T0214331

Lecture Notes in Artificial Intelligence   12797

Subseries of Lecture Notes in Computer Science

More information about this subseries at http://www.springer.com/series/1244

Helmut Degen · Stavroula Ntoa (Eds.)

Artificial Intelligence in HCI

Second International Conference, AI-HCI 2021
Held as Part of the 23rd HCI International Conference, HCII 2021
Virtual Event, July 24–29, 2021
Proceedings

 Springer

Editors
Helmut Degen
Siemens Corporation
Princeton, NJ, USA

Stavroula Ntoa
Foundation for Research and Technology –
Hellas (FORTH)
Heraklion, Greece

ISSN 0302-9743   ISSN 1611-3349   (electronic)
Lecture Notes in Artificial Intelligence
ISBN 978-3-030-77771-5   ISBN 978-3-030-77772-2   (eBook)
https://doi.org/10.1007/978-3-030-77772-2

LNCS Sublibrary: SL7 – Artificial Intelligence

This Springer imprint is published by the registered company Springer Nature Switzerland AG
The registered company address is: Gewerbestrasse 11, 6330 Cham, Switzerland

Foreword

Human-Computer Interaction (HCI) is acquiring an ever-increasing scientific and industrial importance, and having more impact on people's everyday life, as an ever-growing number of human activities are progressively moving from the physical to the digital world. This process, which has been ongoing for some time now, has been dramatically accelerated by the COVID-19 pandemic. The HCI International (HCII) conference series, held yearly, aims to respond to the compelling need to advance the exchange of knowledge and research and development efforts on the human aspects of design and use of computing systems.

The 23rd International Conference on Human-Computer Interaction, HCI International 2021 (HCII 2021), was planned to be held at the Washington Hilton Hotel, Washington DC, USA, during July 24–29, 2021. Due to the COVID-19 pandemic and with everyone's health and safety in mind, HCII 2021 was organized and run as a virtual conference. It incorporated the 21 thematic areas and affiliated conferences listed on the following page.

A total of 5222 individuals from academia, research institutes, industry, and governmental agencies from 81 countries submitted contributions, and 1276 papers and 241 posters were included in the proceedings to appear just before the start of the conference. The contributions thoroughly cover the entire field of HCI, addressing major advances in knowledge and effective use of computers in a variety of application areas. These papers provide academics, researchers, engineers, scientists, practitioners, and students with state-of-the-art information on the most recent advances in HCI. The volumes constituting the set of proceedings to appear before the start of the conference are listed in the following pages.

The HCI International (HCII) conference also offers the option of 'Late Breaking Work' which applies both for papers and posters, and the corresponding volume(s) of the proceedings will appear after the conference. Full papers will be included in the 'HCII 2021 - Late Breaking Papers' volumes of the proceedings to be published in the Springer LNCS series, while 'Poster Extended Abstracts' will be included as short research papers in the 'HCII 2021 - Late Breaking Posters' volumes to be published in the Springer CCIS series.

The present volume contains papers submitted and presented in the context of the 2nd International Conference on Artificial Intelligence in HCI (AI-HCI 2021), an affiliated conference to HCII 2021. I would like to thank the Co-chairs, Helmut Degen and Stavroula Ntoa, for their invaluable contribution to its organization and the preparation of the proceedings, as well as the members of the Program Board for their contributions and support. This year, the AI-HCI affiliated conference has focused on topics related to ethics, trust and explainability, human-centered AI, and AI applications in HCI and Smart Environments.

I would also like to thank the Program Board Chairs and the members of the Program Boards of all thematic areas and affiliated conferences for their contribution towards the highest scientific quality and overall success of the HCI International 2021 conference.

This conference would not have been possible without the continuous and unwavering support and advice of Gavriel Salvendy, founder, General Chair Emeritus, and Scientific Advisor. For his outstanding efforts, I would like to express my appreciation to Abbas Moallem, Communications Chair and Editor of HCI International News.

July 2021   Constantine Stephanidis

List of Conference Proceedings Volumes Appearing Before the Conference

1. LNCS 12762, Human-Computer Interaction: Theory, Methods and Tools (Part I), edited by Masaaki Kurosu
2. LNCS 12763, Human-Computer Interaction: Interaction Techniques and Novel Applications (Part II), edited by Masaaki Kurosu
3. LNCS 12764, Human-Computer Interaction: Design and User Experience Case Studies (Part III), edited by Masaaki Kurosu
4. LNCS 12765, Human Interface and the Management of Information: Information Presentation and Visualization (Part I), edited by Sakae Yamamoto and Hirohiko Mori
5. LNCS 12766, Human Interface and the Management of Information: Information-rich and Intelligent Environments (Part II), edited by Sakae Yamamoto and Hirohiko Mori
6. LNAI 12767, Engineering Psychology and Cognitive Ergonomics, edited by Don Harris and Wen-Chin Li
7. LNCS 12768, Universal Access in Human-Computer Interaction: Design Methods and User Experience (Part I), edited by Margherita Antona and Constantine Stephanidis
8. LNCS 12769, Universal Access in Human-Computer Interaction: Access to Media, Learning and Assistive Environments (Part II), edited by Margherita Antona and Constantine Stephanidis
9. LNCS 12770, Virtual, Augmented and Mixed Reality, edited by Jessie Y. C. Chen and Gino Fragomeni
10. LNCS 12771, Cross-Cultural Design: Experience and Product Design Across Cultures (Part I), edited by P. L. Patrick Rau
11. LNCS 12772, Cross-Cultural Design: Applications in Arts, Learning, Well-being, and Social Development (Part II), edited by P. L. Patrick Rau
12. LNCS 12773, Cross-Cultural Design: Applications in Cultural Heritage, Tourism, Autonomous Vehicles, and Intelligent Agents (Part III), edited by P. L. Patrick Rau
13. LNCS 12774, Social Computing and Social Media: Experience Design and Social Network Analysis (Part I), edited by Gabriele Meiselwitz
14. LNCS 12775, Social Computing and Social Media: Applications in Marketing, Learning, and Health (Part II), edited by Gabriele Meiselwitz
15. LNAI 12776, Augmented Cognition, edited by Dylan D. Schmorrow and Cali M. Fidopiastis
16. LNCS 12777, Digital Human Modeling and Applications in Health, Safety, Ergonomics and Risk Management: Human Body, Motion and Behavior (Part I), edited by Vincent G. Duffy
17. LNCS 12778, Digital Human Modeling and Applications in Health, Safety, Ergonomics and Risk Management: AI, Product and Service (Part II), edited by Vincent G. Duffy

http://2021.hci.international/proceedings

2nd International Conference on Artificial Intelligence in HCI (AI-HCI 2021)

Program Board Chairs: **Helmut Degen,** *Siemens Corporation, USA* **and Stavroula Ntoa,** *Foundation for Research and Technology – Hellas (FORTH), Greece*

- Esma Aimeur, Canada
- Gennaro Costagliola, Italy
- Lynne Coventry, UK
- Ahmad Esmaeili, USA
- Mauricio Gomez, USA
- Jennifer Heier, Germany
- Thomas Herrmann, Germany
- Rania Hodhod, USA
- Sandeep Kaur Kuttal, USA
- Adina Panchea, Canada

- Ming Qian, USA
- Robert Reynolds, USA
- Gustavo Rossi, Argentina
- Carmen Santoro, Italy
- Marjorie Skubic, USA
- Lucio Davide Spano, Italy
- Brian C. Stanton, USA
- Zac Tashdjian, USA
- Roberto Vezzani, Italy
- Giuliana Vitiello, Italy

The full list with the Program Board Chairs and the members of the Program Boards of all thematic areas and affiliated conferences is available online at:

http://www.hci.international/board-members-2021.php

HCI International 2022

The 24th International Conference on Human-Computer Interaction, HCI International 2022, will be held jointly with the affiliated conferences at the Gothia Towers Hotel and Swedish Exhibition & Congress Centre, Gothenburg, Sweden, June 26 – July 1, 2022. It will cover a broad spectrum of themes related to Human-Computer Interaction, including theoretical issues, methods, tools, processes, and case studies in HCI design, as well as novel interaction techniques, interfaces, and applications. The proceedings will be published by Springer. More information will be available on the conference website: http://2022.hci.international/:

General Chair
Prof. Constantine Stephanidis
University of Crete and ICS-FORTH
Heraklion, Crete, Greece
Email: general_chair@hcii2022.org

http://2022.hci.international/

Contents

Ethics, Trust and Explainability

Human-Centered AI

AI Applications in HCI

AI Applications in Smart Environments

Ethics, Trust and Explainability

Can You Trust the Black Box? The Effect of Personality Traits on Trust in AI-Enabled User Interfaces

Martin Böckle(✉), Kwaku Yeboah-Antwi, and Iana Kouris

BCG Platinion, Design and Engineering, Berlin, Germany
{boeckle.martin,yeboah-antwi.kwaku,kouris.iana}@bcgplatinion.com

Abstract. Human-centred artificial intelligence is a fast-growing research stream within the artificial intelligence (AI) and human–computer interaction (HCI) communities. One key focus of this stream is the enablement of trust between end users and the intelligent solution. Although, the current body of literature discusses and proposes a range of best practices for the design of user interfaces for intelligent solutions, there is a dearth of research how such interfaces are perceived by users and especially focusing on trust in these interfaces. In this paper, we investigate how the Big Five personality traits affect trust in AI-enabled user interfaces. We then experimentally verify which design best practices and guidelines proposed by Google enable trust in AI-enabled user interfaces for the different personality types. Initial results (n = 211) reveal that three of the Big Five personality traits – *Extraversion, Agreeableness and Open-Mindedness* – show a significant correlation between the degree of the personality trait and trust in the proposed storyboards. In addition, we identified significant positive relationships between the perception of trust by users and four out of the twelve design principles: *review implicit feedback; connect the feedback to UX changes; create opportunities for feedback; fail gracefully and highlight failure.* This paper is of a highly explorative character and provides first experimental results on designing for trust to the HCI/AI community and also highlights future research directions in the form of a research agenda.

Keywords: Human-centred AI · Personality traits · HCI/AI · Big five

1 Introduction

Human-Centred Artificial Intelligence involves considering human needs when developing intelligent solutions. This approach has attracted a lot of attention in the AI and human–computer interaction (HCI) communities [1]. Although Grudin [3] states that, "both research disciplines have been divided by a common focused goal", current research endeavours are bridging this gap [4, 5]. They are accomplishing this by resolving relevant issues not only from a technological perspective but also by highlighting that humans still represent the most central and critical element in many scenarios to which machine learning (ML) algorithms are applied [4]. AI has already been used to support

© Springer Nature Switzerland AG 2021
H. Degen and S. Ntoa (Eds.): HCII 2021, LNAI 12797, pp. 3–20, 2021.
https://doi.org/10.1007/978-3-030-77772-2_1

the decision-making process in various application domains [6–8]. Examples include the medical domain [6, 11, 12], where AI has been used to classify computed tomography images to aid in the early diagnosis of retinal diseases [6], or in finance, supporting humans in risk management and option pricing decisions [7, 13]. Furthermore, AI [8, 14] has found applications in healthcare where it assists humans with patient administration, patient monitoring and clinical decisions [8]. Although these examples highlight promising early results of the application of AI to support human decision-making in different domains, the design of intelligent systems that provide seamless interaction for individual end users by understanding their different needs is still an ongoing challenge for human-centred AI. In fact, Xu and Riedl [1, 2] argue that while technological factors are of major importance because technology is developing at such a fast pace, nontechnical factors must also receive considerable attention.

From a HCI perspective, the first two waves of AI can be considered failures because they neglected human needs [2]. The focus was predominantly driven by academia and mainly on technological solutions. The third wave of AI, starting in approximately 2006 and characterized by major breakthroughs in the application of deep learning to big data, pattern and speech recognition [2], has been shaping up differently. Very importantly, there is a conversation on how to combine these technological advances with a human-centred approach. There is a clear understanding that intelligent systems need to be designed while taking into account that they form part of a larger ecosystem consisting of human stakeholders including users, operators and clients. This has been defined as human-centred AI [1].

There is increasing awareness that the AI goals of intelligent interfaces would strongly benefit from the application of user-centred design principles, especially the principles concerning user testing defined in the HCI community [9]. HCI needs to tackle several challenges in the design of AI-enabled interfaces. Firstly, AI algorithms deal with a high level of complexity and therefore the interfaces need to provide transparency and better explanations. Secondly, AI-enabled user interfaces are generally designed for more long-term interactions since algorithms initially need to be "trained up" before obtaining the full benefit of the user-experience (UX) design [9]. Consequently, there is a need for a user-centred approach which helps users understand the characteristics and output of the AI algorithm in order to address the AI black-box problem [2], to enable trust and to tackle issues such as social responsibility, fairness and accountability [1].

If systems with AI components are designed to be human centric and enable trust, Riedl [1] highlight two main aspects for consideration. Firstly, AI and ML systems must be able to produce understandable explanations. This is mainly covered by the emerging and fast-growing research stream of explainable AI (xAI), which provides toolkits and taxonomies for AI explainability techniques [16, 17]. Secondly, human-centred AI and ML systems need to consider the behaviour and objectives of different personalities in order to address their individual needs [15]. The current paper investigates this second aspect and examines how it affects trust in AI-enabled user interfaces by addressing the following research question:

RQ 1: Do personality traits affect trust in human-centred AI-enabled user interfaces?

In the current body of HCI/AI literature, there is consensus around a set of proposed best practices and guidelines for building human-centred AI-enabled user interfaces [18, 19]. If personality traits affect trust in these interfaces, then it is important to understand which best practices and guidelines encourage a high level of trust for different personality traits, an area which considerably lacks research. Such knowledge will be very useful in the design of human-centred AI approaches. This paper therefore attempts to answer a second research question:

RQ 2: Which best practices and guidelines for the design of human-centred AI-enabled interfaces instil the highest levels of trust for different personalities?

As such, this study has two main objectives. First, to investigate the influence of personality traits on trust in AI-enabled user interfaces and second, to experimentally verify which design best practices and guidelines enable trust in AI-enabled user interfaces according to personality type. We also propose a research agenda intended as a basis for future research directions and to guide the practical application of human-centred AI design by considering users' personality traits.

To the best of our knowledge, this is one of the first papers to investigate the relationship between personality traits and trust in AI-enabled user interfaces and therefore contributes new design knowledge to the current body of HCI/AI literature and community.

2   Research Background and Related Work

To identify and develop the research gap and the definition of the proposed research design, we studied several streams within the literature on HCI/AI.

2.1   Human-Centred AI

There are several emerging and fast-growing research streams under the umbrella of human-centred AI that highlight different aspects of intelligent solutions. One stream focuses on ethically responsible AI [2], which aspires to avoid discrimination and achieve fairness. Another stream focuses on designing explainable, useful and uscable AI solutions [1, 2], as these features have been neglected in the past. As AI is increasingly applied in multiple end-user applications, most users, especially those with a limited technical background, perceive intelligent systems as a black box. This phenomenon causes end users to ask systems questions such as: "Why did you do that? Why is this the result? Why did you succeed or fail? When can I trust you?" [1, 2, 10]. These questions represent the baseline of explainable AI (xAI), which is currently a popular topic in human-centred AI literature. xAI attempts to address these systems' opacity for the end user by explaining why a certain solution is presented. This approach focuses on understanding and interpreting the AI system output. Recent examples of such work include the classification of explanations of AI solutions for different types of user – developers, AI researchers, domain experts, lay end users – by considering theirs goals [20]. This is a novel framework that comprises relevant theories on human decision-making, e.g.,

how people should and actually reason in order to inform the xAI techniques [21] and a taxonomy focusing on questions about, "What is explained?" (e.g., data or model), "How it is explained?" (e.g., direct/post-hoc, static/interactive) and "At what level?" (e.g., local/global) [22]. These approaches address the research question, "How to and why trust AI-enabled user interfaces and their output?"

2.2 Trust in AI

When end users overcome the perception of uncertainty or risk, they then start to develop trust in technology. This usually happens after an assessment of the technology's performance, reliability, safety and security [23]. Generally, trust is a very complex construct [24] and within the domain of human–machine interaction (HMI) it is defined as, "the attitude that an agent will help achieve an individual's goals in a situation characterized by uncertainty and vulnerability" [25]. The work of Ferrario et al. [26] discussed the concept of e-trust when end users interact with e-commerce platforms, group chats and online communities [27]. McKnight et al. [28] investigated the concept of initial trust formation, which explains the preliminary acceptance of technology, in contrast to continuous trust development, where trust needs to be maintained over time. These approaches deal with the question, "How is trust formed when interacting in a digital context?"

2.3 Best Practices and Guidelines for the Design of Intelligent Solutions

This research paper aims to answer the question, "Who trusts AI-enabled user interfaces and which ones do they trust?", rather than the questions put forward at the end of the last two paragraphs, both having already been discussed extensively in the domain. Jamson et al. [29] showed that AI technology is only effective when it assumes some degree of control for users. To support this effective use of AI technology, interaction with AI-enabled user interfaces should therefore aim to build trust in various ways by considering the factors that influence trustworthiness in such systems, as outlined by Ashoori et al. [30]. Google has created a collection of recommended best practices and guidelines from the UX domain that should be followed in order to achieve a human-centred approach to AI that encourages trust [18]. These include how to introduce the end user to the AI system in a meaningful way and how to set expectations for adaptation and create effective mental models. Other best practices and guidelines show that feedback is crucial to developing trust in AI-enabled user interfaces, as well as explaining how to review, collect and connect implicit and explicit feedback to inform and enhance the user's experience of the product [18]. Google [18] proposed further guidelines on how to define errors and failure and provide a path forward from failure, since AI capabilities can change over time, leading to errors and failures, and the ability to deal with this gracefully is crucial to trust. As explainability is considered one of the major drivers for increasing trust in these systems, there are also best practices that explain how AI systems work by connecting explanations to the end users' actions with the AI system

output and workings of the system's optimisation process [18, 20]. In the current paper, we highlight four sets of best practices and guidelines out of six proposed by Google [18], all of which are widely used and accepted in practice. The four sets of guidelines are:

- *Mental Model:* This set of guidelines concerns the end users' understanding of how AI systems work and how their interactions affect the interface. Generally, mental models aim to set expectations about functionalities and communication limitations.
- *Explainability and Model Confidence:* These guidelines address how the end user receives an appropriate level of explanation regarding how the system works and its degree of confidence in its output. After developing a clear mental model and aware-ness of the system's overall capabilities, these guidelines help end users learn how and when to trust the underlying system. They were originally named "Explainability + Trust" but have been renamed here to avoid confusion.
- *Feedback and Control:* This set of guidelines concerns the design of feedback and control mechanisms that provide a meaningful end-user experience (UX) when sug-gesting personalized content. These mechanisms can also be used to improve the underlying AI model output.
- *Errors and Graceful failures:* These guidelines help identify and diagnose AI context errors and communicate the way forward. Context errors include false starts, misun-derstandings and edge cases that cannot be foreseen within the development process. Google suggests these errors should be seen as opportunities to correct the end user's mental model, encourage the end user to provide feedback and enhance the overall learning process through experimentation and error resolution processes [18].

We focused on these four sets of guidelines from the total of six sets because only these four applied to the context of our evaluation.

Although these UX best practices and guidelines for the design of AI-enabled user interfaces are widely used in the industry, there is a lack of experimental vali-dation of their efficacy in terms of individual user needs that can be associated with different user/personality types [15, 51]. This is very important, since these different user/personality types are known to have distinct needs which should be taken into consideration when implementing a human-centred design approach.

2.4 User-Type Models

User typologies or personality types have a long history of use in the design of person-alised solutions, especially in the domain of learning and persuasion. For instance, Böckle et al. [31, 32] used user/player types to define user-centred design possibilities at the inter-section of gamification and persuasive technology. These typologies are also very useful in HCI [33] for defining boundaries to ensure successful interaction with AI-enabled systems and user interfaces [15, 34]. Stachl et al. [35] reported that user/personality types are better predictors of AI application usage than basic demographic parameters [36]. Although several different user type models have been discussed in the literature, in this paper, we focused on the Big Five personality types proposed by Goldberg [36]. These types are described in the Table 1.

Table 1. Big Five factors of personality (BFI-2) [54]

Extraversion (E)	Active, assertive, energetic, enthusiastic, outgoing, talkative
Agreeableness (A)	Appreciative, forgiving, generous, kind, sympathetic, trusting
Conscientiousness (C)	Planula
Negative Emotionality (N)	Anxious, self-pitying, tense, touchy, unstable, worrying
Open-Mindedness (O)	Artistic, curious, imaginative, insightful, original, wide interests

3   Research Design

To collect data to study how personality traits affect trust in human-centred AI-enabled user interfaces, we created a survey with storyboards covering the four sets of best practices and guidelines for the design of AI-enabled user interfaces proposed by Google [18] and described in Sect. 2.3. The storyboards show a prototype of a mobile application called "Zycle" created using the sets of best practices from Google [18]. Zycle is an application that helps users who cycle for exercise, a common mode of exercise in Germany, the country of residence of the authors. Zycle has various functionalities such as suggesting routes for rides based on the user's profile, as well as ride summaries and music playlists. We designed three storyboards for each set of best practices and guidelines, giving a total of twelve storyboards. The storyboards and their matching sets of guidelines are shown in Table 2.

Users were asked to rate each storyboard on the perceived trust. They were given the following scale and asked to pick one option.

(1)   *The system is reliable*
(2)   *I am confident in the system*
(3)   *I can trust the system*
(4)   *I am not suspicious of the system's intentions, actions or outputs.*

Although the literature discusses several approaches for measuring trust [38, 39], we selected this scale for measuring trust in automated systems [37]. We believe this scale is the most appropriate as it is one of the few that has been empirically tested and used to measure trust in automated systems such as AI-enabled systems.

Table 2. Defined storyboards inspired by [18]

Mental Model Mapping		
S1 Design for experimentation – The interface indicates that the application will average the first few rides together before starting to make recommendations.	**S2 Fail gracefully and highlight failure** – The interface explains that the application could not map the entire bike ride.	**S3 Clearly communicate the limits of AI** – The interface highlights the features of the AI component and sets expectations while also helping the end user.

Explainability and model confidence		
S1 Articulate data sources – The interface reports a lack of data and suggests that the user uses their own judgement.	**S2 Account for situational stakes** – The interface explains why certain recommendations have been made in a certain context (e.g., best for ankles).	**S3 Communicate model confidence** – The interface highlights different recommendations of representing the confidence values as categories (e.g., high/medium/low).

(continued)

Table 2. (*continued*)

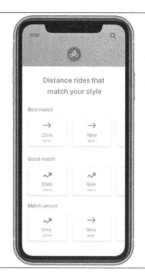

Feedback and control

| **S1 Review implicit feedback** – The interface provides the option of looking at past rides and adjusting data collection settings. | **S2 Connect the feedback to UX changes** – The interface allows users to provide feedback, which will have a direct impact on the UX. | **S3 Consider user preferences** – The interface provides the option to make adjustments if the user's preferences change. |

(*continued*)

Table 2. (*continued*)

Errors and graceful failures		
S1 Define meaningful error states – Use of error states to inform the user what input the AI needs.	**S2 Consider contextual recommendations** – The interface offers proactive recommendations.	**S3 Create opportunities for feedback** - The interface presents the opportunity to receive feedback if the user rejected AI outputs.

4 Results

We used the Amazon Mechanical Turk (MTurk) micro-task market to collect the results of the questionnaires. MTurk is a widely accepted and reliable method [40] for collecting end user responses and has also been applied in a variety of peer-reviewed HCI studies [41–45]. Several studies have confirmed the methodological validity of MTurk in various types of survey, including the issues of sampling and participant representatives and potential biases [46–48]. The major advantage of MTurk is that it taps into a diverse population of potential participants who are more easily accessible than via traditional recruitment methods. To obtain high-quality results, we followed the best practices recommended by Goodman and Kurtis by setting a minimum approval rating of 97% and a minimum number of approved tasks of 5,000. To answer our research questions, we applied a multiple regression analysis to the results of the survey to determine which of the predictors, in this case the Big Five personality traits, showed a high correlation with a perception of trust in each of the storyboards.

4.1 Demographic Information

We received 239 complete responses and discarded 28 responses that we considered invalid, giving a total of 211 valid responses. The discarded responses violated certain rules such as completing the questionnaire in less than the minimum time of 15 min. Demographic data revealed that the participants represented a very diverse population with some differences regarding their home country and level of education. The demographic data are shown in Table 3. It shows that 41% of the subjects were female and 59% male. The largest age group (46%) were between 26 and 35 years old and 49% of all respondents had a degree. Most respondents selected India (51%) as their home country, followed by the United States (41%).

Table 3. Demographic data

Total responses (n = 211)	
Gender	Female (41%), Male (59%), Trans (0%), Other (0%)
Age	15−25 (4%), 26−35 (46%), 36−45 (28%), Over 45(22%)
Education	Less than high school (1%), High school (9%), Graduate (16%), College diploma (7.5%), Degree (49%), Master's degree (15%), Doctoral degree (0), Other (2.5%)
Country	India (51%), USA (41%), Austria (4%), Germany (3%), Singapore (1%)

4.2 Personality Traits and Trust in AI-Enabled User Interfaces

Firstly, we shall address the research question:

RQ 1: Do personality traits affect trust in human-centred AI-enabled user interfaces?

We applied a multiple regression analysis to the results of our survey, focusing on the relationship between the Big Five personality traits and perceived trust in the storyboards and, subsequently, the best practices and guidelines used to design the storyboards. The results are shown in Tables 4, 5, 6 and 7.

Overall, the results reveal that all the user types have differing correlations for forming trust in AI-enabled user interfaces. Three out of five user types, namely those with a high tendency towards *Extraversion, Agreeableness* and *Open Mindedness*, show a statistically significant correlation with trust in AI-enabled user interfaces. We could not identify any significant relationships for users who tend towards the user types *Conscientiousness* and *Negative Emotionality*. Our results unequivocally show that personality traits affect trust in human-centred AI-enabled user interfaces.

Our second aim was to answer the research question:

RQ 2: Which best practices and guidelines for the design of human-centred AI-enabled interfaces instil the highest levels of trust for different personalities?

The following subsections provide a more detailed analysis of the relationship between each Big Five personality trait and trust and attempts to answer RQ1 and RQ2 for each trait.

Extraversion: Individuals with this trait are sociable, gregarious, assertive, talkative, energetic and optimistic [49]. A lower degree of *Extraversion* corresponds to more reserved, independent and quieter users [50]. Our results in Tables 4 and 7 revealed a significant positive relationship between the degree of *Extraversion* and trust in storyboards belonging to the categories *Mental Model* and *Errors and Graceful Failures*. Storyboard S2 on *Mental Model Mapping* highlights the best practice of failing gracefully when the system fails to meet expectations but provides the option to map the run manually. In this case, higher degrees of *Extraversion* correlated positively with an increase in trust when this guideline is followed. Users may see the failure not only as forgivable but something they can help fix, thus increasing their perceived trust in the system. There is a similar positive correlation between trust and higher degrees of *Extraversion* for storyboard S3 on *Errors and Graceful Failures*, which corresponds to the best practice for creating feedback opportunities. These results suggest that extraverted users have a higher degree of perceived trust in systems when they are given chances to supply feedback to fix system failures.

Agreeableness: *Agreeableness* is an important factor in human–computer interactions and is characterised as individuals who are helpful, cooperative, good-natured, sympathetic and tolerant of others [49]. Results in Table 6 show significant positive correlations between the degree of *Agreeableness* and trust in storyboards S1 and S2 for *Feedback and Control*. The first storyboard (S1) deals with the best practice of reviewing implicit feedback and informs end users about an option to opt-out of data collection functionalities. In this context, higher degrees of *Agreeableness* correlated positively with an increase in trust when these guidelines were followed. Generally, since the application of data collection practices may conflict with the level of trust in AI, end users perceive this option positively as a transparent system behaviour. The second storyboard (S2) highlights feedback opportunities by directly influencing the UX of the present ride by asking, "How's the ride going?"; this reveals the balance between control and automation [18].

In this case, individuals can respond with options such as, "Too many hills", "Boring" or "Too long", etc. and the results in Table 6 show that higher degrees of *Agreeableness* correlated positively with an increase in trust when this guideline was applied. Therefore, our results confirm that users have a higher level of trust when they are given chances to opt-out of critical functionalities or providing feedback with a direct impact on the present UX.

Open-Mindedness: *Open-Mindedness* is characterised by intellectual curiosity, a high level of creativity, complex and deep thinking, including the interest in abstract ideas [49]. On the other hand, a low level of *Open-Mindedness* corresponds to conventional, conservative behaviour and a preference for familiar situations [33, 52].

Our results in Tables 4 and 5 show that there is a highly significant inverse relationship between the degree of *Open-Mindedness* and trust for all the storyboards for *Mental Model Mapping* and the third storyboard (S3) for *Explainability and Model Confidence*.

Storyboard S1 in *Mental Model Mapping* demonstrates the best practice of designing for experimentation and reassuring users that these features will not dictate their future experiences [18]. Our results in Table 4 show that higher degrees of *Open-Mindedness* correlate negatively with an increase in trust when this guideline was applied.

Consequently, this shows that these individuals perceive experimentation in which the first few routes will be averaged as untrustworthy. Similarly, when this guideline was followed for the second storyboard (S2), where the route could not be mapped properly, we identified a highly significant negative relationship with increased trust. Users with this personality trait do not tolerate any failures in the system or the option of running the mapping manually. Finally, the last storyboard (S3) aims to communicate the limits of AI by carefully describing the AI features of the system which form the basis of the data used for the proposed recommendations. End users with a higher degree of *Open-Mindedness* did not trust this recommendation. Within this context, Google [18] suggested that there is a risk of integrating a generalised "AI helper", yet the risk of mistrust is even higher if the system limitations are unclear, as it could lead to over-trust or failing to benefit from the real added value [18]. In the category *Explainability and Model Confidence*, we identified a highly significant negative relationship (Table 5) between the degree of *Open-Mindedness* and the third storyboard (S3), which corresponds to the best practice of how to communicate the model confidence in a meaningful way. Although, the interface breaks down model confidence into three levels – best match, good match, match unsure – this personality type does not trust this type of explanation.

Table 4. Mental model mapping

	Storyboard 1			Storyboard 2			Storyboard 3		
	RC	t Stat	p-value	RC	t Stat	p-value	RC	t Stat	p-value
EXT	0.17	1.62	0.10	**0.35***	2.20	0.02	0.24	1.84	0.06
AGR	0.14	0.97	0.33	0.00	0.0	0.99	0.11	0.62	0.53
CON	0.15	1.08	0.28	−0.31	−1.47	0.14	0.03	0.20	0.83
NEG	−0.09	−0.88	0.37	−0.25	−1.47	0.14	−0.11	−0.84	0.40
OPE	**-0.29***	−2.43	0.01	**−0.57***	−3.05	0.00	**−0.52***	−3.47	0.00

Coefficients in bold represent a significant correlation (*p-value ≤ 0.05, **p-value ≤ 0.01), EXT – Extraversion, AGR – Agreeableness, CON – Conscientiousness, NEG – Negative Emotionality (Neuroticism), OPE – Open-Mindedness, RC – Regression Coefficient)

Table 5. Explainability and model confidence

	Storyboard 1			Storyboard 2			Storyboard 3		
	RC	t Stat	p-value	RC	t Stat	p-value	RC	t Stat	p-value
EXT	0.15	1.2	0.21	0.12	1.12	0.26	0.09	0.81	0.41
AGR	0.13	0.75	0.45	0.16	1.00	0.31	0.10	0.63	0.52
CON	0.14	0.91	0.36	0.18	1.24	0.21	0.19	1.28	0.20
NEG	0.08	0.66	0.50	0.10	0.85	0.39	−0.06	−0.55	0.58
OPE	−0.14	−1.04	0.29	−0.23	−1.75	0.08	**−0.31***	−2.37	0.01

Coefficients in bold represent a significant correlation (*p-value ≤ 0.05, **coefficient p ≤ 0.01), EXT – Extraversion, AGR – Agreeableness, CON – Conscientiousness, NEG – Negative Emotionality (Neuroticism), OPE – Open-Mindedness, RC – Regression Coefficient)

Table 6. Feedback and control

	Storyboard 1			Storyboard 2			Storyboard 3		
	RC	t Stat	P-value	RC	t Stat	P-value	RC	t Stat	P-value
EXT	0.15	1.38	0.16	0.15	1.38	0.16	0.16	1.48	0.13
AGR	**0.30***	1.92	0.05	**0.30***	1.92	0.05	0.19	1.21	0.22
CON	0.12	0.88	0.37	0.12	0.88	0.37	0.11	0.76	0.44
NEG	0.14	1.22	0.22	0.14	1.22	0.22	0.08	0.68	0.49
OPE	−0.17	−1.36	0.17	−0.17	−1.36	0.17	−0.18	−1.43	0.15

Coefficients in bold represent a significant correlation (*p-value $\leq$ 0.05, **p-value $\leq$ 0.01), EXT – Extraversion, AGR – Agreeableness, CON – Conscientiousness, NEG – Negative Emotionality (Neuroticism), OPE – Open-Mindedness, RC – Regression Coefficient)

Table 7. Errors and graceful failures

	Storyboard 1			Storyboard 2			Storyboard 3		
	RC	t Stat	P-value	RC	t Stat	P-value	RC	t Stat	P-value
EXT	0.12	0.97	0.33	0.09	0.76	0.44	**0.24***	2.03	0.04
AGR	0.18	1.06	0.29	0.24	1.37	0.17	0.17	1.01	0.31
CON	0.02	0.12	0.89	0.01	0.06	0.94	−0.01	−0.11	0.91
NEG	−0.06	−0.45	0.64	0.02	0.20	0.83	−0.04	−0.36	0.71
OPE	−0.09	−0.65	0.51	−0.11	−0.74	0.45	−0.01	−0.07	0.94

Coefficients in bold represent a significant correlation (*p-value $\leq$ 0.05, **p-value $\leq$ 0.01), EXT – Extraversion, AGR – Agreeableness, CON – Conscientiousness, NEG – Negative Emotionality (Neuroticism), OPE – Open-Mindedness, RC – Regression Coefficient)

5   Conclusion

The results of the present work aim to support researchers in the design of human-centred AI applications with a major focus on trust. Generally, the application of psychology to HCI can be used to assess design effectiveness, as different personality types perceive and use technology differently. Therefore, we looked into best practices from a UX perspective in order to design AI user interfaces that enable trust. These guidelines indicate how to design intelligent solutions in a meaningful way and cover topics such as *Mental Model Mapping, Explainability and Model Confidence, Feedback and Control* and *Errors and Graceful Failures*. The storyboards we developed were designed to encourage trust and followed a human-centred approach to AI.

The human-centred AI design elements applied within the prototype captured a holistic picture of a mobile phone cycling application that informs and enhances the users training habits through feedback and suggestions. Regarding the first research question, we identified significant correlations between three of the Big Five personality traits and the proposed storyboards and therefore conclude that personality types do affect trust in

human-centred AI-enabled user interfaces. For people with high levels of *Extraversion* and *Agreeableness*, we found a positive relationship with the aforementioned best practices, while *Open-Mindedness* presented negative relationships. For the analysis of the second research question, we have shown which of the proposed best practices actually instil higher levels of trust according to personality type. We identified positive correlations between personality types and trust for: *review implicit feedback; connect the feedback to UX changes; create opportunities for feedback; fail gracefully and highlight failure.*

We believe that there are multiple means and opportunities to consider different personality traits when designing AI applications, which is an area that has received very little research interest. For instance, the term cognitive compatibility indicates that the structure of the human–machine interface should actually match the user's cognitive style [53]. Recent research also suggests that users actually perform better when they use an interface that has been matched to their personality type [53].

Unfortunately, there is no complete set of design guidelines to describe preferences for specific design elements, as most of the limited number of studies published to date present results for combined personality traits [53]. Consequently, this paper has a highly explorative character and contributes first results to the HCI/AI literature and community by highlighting future research directions in the form of a research agenda.

6   Limitations and Future Research

The current paper is also subject to several limitations. First, we could not identify an appropriate scale for measuring trust in AI-enabled user interfaces within the literature, so we applied certain items from the work of Jian et al. [37]. Second, each individual's perceived trust was measured through responses to screenshots of a prototype design. Particularly in the case of AI applications, trust can change while using the application in everyday life, the scenario in which behavioural data will be generated, which has an impact on the maturity of the AI functions and therefore on the level of trust. Both initial trust and continuing trust development play a major role throughout these stages. Furthermore, certain AI features of the prototype may reveal practical strengths and weaknesses. Third, since we used MTurk we knew little about the participants, although it is already a well-established method for the design of large-scale studies. Yet many unknowns still remain, for instance the cognitive profile of the MTurk workers as well as how prior experience might influence their responses [40]. Fourth, the prototype cycling application and its presentation to the end user may be perceived differently in certain cultures and regions around the world.

While there are many unexplored issues within this domain, as discussed by Alves et al. [15, 53], we believe that this paper makes a valuable contribution to the HCI/AI community by highlighting the importance of trust in AI-enabled interfaces based on different individual personal types. Finally, based on the initial literature review we identified challenges that are summarized in a proposed research agenda in Table 8. This is intended as a research outlook within the emerging and fast-growing research stream of human-centred AI. While the first challenge (C1) focuses on the difficulties of meaningfully integrating personality traits into the design of intelligent solutions, the

second challenge (C2) aims to shed light on how different personality traits actually affect certain types of intelligent support. The third challenge (C3) does not relate to the work of Google and Amershi et al. [18, 19], but instead deals with UI related elements; for instance how to present intelligent support in a significant manner (e.g., buttons, element style, information density, themes, etc.), as discussed in Alves et al. [53]. The fourth challenge (C4) aims to uncover what the modification techniques might look like when attempting to foster initial trust and constantly improve the level of trust and therefore the overall experience of the AI application. Finally, the fifth challenge (C5) investigates how to define reinforcement strategies if the level of trust starts to decline for the end user.

Table 8. Proposed research challenges

C1	To examine and understand the difficulties of incorporating personality traits into the design of intelligent solutions to increase the level of trust. In particular, to consider how they emerge and connect to different AI layers in a meaningful way, in function of changing needs and demands from the end user.
C2	To explore how personality types affect different design elements and tasks of the intelligent solution, for instance the frequency, type and quality of explanations for different types of information.
C3	To examine how individuals perceive as trustful, different elements (e.g. buttons, information density) of a user interface (UI) for an AI system, with a specific focus on trust.
C4	To design and develop meaningful modification techniques for different types of intelligent solution (e.g., personality engines) in order to maintain and constantly improve the present level of trust (initial trust versus continual trust development).
C5	To design for trust reinforcement in AI-enabled interfaces.

References

1. Riedl, M.O.: Human-centered artificial intelligence and machine learning. Hum. Behav. Emerg. Technol. 1(1) (2019)
2. Xu, W.: Toward human-centered ai: a perspective from human-computer interaction. Interactions 26(4), 42–46 (2019)
3. Grudin, J.: AI and HCI: Two fields divided by a common focus. AI Mag. 30, 48–57 (2009)
4. Inkpen, K., Chancellor, S., Choudhury, M.D., Veale, M. and Baumer E.: Where is the human? bridging the gap between AI and HCI. In: Proceedings of CHI' 19 Extended Abstract, Glasgow, Scotland, UK (2019)
5. Harper, R.H.: The role of HCI in the Age of AI. Int. J. Hum.-Comput. Interact. 35(15), 1331–1344 (2019)
6. Ting, D., Liu, Y., Burlina, P., Xu, X., Bressler, N.M., Wong, T.Y.: AI for medical imaging goes deep. Nat. Med. 24, 539–540 (2018)
7. Aziz, S., Dowling, M.: Machine learning and ai for risk management. In: Lynn, T., Mooney, J., Rosati, P., Cummins, M. (eds.) Disrupting Finance. Palgrave Studies in Digital Business & Enabling Technologies. Palgrave Pivot, Cham (2019)

8. Reddy, S., Fox, J., Purohit, P.M.: Artificial intelligence-enabled healthcare delivery. J. R. Soc. Med. **112**, 22–28 (2018)
9. Liebmann, H.: User interface goals. AI opportunities. AI Mag. **30**(4), 16–22 (2009)
10. Preece, A.: Asking 'WHY' in AI: explainability of intelligent systems – perspectives and challenges. Intell. Syst. Account. Finan. Manage. **1**, 1–10 (2018)
11. Montani, S.: Exploring new roles for case-based reasoning in heterogenous AI systems for medical decision support. Appl. Intell. **28**, 275–285 (2008)
12. Rastgarpour, M., Shanbehzadeh, J.: Application of AI techniques in medical image segmentation and novel categorization of available methods and tools. In: Proceedings of the International MultiConference of Engineers and Computer Scientists (IMECS), vol. 1, Hong Kong (2011)
13. Culkin, R., Das, S.R.: Machine learning in finance: the case of deep learning for option pricing. J. Invest. Manage. **15**, 4 (2017)
14. Lysaght, T., Lim, H.Y., Xafis, V., Ngiam, K.Y.: AI-assisted decision-making in healthcare. Asian Bioeth. Rev. **11**, 299–314 (2019)
15. Völkel, S.T., Schödel, R., Hussmann, H.: Designing for personality in autonomous vehicles: considering individual' s trust attitude and interaction behavior. In: Proceedings of CHI Workshop – Interacting with Autonomous Vehicles: Learning from other Domains (2018)
16. Liao, Q.V., Gruen, D., Miller, S.: Questioning the AI: informing design practices for explainable AI user experiences. In: Proceedings of CHI 2020, Honolulu, USA (2020)
17. El-Essady, M., et al.: Towards XAI: structuring the processes of explanations. In: Proceedings of HCML Workshop at CHI'19, Glasgow, UK (2019)
18. Google PAIR. People + AI Guidebook (2019). https://pair.withgoogle.com/guidebook/
19. Amershi, S., et al.: Guidelines for human-AI iteraction. In: Proceedings of CHI 2019, Glasgow, Scotland, UK (2019)
20. Ribera, M., Lapedriza. A.: Can we do better explanations? A proposal of User-Centered Explainable AI. In: Joint Proceedings of the ACM IUI 2019 Workshop, Los Angeles, USA (2019)
21. Wang, D., Yang, Q., Lim B.: Designing theory-driven user-centric explainable AI. In: Proceedings of the 2019 CHI Conference on Human Factors in Computing Systems, Glasgow, Scotland, UK (2019)
22. Arya, V, et al.: One Explanation Does Not Fit All: A Toolkit and Taxonomy of AI Explainability Techniques. arXiv:1909.03012 (2019)
23. Arnold, M., et al.: FactSheets: Increasing trust in AI services through supplier's declarations of conformity. IBM J. Res. Dev. **63**, 4/5, 6–13 (2019)
24. Culley, K.E., Madhavan, P.: Trust in automation and automation designers: Implications for HCI and HMI. Comput. Hum. Behav. **29**, 2208–2210 (2013)
25. Lee, J.D., See, K.A.: Trust in automation: designing for appropriate reliance. Hum. Fact. **46**, 50–80 (2004)
26. Ferrario, A., Loi, M., Vigano, E.: In AI we trust incrementally: a multi-layer model of trust to analyze human-artificial intelligence interactions. Philos. Technol. **35**, 523–539 (2020)
27. Taddeo, M., Floridi, L.: The case of e-trust. Ethics Inform. Technol. **13**(1), 1–3 (2011)
28. McKnight, D.H., Choudhury, V., Kacmar, C.: The impact of initial consumer trust on intentions to transact with a web site: a trust building model. J. Strateg. Inform. Syst. **11**(3), 297–323 (2002)
29. Jamson, A.H., Merat, N., Carsten, O.M., Lai, F.C.: Behavioural changes in drivers experiencing highly-automated vehicle control in varying traffic conditions. Transp. Res. Part C: Emerg. Technol. **30**, 116–125 (2013)
30. Ashoori, M., Weisz J.D.: In AI we trust? Factors That Influence Trustworthiness of AI-infused Decision-Making Processes. arXiv:1912.02675 (2019)

31. Böckle, M., Yeboah-Antwi, K.: Designing at the intersection of gamification and persuasive technology to incentivize energy-saving. In: Pappas I., Mikalef P., Dwivedi Y., Jaccheri L., Krogstie J., Mäntymäki M. (eds.) Digital Transformation for a Sustainable Society in the 21st Century. I3E 2019. Lecture Notes in Computer Science, vol. 11701 (2019)
32. Böckle, M., Novak, J., Bick, M.: Exploring gamified persuasive system design for energy saving. J. Enterp. Inform. Manage. **33**(6), 1337–1356 (2020)
33. Pillis, E., Green, D.: Personality influences trust differently in virtual and face-to-face teams. Int. J. Hum. Resour. Dev. Manage. **9** (2009)
34. Zhou, X., Mark, G., Li, J., Yang. H.: Trusting virtual agents: the effect of personality. ACM Trans. Interact. Intell. Syst. **9**, 2–3, Article 10 (2019)
35. Stachl, C., Hilbert, S., Au, J., Buschek, D.De, Luca, A., Bischl, B., Hussmann, H., Bühner, M.: Personality, traits predict smartphone usage. Euro. J. Pers. **31**(6), 701–722 (2017)
36. Goldberg, L.R.: An alternative description of personality: the Big-Five factor structure. J. Pers. Soc. Psychol. **59**, 1216–1229 (1990)
37. Jian, J., Bisnatz, A., Drury, C.: Foundations for an empirically determined scale of trust in automated systems. Int. J. Cogn. Ergon. **4**(1), 53–72 (2000)
38. Schaefer, K.E.: Measuring trust in human robot interactions: development of the "trust perception scale-HRI". In: Mittu, R., Sofge, D., Wagner, A., Lawless, W. (eds.) Robust Intelligence and Trust in Autonomous Systems. Springer, Boston, MA (2016)
39. Madsen, M., Gregor, S.: Measuring human-computer trust. In: Proceedings of the 11th Australasian Conference on Information Systems, pp. 6–8 (2000)
40. Paolacci, G., Chandler, J.: Inside the Turk: understanding Mechanical Turk as a participant pool. Curr. Dir. Psychol. Sci. **23**(3), 184–188 (2014)
41. Orji, R., Tondello, G.F., Nacke, L.: Personalizing persuasive strategies in gameful systems to gamification user types. In: Proceedings of CHI 2018, Montreal, Canada (2018)
42. Toomin, M., Kriplean, T., P€ortner, C. and Landay, A.J.: Utility of human-computer interactions: toward a science of preference measurement. In Proceedings of CHI 2011, Vancouver, Canada (2011)
43. Attali, Y., Arieli-Attali, M.: Gamification in assessment: do points affect test performance? Comput. Educ. **83**, 57–63 (2015)
44. Feller, J., Gleasure, R., Treacy, S.: Information sharing and user behavior in internet enabled peer-to-peer lending systems: an empirical study. J. Inform. Technol. **32**, 127–146 (2017)
45. Huang, Y., Li, C., Wu, J., Lin, Z.: Online customer reviews and consumer evaluation: the role of review font. Inform. Manage. **55**, 430–440 (2018)
46. Bartneck, C., Duenser, A., Moltchanova, E., Zawieska, K.: Comparing the similarity of responses received from studies in Amazon's Mechanical Turk to studies conducted online and with direct recruitment. PloS One **10**(4) (2015)
47. Clifford, S., Jewell, R.M., Waggoner, P.D.: Are samples drawn from Mechanical Turk valid for research on political ideology? Res. Polit. **2**(4) (2015)
48. Heen, M.S., Lieberman, J.D., Miethe, T.D.: A comparison of different online sampling approaches for generating national samples. Center Crime Justice Policy, **1**, 1–8 (2014)
49. Soto, C.J., John, O.P.: The next Big Five Inventory (BFI-2): Developing and assessing a hierarchical model with 15 facets to enhance bandwidth, fidelity, and predictive power. J. Pers. Soc. Psychol. **113**, 117–143 (2017)
50. Pillis, E., Green, D.: Personality influences trust differently in virtual and face-to-face teams. Int. J. Hum. Res. Dev. Manage. **9** (2009)
51. Zhou, J., Luo, S., Chen, F.: Effects of personality traits on user trust in human-machine collaborations. Journal of Multimodal User Interfaces **14**, 387–400 (2020)
52. Bruck, C.S., Allen, T.D.: The relationship between Big Five personality traits, negative affectivity, type a behavior, and work-family conflict. J. Vocat. Behav. **63**, 457–472 (2003)

53. Alves, T., Natlio, J., Henriques-Calado, J., Gama, S.: Incorporating personality in user-interface design: a review. Pers. Individ. Differ. **155** (2020)
54. McCrae, R.R., John, O.P.: An introduction to the five-factor model and its applications. J. Pers. **60**, 175–215 (1992)

Towards Design Principles for User-Centric Explainable AI in Fraud Detection

Douglas Cirqueira[1,3](✉) [iD], Markus Helfert[2,3] [iD], and Marija Bezbradica[1,3] [iD]

[1] School of Computing, Dublin City University, Dublin, Ireland
douglas.darochacirqueira2@mail.dcu.ie
[2] Innovation Value Institute, Maynooth University, Maynooth, Ireland
[3] Lero - the Science Foundation Ireland Research Centre for Software,
Dublin City University, Dublin, Ireland

Abstract. Experts rely on fraud detection and decision support systems to analyze fraud cases, a growing problem in digital retailing and banking. With the advent of Artificial Intelligence (AI) for decision support, those experts face the black-box problem and lack trust in AI predictions for fraud. Such an issue has been tackled by employing Explainable AI (XAI) to provide experts with explained AI predictions through various explanation methods. However, fraud detection studies supported by XAI lack a user-centric perspective and discussion on how principles are deployed, both important requirements for experts to choose an appropriate explanation method. On the other hand, recent research in Information Systems (IS) and Human-Computer Interaction highlights the need for understanding user requirements to develop tailored design principles for decision support systems. In this research, we adopt a design science research methodology and IS theoretical lens to develop and evaluate design principles, which align fraud expert's tasks with explanation methods for Explainable AI decision support. We evaluate the utility of these principles using an information quality framework to interview experts in banking fraud, plus a simulation. The results show that the principles are an useful tool for designing decision support systems for fraud detection with embedded user-centric Explainable AI.

Keywords: Explainable AI · Fraud detection · Decision support systems · Artificial intelligence · Design principles · HCI · Human-AI interaction · Human-centered AI

1 Introduction

Digital platforms are convenient for customers in online retail and banking as they allow quick transactions and a choice between multiple E-commerce and

This research was supported by the European Union Horizon 2020 research and innovation programme under the Marie Sklodowska-Curie grant agreement No. 765395; and supported, in part, by Science Foundation Ireland grant 13/RC/2094_P2.

H. Degen and S. Ntoa (Eds.): HCII 2021, LNAI 12797, pp. 21–40, 2021.
https://doi.org/10.1007/978-3-030-77772-2_2

offline channels [8]. However, the convenience is followed by increased fraud cases [30,33]. Companies increasingly use Artificial Intelligence for decision support and fraud detection systems to automatically classify and alert experts of cases where revision is needed. However, fraud experts are not knowledgeable about AI's inner workings and face the black-box problem [35].

Ideally, fraud experts should be able to trust AI partners in such scenarios, as due to the complexity of their work, if decisions those experts make are wrong, this might cause harm and financial loss for institutions and customers [62]. Indeed, trust in AI is recognized as essential by agencies in Europe and worldwide, which develop guidelines for trustworthy and responsible AI for businesses and society [20]. In the meantime, Explainable AI (XAI) research was developed to optimize a diversity of explanation methods (EM) and to enable user's understanding of AI predictions for decision support [2].

Concerning fraud detection specifically, previous studies supported by XAI provide experts with explanations [12], but a user-centric perspective is lacking. Requirements are not elucidated before the deployment of explanations, and they also lack prescriptive principles for the alignment between EM and fraud experts requirements [3,61]. Therefore, this can cause the lack of trust from fraud experts in AI predictions [11]. Since XAI research has a major focus on optimizing EM and AI models, and previous XAI studies for fraud lack a user-centric perspective, in this work, we develop design principles (DP) to prescribe the alignment of fraud expert's tasks to EM for enabling Explainable AI decision support (XAIDSS) in fraud detection. We adopt an Information Systems (IS) lens and follow a design science research methodology to develop and evaluate DP through multiple iterations with fraud experts. We evaluate the utility of our principles through an information quality framework with experts in banking fraud via interviews and also via a simulation on a real transaction fraud dataset.

This paper is structured as follows: Sect. 2 presents the theoretical foundation and related work; Sect. 3 highlights the research methodology; Sect. 4 depicts the developed design principles as the core contribution of this study; Sect. 5 provides the evaluation design and results for design principles; Sect. 6 discusses results and implications followed by Sect. 7 with conclusions.

2   Theoretical Foundation and Related Work

Shopping transactions via digital retail platforms are constantly increasing [10], which opens opportunities for fraudsters to act. This work focuses on transaction fraud cases, which occur when a customer card or online account balance is used without a customer's consent to perform a transaction, for instance, in retail or via bank transfers [45,68]. In order to review and make decisions on fraud cases classified by AI models, experts rely on fraud detection and decision support systems [14].

Explainable AI research applies and develops a diversity of explanation methods to explain AI predictions in particular applications [4]. There is no consensus on how to classify those methods. Currently, well-regarded surveys

classify EM based on their dimensions of scope (local and global explanations), target (to explain the data, model, or features), and explanation type [2,4,5,18,38,39,41,54,67]. Therefore, each method provides the users with an explanation type for decision support, and it analyses particular aspects of AI predictions and models. Recent research in XAI advocates for the importance of user-centric XAI, relying on Human-Computer Interaction (HCI) and interdisciplinary social sciences [1]. The researchers focus on the design of EM following user requirements to enhance decision support and trust in AI [58].

Design principles are well researched in Information Systems and HCI disciplines. They enable prescriptive knowledge on establishing and designing decision support systems to aid user practices [15,52,53]. Design principles have also been developed to guide the design of user interactions with AI, support explanatory data analysis, and debugging AI models [6,65]. Indeed, [19] highlights the need for the development of principles for informed predictions and interactions between users and AI predictions, which can then be mitigated with HCI research support.

Previous studies employing XAI and EM for fraud detection explore the effects and performance of explanations in fraud expert's work. In [31], authors provide a service architecture for security experts with explanations, aiming to introduce more context for the outlier score given anomalous records of network flows. In [12], authors provide experts with Shapley Additive Explanations (SHAP) [34] for why particular warranty claims are marked as anomalies by a machine learning (ML) model. In [61], the authors also work with SHAP explanations for fraud cases, and they observe through experiments that explanations positively impact the decision-making for fraud alerts. The same authors in [60] go further and develop case-based explanations with visualizations for similar fraudulent cases in banking. In [7] authors develop an EM to explain the importance of current and past events and features on sequential data, enabling experts with a temporal perspective on explanations for recurrent ML models such as RNNs and LSTMs [25]. They evaluate their model through experiments and simulations regarding the relevance of features, events, and efficiency for providing explanations that can support debugging AI models for fraud detection. In [26], authors evaluate popular EM and tools focusing on feature importance explanations and their impact on user's accuracy and time to make decisions. However, the assessment of user requirements and exploration of different explanation types are left for future studies.

Existing literature employing EM in fraud detection has not tackled fraud expert's requirements, making it challenging to align explanations and these requirements to establish trustworthy XAIDSS. Furthermore, as used by IS, AI, and HCI studies, design principles are lacking in fraud scenarios for providing prescriptive knowledge on how to deploy explanations for fraud expert's decision support with XAI. This study aims to address this gap by developing and evaluating principles for user-centric XAI and enable XAIDSS in fraud detection.

3   Research Methodology

We follow design science research (DSR) [42], an IS research methodology focused on interactive developing and evaluating artefacts for solving a practitioner's problem and bringing research contributions. We start by identifying the research problem through a literature review presented in the theoretical background section and discussions with experts in fraud detection within a European bank. We identify the problem of fraud experts' lack of trust towards AI predictions due to insufficient alignment between their tasks and explanations to review fraud cases. The next step is to establish the research objective, which is to align expert's tasks to explanations according to their needs for XAIDSS. Adopting HCI and IS theoretical lenses, this study develops an artefact, a set of design principles to guide such alignment.

For the design and development phase of design science research, we first need to establish the kernel theory governing the artefact development process for solving the identified problem [22,56]. Given the poor alignment between fraud expert's tasks and explanations for XAIDSS, our problem relates to an expert's decision-making process when reviewing fraud cases. We then establish the kernel theory to develop the artefact based on two main sets of constructs: 1) fraud expert tasks and 2) design features revealing meta-design principles of EM, which facilitate experts to perform their tasks. Those constructs compose the design knowledge to develop our principles.

Regarding the first set of constructs for artefact development, we rely on our previous study results to elicit 13 fraud expert's tasks when analyzing suspicious fraud cases [51]. In the referred study, we adopt expert interviews with a scenario-based method, and a systematic literature review [11]. Scenario-based elicitation facilitates an HCI and problem-centered perspective to identify stakeholder requirements, goals, tasks and knowledge to develop decision support systems [64]. In the current study, and guided by [51], we extend the previous work by grouping the tasks into requirements. Those requirements should reflect experts' actions and goals when analyzing fraud cases supported by a decision support system. For instance, experts compare, cluster and contrast cases, so those tasks are grouped within a requirement established as similarity and previous pattern matching. We employ the tasks and requirements in our design principles.

In relation to the second set of constructs for artefact development, we perform a systematic literature review following [59] to identify design features of EM, which enable fraud experts to perform their tasks and understand AI predictions. Those features help identify meta-design principles, which are post-instantiated principles found on a class of artefacts [13,57]. In our case, the class of artefacts is an EM within the XAI literature. Then, we start defining the main research question as "What are the design features and meta-design principles of explanation methods for their user's decision support?". To answer the research question through relevant literature, we define a search query as "("explainable ai" OR "explainable artificial intelligence" OR "interpretable machine learning")" and query the databases of Scopus, ACM Digital Library, IEEEXplore,

Science Direct, Springer Link, and arXiv. The database search is performed until July 2020. We obtain 2507 studies and read their abstract, introduction and conclusion to select papers that discuss explanation methods. That process gives 372 papers, which are thoroughly read following inclusion and exclusion criteria.[1] We then obtain 140 papers, for which backward and forward search gives us additional 51 papers. We also include papers from scientific events focused on XAI research following our inclusion criteria and extending our coverage, which adds 43 papers to our pool. Therefore, the total of selected papers was 177. Those are analyzed to extracting design features of explanation methods. The complete data for our systematic literature review is available externally.[2]

To analyze the systematic literature review results and elicit design features of EM, we adopt a classification for explanations following our theoretical foundation [2,4,5,18,38,39,41,54,67]. Given that our DP should align EM to fraud experts tasks to establish XAIDSS, this foundation guides our elicitation of EM design features. Therefore, we focus our analysis on every paper selected based on how the EM employed enables decision support based on the scope, target, and explanation type they provide. We adopt a Concept Centric Matrix [28] to structure the findings and design features of EM from every selected paper. Finally, we establish the DP reflecting the instantiated meta-design principles and design features aligned to fraud expert's tasks.

In the first iteration of our design science research, the design principles are discussed with fraud experts in one major bank and our project partner. We ask the experts about their perceptions regarding the principle's correctness, understandability, and comprehensibility according to their tasks and requirements for analyzing fraud cases. From that iteration, we obtain issues with the terminology adopted to describe each principle, which is deemed ambiguous to describe requirements and their grouping towards design features of EM. After analyzing our kernel knowledge and references discussing templates for DP development, we incorporate the guidelines with the anatomy of DP by [23] bringing principle's aim, implementer, context, mechanism, and rationale based on expert's requirements. The authors developed a systematic template for clear delivery of DP based on rigorous analysis of IS literature, including the development of intelligent decision support systems, which all relate to our context. The next round of iterations was focused on evaluating the tasks, requirements, design features, and DP presented in Sect. 4.

---

[1] Inclusion criteria (IC) for papers: IC1-"Paper focuses on providing explanation methods for supporting understanding of AI predictions or user decision-making in user experiments"; IC2-"Paper presents explanation methods that possess an interface for providing explanations as outputs"; IC3-"Paper focuses on discussing classifications or types of explanation methods". The exclusion criteria (EC) is elaborated as: EC1-"Paper is not written in English"; EC2-"Paper is not a journal, conference, workshop article, or Ph.D. thesis"; EC3-"The paper is not fully available".

[2] https://github.com/dougcirqueira/hcii-design-principles-user-centric-explainable-ai-fraud-detection/tree/main/resources/systematic_literature_review.

4 Design Principles

Figure 1 presents developed design principles from our study. It starts with the 13 fraud expert's tasks when analyzing fraud cases (T1–T13). Those tasks are grouped, and 7 requirements are established (R1–R7), as described in our methodology in Sect. 3. Table 1 presents the descriptions for requirements.

Table 1. Grouped fraud expert tasks and requirements with descriptions

Tasks	Requirements	Description
T1, T2	R1: System confidence and limitations	To provide predicted fraud cases based on a probability ranking, and the limitations for classifying cases based on the current AI model performance on training and validation datasets
T3, T5, T6	R2: Similarity and previous pattern matching	To provide similar and dissimilar classified fraud cases to enable comparative analysis of AI predictions
T3, T12	R3: System interactivity	To provide interactivity and enable experts with a dynamic view on data and detail when comparing fraud cases or investigating the impact of attributes on predictions
T4, T7, T13	R4: Relationships	To provide relationships between attributes in single and multiple classified fraud cases
T8, T9	R5: Importance of attributes	To provide the importance of attributes used by the AI model for classifying fraud cases
T10, T11	R6: Inference path	To provide the reasoning process of the AI model for classifying fraud cases, based on rules and a friendly language for fraud experts
T12, T13	R7: Impact on specific decisions	To provide the impact of attributes used by the AI model on specific classifications given to suspicious fraud cases

The fraud expert's tasks and requirements are aligned with the 8 design features extracted from existing explanation methods (DF1–DF8). Table 2 presents the descriptions for each design feature. According to the explanation scope, target, and type, the dimensions of EM are also highlighted within the description of each design feature. Those design features reveal meta-design principles of EM, from which we derive the five design principles (DP1–DP5) developed for user-centric XAIDSS in fraud detection.

Therefore, our design principles aim to provide utility and information quality to fraud experts and researchers in XAI to clearly understand how to set up explanation methods for decision support in fraud detection. Each principle highlights the users, aim, mechanism, and rationale for supporting experts with explanations to understand AI predictions for fraud. We focus on the alignment

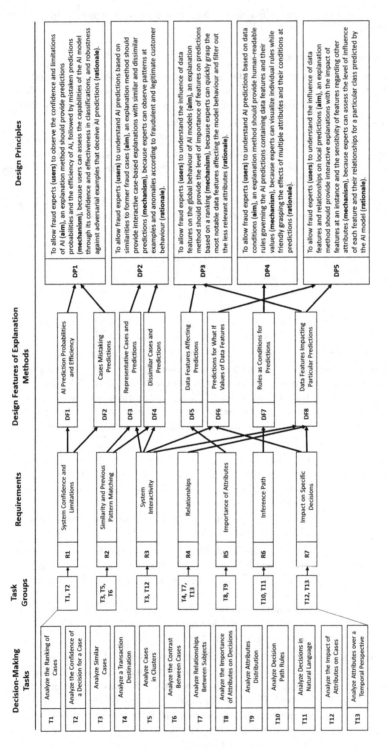

Fig. 1. Design principles for user-centric XAI in fraud detection

Table 2. Descriptions for design features, which reveal meta-design principles of explanation methods

Design features	Description
DF1: AI prediction probabilities and efficiency	The inclusion of prediction probabilities and information on the efficiency of AI may give the user with an explicit visualization of the AI model confidence, and attend R1
DF2: Cases mistaking predictions	Explanations containing local (scope) cases (target) which deceive the AI model predictions, such as those provided by adversarial explanation methods (type). These explanations may support the understanding of weaknesses of AI models and attend to R1
DF3: Representative cases and predictions	Explanations containing local (scope) similar cases (target) to a current prediction, such as those provided by prototypes as explanation methods (type). These explanations may help the comparison between cases by an expert and attend R2 and R3
DF4: Dissimilar cases and predictions	Explanations containing local (scope) dissimilar cases (target) to a current prediction, such as those provided by criticisms as explanation methods (type). These explanations may support the contrast between true positives and outliers, which can attend R2 and R3
DF5: Data features affecting predictions	Explanations containing global (scope) data features (target) considered important by the AI for learning and predicting classes, such as those provided by Feature Importance (type). These explanations may help experts in understanding what data features deserve more attention when deciding on suspicious cases and attend to R5
DF6: Predictions for what if values of data features	Explanations containing global and local (scope) changes on values of data features (target) which can switch the AI model prediction for a case (type), such as those provided by Counterfactual explanation methods. These explanations help in visualizing needed changes to shift AI predictions and enable fine-tuning against errors, which can attend R3 and R7
DF7: Rules as conditions for predictions	Explanations containing rules in natural language for global predictions (scope) which possess data feature values (target) and their combinations for AI model predictions, such as those provided by rule-based explanations and Decision Trees (type). These explanations provide human-readable relationships from the data and AI and attend to R6
DF8: Data features impacting particular predictions	Explanations containing local (scope) data features (target) that impact the AI model predictions for a given class, such as those provided by Feature Impact methods (type). These explanations may enable the user with understanding the important features that impact local cases towards being true positives for the target class and attend to R3, R4 and R7

between explanations and expert's tasks and requirements. Section 5 presents the evaluation of our principles regarding the achievement of the goal established in this study.

5   Evaluation and Results

5.1   Evaluation and Experiment Design

We conduct a naturalistic ex-ante evaluation to assess the extent to which our design principles attend the utility and quality requirements established at the start of the project [42]. The ex-ante evaluation aims to assess the partial design of artefacts before their deployment in real settings. We interviewed three fraud experts (minimum of 3 years of experience) within a bank partner. To structure the evaluation, we adopt the utility and information quality framework of [24]. Those authors there provided a practical framework with semiotic-based pragmatic, semantic, and syntactic levels to establish evaluation criteria for the information quality of DSR artefacts. The framework is suitable for complex design environments, which matches our study context of aligning diverse explanation methods to user requirements for XAIDSS in fraud detection.

To perform an evaluation following [24] framework, we employ a problem-centered interview method [63], which is an approach ratified for ex-ante evaluation in DSR studies [55]. We interviewed each expert for one hour on the matters on how they perceive the correctness of DP's terminology matching their experience and knowledge based on syntactic quality (adequacy, accessibility, consistency), semantic quality (unambiguity, preciseness, understandability, interpretability, and accuracy), and the principle's instantiation helping in everyday work. We also encourage experts to provide reasons for their views and enrich our qualitative data collection. We aim to understand their stressing points and issues worthy of further investigation. Our evaluation strategy matches current research assessing the utility and impact of DP [36]. To structure the feedback collected, we also allow experts to provide their answers based on objective criteria following a Likert scale from extremely unlikely to likely.

We also perform a simulation to evaluate the quality of design principles based on their instantiation. We instantiate the principles by developing an interface mockup to implement explanation methods that reflect the design features within our principles, illustrated in Fig. 2. The mockup instantiates DP1 by showing the AI confidence after the training and testing phases (DP1). It instantiates DP3 and DP5 by presenting the EM of Local Feature Importance (LFI), Global Feature Importance (GFI), and Feature Impact (FI) to provide relevant data features for predictions and their relationships. We instantiate DP2 and DP4 by presenting the EM of Prototypes for providing experts with similar cases to the fraud under analysis, and we provide Anchors for presenting rules governing predictions. The explanation methods described explain the predictions of a Random Forest model trained on a bank partner's dataset with 3269 suspicious cases out of 7653 transactions over three months. The transactions

belong to ten customers of the institution. The set of features adopted for training and testing are: amount, device, anonymous receiver ID, receiver location, sender location, and currency. For each customer, the institution has provided 75% of past transactions for training and 25% for testing. Python programming language version 3.8 is adopted for this implementation and the scikit-learn[3] library for training and testing ML models. Python libraries for LIME [48], Anchors [49] and SHAP [34] are adopted for the implementation of LFI, GFI, FI and Anchor rules. Regarding the EM of Prototypes, we follow the guidelines of [38,60] and train a KNearestNeighbor [44] classifier based on the SHAP values of the training instances. The complete implementation and dataset used for our simulation results are available externally.[4]

An expert would use the described mockup and explanations following our principles. Therefore, this simulation focuses on automatically evaluating the goodness of those explanations to estimate the user confidence in them. The methodology is aligned with the functionally-grounded evaluation established by [17], where an author defines proxy tasks for assessing how good an explanation is in achieving its goal without human participation. Given that our interface has multiple EMs, their goodness is computed according to the explanation types. For Local Feature Importance, Anchors, GFI, and FI, we compute their fidelity to the AI model being explained [40,43,46]. For that, we retrieve the features highlighted in those explanations and change their values in data instances until the prediction for those instances changes. We report the average prediction switching point (ASP). Lower values for switching prediction point indicate that the EM presents the features that contributed most towards the predicted class, which would foster user confidence in the explanation [40]. We compare this result with a random deletion of features. Ideally, the average prediction switching point should be lower than a random deletion switching point to assure the quality of explanations.

Concerning the EM of Anchors, we compute the percentage of instances that change their predicted class when following the Anchor rules to change feature values. That should be above 0.5 to ensure the quality of those explanations. For Prototypes, we compute the number of neighbors which match the correct label prediction for the current transaction under analysis. Therefore, we obtain the explanations for predictions belonging to the last suspicious transactions in the test set from five customers with suspicious fraud cases. Those customers are selected because the AI confidence was diverse, with levels ranging from 66% to 98%. We report the Anchor percentage of instances and percentage of Prototypes to illustrate the refereed explanations' goodness following our design principles.

---

[3] https://scikit-learn.org/.

[4] https://github.com/dougcirqueira/hcii-design-principles-user-centric-explainable-ai-fraud-detection/tree/main/resources/simulation.

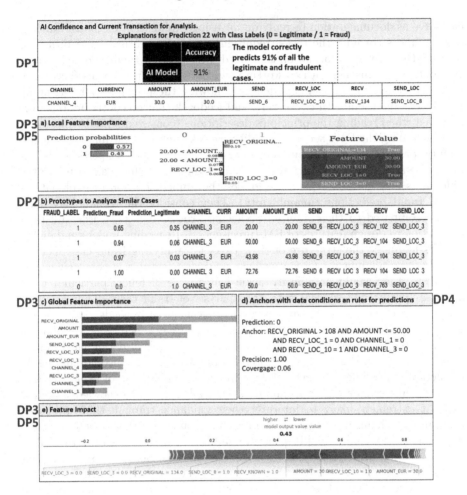

Fig. 2. Mockup with explanation methods following design principles

5.2 Evaluation Results

5.3 Information Quality of Design Principles with Syntactic and Semantic Criteria

Table 3 presents the questions and qualitative feedback from experts during our evaluation and the average rating selected by them regarding the information quality criteria assessed. Concerning the syntactic quality, our interviewed experts report that the terminology to describe design principles is extremely likely to be aligned with their internal practices (adequacy). However, they would like to have included examples of cases and fraudulent customer journeys. We suggested the addition of scenarios, such as developed in [11], and the experts agree those are good examples of additions to the DP for fulfilling that need. They also agree that the DP give a sense of better structuring practices and

can serve as documentation to reply on for using explanation methods. Further-more, experts consider the tasks, requirements, design features, and principles to bring the primary information they need, which enable them to quickly grasp the content and discuss with co-workers and interdisciplinary teams (accessibil-ity). Therefore, experts feel confident in following the principles for setting up EM, discussing them with co-workers, and feedback that no contradictions were observed when reading the principles descriptions (consistency).

Concerning the DP's semantic quality, we ask experts regarding DP's unam-biguity, preciseness, understandability, and interpretability. Experts agree our principles are extremely likely to be not ambiguous, as they can visualize the differences and how each principle guides the implementation of explanations for AI fraud predictions (unambiguity). They consider the terminology quite precise compared to their internal reviewing process of fraud cases and their connection to design features of EM (preciseness). Regarding understandability, experts highlight the structure of their tasks for analyzing fraud cases is clear and that the alignment of requirements to design features is comprehensible. Therefore, they would be able to tackle why to use or follow particular tasks and use specific EM to deal with the analysis of fraud cases. Therefore, the experts agree they are extremely likely to reflect on their needs when reading the DP for fraud detection, what they can do to support their work, and how to use them (understandability). They appreciate the levels of detail which match their practices, and they agree it is understandable that they would analyze AI predictions with different perspectives through a combination of explanations.

Regarding semantic interpretability specifically, experts acknowledge the DP make it extremely likely to interpret their guidance through the established user, aim, mechanism, and rationale in each principle description (interpretabil-ity). Experts conclude the DP structure and layout help understand the DP as it depicts each principle's main goals and what they can do to support their work. Experts are stimulated to give examples of such understanding. They give an example that DP could also help set up separate defense lines for fraud teams, where one team deals with the preliminary filtering of suspicious trans-actions through the first EM, and more complex cases or new schemes of fraud go through further analysis. For instance, one expert mentions he could filter out cases based on Local Feature Importance, and discuss more critical cases using further methods provided by our principles. They stress this aspect would also reduce the workload to fraud teams. Finally, experts agree the principles are extremely likely to be accurate regarding their tasks and requirements for fraud detection and how AI predictions can be explained based on their context and dataset that is daily analyzed (accuracy).

Overall, experts regarded the design principles as quite likely to be supportive in their context, as it can be observed by the average score of 2.33 in Table 3, as well as qualitative data and feedback provided by experts. The results highlight our DP's information quality, which would enable fraud experts to reflect on their fraud detection practices and how to employ XAIDSS.

Table 3. Information quality evaluation results for design principles (Expert rating goes from a –3 to 3 scale, where –3 and 3 represent extremely unlikely and likely, respectively. The letter E stands for Expert)

Theme	Questions and qualitative feedback (A for Answer)	Quantitative average expert rating			
		E1	E2	E3	Average
Syntactic information quality	1) Do you consider the design principles have an adequate representation in accordance with existing instructional material you have seen and used in your organization to understand software tools? (Adequacy) A1: DP terminology is aligned with internal practices. The principles could be accompanied by scenarios to illustrate the fraud detection context they are applicable	3	2	2	2.33
	2) Do you consider that the design principles have a good accessibility to yourself and for your discussion with colleagues regarding your needs for fraud detection, and how Explainable AI and methods can support you? (Accessibility) A1: DP can structure their practices and primary information for interdisciplinary discussions with colleagues that do not have deep knowledge of AI. There is no need for lengthy material given the principles A2: DP can serve as documentation to reply on for using EM to analyze and understand fraud cases	3	3	1	2.33
	3) Do you consider that the descriptions for design principles are consistent and do not bring contradictions? (Consistency) A1: No contradictions were observed when reading the DP descriptions. Each DP description is unique to understand their guidance and capabilities when instantiated A2: DP bring confidence for interdisciplinary discussion with colleagues from various departments, including technical and management levels	3	2	2	2.33
Semantic information quality	4) Do you consider the design principles description is not ambiguous and that there are no principles that could be viewed as the same? (Unambiguity) A1: DP description and template enable the visualization of the differences and how each principle guides the implementation of explanations for AI fraud predictions	3	3	1	2.33
	5) Do you consider the terminology in the design principles is precise to describe your needs for fraud detection tasks and analysis? (Preciseness) A1: DP terminology is precise compared to their internal discussion themes and analysis tasks for fraud cases, as well as design features of EM	3	2	2	2.33
	6) Can you interpret what the design principles can do to support your work based on their description for understanding fraud cases classified by Artificial Intelligence? (Understandability) A1: The flow from tasks to requirements and design principles is comprehensible and reasonable A2: Experts can understand why to use or follow particular tasks and using specific EM, as well as what EM can do to support their work and how to use them A3: DP highlight the different perspectives through a combination of explanations that can be leveraged to understand AI and review fraud cases	3	1	2	2
	7) Do you consider it is easy to understand the description of design principles related to your current work? (Interpretability) A1: Experts can understand the DP as it depicts each principle's main goals, what is behind them, and what they can do to support their work A2: DP can be considered for guiding organizational changes, such as for setting up separate defense lines for fraud teams	3	3	2	2.66
	8) Do you consider the description of design principles is accurate and free of error when you relate it to your current work and tools for fraud detection? (Accuracy) A1: DP description is accurate to reflect experts tasks and requirements for fraud detection, and how AI predictions can be explained A2: DP give the confidence to refer to internal practices and select appropriate tools for fraud detection while interacting with AI predictions	3	2	2	2.33

5.4   Quality of Design Principles Instantiation Through a Simulation

Regarding the simulation results, Table 4 shows the average prediction switching point for the explanations of LFI, GFI, and FI given different AI confidence levels. The explanations are provided with 50 random seeds. For these methods, the lower the ASP reported, the better the estimated user confidence. Columns 3 to 8 show the ASP for the referred methods. Those are compared to a random deletion of features. The numbers highlighted in bold mean that deleting features by importance reported by EM requires lesser deletions than a random selection, which ensures the quality of explanations. Next, Table 5 shows the percentage of cases with prediction changed following Anchor rules and the cases reported by Prototypes explanations belonging to the correct suspicious case classification. For these methods, the higher the reported value, the better the estimated user confidence. Column 3 shows the results for Anchor rules, which should be at least over 0.5 to be considered good in our scenario. Finally, column 4 shows the simulation results for Prototypes.

Table 4. Estimated user confidence based on average prediction switching point for instantiated explanation methods of Local Feature Importance (LFI), Global Feature Importance (GFI) and Feature Impact (FI)

Average AI model confidence for transaction	Customer ID and transaction number	The lower the value, the better the estimated user confidence					
		LFI		GFI		Feature impact (FI)	
		ASP by deleting explanation features	ASP by random deletion	ASP by deleting explanation features	ASP by random deletion	ASP by deleting explanation Features	ASP by random deletion
0.98	SEND_1 - 846	1	1	1	1	**0.63**	1
0.95	SEND_2 - 115	**0.2**	0.8	**0.94**	0.95	**0.61**	0.95
0.89	SEND_4 - 37	**0.2**	0.56	**0.17**	0.22	0.24	0.18
0.82	SEND_1 - 845	1	0.88	1	1	**0.45**	1
0.79	SEND_3 - 89	**0.22**	0.56	**0.1**	0.12	0.28	0.1
0.75	SEND_4 - 36	**0.17**	0.42	**0.24**	0.28	**0.16**	0.27
0.69	SEND_10 - 22	**0.24**	0.49	**0.17**	0.26	**0.23**	0.24
0.66	SEND_4 - 35	**0.11**	0.36	**0.15**	0.17	0.15	0.15

Table 5. Estimated user confidence for instantiated explanation methods of Anchors and Prototypes

Average AI model confidence for transaction	Customer ID and transaction number	The higher the value, the better the estimated user confidence	
		Average percentage of instances affected by anchors rules	Average percentage of prototypes with the same label as the transaction being analyzed
0.98	SEND_1 - 846	0	1
0.95	SEND_2 - 115	0.06	1
0.89	SEND_4 - 37	**0.78**	1
0.82	SEND_1 - 845	0	1
0.79	SEND_3 - 89	**0.98**	1
0.75	SEND_4 - 36	**0.6**	1
0.69	SEND_10 - 22	**0.64**	0.93
0.66	SEND_4 - 35	**0.86**	1

6   Discussion and Implications

6.1   Practical Implications of Design Principles

DP1. Each principle would have practical implications when instantiated by researchers and fraud experts. DP 1 states it is essential to enable fraud experts with the capabilities and limitations of AI. Probabilities provide an explicit visualization of AI models' confidence for predictions based on a trained dataset [32,47]. Moreover, the method of Adversarial explanations enables experts to assess the cases that would affect legitimate predictions. Adversarial transactions are generated in order to deceive the AI, which enables experts to spot weaknesses of the trained models. Adversarial explanations might also be useful for fraud prevention purposes, where simulated fraud schemes can be fed to the AI, which has to determine their legitimacy [66]. Without the instantiation of such a principle, experts might become overconfident in AI predictions and explanations. Overconfidence is not ideal as users might be misled by blindly believing the AI is correct in every prediction, which can cause damage to AI stakeholders and end-users [12].

DP2. The instantiation of DP 2 provides experts with similarities and dissimilarities to a current case under analysis. According to experts during our evaluation, it is essential to analyze customer behaviour from a dynamic perspective, as it changes over time. Such principle supports the analysis of typical patterns within a dataset [21], as experts need to get insights on typical behaviour of legitimate users. Notably, the EM of Prototypes and Criticisms would play a role in the instantiation of DP 2 [27]. Without this principle, experts would rely only on important features and lack the analogy perspective for understanding predictions, which is inherently part of how humans aim to digest explanations [37].

DP3. DP 3 instantiation enables experts to look closely into what data features the AI is considering as the most important when learning legitimate and fraudulent patterns from the whole dataset of customer transactions. EM methods fulfilling this DP are among the most used in extant Explainable AI studies, as the understanding of important features enables not only explanations for the AI behaviour but also to clarify if the model is working correctly, which is valuable for AI engineers aiming to optimize their models [29]. In the scenario of fraud detection, the analysis of important features is among the first tasks performed by experts when reviewing cases [11]. Therefore, without DP 3, experts would lack explanations for understanding whether the AI is focusing on the correct parameters, according to their domain knowledge in fraud cases. The wide adoption of such a principle is reflected by 66% of studies retrieved in our systematic literature review to establish EM design features, which adopt Global Feature Importance as part of explanations.

DP4. Fraud experts are constantly under pressure for protecting customers and being efficient in reviewing fraud cases. DP 4 instantiation enables a friendly and quick overview of fraud cases, as it advocates for the provision of rules and human-friendly explanations. With rules, experts have an overview of multiple data features and values at once. Indeed, experts have highlighted the need to observe illustrations of the effects of multiple features in AI predictions. When discussing DP 4, experts emphasize they get the sense that its instantiation helps them in deciding faster what an AI prediction means, and if it is right, or if the AI is "thinking wrongly". Without DP 4, experts would not rely on a language-friendly explanation to digest AI predictions, which might affect their performance when working with colleagues and communicating to customers the reasons for suspicious fraud [50].

DP5. Transaction fraud and customer datasets have a temporal nature, and experts need to analyze the local impact of data features on single transactions, as well as the influence of past behaviour on current transactions. DP 5 instantiation enables experts to observe the local feature impact and the relationships between features in the dataset. Furthermore, the relationship between multiple instances, as transactions or customers, is also embedded into this principle. This principle attends fraud expert tasks when analyzing complex fraud cases involving multiple actors and comparing past patterns, which is constantly performed during their analyses whether through Feature Impact EM or network and graph visualizations of multiple customer transactions over time [9, 16, 26]. Without DP 5, experts would lack such features for understanding AI with temporal and multiple feature and instances perspectives. Therefore, we perceive our design principles are aligned with the user-centric XAI community and can be useful for instantiating XAIDSS to conducting empirical studies and assess the impact of explanations on fraud experts' work and confidence in AI.

6.2 Simulation Findings

From a quantitative perspective, our simulation results illustrate the quality of DP's instantiations for fraud experts. For the EM of LFI and GFI, the ASP is lower for features deleted based on explanations in 6 out of 8 customer cases, respectively. When changing features based on Anchor rules, the prediction changes for more than 50% of transactions for 5 out of 8 customer cases. Regarding the EM of Prototypes, the method returns at least 93% of similar transactions to a suspicious case that belongs to the same class. Therefore, the reported results obtained based on the ASP computation highlight the data features deemed as important by the AI model implemented, impacting predictions if compared to random features, which is required by expert tasks when analyzing suspicious transactions. Finally, experts compare and contrast fraud cases when analyzing suspicious transactions, and the results are satisfactory to attend those needs as illustrated by Anchor and Prototype reported simulation results.

7   Conclusions

Experts recognize the value of Artificial Intelligence decision support for fraud analysis and detection, despite the lack of transparency of black-box models and, consequently, trust in AI predictions. We develop design principles to align expert requirements and explanation methods for decision support and understanding AI predictions. We adopt an Information Systems perspective and a design science research methodology in the study. Based on the results, we argue that IS theoretical lens is valuable towards user-centric Explainable AI development and worth further investigation. The principles may contribute to user-centric XAI design knowledge and Explainable AI decision support in fraud detection, given their foundation based on industry practices and literature. The developed design principles could impact fraud experts' working processes and guide designing fraud operations based on the information provided by explanation methods through workload splitting. The principles could also be considered by companies developing fraud detection solutions taking into account explainability requirements for the users of such tools.

As limitations, experts highlighted the interest in a prototype with interactivity. However, no interface was deployed at this time, which will be addressed at another iteration of the project. Therefore, as future work, we will conduct an ex-post naturalistic evaluation of the design principles and expand the dataset for experiments with a large bank partner. We can evaluate our artifact's instantiation by assessing an expert's efficiency when using explanation methods instantiated by our design principles. Researchers could also assess the design principles in different fraud detection contexts, including phishing cases. It can be further considered implementing a process perspective for collaboration between experts and explanations for trustworthy and explainable AI decision support, enhancing the efficiency of teams splitting their workload.

References

1. Abdul, A., Vermeulen, J., Wang, D., Lim, B.Y., Kankanhalli, M.: Trends and trajectories for explainable, accountable and intelligible systems: an HCI research agenda. In: Proceedings of the 2018 CHI Conference on Human Factors in Computing Systems, pp. 1–18 (2018)
2. Adadi, A., Berrada, M.: Peeking inside the black-box: a survey on explainable artificial intelligence (XAI). IEEE Access **6**, 52138–52160 (2018)
3. Antwarg, L., Shapira, B., Rokach, L.: Explaining anomalies detected by autoencoders using shap. arXiv:1903.02407 (2019)
4. Arrieta, A.B., et al.: Explainable artificial intelligence (XAI): concepts, taxonomies, opportunities and challenges toward responsible ai. Inform. Fus. **58**, 82–115 (2020)
5. Arya, V., et al.: One explanation does not fit all: a toolkit and taxonomy of ai explainability techniques. arXiv:1909.03012 (2019)
6. Arzate Cruz, C., Igarashi, T.: A survey on interactive reinforcement learning: design principles and open challenges. In: Proceedings of the 2020 ACM Designing Interactive Systems Conference, pp. 1195–1209 (2020)
7. Bento, J., Saleiro, P., Cruz, A.F., Figueiredo, M.A., Bizarro, P.: Timeshap: explaining recurrent models through sequence perturbations. arXiv:2012.00073 (2020)

8. Cakir, G., Iftikhar, R., Bielozorov, A., Pourzolfaghar, Z., Helfert, M.: Omnichannel retailing: digital transformation of a medium-sized retailer. J. Inform. Technol. Teach. ases p. 2043886920959803 (2021). https://doi.org/10.1177/2043886920959803

9. Cheng, D., Wang, X., Zhang, Y., Zhang, L.: Graph neural network for fraud detection via spatial-temporal attention. IEEE Trans. Knowl. Data Eng. **28**(10), 2765–2777(2020)

10. Cirqueira, D., Hofer, M., Nedbal, D., Helfert, M., Bezbradica, M.: Customer purchase behavior prediction in e-commerce: a conceptual framework and research agenda. In: International Workshop on New Frontiers in Mining Complex Patterns, pp. 119–136. Springer (2019)

11. Cirqueira, D., Nedbal, D., Helfert, M., Bezbradica, M.: Scenario-based requirements elicitation for user-centric explainable AI. In: International Cross-Domain Conference for Machine Learning and Knowledge Extraction, pp. 321–341. Springer (2020)

12. Collaris, D., Vink, L.M., van Wijk, J.J.: Instance-level explanations for fraud detection: a case study. arXiv:1806.07129 (2018)

13. Creedon, F., O'Kane, T., O'Donoghue, J., Adam, F., Woodworth, S., O'Connor, S.: Evaluating the utility of the Irish HSE's paper based early warning score chart: a reflective data gathering phase for the design of the reviews framework. In: DSS, pp. 165–176 (2014)

14. Dal Pozzolo, A., Boracchi, G., Caelen, O., Alippi, C., Bontempi, G.: Credit card fraud detection: a realistic modeling and a novel learning strategy. IEEE Trans. Neural Netw. Learn. Syst. **29**(8), 3784–3797 (2017)

15. Dellermann, D., Lipusch, N., Ebel, P., Leimeister, J.M.: Design principles for a hybrid intelligence decision support system for business model validation. Electron. Markets **29**(3), 423–441 (2019)

16. Didimo, W., Liotta, G., Montecchiani, F., Palladino, P.: An advanced network visualization system for financial crime detection. In: 2011 IEEE Pacific visualization symposium, pp. 203–210. IEEE (2011)

17. Doshi-Velez, F., Kim, B.: Towards a rigorous science of interpretable machine learning. arXiv:1702.08608 (2017)

18. Du, M., Liu, N., Hu, X.: Techniques for interpretable machine learning. Commun. ACM **63**(1), 68–77 (2019)

19. Dudley, J.J., Kristensson, P.O.: A review of user interface design for interactive machine learning. ACM Trans. Interact. Intell. Syst. **8**(2), 1–37 (2018)

20. Floridi, L.: Establishing the rules for building trustworthy AI. Nat. Mach. Intell. **1**(6), 261–262 (2019)

21. Gee, A.H., Garcia-Olano, D., Ghosh, J., Paydarfar, D.: Explaining deep classification of time-series data with learned prototypes. arXiv:1904.08935 (2019)

22. Gregor, S., Hevner, A.R.: Positioning and presenting design science research for maximum impact. MIS Q. **3**(2), 337–355 (2013)

23. Gregor, S., Kruse, L.C., Seidel, S.: The anatomy of a design principle. J. Assoc. Inform. Syst. **21**(6), 1622–1652 (2020)

24. Helfert, M., Donnellan, B., Ostrowski, L.: The case for design science utility and quality-evaluation of design science artifact within the sustainable ict capability maturity framework. Syst. Signs Act.: Int. J. Inform. Technol. Act. Communi. Workpract. **6**(1), 46–66 (2012)

25. Hochreiter, S., Schmidhuber, J.: Long short-term memory. Neural Comput. **9**(8), 1735–1780 (1997)

26. Jesus, S., et al.: How can i choose an explainer? an application-grounded evaluation of post-hoc explanations. arXiv:2101.08758 (2021)

27. Kim, B., Koyejo, O., Khanna, R., et al.: Examples are not enough, learn to criticize! criticism for interpretability. In: NIPS, pp. 2280–2288 (2016)
28. Klopper, R., Lubbe, S., Rugbeer, H.: The matrix method of literature review. Alternation **14**(1), 262–276 (2007)
29. Koh, P.W., Liang, P.: Understanding black-box predictions via influence functions. In: International Conference on Machine Learning, pp. 1885–1894. PMLR (2017)
30. Kumari, P., Mishra, S.P.: Analysis of credit card fraud detection using fusion classifiers. In: Computational Intelligence in Data Mining, pp. 111–122. Springer (2019)
31. Laughlin, B., Sankaranarayanan, K., El-Khatib, K.: A service architecture using machine learning to contextualize anomaly detection. J. Database Manage. **31**(1), 64–84 (2020)
32. Le, T., Wang, S., Lee, D.: Why x rather than y? explaining neural model'predictions by generating intervention counterfactual samples (2018)
33. Li, Z., Liu, G., Jiang, C.: Deep representation learning with full center loss for credit card fraud detection. IEEE Trans. Comput. Soc. Syst. **7**(2), 569–579 (2020)
34. Lundberg, S.M., Lee, S.I.: A unified approach to interpreting model predictions. In: Advances in Neural Information Processing Systems, pp. 4765–4774 (2017)
35. Marino, D.L., Wickramasinghe, C.S., Manic, M.: An adversarial approach for explainable AI in intrusion detection systems. In: IECON 2018–44th Annual Conference of the IEEE Industrial Electronics Society, pp. 3237–3243. IEEE (2018)
36. Meier, P., Beinke, J.H., Fitte, C., Teuteberg, F., et al.: Generating design knowledge for blockchain-based access control to personal health records. Inform. Syst. e-Bus. Manage. **19**, 1–29 (2020)
37. Miller, T.: Explanation in artificial intelligence: insights from the social sciences. Artif. Intell. **267**, 1–38 (2019)
38. Molnar, C.: Interpretable Machine Learning. Lulu. com (2020)
39. Mueller, S.T., Hoffman, R.R., Clancey, W., Emrey, A., Klein, G.: Explanation in human-ai systems: a literature meta-review, synopsis of key ideas and publications, and bibliography for explainable AI. arXiv:1902.01876 (2019)
40. Nguyen, D.: Comparing automatic and human evaluation of local explanations for text classification. In: Proceedings of the 2018 Conference of the North American Chapter of the Association for Computational Linguistics: Human Language Technologies, Volume 1 (Long Papers). pp. 1069–1078 (2018)
41. Nunes, I., Jannach, D.: A systematic review and taxonomy of explanations in decision support and recommender systems. User Model. User-Adap. Inter. **27**(3–5), 393–444 (2017)
42. Ostrowski, L., Helfert, M., Hossain, F.: A conceptual framework for design science research. In: International Conference on Business Informatics Research, pp. 345 354. Springer (2011)
43. Papenmeier, A., Englebienne, G., Seifert, C.: How model accuracy and explanation fidelity influence user trust. arXiv:1907.12652 (2019)
44. Peterson, L.E.: K-nearest neighbor. Scholarpedia **4**(2), 1883 (2009)
45. Raj, S.B.E., Portia, A.A.: Analysis on credit card fraud detection methods. In: 2011 International Conference on Computer, Communication and Electrical Technology (ICCCET), pp. 152–156. IEEE (2011)
46. Ramon, Y., Martens, D., Provost, F., Evgeniou, T.: Counterfactual explanation algorithms for behavioral and textual data. arXiv:1912.01819 (2019)
47. Renard, X., Laugel, T., Lesot, M.J., Marsala, C., Detyniecki, M.: Detecting potential local adversarial examples for human-interpretable defense. In: Joint European Conference on Machine Learning and Knowledge Discovery in Databases, pp. 41–47. Springer (2018)

48. Ribeiro, M.T., Singh, S., Guestrin, C.: "why should I trust you?": Explaining the predictions of any classifier. In: Proceedings of the 22nd ACM SIGKDD International Conference on Knowledge Discovery and Data Mining, pp. 1135–1144. San Francisco, CA, USA, 13–17 Aug 2016

49. Ribeiro, M.T., Singh, S., Guestrin, C.: Anchors: high-precision model-agnostic explanations. In: AAAI Conference on Artificial Intelligence (AAAI) (2018)

50. Rosenfeld, A., Richardson, A.: Explainability in human-agent systems. Auton. Agent. Multi-Agent Syst. 33(6), 673–705 (2019)

51. Rosson, M.B., Carroll, J.M.: Scenario based design. Human-Computer Interaction, pp. 145–162. Boca Raton, FL (2009)

52. Seidel, S., Chandra Kruse, L., Székely, N., Gau, M., Stieger, D.: Design principles for sensemaking support systems in environmental sustainability transformations. Eur. J. Inf. Syst. 27(2), 221–247 (2018)

53. Seidel, S., Watson, R.T.: Integrating explanatory/predictive and prescriptive science in information systems research. Commun. Assoc. Inf. Syst. 47(1), 12 (2020)

54. Sokol, K., Flach, P.: Explainability fact sheets: a framework for systematic assessment of explainable approaches. In: Proceedings of the 2020 Conference on Fairness, Accountability, and Transparency, pp. 56–67 (2020)

55. Sonnenberg, C., Vom Brocke, J.: Evaluation patterns for design science research artefacts. In: European Design Science Symposium, pp. 71–83. Springer (2011)

56. Venable, J.: The role of theory and theorising in design science research. In: Proceedings of the 1st International Conference on Design Science in Information Systems and Technology (DESRIST 2006), pp. 1–18. Citeseer (2006)

57. Walls, J.G., Widmeyer, G.R., El Sawy, O.A.: Building an information system design theory for vigilant EIS. Inf. Syst. Res. 3(1), 36–59 (1992)

58. Wang, D., Yang, Q., Abdul, A., Lim, B.Y.: Designing theory-driven user-centric explainable AI. In: Proceedings of the 2019 CHI Conference on Human Factors in Computing Systems, pp. 1–15 (2019)

59. Webster, J., Watson, R.T.: Analyzing the past to prepare for the future: writing a literature review. MIS Q. 26(2), xiii–xxiii (2002)

60. Weerts, H.J., van Ipenburg, W., Pechenizkiy, M.: Case-based reasoning for assisting domain experts in processing fraud alerts of black-box machine learning models. arXiv:1907.03334 (2019)

61. Weerts, H.J., van Ipenburg, W., Pechenizkiy, M.: A human-grounded evaluation of shap for alert processing. arXiv:1907.03324 (2019)

62. West, J., Bhattacharya, M.: Intelligent financial fraud detection: a comprehensive review. Comput. Secur. 57, 47–66 (2016)

63. Witzel, A., Reiter, H.: The Problem-Centred Interview. Sage (2012)

64. Wolf, C.T.: Explainability scenarios: towards scenario-based xai design. In: Proceedings of the 24th International Conference on Intelligent User Interfaces, pp. 252–257 (2019)

65. Yang, Q., Suh, J., Chen, N.-C., Ramos, G.: Grounding interactive machine learning tool design in how non-experts actually build models. In: Proceedings of the 2018 Designing Interactive Systems Conference, pp. 573–584 (2018)

66. Yuan, X., He, P., Zhu, Q., Li, X.: Adversarial examples: attacks and defenses for deep learning. IEEE Trans. Neural Netw. Learn. Syst. 30(9), 2805–2824 (2019)

67. Zerilli, J., Knott, A., Maclaurin, J., Gavaghan, C.: Transparency in algorithmic and human decision-making: is there a double standard? Philos. Technol. 32(4), 661–683 (2019)

68. Zheng, L., Liu, G., Yan, C., Jiang, C.: Transaction fraud detection based on total order relation and behavior diversity. IEEE Trans. Comput. Soc. Syst. 5(3), 796–806 (2018)

Disentangling Trust and Anthropomorphism Toward the Design of Human-Centered AI Systems

Theodore Jensen[✉]

University of Connecticut, Storrs, CT 06269, USA
theodore.jensen@uconn.edu

Abstract. Advances in artificial intelligence will continue to broaden the space of applications of AI-based systems. Trust is a complexity-reducing mechanism that will guide human interactions with AI and anticipation of AI behavior. Similarly, anthropomorphizing, or attributing humanlike qualities to an entity, allows humans to predict and prepare for the complex and motivated behavior of humanlike entities in our environment. The shared role of these two processes is particularly relevant to interactions with complex and humanlike AI systems. This paper firsts describes the nature of trust and anthropomorphism as predictive processes, how they are similar, and how they differ. Then, how each may be considered in the design of human-centered AI systems is discussed, including some future research directions.

Keywords: Ethical and trustworthy AI · Human-centered AI · Anthropomorphism · Trust

1 Introduction

Advances in artificial intelligence (AI) mean that humans will (and often already do) come into contact with complex, AI-based systems on a daily basis. In such interactions, the human user is generally exposed only to small set of information and cues about the underlying technical system. The interface, including system messages and appearance, is consequential for the user's perceptions of and behavior with the system.

In particular, such design features play a critical role in the user's trust in the AI system. Trust is a complexity-reducing mechanism used to limit the space of potential outcomes of interaction with another individual. As with other humans and with automated systems, perceptions of trustworthiness will guide our expectations of AI systems and our anticipation of system behavior. Similarly, anthropomorphizing, or attributing humanlike qualities to a non-human entity, allows us to predict and prepare for the complex and motivated behavior of humanlike entities in our environment. The shared role of these two processes is particularly germane to AI user trust for two reasons: 1) AI trustees

© Springer Nature Switzerland AG 2021
H. Degen and S. Ntoa (Eds.): HCII 2021, LNAI 12797, pp. 41–58, 2021.
https://doi.org/10.1007/978-3-030-77772-2_3

are inherently complex, and 2) the trustee is non-human, but may be perceived as considerably humanlike.

Overall, trust and humanness lie at the heart of safe and effective collaborations between humans and AI. This paper is not intended to be an extensive review of the respective bodies of literature–the reader is referred to the following articles for reviews of the literature on trust in automation [12,18] and AI [5], and anthropomorphism [4]. Instead, this paper focuses on the intertwined relationship between trust and anthropomorphism and its implications for the design of human-centered AI systems. First, the nature of both trust and anthropomorphism as predictive processes are explored, as well as how they have been studied in human-machine interaction. The subsequent section discusses how the processes are similar and how they differ. Then, some suggestions are made for researchers and designers regarding how trust and anthropomorphism can be leveraged toward the design of human-centered AI systems.

2   Trust

Trust has been defined as:

> *"the willingness of a party to be vulnerable to the actions of another party another based on the expectation that the other will perform a particular action important to the trustor, irrespective of the ability to monitor or control that party"* [24]

While this definition was developed with human-human interaction in mind, it accommodates non-human trustees. The "particular action important to the trustor" exists in task-based interactions with computers and automation. Moreover, users' or operators' "ability to monitor or control" is lacking given the degree of uncertainty about a technological system. This latter feature of the definition is the crux of human-machine trust, since the role of system designers is to develop system interfaces that quell the user's uncertainty by accurately and effectively communicating system strengths and limitations such that trust is appropriate and task goals can be achieved.

This section delves into the function of trust and how it has been studied in the context of human-machine interaction.

2.1   The Function of Trust

Luhmann describes trust as a means for reducing complexity in interactions with other actors [21]. The agency of other individuals in our environment means that we are faced with an infinite number of possible behaviors. Trust allows us to deal with this uncertainty by limiting the space of the potential behaviors of those individuals. If we decide that we trust a friend, we do not have to expend energy preparing for that friend's potential negative behaviors. Wicks, Berman and Jones [48] discussion of optimal trust emphasizes these benefits of trust in

firm-stakeholder relationships, and provides an illuminating quote from Baier [2]: "Part of what it is to trust is not to have too many thoughts about possible betrayals." Humans trust in order to live cooperatively.

Meanwhile, Luhmann notes that distrust is not the opposite of trust, but its functional equivalent [21]. If one decides that an individual is not to be trusted, complexity remains. Deciding that there are reasons to distrust, however, reduces complexity, given that cooperative behavior is not expected. Lewicki, McAllister, and Bies [20] note that often the healthiest organizational relationships are characterized by simultaneously high levels of trust and distrust. In such a relationship, the trustor is more certain of the trustee's future behavior and when it can be expected to be negative or positive. So, distrust also helps to reduce complexity. Humans distrust in order to protect themselves.

Game theory researchers have studied what strategies for cooperation with others are the most advantageous and, therefore, can evolve naturally among a population. This can elucidate the advantages of both trust and distrust among humans. Looking at various cooperative strategies, Axelrod and Hamilton [1] found that "Tit for Tat" consistently emerged as the most advantageous–this strategy involves cooperating with other individuals during the first interaction (i.e., trusting), and mirroring the other's previous action for each subsequent interaction. Essentially, this showed that an initial bias toward trusting could naturally emerge among a population. This trust is not blind, however, as individuals learn to defect from (i.e., not cooperate with) those who are not cooperative with them. Trust and distrust serve cooperative needs and protection from untrustworthy individuals, respectively. Wicks, Berman, and Jones [48] note that "prudence with a bias toward trust" permits safe cooperation. While trust is a morally and socially desirable trait, prudence allows one to discern between the trustworthy and untrustworthy in order to not be taken advantage of.

Overall, trust is a mechanism by which humans can cooperate with others. It reduces the uncertainty posed by the freedom of other individuals. Given this core function of trust, it follows that machines exhibiting some degree of independent behavior receive the same treatment as our human peers. In other words, we can trust machines.

2.2 Trust in Human-Machine Interaction

Trust in technology was first considered by Muir [27], who noted that different operators of the same automated system exhibit a variety of responses to that system and also that operators are faced with incomplete knowledge about complex systems. As a result, she suggested that human behavior toward automation is driven by "some intervening or organismic variable that mediates between the automation and the supervisor's responses to the automation," rather than solely the objective properties of that automated system [27]. In other words, while the objective characteristics of automation are influential, it is through operator or user perceptions that they influence behavior. This claim was substantiated as Muir found that experimental participants were able to report on their subjective level of trust in an automated system [27]. Moreover, Muir and

Moray [28] subsequently found that subjective trust was predictive of reliance on the automation. Reliance is a behavioral outcome of trust, akin to Mayer, Davis, and Schoorman's "risk-taking in relationship" [24] and underscores why an understanding of trust is consequential for human-machine trust research–system design influences operator trust and reliance, which influences human-machine performance.

Parasuraman and Riley [34] described the importance of trust in the design of automated systems given the potential for disuse and misuse. Disuse occurs when an operator neglects automation that could assist them in a task, whereas misuse occurs when an operator relies on automation on a task for which it is not suited. Disuse and misuse are the result of undertrust and overtrust, respectively, both conditions of inappropriate trust in automation.

Lee and See's [18] review provides guidance on how trust research can be leveraged in system design to prevent disuse and misuse. Trust calibration refers to the degree of matching between perceived and actual performance, while trust resolution and specificity refer to the degree to which operator trust distinguishes among contextual changes in system performance. System designers are faced with the challenge of communicating effectively and efficiently the information that operators need in order to foster an accurate perception of system capabilities. This is consistent with the function of trust for reducing complexity and uncertainty.

Hoff and Bashir's [12] more recent review categorizes the factors relevant to trust in automation as dispositional, situational, and learned, corresponding roughly to Mayer et al.'s [24] factors related to the trustor, context, and trustee. Hoff and Bashir also note that the relationship between trust and reliance depends on a number of factors. For instance, given a high degree of decisional freedom, operator trust and reliance have a strong relationship. With less decisional freedom, an operator may essentially have no other choice but to rely on the automated system on a given task–trust plays a less significant role in their decision to do so. In line with Lee and See [18], Hoff and Bashir [12]'s review stresses that the design of system interfaces shapes the perceptions of system performance and capabilities on which operator trust is based.

The increasing complexity of automated systems necessitated a consideration of operator trust in their design. Trust, after all, serves to reduce uncertainty in interactions with complex actors. The advent of artificial intelligence further brings trust to the forefront, as technological systems will now be characterized by an even greater degree of behavioral complexity. Compared to traditional automation, AI-based systems are arguably closer to humans in their degree of unpredictability. Madhavan and Wiegmann [22] posit that differences in trust in human and automated decision aids are based on notions of flexibility. Humans are viewed as capable of changing, whereas automation is viewed as static in its behavior. The question arises as to whether perceptions of AI will be more like automation or human. Do notions of machinelike rigidity extend to AI systems? Or does the presence of AI alter perceptions of a system as dynamic actors? This question highlights that technological trust is naturally connected to anthropomorphism.

3   Anthropomorphism

Anthropomorphism has been defined as the process of:

"imbuing the imagined or real behavior of nonhuman agents with human-like characteristics, motivations, intentions and emotions" [4]

While anthropomorphism is commonly associated with animals and supernatural agents, this definition clearly applies to interactions with computers, automation, and AI systems. The label "artificial intelligence" in and of itself is a widely accepted act of anthropomorphizing. It follows that anthropomorphism will play a fundamental role in human perceptions of AI systems.

This section explores the function of anthropomorphism and how it has been considered in research on human-machine interaction, after which we compare and contrast it with trust.

3.1   The Function of Anthropomorphism

Epley, Waytz, and Cacioppo [4] define anthropomorphism as a process of inductive inference that humans use to reason about non-human actors. Inductive inference involves the acquisition of knowledge, the elicitation of stored knowledge, and the application of knowledge to a given target [11]. The application stage also involves the use of alternative knowledge structures activated at the time of perception to correct the applied knowledge. When anthropomorphizing a non-human agent, humans essentially use a wealth of acquired knowledge about other humans and apply this to interpret and anticipate the behavior of the target. Anthropomorphizing determines whether the target agent is treated as "worthy of respect and concern" and, therefore, how that agent is expected to behave [4].

This theory expresses anthropomorphism in terms of three factors: elicited agent knowledge, effectance motivation, and sociality motivation [4]. As a cognitive basis for anthropomorphism, elicited agent knowledge is composed of the knowledge base about humans that is applied to the target, as well as information about non-human agents in general and information about that particular non-human agent. Epley et al. [4] note that, just as knowledge about the self is commonly applied in evaluations of other humans (i.e., egocentrism), knowledge about other humans is applied in the perception of non-human agents. Effectance motivation entails a human's desire to act effectively in their environment: "Attributing human characteristics and motivations to nonhuman agents increases the ability to make sense of an agent's actions, reduces the uncertainty associated with an agent, and increases confidence in predictions of this agent in the future" [4]. The theory suggests that when there is a high degree of uncertainty, or when a human has incentive to predict the target's behavior, anthropomorphism can help to satisfy the perceiver's situational needs. Sociality motivation consists of the human desire for connection with other humans. Anthropomorphism may satisfy the human need for social connection when it is lacking.

The identification of humanlike qualities in other actors is critical given the characteristics and behaviors associated with humanness and personhood. Haslam, Loughnan, Kashima, and Bain [10] describe humanness in terms of uniquely human and human nature traits. Uniquely human traits are those that typically separate humans from other species and are associated with civility and rationality, including secondary emotions like nostalgia and admiration. Human nature includes traits that may be shared with other species, but are typically emphasized as essential to being human, such as emotionality, warmth, and curiosity. While the anthropomorphizing of non-humans has been the focus of our discussion, Haslam [9] emphasizes that dehumanisation, the inverse of anthropomorphism, occurs when a human is denied these attributes of humanness. The lack of civility associated with a denial of uniquely human traits is observed in the likening of outgroups to other animals (i.e., animalistic dehumanisation). The rigidity and lack of emotionality associated with a denial of human nature traits likens outgroup members to machines (i.e., mechanistic dehumanisation).

Mechanistic dehumanisation is pertinent to human perceptions of AI-based systems, which are objectively non-human, but may be attributed or denied human nature traits. Characteristics of the technology may lead a system user to apply knowledge about humans to a greater or lesser degree. Overall, a tendency to socially respond to even minimally social computers has been observed [35], suggesting a general human sensitivity to social characteristics of technology. This may be the result of the particular threat that other humans posed in an evolutionary environment, making it advantageous to recognize the presence of a humanlike intelligence [32]. The function of anthropomorphism for the human species has important implications for human-machine interaction.

3.2   Anthropomorphism in Human-Machine Interaction

Consideration of the "humanness" of computers is not unique to the most recent advances in AI. In 1950, Turing proposed the imitation game to answer the question, "Can machines think?", which tested if an observer could tell whether a subject hidden from view was a machine or a human [39]. His claim was that meeting some objective standard for intelligence was besides the point–machine intelligence would be achieved when the computer was perceived as human. Humanness was implicated in the design of computers from the beginning, and has persisted even as computers have become ubiquitous in our daily lives. Turkle's early work emphasized the reflective role of computing technologies [40]– computers evoke thought about what makes us human, reflecting aspects of Haslam's aforementioned humanness traits [10].

Humans are, after all, perhaps the closest model we have for thinking about "intelligent" computers. The Computers as Social Actors (CASA) paradigm lends to this argument, and has been used by communication researchers to observe whether social interaction phenomena hold when one individual in an interaction is replaced by a computer [35]. For instance, in one prominent experiment, researchers tested whether social politeness norms extend to computers [31]. When evaluating another human, we are more positive when giving

direct feedback than when we give feedback indirectly. In the experiment, participants evaluated a computer's performance either on that same computer or on a separate computer. Results were consistent with human-human politeness behavior, in that indirect feedback was more negative than direct feedback, suggesting that participants in the direct feedback condition were being polite to the computer that they were evaluating. CASA has been applied in various other experiments, showing that gender stereotypes and social rules about praise and criticism [31], as well as reciprocal self-disclosure [26] extend to our interactions with computers.

Researchers have considered various explanations for these social responses, including conscious anthropomorphism [29]. In the CASA experiments, however, participants generally denied that they had treated the computer like another person. Moreover, samples often contained expert computer users who would know that the computers were not actually human. Nass, Steuer, Tauber, and Reeder [30] use the term "ethopoeia" instead of anthropomorphism in order to accommodate these findings–they defined anthropomorphism as a conscious and willing attribution of humanlike qualities to a nonhuman. Nass and Moon [29] propose that social responses to computers occur due to mindlessness–given a sufficient amount of social cues, humans cannot help but to respond socially to computers outside of conscious awareness. Reeves and Nass [35] effectively summarize this human tendency as, "When in doubt, treat it as human".

While these arguments consider anthropomorphism as an active and conscious cognitive process, CASA findings show that people treat computers as if they were other people–this aligns with definitions of anthropomorphism in social psychology that do not specify the degree to which the process is conscious [4,10]. Some researchers have suggested that anthropomorphism may be a mindless response to technology [16]. Epley et al. [4] also note that anthropomorphism may vary from strong to weak attributions of humanness, where the latter does not include an explicit endorsement of those beliefs by the perceiver. Authors note that, "Metaphors that might represent a very weak form of anthropomorphism can still have a powerful impact on behavior, with people behaving toward agents in ways that are consistent with these metaphors" [4]. In some ways, anthropomorphizing is inherently illogical because it involves application of knowledge about humans to nonhuman agents. As such, especially when not knowingly applied, anthropomorphism may have a significant influence on human behavior toward computing technologies, including AI systems.

Throughout the history of computing and up until the latest advances in AI, computing technologies have evoked thoughts about humanness. Anthropomorphism operates as a method of reasoning about nonhumans, including machines and computers, in order to effectively interact with those agents. As such, anthropomorphism and trust play a shared role in human relationships with AI systems. The next section reviews some of the literature that has connected the two, and compares and contrasts them with respect to their functions as human perceptual processes.

4   Trust and Anthropomorphism

Researchers tend to view anthropomorphism as a predictor of trust. Trust is a logical dependent variable because of its concrete implications for behavior toward technology. Anthropomorphism is thus often viewed as a characteristic of the technology. For instance, Gong [7] defines anthropomorphism as "the technological efforts of imbuing computers with human characteristics and capabilities". This approach tends to discount the fact that anthropomorphism is a process by which humans apply anthropocentric knowledge to a nonhuman. Technological features that are considered objectively humanlike are so because they are perceived as humanlike. Experiments finding that even minimally social interfaces elicit social responses [35] as well as the idea of weak anthropomorphism [4] tell us that anthropomorphism can occur more subtly and in the absence of overtly humanlike cues. This section considers how anthropomorphism is similar and different from trust in order to illuminate the distinction between technological features and human perceptions of them.

Nonetheless, the anthropomorphism-predicts-trust approach has contributed greatly to an understanding of the relationship between the two constructs. De Visser and colleagues [41–43] ran a series of experiments manipulating the degree of anthropomorphism of an automated agent that suggests answers to participants in a number-guessing task. The anthropomorphism manipulation involves both the appearance of the agent, which is shown via video throughout the task, as well as a background story which establishes the agent's intent. While each study differs in its other variables of interest, each includes decreasing agent reliability throughout the task and observes changes in participant's trust and willingness to comply with agent advice. These experiments found that greater anthropomorphism was associated with greater "trust resilience"–although not trusted more initially, declines in trust (in response to decreasing reliability) were more shallow for humanlike agents. Other researchers have found that greater anthropomorphism can increase trust and compliance with an agent [33,47]. Others did not find a behavioral effect but found greater subjective trust toward a more humanlike agent [17].

Other studies suggest that perceptions of anthropomorphism are not static. For instance, Salem and colleagues found that, in one study, a more smoothly operating robot was perceived as more anthropomorphic [37] and, in another, that inconsistency between a robot's gestures and spoken instructions lent to greater perceived anthropomorphism [36]. Jensen, Khan, and Albayram [15] found not only that a more reliable automated system was trusted and relied upon more, but that greater system reliability was associated with greater perceived anthropomorphism. These findings highlight an important component of Epley et al.'s [4] theory of anthropomorphism, in that the inductive process can be influenced during the acquisition, application, or correction of knowledge about humans toward non-humans. Perceptions of anthropomorphism toward an entity are not a fixed function of design features, but can change as a result of new information or observed agent behaviors. Technological characteristics such as reliability that are generally associated with perceived trustworthiness may just as well influence the anthropomorphizing process.

This comparison of trust and anthropomorphism aims to broaden perspectives on the role of the two processes in human-AI interaction.

4.1 Similarities

Trust and anthropomorphism align in that both are 1) based in uncertainty, 2) predictive, and 3) used for cooperation and protection.

Based in Uncertainty. Both trust and anthropomorphism are useful for dealing with uncertainty regarding the behavior of other actors. As mentioned, trust's role for complexity reduction is essential–without some notion of trust, we would live in a paralyzing state of fear [21], constantly expecting that other humans intended to hurt or take advantage of us. Anthropomorphism, likewise, gives us a means of accommodating the uncertain and, perhaps, unfamiliar behavior of non-humans in our environment. As will be discussed, however, a greater degree of anthropomorphism may not necessarily lead to a greater degree of certainty.

Predictive & Future-Oriented. The two processes are predictive and future-oriented. The aim of each is to be able to deal with the behavior of other actors that may impact our own well-being. Trust essentially answers the question, "Will this person help me or hurt me?" Anthropomorphism answers the question, "To what extent will this person behave as a human would?" In terms of system design, user perceptions related to both trust and anthropomorphism are critical to understanding the user's ability to accurately anticipate system behavior and, therefore, to overall task performance.

Cooperation & Protection. Lastly, both processes serve not only to facilitate effective cooperation, but to protect individuals. Trust and distrust help to characterize individuals' expected positive or negative behavior, respectively, in order to safely interact and achieve cooperative benefits. Anthropomorphism similarly serves to characterize the complexity of expected behavior and serves as a tool for interacting with non-humans in the absence of information about that specific agent or similar non-humans. It can prepare the perceiver for the potentially detrimental, intelligent behavior of the non-human target.

While these general functions are similar, the means by which trust and anthropomorphism inform interactions differ in some important ways.

4.2 Differences

Trust and anthropomorphism differ in terms of 1) complexity outcomes, 2) type of target actor they are commonly associated with, 3) counterpart processes, 4) nature of the conservative error, and 5) degree of flexibility.

Complexity Outcomes. For one, trust reduces complexity, leading to a simpler consideration of the pertinent actor. If a trustor decides that they trust another actor, they can consider a smaller (i.e., less complex) set of potential future behaviors. A decision to distrust leads to the same outcome (i.e., reduced complexity), although the set is comprised of more negative potential future behaviors. However, complexity is not necessarily reduced by anthropomorphism. When an actor is deemed humanlike, there may be less uncertainty regarding the unfamiliar non-human–the human metaphor applied to the non-human target can make the interaction more manageable. However, given that anthropomorphism is defined by the attribution of humanlike emotion and intentions, it may lead to the perceiver's expectation of more complex behavior. When confronted with a human or something humanlike, the perceiver must prepare for the motivations and intentions of another individual.

Type of Target Actor. In line with this difference in complexity outcomes, each process has been defined with respect to a different kind of actor. Mayer et al. [24] note that trust is not involved in all risk-taking behavior, citing an example of a farmer planting crops with the expectation that sufficient rain will fall. Their argument is that the farmer has no identifiable relationship with the "other party" (i.e., the weather) and, as such, cannot engage in "risk-taking in relationship," the behavioral outcome of trust. Thus, some non-human actors cannot be trusted according to this theory. Nonetheless, research on social responses to computers [35] and trust in automation [27] lend to the construct's viability with respect to machines. Researchers have also specifically suggested greater consideration of automated and autonomous systems as actors with which humans have relationships, rather than one-shot interaction partners [44]. Anthropomorphism is defined as the attribution of humanlike characteristics to non-humans. This precludes the possibility of anthropomorphizing another human, although researchers have referred to dehumanization as the inverse process by which human individuals or groups are denied humanlike traits [9].

Distrust vs. Dehumanization. The counterparts to trust and anthropomorphism are associated with different outcomes. Again, distrust is not an opposite of trust but a functional equivalent [21]–both allow the trustor to be more certain about the behavior of another actor. Trust and distrust lead the perceiver to the same place of greater certainty, although positive and negative behaviors are respectively expected. Dehumanization is the inverse process of anthropomorphism [9] and involves the denial of humanlike attributes to another human. According to this definition, a technological entity cannot be dehumanized, although it could be anthropomorphized to a lesser extent. A lesser degree of perceived anthropomorphism is not necessarily associated with less certainty, as with a lesser degree of trust. On the contrary, it may be the product of ample information about the non-human agent, such that anthropocentric knowledge is not needed. Research into particular features or behaviors that make a technology appear less humanlike (i.e., more machinelike) would illuminate how user

perceptions and behavior are influenced when anthropomorphism is reduced. For instance, one study found that blaming developers for an automated system's errors altered perceptions of trustworthiness compared to self-blame by the automation [14]. It is possible that mention of developers acts as a cue to diminish perceived anthropomorphism.

Conservative Errors. As protective perceptual mechanisms, trust and anthropomorphism may be biased in different ways. Error Management Theory (EMT) suggests that asymmetry between the costs of two errors in the human evolutionary environment can lead to a systematic deviation or bias toward making the less costly error [8]. With respect to trust, game theory researchers have shown that "Tit for Tat" is an evolutionarily stable strategy for interaction–individuals are predisposed to trust and cooperate with new interaction partners, but to subsequently punish and defect from those who prove to be untrustworthy [1]. Wicks, Berman, and Jones [48] notion of a bias toward trusting shaped by prudence mirrors this, highlighting the desirability of a state of trust alongside the protective role of distrust. A bias toward trust may occur because it is more costly to lose opportunities for cooperation (i.e., not trust) than to cooperate with some untrustworthy individuals (i.e., trust), at least initially. In applying anthropocentric knowledge or social treatment to non-humans, perceivers may be making the, "When in doubt, treat it as human" conservative error noted by Reeves and Nass [35]. To decide that some non-humans are humanlike is likely less costly than misidentifying an intelligent and motivated individual as non-human–use of more cognitive resources is a lesser risk than the harm another human may cause [32]. While these perceptual biases serve a similar purpose, the bias toward trusting is characterized by benefits of cooperation, while the bias toward anthropomorphizing is characterized by the risk of interacting with another human.

Degree of Flexibility. The degree to which levels of trust and anthropomorphism are flexible or change with respect to a given target may also differ. Trust is well-defined by its dynamic nature. Existence of terms such as "building" and "breaking" trust imply that it is an ongoing and fluid property of relationships, and underlies calls by researchers to consider trust repair by autonomous systems [44]. Likewise, "trust resilience" reflects the degree to which new information via observed behaviors is able to change a trust in an agent [43]. Anthropomorphism is generally portrayed as a static property. However, a change in the acquisition, application, or correction of anthropocentric knowledge that characterize the inductive inference process could alter the degree to which a target is anthropomorphized [4]. Recent evidence suggests that perceived anthropomorphism is influenced by system reliability [15], a known factor in human-machine trust, implying that system performance may alter anthropomorphism perceptions. Still, the differences mentioned above suggest that the dynamic profile of perceived anthropomorphism may look different than that of perceived trustworthiness. Trust by nature may need to be more flexible in order to continuously

predict positive or negative future behavior. Mayer et al. [24] include a feedback loop in their model to account for the effect of the observed outcomes of reliance on perceived trustworthiness and future trust. Changes in anthropomorphism reflect either cognitive changes via acquired agent knowledge or motivational changes via effectance or sociality motivations as a product of task demands [4]. More research is needed to determine the extent to which anthropomorphism perceptions of a target change over time.

5   Toward Human-Centered Design

The complexity and unpredictability inherent to AI systems places great importance on design for the human users of those systems. Because trust and anthropomorphism are predictive perceptual processes used to confront complexity, they are two highly important variables to consider during human-AI system design. Human-centered design has been defined in ISO 9241-210:2019 as the "approach to systems design and development that aims to make interactive systems more usable by focusing on the use of the system and applying human factors/ergonomics and usability knowledge and techniques" [13]. The human-centered design process is focused on consistently incorporating input from human stakeholders of a technological system when establishing context of use, user requirements, design solutions, and evaluation of designs. In general, an AI system that is human-centered will serve the human user or other stakeholder's task needs.

This final section integrates the prior discussion of trust and anthropomorphism toward some suggestions for areas of needed research to design human-centered AI.

5.1   Designing for Trust Repair and Trust Dampening

Due to the increasing autonomy of technological systems, researchers have highlighted the need for an understanding of trust as a dynamic property of human-autonomy systems and teams, toward the incorporation of trust repair capabilities into those systems [44,45]. Because trust helps individuals to anticipate an interaction partner's behavior, human-human trust repair, via social accounts such as apologies and attribution of blame [19], can help the trustee to manage the trustor's expectations of future behavior. Because of the increasing behavioral flexibility of technological systems, de Visser, Pak, and Shaw [44] suggest that systems engage in these relationship regulation acts in order to maintain appropriate trust. In other words, designers may incorporate messages or other features that dynamically facilitate users' understanding of not only the system's strengths, but its limitations. If a system does something detrimental to the user or the task, a *trust repair act* such as an apology or explanation can prevent trust from becoming substantially lower. Meanwhile, if a system does something beneficial for the user, a *trust dampening act*, such as a message regarding system limitations, can prevent trust from becoming inappropriately high. The degree

to which a user's perception of system trustworthiness aligns with actual system trustworthiness has been referred to as trust calibration [18].

Human-centered design can give designers information about stakeholders' needs in terms of trust repair and trust dampening. For example, in a given context of use, users may be prone to overtrust following an instance of good system performance, which can lead to unrealistic expectations of the system in situations where it is not capable. In such a circumstance, trust dampening by the system is necessary to calibrate user trust. Involvement of users can determine both when trust dampening is necessary and which types of messages or information can effectively dampen trust and improve performance. McDermott and ten Brink suggest that human factors practitioners evaluate user trust at "Calibration Points," or circumstances where system reliability changes, in order to observe whether trust is appropriately calibrated [25]. This practice can be extended throughout the stages of the human-centered design process, guiding designers on how to facilitate appropriate trust, and ensuring that AI systems serve human stakeholder needs.

Overall, fostering appropriate trust can improve task performance by contributing to a user's accurate perceptions of a technical system's strength and limitations. The next section gives some suggestions for appropriately incorporating anthropomorphism into system design in order to benefit users' understanding of a system.

5.2   Anthropomorphism as a Tool for User Understanding

Hoff and Bashir's [12] assertion about the role of interface design in perceptions of system performance emphasizes how, alongside trust, anthropomorphism is a critical consideration in the representation of an AI system and the delivery of information to users. Anthropomorphism has been largely viewed as an antecedent to trust in technology, and for good reason–it relates to the expectation of humanlike behavior, whereas trust relates more specifically to the valence of expected behavior. The literature also suggests that anthropomorphism itself can be dynamically influenced by factors in trust. Measurement of perceived anthropomorphism or humanlikeness in the design process can elucidate how these perceptions may fluctuate just as trustworthiness perceptions do.

As with trust appropriateness, anthropomorphizing should be considered as a means to performance consequences. Epley et al. [4] suggest that anthropomorphizing can be a useful analog for interacting with a non-human in the absence of other information, regardless of the objective accuracy of the anthropomorphic beliefs. In other words, it is possible that the human metaphor can assist user understanding of an AI-based system and lead to more appropriate expectations for system behavior. However, Culley and Madhavan [3] express concern for the role of anthropomorphic agents in eliciting inappropriate expectations for system behaviors. Even if qualities such as liking and user satisfaction are improved, if an objectively non-human system is perceived as having human qualities like motivation and emotion, those interacting with the system may expect it to do things it cannot. The general human tendency to respond

socially to computers [35] shows that anthropomorphizing can occur even as an unintended design consequence, which may represent a human vulnerability when interacting with complex AI systems. It is the responsibility of designers to protect human needs and interests with calculated and research-based incorporation of humanlike agents and features in their non-human systems. This includes human-centered evaluation of both perceptions of anthropomorphism and expectations of system capability even when systems are not deliberately anthropomorphic.

In this vein, research is needed into the factors that make anthropomorphic agents an effective interface tool in terms of task performance. Situating evaluation of system design in the context of task performance can help researchers and designers to conclude whether certain features help or hurt the user and their task goals. At the time of this writing, at least one study explicitly considered the role of humanlike features in user trust calibration [15]. Although a behavioral effect was not found, an effect of humanlike communication style on perceived benevolence of the automated system lends to the relationship between anthropomorphism and perceived trustworthiness found in other work [17,33,38,41–43]. In studies of human-AI interaction, we urge researchers to measure perceived anthropomorphism, in order to understand anthropomorphism as a consequential psychological phenomenon rather than as a direct consequence of design features.

Much like trust, evaluating anthropomorphism perceptions as a product of various system behaviors, features, or messages throughout the stages of the human-centered design process can improve understanding of how it functions as a useful perceptual process for the human users of AI systems. This can be leveraged toward design that improves safety and performance. For instance, consider a virtual assistant that helps its user maintain a calendar via voice command. The conversational interface may elicit the perception of certain humanlike capabilities–perhaps it leads the user to believe that the system "cares" enough to warn them when the timing of a newly scheduled event conflicts with another. In this case, the user's expectation of communication about the conflict by the virtual assistant may lead them to miss one of the overlapping events. Moreover, this violation of trust may lead to subsequent disuse of the virtual assistant. Evaluating perceptions of anthropomorphism and expectations of performance ahead of time would give designers information about the user's unrealistic perceptions of system capabilities arising as a result of the conversational interface, allowing them to appropriately communicate information or to re-design based on the potential for such an outcome.

Research on user- and context-based variability provides insights into potential applications of trust- and anthropomorphism-driven human-centered design. First, the use of anthropocentric knowledge to reason about an AI system may differ across users. Waytz, Cacioppo, and Epley [46] pursued this line of inquiry and established that there do exist individual differences in the tendency to anthropomorphize. The Individual Differences in Anthropomorphism Questionnaire (IDAQ) developed by these authors has been shown to explain variance

in perceptions of anthropomorphism of an automated system [15] and in trust in technology [46]. Likewise, Marakas, Johnson, and Palmer [23], drawing from CASA and the notion of mindless social responses to technology [29], suggest a spectrum of social attributions to computing technologies, where some individuals are more prone to treat computers as social actors and others to treat computers as tools. These perspectives suggest that not only different individual users, but groups of users may respond differently to the intended or unintended presence of humanlike qualities in an AI interface. Measurement of individual tendency to anthropomorphize throughout the design process can help to tailor system design to individuals or groups of end users.

Second, the value of a humanlike presence may differ across contexts of use. Smith, Allaham, and Wiese [38] confirmed that trust in a humanlike agent was greater on a task that was social in nature, while machinelike agents were preferred on a more analytical task. Similarly, Goetz, Kiesler, and Powers [6] showed that robot appearance interacted with task sociability, such that humanlike robots were preferred for more social tasks. Again, anthropomorphism may provide a suitable metaphor for interaction with non-human technical systems in some cases, but be a distraction or set unrealistic expectations in other circumstances. The degree to which humanlike features in an interface are useful toward a user's understanding of the AI system and task performance certainly should vary with characteristics of the task. The relationships between AI task characteristics, perceptions of anthropomorphism and trustworthiness, and task performance are important areas of research that can inform when certain design features should or should not be employed.

6   Conclusion

Trust and anthropomorphism, as perceptual mechanisms for dealing with the uncertain behavior of autonomous actors, will naturally extend to AI technologies. Both are based in uncertainty, predictive, and used toward cooperation and protection. Meanwhile, they differ in terms of complexity outcomes, the type of target actor they are commonly associated with, their counterpart processes, the nature of their conservative error, and their degree of flexibility. Measurement of perceived trustworthiness and perceived anthropomorphism throughout the design process, alongside performance indicators or metrics of user understanding, can ensure that these ingrained perceptual processes are being sufficiently considered in AI system design. As researchers and system designers, an understanding of the two can contribute to systems that accommodate trust and anthropomorphism, that protect and promote human stakeholders' interests, and that ultimately facilitate safe and effective collaborative outcomes, ensuring that AI systems serve human needs.

References

1. Axelrod, R., Hamilton, W.D.: The evolution of cooperation. Science 211(4489), 1390–1396 (1981)

2. Baier, A.: Moral Prejudices. Harvard University Press, Cambridge (1994)
3. Culley, K.E., Madhavan, P.: A note of caution regarding anthropomorphism in HCI agents. Comput. Hum. Behav. **29**(3), 577–579 (2013)
4. Epley, N., Waytz, A., Cacioppo, J.T.: On seeing human: a three-factor theory of anthropomorphism. Psychol. Rev. **114**(4), 864 (2007)
5. Glikson, E., Woolley, A.W.: Human trust in artificial intelligence: review of empirical research. Acad. Manage. Ann. (JA) **14**, 627–660 (2020)
6. Goetz, J., Kiesler, S., Powers, A.: Matching robot appearance and behavior to tasks to improve human-robot cooperation. In: Proceedings of the 12th IEEE International Workshop on Robot and Human Interactive Communication, pp. 55–60. IEEE (2003)
7. Gong, L.: How social is social responses to computers? The function of the degree of anthropomorphism in computer representations. Comput. Hum. Behav. **24**(4), 1494–1509 (2008)
8. Haselton, M.G., Nettle, D., Murray, D.R.: The evolution of cognitive bias. In: The Handbook of Evolutionary Psychology, pp. 1–20 (2015)
9. Haslam, N.: Dehumanization: an integrative review. Pers. Soc. Psychol. Rev. **10**(3), 252–264 (2006)
10. Haslam, N., Loughnan, S., Kashima, Y., Bain, P.: Attributing and denying humanness to others. Eur. Rev. Soc. Psychol. **19**(1), 55–85 (2008)
11. Higgins, E.T.: Knowledge activation: accessibility, applicability, and salience. In: Social Psychology: Handbook of Basic Principles, pp. 133–168 (1996)
12. Hoff, K.A., Bashir, M.: Trust in automation: integrating empirical evidence on factors that influence trust. Hum. Factors **57**(3), 407–434 (2015)
13. Ergonomics of Human-System Interaction - Human-Centred Design for Interactive Systems. Standard. International Organization for Standardization, Geneva, CH (2019)
14. Jensen, T., Albayram, Y., Khan, M.M.H., Fahim, M.A.A., Buck, R., Coman, E.: The apple does fall far from the tree: user separation of a system from its developers in human-automation trust repair. In: Proceedings of the 2019 on Designing Interactive Systems Conference, pp. 1071–1082 (2019)
15. Jensen, T., Khan, M.M.H., Albayram, Y.: The role of behavioral anthropomorphism in human-automation trust calibration. In: Degen, H., Reinerman-Jones, L. (eds.) HCII 2020. LNCS, vol. 12217, pp. 33–53. Springer, Cham (2020). https://doi.org/10.1007/978-3-030-50334-5_3
16. Kim, Y., Sundar, S.S.: Anthropomorphism of computers: is it mindful or mindless? Comput. Hum. Behav. **28**(1), 241–250 (2012)
17. Kulms, P., Kopp, S.: More human-likeness, more trust? The effect of anthropomorphism on self-reported and behavioral trust in continued and interdependent human-agent cooperation. Proc. Mensch Comput. **2019**, 31–42 (2019)
18. Lee, J.D., See, K.A.: Trust in automation: designing for appropriate reliance. Hum. Factors **46**(1), 50–80 (2004)
19. Lewicki, R.J., Brinsfield, C.: Trust repair. Ann. Rev. Organ. Psychol. Organ. Behav. **4**, 287–313 (2017)
20. Lewicki, R.J., McAllister, D.J., Bies, R.J.: Trust and distrust: new relationships and realities. Acad. Manage. Rev. **23**(3), 438–458 (1998)
21. Luhmann, N.: Trust and Power. Wiley, Hoboken (1979)
22. Madhavan, P., Wiegmann, D.A.: Similarities and differences between human-human and human-automation trust: an integrative review. Theor. Issues Ergon. Sci. **8**(4), 277–301 (2007)

23. Marakas, G.M., Johnson, R.D., Palmer, J.W.: A theoretical model of differential social attributions toward computing technology: when the metaphor becomes the model. Int. J. Hum. Comput. Stud. **52**(4), 719–750 (2000)
24. Mayer, R.C., Davis, J.H., Schoorman, F.D.: An integrative model of organizational trust. Acad. Manage. Rev. **20**(3), 709–734 (1995)
25. McDermott, P.L., Brink, R.N.t.: Practical guidance for evaluating calibrated trust. In: Proceedings of the Human Factors and Ergonomics Society Annual Meeting, vol. 63, pp. 362–366. SAGE Publications, Los Angeles (2019)
26. Moon, Y.: Intimate exchanges: using computers to elicit self-disclosure from consumers. J. Consum. Res. **26**(4), 323–339 (2000)
27. Muir, B.M.: Trust in automation: Part I. Theoretical issues in the study of trust and human intervention in automated systems. Ergonomics **37**(11), 1905–1922 (1994)
28. Muir, B.M., Moray, N.: Trust in automation. Part II. Experimental studies of trust and human intervention in a process control simulation. Ergonomics **39**(3), 429–460 (1996)
29. Nass, C., Moon, Y.: Machines and mindlessness: social responses to computers. J. Soc. Issues **56**(1), 81–103 (2000)
30. Nass, C., Steuer, J., Tauber, E., Reeder, H.: Anthropomorphism, agency, and ethopoeia: computers as social actors. In: INTERACT'93 and CHI'93 Conference Companion on Human Factors in Computing Systems, pp. 111–112 (1993)
31. Nass, C., Steuer, J., Tauber, E.R.: Computers are social actors. In: Proceedings of the SIGCHI Conference on Human Factors in Computing Systems, pp. 72–78 (1994)
32. Nowak, K.L.: Examining perception and identification in avatar-mediated interaction. In: The Handbook of the Psychology of Communication Technology, pp. 89–114 (2015)
33. Pak, R., Fink, N., Price, M., Bass, B., Sturre, L.: Decision support aids with anthropomorphic characteristics influence trust and performance in younger and older adults. Ergonomics **55**(9), 1059–1072 (2012)
34. Parasuraman, R., Riley, V.: Humans and automation: use, misuse, disuse, abuse. Hum. Factors **39**(2), 230–253 (1997)
35. Reeves, B., Nass, C.I.: The Media Equation: How People Treat Computers, Television, and New Media Like Real People and Places. Cambridge University Press, Cambridge (1996)
36. Salem, M., Eyssel, F., Rohlfing, K., Kopp, S., Joublin, F.: To err is human (-like): effects of robot gesture on perceived anthropomorphism and likability. Int. J. Soc. Rob. **5**(3), 313–323 (2013)
37. Salem, M., Lakatos, G., Amirabdollahian, F., Dautenhahn, K.: Would you trust a (faulty) robot? Effects of error, task type and personality on human-robot cooperation and trust. In: 2015 10th ACM/IEEE International Conference on Human-Robot Interaction (HRI), pp. 1–8. IEEE (2015)
38. Smith, M.A., Allaham, M.M., Wiese, E.: Trust in automated agents is modulated by the combined influence of agent and task type. In: Proceedings of the Human Factors and Ergonomics Society Annual Meeting, vol. 60, pp. 206–210. SAGE Publications, Los Angeles (2016)
39. Turing, A.: Computing machinery and intelligence. Mind **59**(236), 433–460 (1950)
40. Turkle, S.: The Second Self: Computers and the Human Spirit. MIT Press, Cambridge (2005)
41. de Visser, E.J., et al.: The world is not enough: trust in cognitive agents. In: Proceedings of the Human Factors and Ergonomics Society Annual Meeting, vol. 56, pp. 263–267. Sage Publications Sage CA, Los Angeles, CA (2012)

42. de Visser, E.J., et al.: A little anthropomorphism goes a long way: effects of oxytocin on trust, compliance, and team performance with automated agents. Hum. Factors **59**(1), 116–133 (2017)
43. de Visser, E.J., Monfort, S.S., McKendrick, R., Smith, M.A., McKnight, P.E., Krueger, F., Parasuraman, R.: Almost human: anthropomorphism increases trust resilience in cognitive agents. J. Exp. Psychol. Appl. **22**(3), 331 (2016)
44. de Visser, E.J., Pak, R., Shaw, T.H.: From 'automation' to 'autonomy': the importance of trust repair in human-machine interaction. Ergonomics **61**(10), 1409–1427 (2018)
45. de Visser, E.J., Peeters, M.M., Jung, M.F., Kohn, S., Shaw, T.H., Pak, R., Neerincx, M.A.: Towards a theory of longitudinal trust calibration in human-robot teams. Int. J. Soc. Rob. **12**(2), 459–478 (2020)
46. Waytz, A., Cacioppo, J., Epley, N.: Who sees human? The stability and importance of individual differences in anthropomorphism. Perspect. Psychol. Sci. **5**(3), 219–232 (2010)
47. Waytz, A., Heafner, J., Epley, N.: The mind in the machine: anthropomorphism increases trust in an autonomous vehicle. J. Exp. Soc. Psychol. **52**, 113–117 (2014)
48. Wicks, A.C., Berman, S.L., Jones, T.M.: The structure of optimal trust: moral and strategic implications. Acad. Manage. Rev. **24**(1), 99–116 (1999)

Designing a Gender-Inclusive Conversational Agent For Pair Programming: An Empirical Investigation

Sandeep Kaur Kuttal[(⊠)], Abim Sedhain, and Jacob AuBuchon

University of Tulsa, Tulsa, OK 74104, USA
{sandeep-kuttal,abs5423,jsa6790}@utulsa.edu

Abstract. Recently, research has shown that replacing a human with an agent in a pair programming context can bring similar benefits such as increased code quality, productivity, self-efficacy, and knowledge transfer as it does with a human. However, to create a gender-inclusive agent, we need to understand the communication styles between human-human and human-agent pairs. To investigate the communication styles, we conducted gender-balanced studies with human-human pairs in a remote lab setting with 18 programmers and human-agent pairs using Wizard-of-Oz methodology with 14 programmers. Our quantitative and qualitative analysis of the communication styles between the two studies showed that humans were more comfortable asking questions to an agent and interacting with it than other humans. We also found men participants showed less uncertainty and trusted agent solutions more, while women participants used more instructions and apologized less to an agent. Our research results confirm the feasibility of creating gender-inclusive conversational agents for programming.

Keywords: Gender · Conversational agents · Communication style

1 Introduction

Conversational agents—such as automated customer support, personal virtual assistants, and social chatbots—have transformed human interactions with computers [40–46]. Despite phenomenal progress in conversational agent's research there does not exist any such agent for programming tasks. To understand the design space of such an agent, we prototyped an interactive pair programming partner agent based on research from conversational agents, software engineering, education, human-robot interactions, psychology, and artificial intelligence [1, 3–5].

In pair programming, two programmers work simultaneously on one design, algorithm, code, or test [14–17]. Programmers switch between the roles of driver (writing code) and navigator (making suggestions). Pair programming provides various benefits, including increased code quality, productivity, creativity, knowledge management, and self-efficacy [18–30].

Our agent provided similar benefits as of pair programming with another human[4]. It's active application of social skills as a navigator increased participants' confidence

© Springer Nature Switzerland AG 2021
H. Degen and S. Ntoa (Eds.): HCII 2021, LNAI 12797, pp. 59–75, 2021.
https://doi.org/10.1007/978-3-030-77772-2_4

in the code and trust in the agent, while its technical skills as a driver helped participants realize their own solutions better [5].

A key facet in the design of a gender-inclusive conversational agent is how it should account for different communication styles; an area where women and men are known to differ [9–12]. To reduce the possibility of gender biases in our agent, we need to consider communication styles of each gender. Therefore, we formulated the following research questions:

R1. What communication styles are used by programmers when interacting with a human vs. an agent?

R2. How do men's communication styles differ when interacting with a human vs. an agent?

R3. How do women's communication styles differ when interacting with a human vs. an agent?

2   Methodology

A human-human study was conducted in a remote lab setting followed by a human-agent study using Wizard of Oz technique to investigate the similarities and differences between human-human and human-agent interactions.

2.1   Human-Human Study

A human-human study was conducted to analyze interactions between humans in a pair programming environment. 18 computer science students were conducted on a first-come first-serve basis, who each had at least a year of experience and some knowledge in java programming. Based on the background questionnaire, we identified only binary genders (men and women) from their own self-identification, though there are other genders [31, 32]. Hence, we paired students into gender balanced pairs (3 man-man, 3 woman-woman, and 3 man-woman). Gender balanced data is essential for discovering gender biases in designs to avoid unintentional gender bias and to support gender specific problem solving [38, 39], communication techniques, and leadership styles [31–39]. Gender was the focus of the study as opposed to other demographics because of the well-documented negative effects of the gender gap in the STEM fields.

We refer to each pair with the label HH-X with X being the gender of the individual: i.e., HH8-M7 and HH8-W9 refers to the seventh man participant and ninth woman participant who were in pair 8.

Study Design

The study was conducted in a lab setting to emulate a remote pair programming environment. Remote pair programming is known to have benefits in the likeness of collocated pair programming [54–56]. Every participant was required to complete a consent form, background questionnaire, and pre-self-efficacy questionnaire before the study [57]. Participants watched instructional video tutorials to teach them concepts of test-driven development (e.g., writing test cases, implementing code), pair programming concepts

(e.g., driver and navigator roles), and think-aloud study (e.g., vocalize any thoughts and feelings as they program [58]). Each pair communicated remotely using TeamViewer [117] and implementing the Eclipse IDE [118] to complete the task. Before the task was performed, the pairs were given a warm-up task to allow pair jelling, a period that allows them to adjust to their partner and work more efficiently [59].

Participants were given the task of implementing a tic-tac-toe game in Java. The game was selected for the study due to its simplicity and popularity. In tic-tac-toe, two players take turns placing marks in a 3 × 3 grid until either one player successfully gets three marks consecutively, or winning becomes impossible, causing a tie. Participants needed to write methods and test cases to complete the game; however, methods for a board and the ability to place marks on it were already provided to them. User stories and acceptance criteria for the task were provided regarding win conditions, full board, taking turns, and a tie. Participants determined their own roles as driver and navigator. They were given 40 min to complete the task to prevent fatigue. Study sessions were recorded using the Morae screen capture tool [60]. After the session was completed, participants answered questionnaires on post-self-efficacy and their pair programming preferences.

2.2 Human-Agent Study

A Wizard of Oz lab study was conducted to identify the interactions between a human and an agent. The study design, and the data analysis, was like our previous human-human study.

Wizard of Oz

Our study followed the basic components of a Wizard of Oz design. Wizard of Oz design helps to replicate a real virtual agent and effectively identify human interactions with an AI software [61–64]. Two wizards maintained the illusion of the agent, using dialogue options from a templated script. Participants had their face, voice, and screen were shared with the agent (the wizards). The wizards simulated a conversational agent through speech recognition, intent understanding, dialogue state tracking, dialogue policy, and response generation, using a constraint called the Wizard of Oz protocol [66]. For example, if the participant asked, *"How to write code for the win game?"*, the wizard would identify this question as *"implementation help"* and would subsequently choose the appropriate response from the wizard's templated script: *"I can make a recommendation from GitHub, would you like me to do so?"*

Participants pair programmed using the Saros plugin for the Eclipse IDE to facilitate remote collaboration with the agent (wizard). The agent (wizard) directly edited participants code using the Saros plugin for Eclipse [77]. Participants interacted with the agent, embodied by a 3D avatar, which was synchronized with the wizard's face using the Facerig software [65]. They communicated directly with the agent using voice and it responded with voice-synthesized messages to reduce switching the context of the participants and to increase interactions with the wizard. The voice-synthesized messages were generated using Google Text-to-Speech. Text communication was used exclusively

for sending links and pictures. Skype, Discord, and Google Hangouts were used to facilitate video and audio communication of the participants with the wizard to give the illusion of a real programming conversational agent.

Agent Design

The agent's design was inspired by multi-disciplinary research on human-computer interactions, conversational agents, human-robotic interactions, education, intelligent tutoring systems, psychology, and management science.

The agent was designed to interact with its partners by using a dynamic 3D avatar, voice, and text chat that enhanced human-computer interactions [3, 67–70]. The inclusion of an avatar makes the agent more human-like, improving understanding, engagement, and trust in novice programmers [61, 71–76].

The agent built rapport with participants as it greeted and introduced itself to them at the beginning of the study [78]. Further, it attributed success to the group and took personal responsibility for its mistakes. To increase participants' trust [67, 79–81] the agent showed uncertainty about its work, asking for verification; for example, after adding code through the IDE, the agent would say *"This might do the trick. I'm not sure though"*.

Motivation has a significant effect on performance of programmers and their productivity, also helping them to increase their creativity and their innovative outcomes [82–86]. Motivation was implemented in the study with motivational statements like *"I think this looks good,"* and *"I'm not sure what we're doing here, but we can always test it."* being given to participants upon both success and failure.

The agent's ability to contribute code (driver role) and give feedback (navigator role) were based on automated code/feedback techniques that require past solutions and search-based feedback [87–92]. The agent was designed to identify unnecessary code found in variables, functions, and classes based on automated tools like UCDetector [93]. If the participant asked for the location of variables/classes/methods within the code, the agent's ability to respond was based on both static and dynamic feature location techniques [94, 95]. Generating test cases automatically (i.e., without past solutions) was done by either code search algorithms or converting user stories to scenarios and then to test cases [96–99]. The agent could identify missing code using a technique investigated in the Haskell programming language [100].

Creative thinking is essential to a programmer's success especially when solving open-ended problems [18, 101–108]. To encourage diversity in thinking the agent offered abstract code templates, code examples, and alternate implementations. This decision was motivated by Tsuei's research that imperfect guidance enhances creativity and encourages exploration of new ideas [109]. Creativity theory suggests the production of a large number of ideas to arrive at creative ones through ideational fluency, since agents themselves cannot ideate [110]. Thus, the agent may ask things like, *"Are there other ways to do this?"*

The agent's responses were tuned to its performance with the partner giving just-in-time help when needed; it also apologized for incorrect or unknown answers, expressed uncertainty, and gave who/what/when/why/where/how answers accompanied with directions or suggestions [111, 112]. The agent also provided verbal feedback by presenting content and code templates to help the programmer's memory. For example, the agent

corrected a participant by highlighting code and giving verbal suggestions. Doing so addressed a self-presentation bias that induces a lack of memory retention on the human's part [113–116].

Study Design
14 participants (7 men and 7 women) were recruited for the study through advertisements and a recruitment site called Upwork. Eight of our participants were university students (4 men and 4 women) and six were professional programmers (3 men and 3 women). This study was conducted during the COVID-19 pandemic. Therefore, virtual lab studies were conducted from participants throughout United States.

The study design was like the human-human study, except (1) the participant completed the task with an agent, (2) they were given an instructional tutorial on the agent, (3) no pair jelling was incorporated, and (4) the agent changed gender presentation halfway through the study. We split the gender embodiment of the agent equally among the study sessions to counterbalance. Each participant signed a consent form at the end of the study, as it was a deception study.

2.3 Analysis of Data

Audio and Video of each session was recorded in both human-human and human-agent studies. These recording were then transcribed into individual snippets of dialogue which were subsequently sorted by their intent. The intent of each was represented by dialogue acts as described in Table 1. "Dialogue acts" have been used to classify human utterances using criteria based on a combination of pragmatics, semantics, and syntaxes [47]. Dialogue acts enable us to understand verbal communication as well as conversations via Hidden Markov Models [47], which are necessary for a conversational agent. 20% of the transcripts were coded by three researchers. Once they reached the inter-rater reliability of 90% (using the Jaccard measure [48]), two researchers independently coded the rest of the transcripts. We analyzed the data qualitatively and quantitatively to understand the differences of communication styles between humans with humans and humans with an agent.

3 Results

The results of the studies: both similarities and differences in the communication styles (dialogue acts) are summarized in Table 2. The differences between the studies in regard to the research questions are discussed subsequently.

RQ1: What communication styles are used by programmers when interacting with a human vs. an agent?
To answer RQ1, we compared the dialogue acts used by participants (both men and women) with another participant and with our agent. Table 2 shows (in yellow) the frequencies of different dialogue acts for human-human vs. human-agent. The different communication styles found between human-human vs. human-agent were:

Table 1. The dialogue acts and their definitions as used to code the study transcriptions.

Dialogue acts	Definition
Abandoned	An unfinished remark
Acknowledgement	Acceptance of the existence of something
Answer no	"No" responses
Apology	A regretful acknowledgement of failure
Answer W/H questions	Answering who/what/when/where/why/how questions
Answer yes	"Yes" Responses
Direct instruction	An explicit instruction
Feedback non-positive	A non-positive response or comment
Feedback positive	A positive response or comment
Indirect instruction	Implicit or polite instruction
Other (filler words)	Meaningless words
W/H/QUESTION	A who/what/when/where/why/how question
Questions yes/no	Questions asking for a yes/no answer
Statement	A declaration or remark
Uncertainty	Dialogue that indicates uncertainty

More Filler Words with Humans vs. Fully Articulated Thoughts with an Agent.
Participants in the human-human study used 55% more filler words (see Table 2) to give their verifications or to fill the gaps in the communication. For example, when HH1-M2 was implementing the horizontal win condition, M1 responded with *"Um" "Ok,"* or *"Huh"* while responding to M2's thought process. Conversely, while communicating with agent partners, participants fully formulated their thoughts before articulating them. The agent themselves never used filler words, as they followed a script.

Non-positive Feedback and Acknowledgment to Other Humans vs. Agents
In the human-human study, participants used non-positive feedback 70.3% more than their counterparts in the human-agent study. For example, HH9-M8 gave non-positive feedback for implementation of the win method with a sarcastic comment: *"okay, so it's a failure, awesome."* Similarly, HH7-W8 gave non-positive feedback to her partner's idea when she stated *"Yeah, I'm not sure about that one."* The decreased usage of non-positive feedback in the human-agent study stemmed from our design choice of agent being motivational, empathetic, and a rapport-builder.

The human-human study participants acknowledged their partners' ideas 52.5% more than human-agent participants. For example, in HH9, M8 shared a possible solution for checking diagonal win conditions and suggested the usage of multiple 'loops', to which his partner M9 responded *"Yeah we could do something like that."* Later in the same study, they switched roles, describing a test method he (M9) might want to implement and M8 confirming *"Yeah, I think that'll work."* These types of dialogue

Table 2. The dialogue acts for human-human and human-agent studies. The most prominent difference in dialogue acts is highlighted. Yellow for humans with humans vs. agent (RQ1), peach for women with human vs. agent (RQ2), and blue for men with human vs. agent (RQ3).

Dialogue Acts	Human-Human				Human-Agent			
	M[#]	W[#]	Total	%age	M[#]	W[#]	Total	%age
Abandoned	97[9]	108[8]	205	6.01	104[7]	101[7]	205	6.56
Acknowledgement	295[9]	326[9]	621	18.22	129[7]	166[7]	295	9.43
Answer No	5[6]	16[6]	21	0.62	6[7]	3[3]	9	0.29
Apology	7[6]	10[5]	17	0.50	5[5]	4[4]	9	0.29
Answer W/H Questions	21[9]	19[5]	40	1.17	65[7]	62[7]	127	4.06
Answer Yes	52[9]	51[9]	103	3.02	70[7]	55[7]	125	4.00
Direct Instruction	87[9]	72[9]	159	4.66	46[7]	81[7]	127	4.06
Feedback Non-Positive	23[8]	14[5]	37	1.09	7[5]	4[4]	11	0.35
Feedback Positive	58[9]	26[7]	84	2.47	23[6]	49[7]	72	2.30
Indirect Instruction	129[9]	89[9]	218	6.39	78[7]	120[7]	198	6.33
Other (Filler Words)	126[9]	112[9]	238	6.98	46[7]	61[7]	107	3.42
W/H Questions	48[9]	62[8]	110	3.23	82[7]	96[7]	178	5.69
Questions Yes/No	97[9]	119[9]	216	6.34	84[7]	125[7]	209	6.68
Statement	688[9]	592[9]	1280	37.55	657[7]	753[7]	1410	45.09
Uncertainty	27[9]	33[9]	60	1.76	16[7]	29[5]	45	1.44
Total	**1760**	**1649**	**3409**		**1418**	**1709**	**3127**	

acts did not appear as frequently in the human-agent studies. One possible reason for this could be that when conversing with another person acknowledgement is key to effectively communicating ideas [49] and hence, in the absence of another human the acknowledgement was decreased.

Asked More WH Questions From the Agent vs. Human

Participants from the human-agent study also asked 38.2% more W/H (who/what/when/why/where/how) questions to an agent as compared to the human-human study. For example, after getting stuck, HA-M6 prompts the agent by commenting, *"What do you think?"* and later with *"How about you drive."* Additionally, HA-W13 asked questions such as *"Why would you return true?"* *"How do we check?"* The WH question were asked by participants to clarify, understand the code, prompt for feedback, or ask for help.

Human participants were more reliant on the agent as they had more confidence in the agent's responses and solutions. For example, HA-M3 asked the agent, *"What do you think the mistake is?"* The agent replied with, *"You have a missing bracket on line 66"*. This helped participants quickly find out the error and save time. HA-M3 also asked, *"What's the next story?"* or *"What's story number four?"* This helped the

participants navigate quickly between tasks. Participants in human-agent study were more comfortable with the agent and saw it as a non-judgmental partner [4].

RQ2: How do men's communication styles differ when interacting with a human vs. an agent?

We found the following differences (highlighted in peach in Table 2) in men's communication style with another human (man/woman) vs. an agent.

Men Showed More Uncertainty with Another Human vs. an Agent

Men participants in the Human-Human study were 40.7% more uncertain about the task. For example, in HH2, both participants were uncertain about the next step, as M3 commented "*Ya I don't know*", to which M4 responded "*I'm not sure how we can write a test for this because it's not going to evaluate to true or false you know.*" Humans trust agents more than other humans [50], and this was evident as HA-M8 commented "*the computer knows more ... than a human knows... I would trust a computer more.*"

Men Gave More Positive Feedback to Their Human Partners

Men participants of the human-human study showed 60.3% more positive feedback, listened to their partners, and acknowledged their suggestions. All nine men participants showed positive feedback during the human-human study. For example, in HH1, while working on method to place marks on a board, M1 laughed and commented "*It's ugly but it should work*" to which M2 responded with a laugh and commented "*Yeah, that's how I feel sometimes*". The frequency of acknowledgement was twice as high in the human-human study than in the human-agent study. For example, in HH2, M4 was describing how to keep track of current player, placing the mark. M3 listened carefully and acknowledged HH2-M4's thought process without abandoning it mid-point.

Men Gave More Direct and Indirect Instructions to Another Human Partner Than an Agent

Men participants gave 51.8% more direct and indirect instructions to another human partner than to an agent. Direct instructions are an explicit way of expressing what needs to be done while indirect instructions are a subtle way of suggesting things that need to be done. Direct instructions were used to express what aspect of the problem needs to be done and how to accomplish it. For example, in HH3, when W1 tried to explain the logic for checking if marks are in the same row verbally, M5 explicitly said "*Uh, write it out for me. I, I can visualize it a bit better if I can see it,*" Indirect instructions were used for directions, suggestions, agreeing to tasks, and giving control. HH1-M1 gave indirect instructions to M2 and commented "*We could probably use the code for tie winner, but that's my guess*".

RQ3: How do women's communication styles differ when interacting with a human vs. an agent?

We found the following differences (highlighted blue in Table 2) in women's communication style with another human (man/woman) vs. an agent. Women participants talked more with an agent, as the number of statements was 21.4% more than with another human.

Women used More Direct and Indirect Instructions with an Agent Than with Another Human

Women participants used 19.9% more instructions (both direct and indirect) to direct the flow of the task process with an agent. For example, HA-W4 was more vocal and direct with the agent. She was controlling the flow of dialogue with *"Do it again"*, *"Sure, let's do that"*, and *"Try it"*. She also used indirect instructions to implicitly direct the flow as she commented *"Okay, Um, so we need to add test"*, *"I think there needs to be more code in check ties"*, and *"Should we just try placing"*.

Women Apologized More with a Human Partner Than with an Agent

Women participants tended to apologize with a human partner when an answer was unknown or wrong. While working with an agent, women were less self-conscious when they made mistakes and were more confident in their approaches. For example, in HH4, W3 apologized to her partner for simple things like clicking by mistake as she commented *"Sorry, I'm clicking"*. Likewise, HH4-W2 read the task wrong and she apologized immediately *"Oh, sorry,…oh my gosh. I'm so sorry. I read that wrong."* Women build rapport with their human partner, not wanting to let their partner down [52, 53].

Women Disagreed with Another Human With a No

Women participants disagreed (with a "No") 81% more with a human partner than an agent. For example, in HH4, W2 asked if they need to write a test case for mark placed, to which her partner W3 responded *"No no no…"*

Women Positively Responded to an Agent

Women participants gave 46.9% more positive feedback to an agent than to a human partner. For example, HA-W4 responded positively to agent's code by commenting, *"It does make more sense."*

4 Discussions

Our results shed light on the conversational styles of humans, both men and women, with other humans and agents. As seen in RQ2 and RQ3 the communication styles were different for both men and women, hence, to create a gender-inclusive programming agent we should integrate styles from both genders. Some of the implications for programming conversational agents are detailed as follows.

Conversational agents for problem solving tasks should support WH questions and answers, WH answers should be accompanied by directions and suggestions. In our study, while answering, humans tended to answer yes/no more frequently but the agent should explain the logic behind the code or explanations regarding the decisions made by the agent. HA-M6 mentioned that the agent should give a *"more specific, explanatory approach like a human."* Though generating human-like explanation and discussions regarding programming may be hard to implement with the current technology, utilizing the static and dynamic testing approaches, generating visually explainable solutions can make the agent more human-like. Further, such plausible explanations can help to build

trust of both professionals and students with an agent and in its solutions [50], while a lack of such explanations may affect losing trust. The agent should show vulnerabilities of being a machine and expose its limitations to increase trust with a human partner. When a human shows uncertainty, the agent should better explain using visualizations and the underlying concepts of how it arrived at a specific solution. Additionally, the agent should provide better verbal feedback and present content and code templates to jog the programmer's memory. For example, to correct a programmer, the agent should highlight the code segments while giving verbal and visual suggestions about the code. One approach to generate such explanations could be using deep learning techniques and training them on WH questions. However, such training will require tremendous amount of data on pair programming conversations that currently is unavailable. Further, supporting gender specific problem solving [37–39], and leadership styles [2] should also be facilitated by a gender-inclusive agent.

Engaging the human partner with both positive and negative feedback was important for the success of the task. The feedback from human as well as agent was regarding (1) status of the code i.e., pass/fail/unexpected results, (2) code reviews i.e., correct/incorrect/coding style, (3) idea implemented work/fails, (4) agreement, and (5) motivation. The positive feedback helped human partners to stay motivated and engaged in the task, while negative feedback was on mistakes in the code and helped them to improve quality of the code. Both positive and negative feedback are important towards problem solving tasks and hence should be used as integral parts of the conversational agents. Since there were differences in how each gender used feedback, it will be important that a gender-inclusive agent is able to adapt itself based on past conversations. This is especially important regarding negative feedback, which should be dealt more carefully. For-example, if there is a typo, the agent can correct the typo by itself but if there is a logical error or the problem-solving approach is wrong, the agent should give suggestions.

One implication for machine learning algorithms is to train them on different genders (equal number of men, women and non-binary) to capture the diversity of conversational styles. The differences between the communication styles between different genders necessitate inclusion of features in the machine learning algorithms that can capture these differences. Hence, avoiding agents that support misogynistic ideologies in machine learning [33–36] and support more gender-inclusive machine learning algorithms.

5   Conclusion

We take the first step towards creating a gender-inclusive Alexa-like programming partner. This paper contributes to understanding the differences between human-human and human-agent conversations, and how gender effects conversations.

1. **RQ1: Communication Styles: Human vs. Agent.** Participants in human-human study gave more directions to complete the task and said more filler words. They also expressed non-positive feedback and acknowledged their partners' ideas and interjected when they thought they could help. They also asked more W/H questions as they were comfortable with the agent.

2. **RQ2: Men's Communication Styles: Human vs. Agent.** Men participants motivated their human partner through positive feedback and giving direction using both direct and indirect instructions. Men participants also showed more trust with the agent.
3. **RQ3: Women's Communication Styles: Human vs. Agent.** Women participants disagreed and apologized more with a human partner while making rapport with them. They instructed the agent and gave more positive feedback. Hence, women participants were hesitant to communicate with human partners.

Finally, we discuss the implications for a gender-inclusive conversational agent. The implications include interface design of such an agent as well as for training machine-learning algorithms.

References

1. Robe, P., Kuttal, S.K., Zhang, Y., Bellamy, R.: Can machine learning facilitate remote pair programming? Challenges, insights & implications. In: Proceedings of Visual Languages and Human-Centric Computing (2020)
2. Kuttal, S.K., Gerstner, K., Bejarano, A.: Remote pair programming in online CS education.: investigating through a gender lens. In: Proceedings of Visual Languages and Human-Centric Computing (2019)
3. Kuttal, S.K., Myers, J., Gurka, S., Magar, D., Piorkowski, D., Bellamy, R.: Towards designing conversational agents for pair programming: accounting for creativity strategies and conversational styles. In: Proceedings of Visual Languages and Human-Centric Computing (2020)
4. Kuttal, S.K., Kwasny, K., Ong, B., Robe, P.: Understand the tradeoffs for substituting humans with an agent - good, bad, and ugly. Submitted to CHI 2021 found at https://drive.google.com/drive/folders/14_0zkttwbVr6pJnB_U4YIReGDLI6mCTX?usp=sharing
5. Robe, P., Kuttal, S.K.: Designing an interactive pair programming partner submitted to TOCHI 2021 found at https://drive.google.com/drive/folders/1vIOdro0pg8C1jSB42KzYrDRKO0PVhqZ1?usp=sharing
6. Stolcke, A., et al.: Dialogue act modeling for automatic tagging and recognition of conversational speech. Comput. Linguist. **26**(3), 339–373 (2000)
7. Tugend, A.: Why is asking for help so difficult? N. Y. Times (2007)
8. PairBuddy Github. https://github.com/grubtub19/pairbuddy
9. Abraham, A.: Gender and creativity.: an overview of psychological and neuroscientific literature. Brain Imaging Behav. **10**(2), 609–618 (2016)
10. Baron-Cohen, S., Knickmeyer, R.C., Belmonte, M.K.: Sex differences in the brain: implications for explaining autism. Science **310**(5749), 819–823 (2005)
11. LeClair, A., Eberhart, Z., McMillan, C.: Adapting neural text classification for improved software categorization. In: IEEE International Conference on Software Maintenance and Evolution (ICSME), Madrid, pp. 461–472 (2018)
12. Lin, W.-L., Hsu, K.-Y., Chen, H.-C., Wang, J.-W.: The relations of gender and personality traits on different creativities: a dual-process theory account. Psychol. Aesthet. Creativity Arts **6**(2), 112–123 (2012)
13. Wood, A., Rodeghero, P., Armaly, A., McMillan, C.: Detecting speech act types in developer question/answer conversations during bug repair. In: Proceedings of the 26th ACM Joint Meeting on European Software Engineering Conference and Symposium on the Foundations of Software Engineering (ESEC/FSE 2018), pp. 491–502 (2018)

14. Woolley, W., Aggarwal, I., Malone, T.W.: Collective intelligence and group performance. Curr. Dir. Psychol. Sci. **24**(6), 420–424 (2015)
15. Palmieri, D.W.: Knowledge management through pair programming, Master's Thesis, Department of Computer Science, North Carolina State University, Raleigh, NC (2002)
16. Williams, L., McDowell, C., Nagappan, N., Fernald, J., Werner, L.: Building pair programming knowledge through a family of experiments. In: 2003 International Symposium on Empirical Software Engineering, pp. 143–152 (2003)
17. Williams, L., Kessler, R.: Pair Programming Illuminated. Addison-Wesley Longman Publishing Co., Inc., Boston (2002)
18. de la Barra, C.L., Crawford, B.: Fostering creativity thinking in agile software development. In: Holzinger, A. (ed.) USAB 2007. LNCS, vol. 4799, pp. 415–426. Springer, Heidelberg (2007). https://doi.org/10.1007/978-3-540-76805-0_37
19. Belshee, A.: Promiscuous pairing and beginner's mind: embrace inexperience, pp. 125–131 (2005)
20. Cockburn, A., Williams, L.: Extreme Programming Examined. Addison-Wesley Longman Publishing Co., Inc., Boston. Ch. The Costs and Benefits of Pair Programming, pp. 223–243 (2001)
21. DeMarco, T., Lister, T.: Peopleware: Productive Projects and Teams. Dorset House Publishing Co., Inc., New York (1987)
22. Zieris, F., Prechelt, L.: On knowledge transfer skill in pair programming. In: Proceedings of the 8th ACM/IEEE International Symposium on Empirical Software Engineering and Measurement, ESEM 2014, pp. 11:1–11:10. ACM, New York (2014)
23. McDowell, C., Werner, L., Bullock, H., Fernald, J.: The effects of pair-programming on performance in an introductory programming course. In: Proceedings of the 33rd SIGCSE Technical Symposium on Computer Science Education. SIGCSE, pp. 38–42. ACM, New York (2002)
24. Katira, N., et al.: On understanding compatibility of student pair programmers. SIGCSE Bull. **36**(1), 7–11 (2004)
25. McDowell, C., Werner, L., Bullock, H.E., Fernald, J.: The impact of pair programming on student performance, perception and persistence. In: Proceedings of the 25th International Conference on Software Engineering, ICSE 2003, pp. 602–607. IEEE Computer Society, Washington, DC (2003)
26. Williams, L., Wiebe, E., Yang, K., Ferzli, M., Miller, C.: In support of pair programming in the introductory computer science course. Comput. Sci. Educ. **12**, 197–212 (2002)
27. Ruvalcaba, O., Werner, L., Denner, J.: Observations of pair programming: variations in collaboration across demographic groups. In: Proceedings of the 47th ACM Technical Symposium on Computing Science Education, SIGCSE, pp. 90–95. ACM, New York (2016)
28. Werner, L.L., Hanks, B., McDowell, C.: Pair-programming helps female computer science students. J. Educ. Resour. Comput. **4**(1) (2004)
29. Celepkolu, M., Boyer, K.E.: Thematic analysis of students' reflections on pair programming in CS1. In: Proceedings of the 49th ACM Technical Symposium on Computer Science Education, SIGCSE, pp. 771–776. ACM, New York (2018)
30. Rodríguez, F.J., Price, K.M., Boyer, K.E.: Exploring the pair programming process: characteristics of effective collaboration. In: Proceedings of the 2017 ACM SIGCSE Technical Symposium on Computer Science Education, SIGCSE 2017, pp. 507–512. ACM, New York (2017)
31. Butler, J.: Revisiting bodies and pleasures: theory. Cult. Soc. **16**(2), 11–20 (1999)
32. West, C., Zimmerman, D.H.: Doing gender. Gend. Soc. **1**(2), 125–151 (1987)
33. Burnett, M., Peters, A., Hill, C., Elarief, N.: Finding gender-inclusiveness software issues with GenderMag: a field investigation. In: Proceedings of the 2016 CHI Conference on Human Factors in Computing Systems, pp. 2586–2598. ACM (2016)

34. Charness, G., Gneezy, U.: Strong evidence for gender differences in risk taking. J. Econ. Behav. Organ. **83**(1), 50–58 (2012)
35. Mendez, C., et al.: Open-source barriers to entry, revisited: a sociotechnical perspective. In: Proceedings of the 40th International Conference on Software Engineering, pp. 1004–1015. ACM (2018)
36. Shekhar, A., Marsden, N.: Cognitive walkthrough of a learning management system with gendered personas. In: Proceedings of the 4th Conference on Gender & IT, pp. 191–198. ACM (2018)
37. Leavy, S.: Gender bias in artificial intelligence: the need for diversity and gender theory in machine learning. In: Proceedings of the 1st International Workshop on Gender Equality in Software Engineering, GE 2018, Gothenburg, Sweden, pp. 14–16. Association for Computing Machinery, New York (2018)
38. Arisholm, E., Gallis, H., Dybå, T., Sjoberg, D.I.K.: Evaluating pair programming with respect to system complexity and programmer expertise. IEEE Tran. Softw. Eng. **33**(2), 65–86 (2007)
39. Falkner, K., Falkner, N., Vivian, R.: Collaborative learning and anxiety: a phenomenographic study of collaborative learning activities. In: Proceedings of the 44th ACM Technical Symposium on Computer Science Education, pp. 227–232 (2013)
40. Virtual Assistant [n.d.]. Amazon Alexa. https://developer.amazon.com/en-US/alexa
41. Virtual Assistant [n.d.]. Apple Siri. https://www.apple.com/siri/
42. Virtual Assistant [n.d.]. Google Assistant. https://assistant.google.com/
43. Social Bot [n.d.]. Cleverbot. https://www.cleverbot.com/
44. Social Bot [n.d.]. Mitsuku. https://www.pandorabots.com/mitsuku/
45. Social Bot [n.d.]. SAP Conversational AI. https://www.sap.com/products/conversational-ai.html
46. Social Bot [n.d.]. Xiaoice AI Assistant. https://www.digitaltrends.com/cool-tech/xiaoice-microsoft-future-of-ai-assistants/
47. Stolcke, A., et al.: Dialogue act modeling for automatic tagging and recognition of conversational speech. Comput. Linguist. **26**(3), 339–373 (2000)
48. Jaccard, P.: Étude comparative de la distribution florale dans une portion des Alpes et des Jura. Bulletin de la Société vaudoise des sciences naturelles **37**, 547–579 (1901)
49. Wentzel, K.R., Watkins, D.E.: Peer relationships and collaborative learning as contexts for academic enablers. Sch. Psychol. Rev. **31**(3), 366–377 (2002)
50. Fiske, S.T., Fiske, E.H.P.P.S.T., Taylor, S.E.: Social Cognition. McGraw-Hill, New York City
51. Kuttal, S.K., Ong, B., Kwasny, K., Robe, P.: Trade-offs for substituting a human with an agent in a pair programming context: the good, the bad, and the ugly. In: Proceedings of the conference on Human Factors in Computing, CHI (2021)
52. Cuadrado, I., Navas, M.M.D., Molero, F., Ferrer, E., Morales, J.F.: Gender differences in leadership styles as a function of leader and subordinates sex and type of organization. J. Appl. Soc. Psychol. **42**, 3083–3113 (2012)
53. Yang, T., Aldrich, H.E.: Whos the boss? Explaining gender inequality in entrepreneurial teams. Am. Sociol. Rev. **79**(2), 303–327 (2014)
54. Baheti, P., Gehringer, E., Stotts, D.: Exploring the efficacy of distributed pair programming. In: Wells, D., Williams, L. (eds.) XP/Agile Universe 2002. LNCS, vol. 2418, pp. 208–220. Springer, Heidelberg (2002). https://doi.org/10.1007/3-540-45672-4_20
55. Duque, R., Bravo, C.: Analyzing work productivity and program quality in collaborative programming. In: Proceedings of the 2008 The Third International Conference on Software Engineering Advances, pp. 270–276. IEEE Computer Society, Washington, DC (2008)
56. Hanks, B.: Empirical evaluation of distributed pair programming. Int. J. Hum Comput Stud. **66**, 530–544 (2008)

57. Compeau, D.R., Higgins, C.A.: Computer self-efficacy: development of a measure and initial test. MIS Q. **19**(2), 189–211 (1995)
58. Lewis, C.: Using the "Thinking-Aloud" Method in Cognitive Interface Design. IBM T.J. Watson Research Center, Yorktown Heights (1982)
59. Jones, D.L., Fleming, S.D.: What use is a backseat driver? A qualitative investigation of pair programming. In: Proceedings of IEEE Symposium on Visual Languages and Human-Centric Computing, VL/HCC, pp. 103–110 (2013)
60. Morae 2019. Morae. http://www.techsmith.com/morae.asp
61. Bickmore, T., Cassell, J.: Relational agents: a model and implementation of building user trust. In: Proceedings of the SIGCHI Conference on Human Factors in Computing Systems (CHI 2001), Seattle, Washington, USA, pp. 396–403. ACM, New York (2001)
62. Bradley, J., Benyon, D., Mival, O., Webb, N.: Wizard of Oz experiments and companion dialogues. In: Proceedings of the 24th BCS Interaction Specialist Group Conference, pp. 117–123. British Computer Society (2010)
63. Dahlbäck, N., Jönsson, A., Ahrenberg, L.: Wizard of Oz studies—why and how. Knowl. Based Syst. **6**(4), 258–266 (1993)
64. Wargnier, P., Carletti, G., Laurent-Corniquet, Y., Benveniste, S., Jouvelot, P., Rigaud, A.-S.: Field evaluation with cognitively-impaired older adults of attention management in the embodied conversational agent Louise. In: 2016 IEEE International Conference on Serious Games and Applications for Health (SeGAH), pp. 1–8. IEEE (2016)
65. Software Application [n.d.]. Facerig. https://facerig.com/
66. Riek, L.D.: Wizard of Oz studies in HRI: a systematic review and new reporting guidelines. J. Hum. Robot Interact. **1**(1), 119–136 (2012)
67. Ashktorab, Z., Jain, M., Liao, Q.V., Weisz, J.D.: Resilient chatbots: repair strategy preferences for conversational breakdowns. In: Proceedings of the 2019 CHI Conference on Human Factors in Computing Systems (CHI 2019), Glasgow, Scotland, UK. Association for Computing Machinery, New York (2019). Article no. 254, 12 pages
68. Lopatovska, I., Williams, H.: Personification of the Amazon Alexa: BFF or a mindless companion. In: Proceedings of the 2018 Conference on Human Information Interaction & Retrieval (CHIIR 2018), New Brunswick, NJ, USA, pp. 265–268. Association for Computing Machinery, New York (2018)
69. Sproull, L., Subramani, M., Kiesler, S., Walker, J.H., Waters, K.: When the interface is a face. Hum. Comput. Interact. **11**(2), 97–124 (1996)
70. Zalake, M., Woodward, J., Kapoor, A., Lok, B.: Assessing the impact of virtual human's appearance on users' trust levels. In: Proceedings of the 18th International Conference on Intelligent Virtual Agents (IVA 2018), Sydney, NSW, Australia, pp. 329–330. Association for Computing Machinery, New York (2018)
71. Gratch, J., Wang, N., Gerten, J., Fast, E., Duffy, R.: Creating rapport with virtual agents. In: Pelachaud, C., Martin, J.-C., André, E., Chollet, G., Karpouzis, K., Pelé, D. (eds.) IVA 2007. LNCS (LNAI), vol. 4722, pp. 125–138. Springer, Heidelberg (2007). https://doi.org/10.1007/978-3-540-74997-4_12
72. Hasegawa, D., Cassell, J., Araki, K.: The Role of Embodiment and Perspective in Direction-Giving Systems (2010)
73. Shamekhi, A., Liao, Q.V., Wang, D., Bellamy, R.K., Erickson, T.: Face value? Exploring the effects of embodiment for a group facilitation agent. In: Proceedings of the 2018 CHI Conference on Human Factors in Computing Systems, CHI 2018, Montreal QC, Canada, pp. 1–13. Association for Computing Machinery, New York (2018)
74. Takeuchi, A., Naito, T.: Situated facial displays: towards social interaction. In: Proceedings of the SIGCHI Conference on Human Factors in Computing Systems (CHI 1995), Denver, Colorado, USA, pp. 450–455. ACM Press/Addison-Wesley Publishing Co., New York (1995)

75. van Mulken, S., André, E., Müller, J.: The persona effect: how substantial is it?. In: Johnson, H., Nigay, L., Roast, C. (eds.) People and Computers XIII. Springer, London (1998). https://doi.org/10.1007/978-1-4471-3605-7_4

76. Yee, N., Bailenson, J.N., Rickertsen, K.: A meta-analysis of the impact of the inclusion and realism of human-like faces on user experiences in interfaces. In: Proceedings of the SIGCHI Conference on Human Factors in Computing Systems (CHI 2007), San Jose, California, USA, pp. 1–10. ACM, New York (2007)

77. Saros [n.d.]. Saros Project. https://www.saros-project.org/

78. Kahn, P.H., et al.: Design patterns for sociality in human-robot interaction. In: Proceedings of the 3rd ACM/IEEE International Conference on Human Robot Interaction (HRI 2008), Amsterdam, The Netherlands, pp. 97–104. Association for Computing Machinery, New York (2008)

79. Jain, M., Kumar, P., Bhansali, I., Liao, Q.V., Truong, K., Patel, S.: Farm chat: a conversational agent to answer farmer queries. In: Proceedings of the ACM Interactive, Mobile, Wearable and Ubiquitous Technologies, vol. 2, no. 4 (2018). Article no. 170, 22 pages

80. Jain, M., Kumar, P., Kota, R., Patel, S.N.: Evaluating and informing the design of chatbots. In: Proceedings of the 2018 Designing Interactive Systems Conference, pp. 895–906 (2018)

81. Luger, E., Sellen, A.: "Like having a really bad PA" the gulf between user expectation and experience of conversational agents. In: Proceedings of the 2016 CHI conference on human factors in computing systems, pp. 5286–5297 (2016)

82. Amabile, T.M., Pratt, M.G.: The dynamic componential model of creativity and innovation in organizations: making progress, making meaning. Res. Organ. Behav. **36**, 157–183 (2016)

83. Armstrong, M.: Armstrong's Handbook of Reward Management Practice: Improving Performance Through Reward, 12 edn. Kogan Page Publishers (2012)

84. Cerasoli, C.P., Nicklin, J.M., Ford, M.T.: Intrinsic motivation and extrinsic incentives jointly predict performance: a 40-year meta-analysis. Psychol. Bull. **140**(4), 980 (2014)

85. Deci, E.L., Olafsen, A.H., Ryan, R.M.: Self-determination theory in work organizations: The state of a science. Ann. Rev. Organ. Psychol. Organ. Behav. **4**, 19–43 (2017)

86. Fischer, C., Malycha, C.P., Schafmann, E.: The influence of intrinsic motivation and synergistic extrinsic motivators on creativity and innovation. Frontiers Psychol. **10**, 137 (2019)

87. Day, M., Penumala, M.R., Gonzalez-Sanchez, J.: Annete: an intelligent tutoring companion embedded into the eclipse IDE. In: 2019 IEEE First International Conference on Cognitive Machine Intelligence (CogMI), pp. 71–80 (2019)

88. Keivanloo, I., Rilling, J., Zou, Y.: Spotting working code examples. In: Proceedings of the 36th International Conference on Software Engineering, Hyderabad, India, pp. 664–675. Association for Computing Machinery, New York (2014)

89. Kim, K., Kim, D., Bissyandé, T.F., Choi, E., Li, L., Klein, J., Traon, Y.L.: FaCoY: a code-to-code search engine. In: Proceedings of the 40th International Conference on Software Engineering, ICSE 2018, Gothenburg, Sweden, pp. 946–957. Association for Computing Machinery, New York (2018)

90. Niu, H., Keivanloo, I., Zou, Y.: Learning to rank code examples for code search engines. Empirical Softw. Eng. **22**(1), 259–291 (2017)

91. Raghothaman, M., Wei, Y., Hamadi, Y.: SWIM: synthesizing what i mean - code search and idiomatic snippet synthesis. In: 2016 IEEE/ACM38th International Conference on Software Engineering, ICSE, pp. pp. 357–367 (2016)

92. Zhi, R., Marwan, S., Dong, Y., Lytle, N., Price, T.W., Barnes, T.: Toward data-driven example feedback for novice programming. In: Proceedings of the 12th International Conference on Educational Data Mining (2019)

93. Jörg Spieler. [n.d.]. UCDetector. http://www.ucdetector.org/

94. Liu, D., Marcus, A., Poshyvanyk, D., Rajlich, V.: Feature location via information retrieval based filtering of a single scenario execution trace. In: Proceedings of ASE 2007 - 2007 ACM/IEEE International Conference on Automated Software Engineering, pp. 234–243 (2007)

95. Savage, T., Revelle, M., Poshyvanyk, D.: FLAT3: feature location and textual tracing tool. In: Proceedings of 2010 ACM/IEEE 32nd International Conference on Software Engineering, vol. 2. pp. 255–258 (2010)

96. Ali, S., Briand, L.C., Hemmati, H., Panesar-Walawege, R.K.: A systematic review of the application and empirical investigation of search-based test case generation. IEEE Trans. Softw. Eng. **36**(6), 742–762 (2010)

97. Meiliana, Septian, I., Alianto, R.S., Daniel, Gaol, F.L.: Automated test case generation from UML activity diagram and sequence diagram using depth first search algorithm. Procedia Comput. Sci. **116**, 629 – 637 (2017). http://www.sciencedirect.com/science/article/pii/S18 77050917320732. Discovery and innovation of computer science technology in artificial intelligence era: The 2nd International Conference on Computer Science and Computational Intelligence (ICCSCI 2017)

98. Mariano, M.M., Souza, É.F., Endo, A.T., Vijaykumar, N.L.: Analyzing graph-based algorithms employed to generate testcases from finite state machines (2019)

99. Rane, P.: Automatic Generation of Test Cases for Agile using Natural Language Processing (2017)

100. Gerdes, A., Heeren, B., Jeuring, J., van Binsbergen, L.T.: Ask-Elle: an adaptable programming tutor for Haskell giving automated feedback. Int. J. Artif. Intell. Educ. **27** (2016)

101. Brown, T.: Change by Design: How Design Thinking Transforms Organizations and Inspires Innovation. Harper Business. (2009)

102. Berland Edelman and Inc. 2010. Creativity and education.: Why it matters. http://www. adobe.com/aboutadobe/pressroom/pdfs/Adobe_Creativity_and_Education_Why_It_Mat ters_study.pdf. Accessed 18 Sept 2019

103. Levine, M.: Effective Problem Solving. Prentice Hall, Hoboken (1988)

104. Liu, Z., Schonwetter, D.J.: Teaching creativity in engineering. Int. J. Eng. Educ. **20**(5), 801–808 (2004)

105. Polya, G.: How to Solve It.: A New Aspect of Mathematical Method, vol. 85. Princeton University Press (2004)

106. Tony, W., Robert, A.C.: Creating Innovators: The Making of Young People Who Will Change the World. Simon and Schuster, New York (2012)

107. Wickelgren, W.A.: How to Solve Problems: Elements of a Theory of Problems and Problem Solving. WH Freeman, San Francisco (1974)

108. Zhao, Y.: World Class Learners: Educating Creative and Entrepreneurial Students. Corwin Press, Thousand Oaks (2012)

109. Tsuei, M.: Learning behaviours of low-achieving children's mathematics learning in using of helping tools in a synchronous peer-tutoring system. Interact. Learn. Environ. **25**(2), 147–161 (2017)

110. Guilford, J.P.: Intelligence, Creativity, and Their Educational Implications. R. R. Knapp (1968) https://books.google.com/books?id=WE8kAQAAMAAJ

111. Robertson, T., et al.: Impact of interruption style on end-user debugging. In: ACM Conference on Human Factors in Computing Systems, pp. 287–294 (2004)

112. Wilson, A., et al.: Harnessing curiosity to increase correctness in end-user programming, pp. 305–312 (2003)

113. Knutsen, D., Le Bigot, L.: The influence of reference acceptance and reuse on conversational memory traces. J. Exp. Psychol. Learn. Mem. Cogn. **4** (2014)

114. Knutsen, D., Le Bigot, L., Ros, C.: Explicit feedback from users attenuates memory biases in human-system dialogue. Int. J. Hum. Comput. Stud. **97**, 77–87 (2017). http://www.scienc edirect.com/science/article/pii/S1071581916301045

115. Knutsen, D., Ros, C., Le Bigot, L.: Generating references in naturalistic face-to-face and phone-mediated dialog settings. Top. Cogn. Sci. **8** (2016)

116. Sharma, R., Gulia, S., Biswas, K.K.: Automated generation of activity and sequence diagrams from natural language requirements. In: 2014 9th International Conference on Evaluation of Novel Approaches to Software Engineering, ENASE, pp. 1–9 (2014)

117. TeamViewer 20219. Teamviewer. https://www.teamviewer.com/

118. Eclipse 2019. Eclipse Foundation https://www.eclipse.org/ide

The Challenge of Digital Education and Equality in Taiwan

Shin-yi Lee[✉]

China Medical University, Jingmao Rd., Beitun Dist., Taichung City 406040, Taiwan
sylee@mail.cmu.edu.tw

Abstract. 2020 has been a year full of changes and challenges to all teachers, students, and even parents. Because of the outbreak of COVID-19, school systems at all levels have been forced to modify their ways of educating students and many turned in-person teaching to online education. Before the pandemic, distance or online learning was treated by many teachers as an alternative or complementary method for the conventional classroom education; however, it has now become a necessity and seemingly the most feasible way to continue education, as long as schools remain closed because of the pandemic. Yet, the more the demand for distance or online learning grows, the more apparent the problem of digital divide would be. A nationwide coverage of the network transmission or the digital infrastructure does not guarantee digital equality among all citizens. Social, economic, political and even racial discrepancies should also be taken into consideration. This paper will begin by discussing the challenge of education in this digital era we have in Taiwan, and how to "bridge" the digital divide to achieve digital equality in our society.

Keywords: Digital education · Digital equality · Digital divide · COVID-19

1 Introduction

In 2020, an unexpected pandemic, COVID-19, just caught everyone off guard and revolutionized the conventional classroom management. A great number of schools around the world were forced to be closed and people were quarantined or self-isolated at home. The situation turned distance or online learning a daily necessity, if people wish to continue education. However, as the demand for distance or online learning grows, the subsequent problems and challenges are to be found even more distinct and imperative.

First of all, the traditional in-person classroom management has been converted into an interactive and quick-pace communication. Without the limitation on time and space, teachers could teach and students could learn anywhere as long as the Internet connection is functioning and well at hand. However, without teachers being in sight, students have to be more self-disciplined and responsible for their own learning. Before the pandemic, home-schooling has merely been an option for education, and now it has become the norm. A great challenge has equally fallen upon countless families: do students know how to carry on their education without teachers' discipline? Are parents capable of accessing to resources and support to help their children?

© Springer Nature Switzerland AG 2021
H. Degen and S. Ntoa (Eds.): HCII 2021, LNAI 12797, pp. 76–83, 2021.
https://doi.org/10.1007/978-3-030-77772-2_5

Yet another greater challenge in education is the lack of a more stable and friendly digital environment for all and the equal distribution of digital resources. Achieving digital equality has always been one of our governmental policies, but COVID-19 was so unexpected to everyone that all decisions were made in haste, including curfew, lockdown or even closing the border. This is the very moment to test if the digital infrastructure could fully satisfy the basic need of all citizens. When people are quarantined and prevented from having any physical contact with others, a stable network transmission and sufficient digital education would be the window for them to reach out to the world and continue learning. However, we learn from news reports (and my personal experiences as well) that the network connection would be jammed or disconnected when overloaded; not to mention that not all people have adequate equipment or devices to access the network. In addition, not all the schools have enough information technology (IT) to satisfy various needs of teachers and students. In fact, teachers may be limited by the teaching environment, if they do not receive full support from the school. Moreover, the socio-economically disadvantaged students might not be able to compete in this "digital arms race" since they do not have sufficient hardware and software at hand, when they need to conduct online learning.

Thus, digital equality concerns not just people's equal access to digital resources *physically*, but, more significantly, the social justice to enable and empower—*ethnically, socially, politically and economically*—all people to access the digital world with sufficient skills and equipment to face this fast-changing world. Then, an ethical and equal digital society could be promised and realized.

2 Digital Divide: A Daily Scenario of Regular Life

Let me begin by one of my personal experiences: a student came to me after class, right after I announced an online-quiz, a safer way to avoid any physical contact, would be held the next time we had class. That was a regular school day after the outbreak of COVID-19. We had class both online and at the physical classroom, and students could choose either way to participate. Most of my students were encouraged to self-isolate at home, but some still would choose to come to class as long as they wore masks and maintained social distancing, since the university where I work was not closed completely. Then the student spoke in a timid way asking me if I could give her a written quiz instead of an online one, for her laptop was a bit old and might not run steadily enough when she took an online exam. All students were given the same amount of time to take the quiz, so if her laptop needed more time to process data, she might not have enough time to finish the task. What she said truly ignited my attention and reflection on how digital equality has been practiced on the teaching sites. It seemed that much of our attention was placed on the equal physical access to the digital resources; that is, students are free to access WIFI at school or basic IT facilities, like computers. However, we did not "bridge" enough or at least cut less in the social and economic sense the digital divide between the advantaged and disadvantaged students.

Another incident also kindled much discussion online. In March 2020, a video selfie taken by Shiri Kenigsberg Levi[1], a mother of four from Israel, was broadcasted on the

---

[1] See https://www.youtube.com/watch?v=MrpYIv0oGhk, last accessed 2021/2/10.

Taiwan local news channel and surprisingly caught much attention. Levi complained with full rage that she was not capable of handling all the assignments her children had to accomplish at the same time and that her four kids fought over for two computers everyday also frustrated her. Under the YouTube link, some parents from Taiwan could relate to Levi's pain and frustration, and struggled to fulfill all the needs of their children to accomplish their online homework and distance learning. Yet, there were still some others thinking Levi should prepare "in advance" to buy each kid a laptop, iPad or appropriate equipment, and prompt and train her children to learn actively. No matter what comments the viewers put, one thing taken for granted behind is that people believe the coverage of the network connection and digital equipment we have are sufficient to conduct distance and online education, and if we do not have enough resources, we could *buy* or at least *borrow* from others.

My student's and Levi's cases both demonstrate a daily scenario during the outbreak of COVID-19. Many people, if prevented from going to school, could naturally turn online to continue learning and finish all the tasks as they do in the physical classroom. However, my student's and Levi's cases also point out a blind spot we commonly share; that is, digital equality will not be achieved nor the digital divide among all groups of people be mended, when we add more computers or free WIFI access to students. There are more issues needed to be considered.

The term of "digital divide" is not a novel idea. Starting from 1994, National Telecommunications and Information Administration (NTIA) has collected data on Americans' use of computer and Internet use and devices.[2] The coverage of the Internet connection and computer ownership rate are constantly collected and studied, which implies that the advantaged—the ones who can afford the access to the technologies—could manage to be benefited from the Internet services, while the disadvantaged could only fall far behind. The ones who own the latest technologies, such as IT equipment, could enjoy the most benefit from the Internet services. Likewise, Taiwan Network Information Center (TWNIC) also collects data on Taiwanese people's use on the Internet services, and divides users into specific groups, such as users in the developed urban areas, users from the rural areas, also based on their age and education differences.[3] Moreover, in 2004, Taiwan established "Digital Opportunity Center" and started a project, "Digital Application Promotion Project in Remote Areas"[4] in hopes of abridging the digital gaps between the remote places and urban areas. The government intends to create more digital opportunities in the remote places, help promote economy and balance the development and application of information technologies in all regions. Yet, at the same time, the government has already classified people and the Internet users into several classes.

NTIA, TWNIC and even Digital Opportunity Center suggest digital accessibility is deeply related to the Internet users' social status and economic power. Higher digital

---

[2] See National Telecommunications and Information Administration (NTIA), U. S. Department of Commerce, https://www.ntia.doc.gov/, last accessed 2021/2/10.

[3] See 2020 Taiwan Internet Report, Taiwan Network Information Center (TWNIC), https://report.twnic.tw/2020/en/index.html, last accessed 2021/2/10.

[4] See Digital Application Promotion Project in Remote Areas, Digital Opportunity Center, Ministry of Education, Taiwan, https://itaiwan.moe.gov.tw/english/, last accessed 2021/2/10.

accessibility implies not only people's economical capability of exploiting the information technologies, but also their information literacy to process and execute the knowledge and resources they acquire online. Then we could say that the Internet and information technology are not a great "equalizer" to close all the gaps among different groups of people, but unveil the urgency that the digital divide is now widening. What's worse, the more the new technologies are invented and expanded, the more gaps and digital divide will be created. If we really would like to close up the digital divide and fulfill the ideal of digital equality, we are obliged to take all users' need socially, economically and culturally into consideration.

3 The Challenges: IT Arm Races or Digital Equality

According to Yuval Noah Harari, globalization and new technologies supposedly should help achieve equality among all people around the world; in other words, they would equally treat everyone as a "cosmopolitan," a global citizen without ethnic or social differences. Yet the truth is they actually widen the gaps of human classes and create the inequality of all time (2019: 94). The ones who have the access to technologies and resources could gain the upper hand to better their power and influence over the socio-economic underprivileged groups.

Moreover, there are the IT oligopolies, such as Google, Facebook, or Baidu, that dominate and control the majority of the corporate wealth, and overwhelm the majority of people. For example, at the beginning of the pandemic in 2020, the Ministry of Education (MOE) encouraged schools in Taiwan to apply free online classroom systems to help students who were not able to go to school. We did not have a local online classroom system suitable for all learners nationwide, so schools at all levels tried and applied various systems to fulfill demands from different teaching sites, finding the most appropriate one to use. At the first place, Zoom was once officially promoted and encouraged to use, but as the concern of information security rose, MOE issued an order that Zoom should be banned as the medium for teaching.[5] This process of "trial and error" showed that the government did not prepare themselves for the coming of the digital age, and led most of the Taiwanese schools to highly rely on the systems provided by foreign companies to conduct online courses and video conferences, such as Google Classroom, Zoom, Microsoft Teams, Facebook, YouTube or Line. These systems are distinctive with assorted features, yet the maintenance and management are in the hand of the international corporations. A demand for a local system which protects privacy and feasible for the teaching sites in Taiwan grows, but none could virtually achieve high penetration rate to cover all needs.

The abovementioned incident points out the truth and the predicament we have today: people are not just classified based on their social and economic power; moreover, the government might also be the helping hand to prompt the social inequality. Since there are so many different free resources with high penetration rate at hand, the government would be very much likely to cut short of the investment in education, and leave education as a commodity controlled by the market. By doing so, education is seen as a

---

[5] See News Release of MOE, https://www.edu.tw/News_Content.aspx?n=9E7AC85F1954 DDA8&s=868B3A6EDF9BA52D, last accessed 2021/02/11.

capitalist property: only those who are capable of affording new technologies could receive education. Take banning Zoom as an example, the teachers and students who were familiar with new technologies might find it easy to switch to other software and continued teaching and learning; on the contrary, the underprivileged might need extra time and effort, and even money, to catch up with others. Thus, the government must not see the online learning system as an equalizer to close the gaps among students; sometimes, it could be an oppressive force to create more gaps and inequality.

Furthermore, according to TWNIC, the top five Internet services people use are "Instant Messaging," "Internet News," "Video/Live," "Email/Search," and "Community Forum." The amount of time spent on the Internet services during off-work are 14.2% for "Instant Message," 13.0% for "social media," 11.1% for "recreation," 9.7% for "Email and search" and 8.7% for "News." Only 1.9% time is spent on online learning. The survey points out that people see technologies as part of daily necessity for communication and fun; most of them are for personal purposes, instead of pedagogic purposes. In addition, broadband rate in Taiwan is almost 100% coverage rate[6], which means it would be easy for people in Taiwan to access the network connection and online resources. Access to the Internet is so "everyday life" that Taiwanese people tend to ignore the possibilities that are caused by social and economic differences among people. Just like my student's and Levi's cases I mention earlier in this paper, many people, including myself, tend to overlook the truth that people actually do not share an equal economic status and power, but assume that accessing the network and enjoying the latest technologies are so common just like drinking a cup of clean water.

Another challenge that comes along with the pandemic is the teachers' information literacy. As a teacher, I also struggled to find the suitable way to teach students on the spot and online simultaneously since the outbreak of COVID-19. Many of my colleagues were busy acquainting themselves with various systems and eager to find the most con-venient and safe one to use. Apart from the regular teaching load, many teachers were exhausted by the extra effort to acquire new digital skills. My university offered online courses to help teachers, yet those were just hasty work to temporize the circumstances, and did not help me much. This incident shows that the technology-related education for teachers may not be sufficient. The devices that have been mostly used are "Pow-erPoint" and "Word," and some may also add audio-video aids to enrich their lectures. This points out that teachers tend to use computer and other application programs sim-ply for basic usage. Moreover, not all schools would offer sufficient IT help for their teachers. Take my university for example, not all E-classroom services are upgraded to the equal level, which limits teachers from doing more active and dynamic teach-ing experiments. Without support from schools and MOE, teachers might not exploit resources and access the latest technologies; consequently, students under this circum-stance may not develop enough creative and critical thinking. Therefore, the challenge lies on that schools should not merely give full support on the use of computers and the

---

[6] According to Taiwan Internet Report, in non-rural areas, the household Internet rate is 83%, and in rural areas, the household rate is 70.3%. The broadband rate in the non-rural areas is 99.7%, and it is 100% in rural areas. The figures show that the Internet is quite accessible all over Taiwan. See Digital Divide Analysis, 2020 Taiwan Internet Report, TWNIC, https://report.twnic.tw/2020/en/index.html, last accessed 2021/02/12.

Internet in classrooms without giving IT help for teachers. Teachers should also build up their interest in technologies and information knowledge, and adapt themselves to the change of times. Teachers' "resourcefulness" would definitely cultivate students in a more pedagogically sound way.

Then the other challenge is to improve all users' information literacy and create a real user-friendly digital environment. It cannot be denied that the innovation and revolution of technologies are only meaningful for those who can access them. However, if we really wish to "equalize" all users online, first of all, we need to recognize digital inequalities and divide do exist; they are never personal or occasional, but systematic and customary. For example, most web pages in Taiwan only offer Mandarin Chinese versions, some would provide English translation, but not all the web sites or portals offer language choices other than Mandarin Chinese or English. Under this condition, speakers of different languages may not find relevant and adequate information online. In addition, the application programs and web pages could be designed in a friendly way for all users, including those who are not familiar with new technologies. During the pandemic, National Health Insurance Administration, Ministry of Health Welfare in Taiwan, developed an App and mask-map for people to purchase masks online, and this name-based system guaranteed that everyone has equal chance to buy masks. However, many people, especially the elders, still chose to go to the pharmacies in person and exposed themselves to the hazard of being infected by the unknown crowd to buy masks, just because they did not know how to operate the App. In a survey inquiring Taiwanese people's idea about what they think if the government should stop distributing masks to pharmacies, but have pharmacies decide what to sell and how to sell. In the meantime, people could still order masks online or through Apps. Surprisingly, about 85% of people still consider it necessary for the government to distribute masks to physical pharmacies, for the elders do not have the habit to order and inquire things online.[7] The abovementioned examples just demonstrate the digital divide does not merely concern the *physical* differences—the infrastructure, the number and quality of computers, the speed of broadband and so on, but the differences resulting from the social, cultural, economic inequalities among people. Speakers of different languages, users of different ages or digital skills and even the able and disabled should all be cared and tended.

Last but not the least, a non-hostile web content is even more significant than technology-based education for the young generation. According a survey done by Child Welfare League Foundation in 2019, 72% children and teenagers admit that they have heavy reliance on the Internet services, and 61% admit they used to do online voice or video chatting until midnight. About 70% are exposed to the content concerning violence, blood, horror or pornography.[8] Moreover, many video games offer sexist or racial

---

[7] About 92% of people are against that the government stops distributing masks, 85% of people think it's more convenient for the elders and the ones who don't use Apps or online shopping. Only 8% of people think the government could stop distributing. See UDN news report (2020), https://udn.com/news/story/7266/4537914, last accessed 2021/02/12.

[8] According to Child Welfare League Foundation's report, 82.7% of children and teenagers have their own mobile phones and 87% have accounts of social media. These figures show our children and teenagers have more chance to assess the Internet services than we think; therefore, it is urgent and imperative for us to teach them how to use the services properly and protect themselves when accessing the online resources. See Child Welfare League Foundation (2019), https://www.children.org.tw/research/detail/67/1525, last accessed 2021/02/12.

content and appeal to users' sensory excitement and pleasure without ethic concerns. In 2005, an act was passed to regularize the digital web site content rating system in Taiwan, but now it has been abolished in 2012.[9] An appropriate content rating system set by the government is by all means essential and compulsory, but what helps more is that the service providers do not post any hostile content and the parents guide the young in an appropriate way. The truth is, at most of the time, children and teenagers could only rely on the self-discipline of the service providers and themselves, and under the parents' guidance. By all means, the young generation could definitely benefit more if the service providers offer friendly and non-hostile content and the parents' information literacy improves and adequate to be a model.

Above all, the idea of technology progress is not simply the improvement of the physical devices or infrastructure, but the *social, cultural and economic* progress for all humans. How all people benefit from new technologies is truly a core mission for developing new technologies.

4 Conclusion

Paul C. Gorski has pointed out, multicultural education should go beyond celebrating the joys of diversity, and push further to think about how to apply the technologies to fulfil equality and justice in school and society (2007: 10). This is also true for the digital education. At this unprecedented digital era, the role of teachers and adults are not the knowledge providers for the young generation anymore, but all people could access information and knowledge if given equal chance. The exchanges of knowledge and information are always on the move, and the authorities and conventions could always be challenged and subverted. The digital resources are diverse and renewed every day. Apart from catching up with the trend and celebrating the joys of *resourcefulness* of the Internet, but the Internet users should also learn how to apply the latest technologies to promote equality and ethics online.

By recognizing the digital divide existing in our society, we could have a better idea about what to do and where to start. We could start by encouraging teachers' taking more technology-based training course to improve their information literacy, then improving the friendly cyber culture by eliminating the hostile content against children and teenagers and taking the needs people from different backgrounds into consideration, and finally, by always bearing in mind that new technologies are not for creating a privileged class, but help equal distribution of power and resources.

Just as Gorski has put it, "multicultural education's chief concerns are equality and justice" (2007: 2), and it is also true about digital education. The reason why we emphasize much on digital equality is that it is an act to promote social justice: to assure everyone shares equal chance freely to access information and benefit from the advantages of new technologies. The network connection is not an ultimate equalizer, but it could be exploited in a sound way to care for the disadvantaged and the underprivileged class.

---

[9] See Regulation for the Rating of the Internet Content, Laws and Regulations Database of the Republic of China, https://law.moj.gov.tw/LawClass/LawAll.aspx?pcode=P0050021, last accessed 2021/02/12.

References

1. Blackman, R.: A practical guide to building ethical AI. Harv. Bus. Rev. (2020). https://hbr. org/2020/10/a-practical-guide-to-building-ethical-ai. Accessed 12 Feb 2021
2. Child Welfare League Foundation (2019). https://www.children.org.tw/research/detail/67/ 1525. Accessed 12 Feb 2021
3. Digital Opportunity Center, Ministry of Education, Taiwan. https://itaiwan.moe.gov.tw/eng lish/. Accessed 10 Feb 2021
4. Dobbs, L.: War on the Middle Class: How the Government, Big Business, and Special Interest Groups Are Waging War on the American Dream and How To Fight Back. Trans. Zhen-ru Liu. Domain Publishing Company, Taipei (2007)
5. Gorski, P.: Insisting on Digital Equality: Reframing the Dominant Discourse on Multicultural Education and Technology. http://edchange.com/publications/digital-equity.pdf. Accessed 05 Feb 2021
6. Harari, Y.: 21 Lessons for the 21st Century. Trans. Jun-hung Lin. Global Views-Commonwealth Publishing, Taipei (2019)
7. Li, M.-H., Tseng, S.-F.: Redefining the digital divide and its measurement. J. Cyber Cult. Inf. Soc. 9, 89–124 (2005). http://ccis.nctu.edu.tw/word/9-02.pdf. Accessed 09 Feb 2021
8. National Telecommunications and Information Administration (NTIA). U. S. Department of Commerce. https://www.ntia.doc.gov/. Accessed 10 Feb 2021
9. News Release of MOE. https://www.edu.tw/News_Content.aspx?n=9E7AC85F1954DDA8& s=868B3A6EDF9BA52D. Accessed 11 Feb 2021
10. Putnam. R.: Our Kids: The American Dream in Crisis. Trans. Cheng-yi Li and Ya-shu Xu. Acropolis publishing, Taipei (2016)
11. Taiwan Network Information Center (TWNIC), 2020 Taiwan Internet Report. https://report. twnic.tw/2020/en/index.html. Accessed 10 Feb 2021
12. The Committee for the Coordination of Statistical Activities (CCSA). How Covid-19 is Changing the World: A Statistical Perspective, vol. I. https://unstats.un.org/unsd/ccsa/doc uments/covid19-report-ccsa.pdf. Accessed 31 Jan 2021
13. The Committee for the Coordination of Statistical Activities (CCSA). How Covid-19 is Changing the World: A Statistical Perspective, vol. II. https://unstats.un.org/unsd/ccsa/doc uments/covid19-report-ccsa_vol2.pdf. Accessed 31 Jan 2021
14. UDN news report (2020). https://udn.com/news/story/7266/4537914. Accessed 12 Feb 2021
15. United Nations Human Rights Office of the High Commissioner (OHCHR). Human Rights Indicators Tables. https://www.ohchr.org/Documents/Issues/HRIndicators/SDG_Ind icators_Tables.pdf. Accessed 31 Jan 2021
16. United Nations Human Rights Office of the High Commissioner (OHCHR). Human Rights Indicators: A Guide to Measurement and Implementation. New York and Geneva: United Nations (2012). https://www.ohchr.org/Documents/Publications/Human_rights_ind icators_en.pdf. Accessed 31 Jan 2021
17. Regulation for Rating of the Internet Content, Laws and Regulations Database of the Republic of China. https://law.moj.gov.tw/LawClass/LawAll.aspx?pcode=P0050021. Accessed 12 Feb 2021
18. YouTube page. https://www.youtube.com/watch?v=MrpYIv0oGhk. Accessed 10 Feb 2021

Morality Beyond the Lines: Detecting Moral Sentiment Using AI-Generated Synthetic Context

Ming Qian[1]($\boxtimes$), Jaye Laguardia[1], and Davis Qian[2]

[1] Barnstorm Research, Boston, MA, USA
{ming.qian,jaye.laguardia}@barnstormresearch.com
[2] School of Information Science, University of North Carolina, Chapel Hill, USA
davisq@live.unc.edu

Abstract. Moral rhetoric is defined as the language used for advocating or taking a moral stance towards an issue by invoking or making salient various moral concerns. The Moral Foundations Theory (MFT) can be used to evaluate expressions of moral sentiment. MFT proposes that there are five innate, universal moral foundations that exist across cultures and societies: care/harm, fairness/cheating, loyalty/betrayal, authority/subversion, and sanctity/degradation. We investigate the case in which texts containing MFD keywords are not expressed explicitly — hidden context. While members of high-context groups can read "between the lines" meanings, word counting methods or other NLP methods for moral sentiment detection and quantification cannot happen if the related keywords are not there to be counted. To explore the hidden context, we leverage a pretrained generative language model such as Generative Pre-trained Transformer (GPT-2) that uses deep learning to produce human-like text—to generate a new story. A human writer would usually provide several prompting sentences, and the GPT model would produce the rest of the story. To customize the GPT-2 model towards a specific domain—for this paper we studied local population's attitudes towards US military bases located in foreign countries—a training dataset from the domain can be used to finetune the GPT-2 model. Finetuning means taking weights of a trained neural network and using it as initialization for a new model being trained on the finetuning dataset. Restricted language codes (meanings are not expressed explicitly) can be used as prompting sentences, and finetuned GPT models can be used to generate multiple versions of synthetic contextual stories. Since the GPT-2 model was trained using millions of examples from a huge text corpus, the generated context contents reflect the cultural-related knowledge and common sense in the culture. In addition, since finetuned models were trained using finetuned dataset, the generated context contents reflect the local people's reaction for that specific domain—which is attitudes towards US military bases in regards to this paper. After using or fine-tuning the GPT-2 model to generate multiple versions of synthetic text, some versions might contain keywords defined in the MFD. Our hypothesis is that the percentage keywords related to the five morality domains can serve as statistical indicators for the five domains. Our experiment shows that the top five morality domain types experiencing significant percentage changes between positive and negative stories, generated by fine-tuned training

© Springer Nature Switzerland AG 2021
H. Degen and S. Ntoa (Eds.): HCII 2021, LNAI 12797, pp. 84–94, 2021.
https://doi.org/10.1007/978-3-030-77772-2_6

models, are HarmVice, AuthorityVirtue, InGroupLoyalty, FairnessVirtue and Fairness Vice. The results are in line with several major issues identified between US oversea military bases and local populations by well-known existing studies. The main contribution of this research is to use AI-generated synthetic context for detecting moral sentiment and quantification.

Keywords: Human-machine teaming · Human-machine interaction · Moral sentiment detection · Moral sentiment quantification · Moral foundations theory · Elaborated and restricted language codes · Generative pre-trained language model · Artificial intelligence · AI-generated synthetic context

1 Introduction

The concept of elaborated and restricted language codes was introduced by sociologist Basil Bernstein [1]. Restricted codes are usually phrased with collapsed and shortened sentences, requiring listeners to share a great deal of common perspective to understand the implicit meanings and nuances of a conversation. Restricted language codes are commonly used in high context culture groups, where group members share the same cultural background and can easily understand the implicit "between the lines" meaning without further elaboration. This is especially true for social media content because most social media texts are short conversational posts or comments that do not contain enough explicit information for natural language processing tools [2]. There exists an intellectual 'great divide' between 'insiders' and 'outsiders' of a cultural group.

Moral rhetoric is defined as the language used for advocating for or taking a moral stance towards an issue by invoking or making salient various moral concerns. It has been argued that moral rhetoric affects people's moral and political worldviews and increases levels of political intensity [3].

Previous works exist using word count methods, the Moral Foundations Theory (MFT), and pre-trained distributed representations [4, 5, 6, 7] to evaluate expressions of moral sentiment. MFT proposes that there are five innate, universal moral foundations that exist across cultures and societies: Care/harm (e.g. intuitions of sympathy, compassion, and nurturance), Fairness/cheating (e.g. notions of rights and justice), Loyalty/betrayal (e.g. supporting moral obligations of patriotism and "us vs. them" thinking), Authority/subversion (e.g. concerns about traditions and maintaining social order), and Sanctity/degradation (e.g. moral disgust and spiritual concerns related to the body). For example, Reference [4, 5] demonstrates that liberals and conservatives attend to different moral intuitions: while liberals focus exclusively on the notions of harm and fairness when making moral judgments, conservatives also attend to ideas of loyalty to in-group members, authority, and purity.

In this paper, we investigate the case in which text containing MFD-related keywords are not expressed explicitly. While members of a high-context group can read the "between the lines" meaning, word counting methods or other NLP methods for moral sentiment detection and quantification cannot happen if the related keywords are not there to be counted.

To explore the hidden context, we leverage a pretrained generative language model such as GPT2. Generative Pre-trained Transformer (GPT) is an autoregressive language model that uses deep learning to produce human-like text [8, 9]. OpenAI released GPT-2 in February 2019. It is trained to predict the next word in a sentence based on a 40GB training corpus. To generate a new story, a human writer would usually provide several prompting sentences, and the GPT model would produce the rest of the story. To customize the GPT-2 model towards a specific domain—for this paper we studied local attitudes towards US military bases located in foreign countries—a training dataset from the domain can be used to finetune the GPT-2 model. Finetuning means taking weights of a trained neural network and using it as initialization for a new model being trained on the finetuning dataset.

We propose a new approach to use the restricted language codes as prompting sentences (meanings are not expressed explicitly), and the finetuned GPT model can be used to generate multiple versions of synthetic contextual stories. Since the GPT model was trained using millions of examples from a huge text corpus, the generated context contents reflect the cultural-related knowledge and common sense in the culture. In addition, since the finetuned model was trained using the fine-tuned dataset, the generated context contents reflect the local people's reaction for that specific domain—attitudes towards US military bases for this paper.

After using or fine-tuning the GPT-2 model to generate multiple versions of synthetic text, some versions might contain keywords defined in the MFD. Our hypothesis is that the percentage of files containing keywords related to the five morality domains — care/harm, fairness/subversion, loyalty/betrayal, authority/subversion and purity/degradation — can serve as statistical indicators for the five domains.

2   Methodologies

2.1   Public Opinion Problem Under Study: Local Attitudes Toward US Oversea Bases and US Troop Presence

After World War II, hundreds of thousands of military personnel have been deployed to countries around the globe, as the United States began to acquire overseas bases. Based on treaties between the U.S. and host countries during the Cold War, US military personnel have been stationed permanently.

The local population's perceptions regarding the United States, its government, and people have been heavily influenced by the nature of their interactions with the US bases and military personnel. A typical social science approach is to conduct surveys which consume a lot of resources in terms of time and resources. For example, reference [10] describes a survey sampling about 14,000 individuals in multiple countries, and each individual was asked 50 questions related to their demographic characteristics, the U.S. military deployments in their country, and their views on several local and global issues.

An alternative and cheaper way is to collect news stories and social media messages and evaluate expressions of moral sentiment embedded in those messages, since moral rhetoric reflects people's moral and political worldviews and levels of political intensity. One challenge of this approach is that since the messages were not answers towards well-defined questions, they would be in different forms and meanings could be implicit. This

was especially true for high-context cultural groups where people can read the "between the lines" meaning based on shared background. We leverage the GPT language model to generate synthetic texts in order to explicate and contextualize the implicit meaning. Since the GPT language model was trained using a large corpus of Internet text, it captures not only linguistic patterns but also cultural knowledge.

To customize the synthetic text generation towards our unique task of evaluating local attitudes toward US oversea bases and US troop presence, we fine-tuned the raw 124M GPT-2 model with collected task related text from Reddit and Quora forums.

2.2 Fine-Tuning Data Collection

For data collection, we looked for answers to the prompting question "what is your view/opinion towards US military bases" from people outside of the US.

On Reddit, there were many forums dedicated to this topic. As much as possible, we avoided subreddits such as "r/AskAnAmerican", and we instead tried to gather answers from users whose locations were public, as again we wanted opinions from the local people of other countries, not the US. We found 23 good subreddits fitting these requirements. A simple Reddit scraper was made using PRAW (Python Reddit API Wrapper) that automated the data collection process. We fed a selected forum's unique ID into the scraper and it gathered all the top-level comments, formatted the query into question/answer style, and saved the data in an output text file. We formatted it this way because we would use this data to train the GPT-2 model after data collection. After building the scraper and scouring Reddit for good forums, we were able to gather 537 unique answers to the above prompting questions from people all over the world.

We also gathered data from Quora forums. We used a 3rd party software called 'webscraper.io' to scrape the data. A similar process used for Reddit was followed and then all outputs were combined into a single text file with the same format. The final output ended up with 977 tokens (stories).

Table 1 shows two sample examples of collected Reddit and Quora data.

2.3 Fine-Tuning GPT-2 Language Model

Next, we followed instructions (described in [11]) to set up and fine-tune the 124M gpt2 model using the 1514 collected stories. After fine tuning, we saved this fine-tuned version of 124M gpt2 so we can always reload and run this specific fine-tuned version.

2.4 Synthetic Context Generation

We picked out 3 prompts from the final output: one with a neutral outlook on military bases, one with a slightly positive outlook, and another with a slightly negative outlook (Table 2).

We put each of the three prompts through the fine-tuned gpt2 model text generator and the generic AI text generator (i.e., default gpt2 model without fine-tuning) and generated 10 stories for each generator.

Table 1. Sample Reddit and Quota forum data

Sample Reddit forum data	< |startoftext|> [Question] for quora user Nanashi no Gonbei, former Semi retired, consultant: What do the Japanese think of the American military base in their country? [Answer]: They are necessary to protect Japan When some aeroplane accidents and incidents like traffic accidents & rapes highlights the media from time to time, but they are small matter compared to the nation's security and safety. Although they are regrettable those things happen if you have an American base or not Those things are unfortunate, but you cannot have everything positive. We have to accept some negative points, too. In Okinawa, those people demonstrating and complaining are mixed people, some Chinese, Korean and the old people paid for the demonstration and some of them are airlifted from mainland Japan. US bases in Okinawa was promised by the previous so-called Liberal administration, namely then Prime Minister Hatoyama, to move the base at least out of Okinawa, but in the end, failed to do so and reversed the previous decision and went back to the original plan to move to less populated area in Okinawa. That is the Hatoyama's legacy the present administration is battling US Bases are necessary in Okinawa, simply because Okinawa is strategically very important to protect many Japanese islands around that area, as the Chinese navy and their smaller ships disguised as fishing boats violate Japanese territorial waters every day < |endoftext|>
Sample Quota forum data	< |startoftext|> [Question] for quora user John Doe, former counselor at Government of Iraq: What do other nations think of US Soldiers? [Answer]: Judging how US soldiers are perceived in Iraq based on first-hand experience in Baghdad. I'm not talking about any army branch or unit in particular, just GIs. They are very well equipped. Nice weapons, uniform and vehicles, good communication, so logistically GIs are well taken care of. In particular, the creature comforts in their basis such as Union-III in Baghdad. It has not just a well catered mess run by contractors providing them with a choice of hot meals three times a day, and at least three sorts of ice-cream, but also a free 24/7 shop for food during periods not covered in between plus there are commercial fast-food restaurants as can be found in the USA. On the flipside: US soldiers appear not just uninterested in Iraq but even without respect for the local culture and ignorant of Iraqi customs and language. This makes them embarrassingly bad at interacting with the local population, which in turn prevents them from winning hearts and minds. One small gesture that would make a lot of difference is to take off their sunglasses when talking to people, as they make them look like big bugs. As a result, Iraqis don't go out of their way to tell them anything. This bad interaction with the local population is in remarkable contrast with e.g. Italian or British soldiers who because of their better interaction with the locals have a correspondingly better situational awareness < |endoftext|>

Table 2. Three types of prompts (negative, positive and neutral).

Prompt negative	US soldier squads on the battlefield are walking battle fortresses, heavy equipments heavy machine guns, night vision, thermal vision, air support, artillery support evacuation support, etc.... But when it comes to close quarters combat, where equipment is ineffective, even US special forces cannot match endoctrinated and decisive fighters, beginning from vietnam war, where communist peasants with AK47 and bolt action rifles caused havoc to US army, until the battle of Fallujah, where militia fighters strained the US forces, until they were forced to withdraw
Prompt positive	Most "nations" don't have opinions about US soldiers. I've worked with American soldiers and my opinion of them varies from [a] a waste of rations to [z] !!W!O!W!! depending on whom I was working with and what was being done. As a general rule, however, members of the USMC are almost as good as they think that they are - and that's pretty damn good
Prompt neutral	As long as European countries like our bases as fountains of money, and as long as people don't like to target nations with US military bases for offensive wars, I think they're a reasonable use of military funds. As someone else mentioned, how would the nations of Ukraine or Georgia be today if treaties had opened US military bases there five or ten years ago?

2.5   Keyword Identification and Counts

For each generated story, we color coded any instance of morality domain types defined in MFD and calculated the stats on each morality type by dividing specific MFD-defined keyword counts by total word counts in the story.

In addition, we expanded the initial keyword lists based on their embedded meanings. For example, based on the keywords "authority", "duty", "lawful" and "serve", we inferred AuthorityVirtue as an embodiment that enforces the law / upholds authority, law, and order. Towards the embodiment, groups of people or individuals who have authority, such as the military or a police officer, also meet this criterion. On the other hand, the word "militia" has a more negative connotation, that word would be categorized as AuthorityVice instead. PurityVice had keywords such as "slut", "whore", "filth", "defile" in the MFD. Therefore, whenever a speaker uttered a degrading word, the intent was to degrade the military through name calling, and we can categorize the word as PurityVice.

We also add another category called morality general and we labeled anything subjective to morals under that whether it was good, bad or neutral. For example, if a speaker believes the military is a bad influence which should be interpreted as a moral judgment, then we can label it as morality general.

Another category defined is concern, which is an emotion-based category. For example, if a majority of the military was not qualified to do their job, that should be regarded as a major concern and was thus labeled so.

Agenda was also labeled based on human judgements. Agenda is like having a plan of action. Usually the speaker talks in the first person about how they want an event to

happen—e.g. building a resort on the island where the existence of a military base could become a major obstacle towards such an agenda.

Emotion was also defined as a separate category. Any words relating to emotion such as "anger", "sadness", "joy" was labeled as emotion. Emotions greatly represent local populations' perceptions on US bases and military.

3   Results and Discussions

3.1   Number of MFD-Related Keywords Before and After Fine-Tuning

We count the total number of MFD-related keywords in 10 generated stories based on the default GPT2 model and 10 stories based on the fine-tuned GPT2 model for each prompt type. The count comparison is listed in Table 3.

For the neutral and negative prompts, the fine-tuned model generated a larger number of MFD keywords in the stories, whilst for the positive prompt, the fine-tuned model generated a smaller number of MFD keywords in the stories. This might be due to the asymmetric morality described in [12, 13]: blame is more differentiated and more extreme than praise. Features of a moral act like causality, intentionality, and magnitude of outcome influence blame more than praise. A functional perspective was proposed: while blame is primarily for punishment and signaling one's moral character, praise is primarily for relationship building. If we believe in this perspective, for moral judgments towards the US bases, local populations showed strong negative moral judgments and weak desire for relationship building.

We made a table representing the MFD statistics before and after fine tuning (Table 4). We wanted to compare the differences between positively and negatively prompted texts, and differences between texts generated before and after fine-tuning.

To compare positively and negatively prompted texts, we compared every category (for both before and after fine-tuning) and found that the following categories are significant (changes are more than 2% between positive and negative): HarmVice (up 2.1% and up 2.8%), IngroupVirtue (up 2.2% and up1.6%), AuthorityVirtue (up 8.6% and down 2.3%), and Morality General (up 6.1% and up 7.9%). HarmVice is related to the concerns about caring and protecting individuals from harm (e.g. attack, kill). IngroupVirtue is related to loyalty to the group (e.g. country names, the community, the public, the people). AuthorityVirtue is related to the obedience and subversion for authority (e.g. US-led coalition). Negative texts related to the local attitudes towards US bases, compared with positive texts, are associated with more harm, more group loyalties such as local culture or society, and more/less obedience for authority.

Before fine-tuning, by comparing the statistics, the significant factors differentiating the texts generated by the positive and negative prompts are HarmVice (up 2.1%), IngroupVirtue (up 2.2%), AuthorityVirtue (up 8.6%) and Morality General (down 6.1%). So, for the baseline language modeling, the negative prompted texts are associated with more harm, more group loyalties, and more obedience for authority.

After fine-tuning, the significant factors differentiating the texts generated by the positive and negative prompts are HarmVice (up 2.8%), AuthorityVirtue (down 2.3%), Morality General (down 7.9%), Agenda (down 2.5%), Emotion (up 7%) and concerns

(down 8.6%). So, for texts generated by the fine-tuned model, negatively prompted texts are associated with more harm, reduced obedience for authority (e.g. US-led coalition), significantly more emotion, and having fewer concerns and agendas.

Table 3. Number of MFD-related keywords before and after fine-tuning.

	Number of MFD-related keywords before fine-tuning	Number of MFD-related keywords after fine-tuning
Positive prompt	864	570
Negative prompt	832	1520
Neutral prompt	806	992

Table 4. Statistics for MFD and other metrics before and after fine tuning.

Morality domain types and other metrics	Positive prompt		Negative prompt		Neutral prompt	
	Before	After	Before	After	Before	After
Harm virtue	0.1%	0%	0.2%	0.2%	0.3%	1.5%
Harm vice	0.5%	0.3%	2.6%	3.1%	2.4%	0.8%
Fairness virtue	0%	2.6%	0%	1.7%	0%	0%
Fairness vice	0%	0%	0%	1.1%	0%	0%
Ingroup virtue	0.9%	0.5%	3.1%	2.1%	4%	5.2%
Ingroup vice	0%	0%	0.6%	0.1%	0%	0%
Authority virtue	2.3%	9.2%	10.9%	6.9%	5.5%	6.8%
Authority vice	0.2%	0.3%	0.1%	0.6%	0.2%	1%
Purity virtue	0%	0%	0%	0%	0%	0%
Purity vice	0%	2.4%	0.4%	0%	0%	0%
Morality general	11%	11.2%	4.9%	3.3%	1.2%	8.1%
Agenda	4.5%	5%	5.7%	2.5%	10%	11.4%
Emotion	0.1%	2%	0%	9%	0%	1.5%
Concerns	3.8%	11.9%	3.4%	3.3%	7.3%	6.8%

3.2 Discussion

Based on extensive research, reference [14] listed several major issues between the local people/government and US bases/personnel.

One main issue is that poor behavior or illegal actions by US personnel could taint local perceptions of the U.S. military. In terms of morality foundations, it belongs to the category of care/harm—poor behaviors and illegal actions harm the local population. Also, some local communities have worries that US personnel would disrupt local social relations (e.g. fraternize with local women). In terms of morality foundations, it belongs to the category of loyalty/betrayal and purity/degradation—the 'us' versus 'them' thought.

Another major concern is the presence of a foreign military force on the territory goes against the most fundamental principles of sovereignty. Some sovereignty-related issues are the sovereign status of bases, the host country's right to restrict the use of the base, and the criminal jurisdiction procedures that govern the legal status of U.S. soldiers accused of crimes. In terms of morality foundations, it belongs to the category of loyalty/betrayal and purity/degradation—the 'us' versus 'them' thought, and the obligation of patriotism.

One other issue is bargains and concessions, in terms of political and economic stakes, offered by the sending country to the receiving country in exchange for securing basing rights. In terms of morality foundations, it belongs to the category of fairness and authority—the sense of fairness and justice (both parties shoulder some shares of responsibility), and the sense of belonging (e.g. US-led coalition).

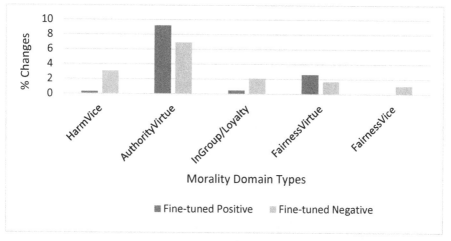

Fig. 1. Morality domain types experiencing significant percentage changes between positive and negative stories, generated by fine-tuned training models.

Figure 1 shows the top five morality domain types experiencing significant percentage changes between positive and negative stories, generated by fine-tuned training models. They are HarmVice (going up by 2.8%), AuthorityVirtue (going down by 2.3%), InGroupLoyalty (going up by 1.6%), FairnessVirtue (going down by 0.9%) and FairnessVice (going up by 1.1%). These results seem to agree with the major negative issues

described in Reference [14]: poor behaviors or illegal actions by US personnel harm the local population; the virtue of authority (e.g. US-led coalition) drops; group loyalty (obligations of patriotism and "us vs. them" thinking) rises; fairness virtue drops; and fairness vice rises (local people felt that they did not get a fair deal in terms of political and economic prices).

4 Conclusion

Moral rhetoric is defined as the language used for advocating for or taking a moral stance towards an issue by invoking or making salient various moral concerns. The Moral Foundations Theory (MFT) can be used to evaluate expressions of moral sentiment. MFT proposes that there are five innate, universal moral foundations that exist across cultures and societies: care/harm, fairness/cheating, loyalty/betrayal, authority/subversion, and sanctity/degradation. We investigate cases in which texts containing MFD keywords are not expressed explicitly. While members of a high-context group can read the "between the lines" meaning, word counting methods or other NLP methods for moral sentiment detection and quantification cannot happen if the related keywords are not there to be counted.

To explore the hidden context, we leverage the pretrained generative language model such as Generative Pre-trained Transformer (GPT-2) that uses deep learning to produce human-like text. To generate a new story, a human writer would usually provide several prompting sentences, and the GPT model would produce the rest of the story. To customize the GPT-2 model towards a specific domain—for this paper we studied local attitudes towards US military bases located in foreign countries—a training dataset from the domain can be used to finetune the GPT-2 model. Finetuning means taking weights of a trained neural network and using it as initialization for a new model being trained on the finetuning dataset.

Restricted language codes can be used as prompting sentences (meanings are not expressed explicitly), and finetuned GPT models can be used to generate multiple versions of synthetic contextual stories. Since the GPT model was trained using millions of examples from a huge text corpus, the generated context contents reflect the cultural-related knowledge and common sense in the culture. In addition, since finetuned models were trained using a fine-tuned dataset, the generated context contents reflect the local people's reaction for that specific domain—attitudes towards US military bases for this paper.

After using or fine-tuning the GPT-2 model to generate multiple versions of synthetic text, some versions might contain keywords defined in the MFD. Our hypothesis is that the percentage keywords related to the five morality domains can serve as statistical indicators for the five domains.

Our experiment shows that the top five morality domain types generated by the fine-tuned training models, experiencing significant percentage changes between positive and negative stories, are HarmVice, AuthorityVirtue, InGroupLoyalty, FairnessVirtue, and FairnessVice. The results agree with major issues identified between US oversea military bases and local populations by well-known existing studies.

References

1. Bernstein, B.: Elaborated and restricted codes: their social origins and some consequences. Am. Anthropol. **66**(6), 55–69 (1964)
2. Benamara, F., Inkpen, D., Taboada, M.: Introduction to the special issue on language in social media: exploiting discourse and other contextual information. Comput. Linguist. **44**(4), 663–681 (2018)
3. Sagi, E., Dehghani, M.: Measuring moral rhetoric in text. Soc. Sci. Comput. Rev. **32**(2), 132–144 (2013)
4. Graham, J., Haidt, J., Nosek, B.A.: Liberals and conservatives rely on different sets of moral foundations. J. Pers. Soc. Psychol. **96**(5), 1029 (2009)
5. Garten, J., Boghrati, R., Hoover, J., Johnson, K.M., Dehghani, M.: Morality between the lines: detecting moral sentiment in text. In: Proceedings of IJCAI 2016 Workshop on Computational Modeling of Attitudes (2016)
6. Moral foundations theory: https://moralfoundations.org/
7. Moral foundations dictionary: https://moralfoundations.org/wp-content/uploads/files/downlo ads/moral%20foundations%20dictionary.dic
8. Haidt, J., Graham, J., Joseph, C.: Above and below left–right: Ideological narratives and moral foundations. Psychol. Inquiry **20**(2–3), 110–119 (2009)
9. Radford, A., Wu, J., Child, R., Luan, D., Amodei, D., Sutskever, I.: Language models are unsupervised multitask learners. OpenAI Blog **1**(8), 9 (2019)
10. Allen, M.A., Flynn, M., Machain, C.M., Stravers, A.: Understanding How Populations Perceive U.S. Troop Deployments, The Owl in the Olive Tree, the blog of Minerva Research Initiative (2019). https://minerva.defense.gov/Owl-In-the-Olive-Tree/Owl_View/Article/179 7784/understanding-how-populations-perceive-us-troop-deployments/
11. https://medium.com/@stasinopoulos.dimitrios/a-beginners-guide-to-training-and-genera ting-text-using-gpt2-c2f2e1fbd10a
12. Guglielmo, S., Malle, B.F.: Asymmetric morality: blame is more differentiated and more extreme than praise. PLOS ONE (2019). https://doi.org/10.1371/journal.pone.0213544
13. Anderson, R.A., Crockett, M.J., Pizarro, D.A.: A theory of moral praise. Trends Cogn. Sci. **24**(9), 694–703 (2020)
14. Cooley, A.: Base Politics: Democratic Change and the US Military Overseas. Cornell University Press, Ithaca (2012)

Whoops! Something Went Wrong: Errors, Trust, and Trust Repair Strategies in Human Agent Teaming

Summer Rebensky⬤, Kendall Carmody⬤, Cherrise Ficke⬤, Daniel Nguyen⬤,
Meredith Carroll$^{(\boxtimes)}$⬤, Jessica Wildman⬤, and Amanda Thayer⬤

Florida Tech, Melbourne, FL 32901, USA
slindsey2013@my.fit.edu, mcarroll@fit.edu

Abstract. Human interactions with computerized systems are shifting from using computers as tools, into collaborating with them as teammates via autonomous capabilities. Modern technological advances will inevitably lead to the integration of autonomous systems and will consequently increase the need for effective human agent teaming (HAT). One of the most paramount ideals human operators must discern is their perception of autonomous agents as equal team members. In order to instill this trust within human operators, it is necessary for HAT missions to apply the proper trust repair strategies after a team member commits a trust violation. Identifying the correct trust repair strategy is critical to advancing HAT and preventing degrading team performance or potential misuse. Based on the current literature, this paper addresses key components necessary for effective trust repair and the numerous variables that can further improve upcoming HAT operations. The impacting factors of HAT trust, trust repair strategies, and needed areas of future research are presented.

Keywords: Human-agent teaming · Trust violations · Trust repair

1 Introduction

Due to the exponential growth in technological advancements, autonomous systems have presented promising results for the future of numerous applications in business, military and medical environments. As the United States Air Force (USAF) is expecting 60% of the USAF fleet to be unmanned or optionally manned by 2035, other military branches are projected to follow a similar trend as well [1]. In the Army, Military Occupational Speciality (MOS) codes are used to identify a soldier's specific skill set and ability to perform a distinct task. By using these specific set of skills as the framework for a computational agent, intelligent agents can be programmed to complement a soldier's abilities during a mission [2]. These autonomous agents can provide assistance in circumstances where human operators are unable to perform a task due to physical difficulty, cognitive load or high risk [2].

The artificial intelligence (AI) used in autonomous agents is also becoming more prominent in contemporary society, beyond the military. Autonomous agents can be used

© Springer Nature Switzerland AG 2021
H. Degen and S. Ntoa (Eds.): HCII 2021, LNAI 12797, pp. 95–106, 2021.
https://doi.org/10.1007/978-3-030-77772-2_7

in business environments to optimize productivity in the workplace. A study by Dubey et al. (2020), assembled an AI framework for a team of human and intelligent call-center agents to interact with customers and address customer concerns. Autonomous agents were able to collect customer purchase information, company policy documentation, and help human operators upsell company products to the customer by assisting in formulating a response for the customer. Through the collaboration between the autonomous agents and the call-center operator, more effective and timely responses for the customer may be produced resulting in profit maximization [3].

The dynamics of human agent teaming (HAT) plays a substantial role in AI operations. As we explore techniques to strengthen the dynamic between the human operator and AI-driven autonomous agent, appropriate levels of trust must be developed to achieve optimal mission performance [4]. To address this need, human operators may need to begin viewing autonomous agents as team members, rather than tools. Moving forward, many more applications of HAT across industries will be prevalent as technology continues to advance in contemporary society.

In a HAT context, trust is defined as "the attitude that an agent will help achieve an individual's goals in a situation characterized by uncertainty and vulnerability" ([5], p. 54). When a human operator is familiar with the AI and has an understanding of how and why the agent is operating, trust is developed, promoting a symbiotic relationship between the operator and the system [6]. However, if trust is not attained, the operator may choose to turn off or disregard the agent and take over responsibilities assigned to the agent. This could result in higher workload levels for the operator who may become overwhelmed with tasks, ultimately resulting in performance degradation [7]. In the autonomous driving domain, those who have experienced incidents with their autonomous driving features, such as unexpected speed changes or turns, have reported lower trust levels and lower frequencies of use [8]. Therefore, the development of appropriate levels of trust is necessary to optimize HAT operations. Appropriate levels of trust are the foundation for operators to begin viewing the AI-driven agent as a teammate, leading to the potential for operators to take advantage of the system's functionalities available to support them in more effectively meeting their performance goals.

Research is needed to examine the unique challenges HATs face compared to human-human teams (HHT) and the impact that these challenges have on trust processes and the effectiveness of trust repair strategies. This includes studying interactions with variables such as degree of risk, time criticality, and environmental surroundings. This paper will provide an overview of the research in HAT conducted to date, along with recommendation for industry and future research. The focus of this paper is primarily on the aspects of the system and what the system does and can do to enhance the trust relationship with its human counterpart. By evaluating HAT trust processes and trust repair strategies, the boundaries of HAT can be more effectively understood, resulting in improvements to the safety and efficiency of future autonomous system operations.

2   Trust Development in HAT

Trust in HATs is different than trust in HHT in several key ways, including the lack of: (a) intentionality, which makes it difficult for humans to understand or gather what the

agent intends, (b) a fluid social exchange that might provide an opportunity for trust to form through interactions, and (c) the attribution process in which trust builds to higher levels (in fact, in HAT, trust often begins high and degrades with mistakes or perceived mistakes; [5]. There has been a great deal of empirical research regarding the changing dynamics of trust associated with human-agent dyads (one agent working with one human) and AI [9]. Meta-analytic reviews centered around system trust in AI have outlined various impacting factors relative to the agent capabilities and attributes including: (a) the tangibility or presence of the autonomous agent, (b) the immediacy or the ability for the autonomous agent to interact with humans in a quick and natural manner, (c) how transparent the system is regarding its processes, (d) the reliability of the automation, and (e) the task the autonomous agent is given [9]. Each of these concepts and their impacts on trust are discussed in detail below.

2.1 Tangibility and Immediacy

Tangibility or the ability to see or touch the agent can bring feelings of trust—and as a result, humans are more likely to trust and comply with physical agents as opposed to virtual agents [9]. Research has demonstrated that humans interacting with anthropomorphic agents experience lesser degradations in trust following a violation compared to non-anthropomorphic agents [10]. Other human-like behaviors, such as communication, are essential for a team to function effectively. Human teams are able to manage one another, communicate their individual actions to ensure shared situation awareness, and coordinate with one another dynamically [11]. With an autonomous agent, communication similar to HHT is necessary to achieve mutual predictability [11]. However, communication is currently a challenge due to the lack of social cues and contexts that humans typically use when communicating in HHT that allows for natural and immediate conversation [6]. It is important for the robot to explain why it is doing a specific task and support two way communication among teammates [6]. In order to further this collective trust, shared mental models, and situational awareness, it is essential that the autonomous teammate attempt to read and incorporate social cues such as gestures in its communication repertoire [12].

2.2 Transparency

Of particular importance is transparency, which allows human agents to understand the system and what it will do next; whilst team members use shared awareness to mutually acknowledge the potential and current actions a team member may execute during an operation [4]. For example, an autonomous car may inform the user it is exiting the highway as congestion is detected ahead. Doing so ensures the human fully understands what is about to happen and why, keeping trust in the vehicle's actions high. Transparency, in conjunction with communication, and customizability are seen as influential factors in creating trust between the human operator and autonomous agent [13]. However there is a careful balance between too little information and too much. Wright, Chen, Barnes, and Hancock (2016) explored the impacts of providing reasoning to reroute suggestions in a convoy managing simulated study. It was revealed that participants were likely to comply with the automation suggestions—even if it was wrong because it offered no

reasoning. However, being transparent in its reasoning behind suggestions led to more correctly rejected advice. In contrast, offering too much information, which may not be clear or fully understood by the participant, can cause participants to be as complacent to automation as when no reasoning was offered. Therefore, too much information or unclear reasoning can lead to a position of over-trusting automation or agents [14]. As autonomous technology starts to perform tasks, like complex decision making, ensuring human agents can understand and deduce what the automation is doing and why, will help establish an appropriate level of trust [6]. However, if the autonomous agent fails to meet human expectations, team members will question the effectiveness and accuracy of the system, resulting in degraded trust, regardless of whether the system is acting appropriately [6]. For example, if a drone is set to fly straight over buildings, and alters its course in order to ensure adequate collision avoidance margins, and does not inform the human, the human may believe the drone is acting inappropriately and take over manual flight. It is essential that the correct calibration of trust be identified in order to create an efficient and effective HAT [15, 16]. Particularly once an autonomous agent acts in err or against the human's expectations. Then, providing the correct level of information to the operator to adjust human's mental models of the agent's abilities and decision making will aid in the correct calibration of trust.

2.3 Reliability

Reliability has served as a common predictor in a human's initial trust towards an autonomous agent [17, 18]. This positive correlation between initial trust and reliability significantly influences the dynamic between operators in HAT environments [18–20]. For example, when a UAS exhibits a decrease in performance from its expected level of reliability, the human's trust will decline, and more so than with a human team member who exhibited a lower than expected level of performance [11]. Further, under high workload circumstances, operators are more likely to comply with false alarms from automated agents when there is high reliability [21]. Multi-unmanned aircraft vehicle (multiUAV) studies have shown that lower reliability led to higher stress and poorer performance in surveillance tasks [22]. In HAT research, negative outcomes could pose relational risks, or negative impacts on other relationships where humans set their trust based on past experiences and history with automation [23, 24]. For example, a negative interaction with one brand of robot could have negative trust implications for other robots of the same or similar brands. Further, Huang et al. (2021) posit potential impacts of degraded trust in future similar autonomous agents based on previous unreliable agents with early versions of technology. This degraded trust could transfer to different agents on the same team, or even degraded trust in a new agent after a trusted agent was destroyed (e.g., a soldier who distrusts a new robot of the same model, after losing a trusted robot the soldier was attached to on the battlefield; [25, 26]). To continue improving HAT performance, future research must investigate the impacts of agent reliability in order to optimize mission performance for future operations. Other variables such as work environment, training experience, and cognitive demands are also significant factors that increase reliability in HAT environments [27].

2.4 Task

Humans may exhibit low initial trust if they perceive the agent's task as having high perceived risk, such as self parking features in an autonomous car (Dikmen and Burns 2017). When considering these risks, one must weigh the responsibilities and risks associated with putting faith in the agent. An individual may trust an agent to control their music or phone calls, but not to drive their car. Furthermore, one single mistake of an autonomous agent could have various negative outcomes. An autonomous car that collides with a street post could result in financial, physical, and time losses that could extend to potential social or psychological risks as well [23]. A study examined how participants' responded to robot recommendations related to social and functional tasks. Results suggest that participants are more willing to uphold the robot's advice on functional issues, such as which item is heavier, rather than social tasks such as which dress looks better [28]. Additionally, other pressures, such as time pressures, can affect human's willingness to trust. The demands of tasks can stress the ability for humans to maintain accurate mental models and awareness with their agent counterparts, which could lead to complacency and following agent direction during times of high workload, even if the agent is wrong [21].

3 Trust Violation in HAT

With an autonomous teammate, trust is never freely provided by the human operator. It instead must be built and well maintained over a period of time. Glikson and Woolley (2020) posit that trust in agents depends on the type of agent, but generally trust starts low for physical agents and starts high for virtual and embedded agents [9]. As time and interactions occur, virtual and embedded AI trust drops rapidly with negative interactions, whereas physical robot trust slowly grows with positive interactions. However, regardless of agent type, trust is fragile and can quickly degrade during early use [9]. Establishment of trust results from consistent and predictable interaction over time [13]. However, a break in predictability can cause the trust between the two teammates to dwindle [13].

Marinaccio et al. (2015) proposed a framework for HAT trust repair in which the violations are either integrity-based or competency-based [29]. Competency-based errors can be defined as errors that occur when the autonomous agent lacks the ability to respond to certain situations (e.g. malfunctions, reliability issues). In other words, failures of the system [30]. An integrity-based error occurs when the agent mis-prioritizes its tasks (e.g., misalignment with human goals) and essentially acts on a set of unpredictable principles [31]. The following sections will discuss the different strategies to repair both of these types of violations. Particularly, what has been found in HHT that could be leveraged in HAT, the limited research that has been conducted with HAT trust repair, and interactions to consider based on the context of the violation.

4 Trust Repair in HAT

As agent errors will inevitably occur, potentially violating the operators' trust in their system compatriot, understanding how to repair this trust is critical. An agent simply

correcting it's behavior is not sufficient as a trust repair strategy and will still negatively affect trust [32]. An understanding of the trust repair process [33] and the impact of different trust-repair strategies after a trust violation occurs in HAT operations. is essential in developing effective relationships between teammates. Different types of errors require different trust repair strategies, and different relationships present idiosyncratic traits that consequently result in different trust repair methods. Baker et al. (2018) identified four common trust repair types in HHT teams including: (a) apology, or apologizing for the mistake, (b) denial, or explanation that one was not at fault, (c) trustworthy action, or performing better for an extended period of time after the violation, or (d) commitment to change, or promising to do better next time [33]. However, HHT trust processes differ from those within HAT operations which occur in a more complex paradigm where multiple variables interact. Marinaccio et al. (2015)'s framework provides a solid foundation for understanding the most effective trust repair strategies that can be utilized within HATs [29, 33]. The strategies for trust repair in HATs depend greatly on the type of violation committed by the teammate, and depending on the context of the violation, an effective repair would either be an (a) apology, (b) denial or (c) context dependent [33].

4.1 Apologies

HHT trust research offers additional options regarding apology strategies. Beyond indicating regret, apologies may include specific components such as an expression of regret, an explanation of the violation, an acknowledgement of the violator's responsibility, a declaration promising not to repeat the offense, an offer of repair, or a request for forgiveness [34]. An apology may include some or all of these components (or variations thereof, e.g., apologizing but attributing the blame to a different party). However, research suggests that explanations of the violation, acknowledgement of responsibility, and offers of repair were the most effective components of apologies [34]. If the violation is perceived to be competency-based, apologies have been found to be the most effective strategy for repairing trust [35]. In HAT, integrity violations will result from agent priorities, which do not align with the human teammates. A proposed study by Quinn, Pak and de Visser (2017) describes a HAT example scenario in which an automated AI taxi-fare program makes a competency- or integrity-based trust violation when suggesting the most effective taxi-fares and routes [30]. In the competency-violation condition, a computer program will suggest that the system could not cope with the current situation due to system inability when selecting the best route or fare. In the integrity-based condition, the system will deny it's error as it's prioritization of tasks were not in-line with the humans. Although it has not been tested, researchers hypothesize that apology-based trust repair will work best for competency issues [30].

4.2 Denial

Research on trust in HHT also identifies when certain strategies may be more or less effective. Often, the success of the repair strategy depends on the perceived nature of the violation. When violations are perceived to be integrity-based, denial has been shown to be more effective than apologies as long as fault is not evidently established [36].

Few studies have tested to see whether these findings hold true in HAT. Quinn, Pak, & de Visser (2017) hypothesized that denial would repair trust best when an agent commits an integrity-based error in a HAT, paralleling HHT trust repair findings from Kim and colleagues [30, 36]. Kohn, Quinn, Pak, & de Visser (2018) conducted an exploratory study to compare the effectiveness of various forms of apology and denial trust repair strategies [37]. In study 2, participants viewed a self-driving car coming close to crashing, then followed up with either no repair strategy, an apology-based strategy, or a denial-based strategy. Overall, apology-based strategies were more effective than denial-based strategies. Kohn and colleagues (2018) note that only one form of trust violation was employed in the study, but that participants perceived the near-crash violation as a competence error [37]. Thus, similar to trust in HHT, apologies appear to be more effective than denials for competency-based trust violations in HATs [37].

4.3 Context Specific Strategies

Compensation is an additional trust repair strategy in HHTs that involves giving a tangible item of equal value to the cost of the loss incurred by the violation (e.g., reimbursement, vouchers). Compensation is only an option in situations when there is a concrete valuation that can be placed on the loss, and its effectiveness has shown mixed results [38]. It is unclear whether it is more or less effective than other common repair strategies such as apologizing; however, there is evidence that it is effective when done in conjunction with an apology [39], it fully recovers the value lost from the violation [40], or is perceived as the violator fully understanding the extent of their damage [41]. Compensation as a repair strategy has been explored in service staff robots in a restaurant. In a study by Lee et al. (2010), humans watched videos of food service performed by a robot that made errors by bringing the customer a Sprite instead of a Coke [42]. The different conditions were repair strategies of apologies, compensation, or options (methods to help the robot correctly get a Coke). It was found that compensation was more effective for participants who put greater emphasis on receiving good service than the social aspects of a server. However those who view service as a social and relational job were more satisfied with apologies from the robot. In emergency situations, agents offering promises to do better next time can be just as effective as apologies—particularly if they immediately follow the mistake [43].

It is also worth noting that these repair strategies are short-term options for restoring trust, and that long-term strategies are also available to repair trust in HHTs. For example, teams may rearrange their operating structure to control how its members interact [44] or reframe the violation to influence how members perceive the long-term effects of the violation [45]. HHT trust research also identifies that victims of the violation may either forgive the violator for their offense [46] or remain silent about the offense [47]. However Baker et al. (2018) notes "only a handful of authors have addressed the issue of [HAT] trust repair" [33]. More research is needed to determine effects over time with violations in HAT.

4.4   Interactions Between Trust Violation and Trust Repair

Factors such as degree of risk, time criticality, environmental surroundings, teammate perception and mission context may impact the effectiveness of HAT trust repair strategies [48]. The literature regarding trust repair strategies for HHT trust provides a basis for understanding human agent trust repair. Some factors that can impact the effectiveness of a trust repair strategy include elements associated with the situation, the nature and length of the relationship, and the type of violation. In HHT, the phase and nature of the relationship can impact the effectiveness of trust repair strategies [49]. Trust violations that occur early in relationships can be more detrimental as trust is fragile; this is also the case for violations that appear hypocritical in nature [49]. For example, if trust violations occur with managers who set rules for subordinates that they themselves violate, this can be more detrimental as the subordinate has preset expectations of their managers. Similar to HHT, in HAT perceptions of their agent teammate can also impact trust repair effectiveness [50], including human's perceptions of the agent's capabilities and their predisposition towards automation in general [50]. If there is a mismatch between what the human expects and what the agent does, there is potential for trust degradation, regardless of whether the agent's actions were accurate.

Relative to the violation itself, HHT trust repair strategy effectiveness can depend on the severity of the violation as well as the history and frequency of these types of violations [49]. For HAT, risk and time criticality must be considered when exploring trust repair strategies. Once a human develops trust in an autonomous agent, the trust is weighed against the potential risk of engaging with the agent [23]. In HAT research, negative outcomes could pose relational risks such as performance or financial risks, or negative impacts on other relationships where humans set their trust based on past experiences and history with automation [23, 24]. For example, a negative interaction with one brand of robot could have negative trust implications for other robots of the same or similar brands. Recent studies have found that human operators are more forgiving when autonomous agents fail during a low risk mission versus a high risk mission [51]. Further, the more time critical a mission is under fatal circumstances, the less likely the human operator is to exhibit trust in the future [50]. As agent capabilities expand from simple and manual tasks to complex decision making tasks with potentially life-threatening outcomes, it is important to consider risk.

4.5   Moving Beyond Repair and into Calibration

Explanations after agent errors in decision making are key to mitigate trust degradation [52]. Findings in dyadic teams reveal that providing explanations after an error can be effective in repairing trust, even if the information in the explanation is inaccurate [43]. This means that in some cases, humans may put blind faith in agents simply due to their attempt at trust repair, regardless of its accuracy. Therefore, trust calibration is key for HAT performance effectiveness [54]. Further, the level of information or transparency that is provided in an apology or denial will determine the human's acceptance of that trust repair strategy [14]. de Visser et al. (2020) have addressed trust repair from the concept of trust calibration—or only using trust repair when trust becomes too low after a trust violation [53]. de Visser et al. posits that trust repair strategies, such as

apologies, denial, and acknowledging one's error, are strategies that could be used to repair trust in an agent when a violation has occurred that is misaligned with a human's mental model. In this case, these methods aim to improve trust when a human's mental model is indicative of under-trust in an agent's capabilities after a violation. Given the inevitability of error occuring due to agent limitations, if a human has expectation that are too high or an inaccurate mental model of an agent's capabilities, an agent could engage in trust dampening strategies designed to lower expectations (e.g., requesting review, or conveying agent limitations). Finally, de Visser et al. suggests transparency during agent use to either provide states and mode parameters, or post-task reviews, in which the agent discusses its own performance with the user, to ensure the user's mental model for how the agent operators is accurate. These can also include immediacy behaviors such as conveying uncertainty in tone or vocabulary, or adjusting behaviors based on the human actions, to be more in-line with human intents [53].

These findings shed light on the need to further understand trust repair strategies and their impact on trust and future use of agents. Ineffective trust repair strategies could lead to under-trust and abandonment of the agent teammate or over-trust of the agent leading to complacency and potential detrimental outcomes [32, 43]. This includes exploring initial attempts into preserving trust with one agent towards the impacts of trust repair strategies on other agent team members, swarms of agents, and future agents [25, 54].

5 Conclusion

HAT trust is dependent on many human, environmental, and agent aspects. Relative to the agent, its ability to be transparent, human-like, and predictable are key. As we put more responsibilities in the hands of agents, we require the human to be more willing to expose themselves to risky situations and instill trust in their agent counterpart. Trust degradation can not only impact the short term HAT effectiveness, but also long term, future, and peripheral HAT effectiveness. Although HHT trust repair strategies provide a substantial foundation for understanding the HAT repair strategies, it is important to recognize HHT interaction cannot be completely emulated by HAT interactions. We as humans know what to expect of our human comrades, but agents are much less clear. The ultimate goal is to achieve the "sweet spot" of agent trust that ensures agents are utilized to their fullest potential, in appropriate ways, without humans becoming complacent. Research on the most effective methods to ensure trust in HAT is emerging; however, many questions have yet to be answered. Limited research has explored trust, and especially trust violation and repair, in larger multi-agent team settings, and this research has primarily focused on controlled experiments or theoretical propositions. More research with a range of different types of agents and situations is necessary, including virtual/physical, low/high intelligence, predictable/unpredictable, and in low/high-risk situations. We must determine how to effectively manage the balance between appropriate levels of trust and risk based on the human's mental model and the agent's actual capabilities. In a HAT, it simply boils down to your ability to answer this question: "What would your agent do?"—and do you trust it?

References

1. OSD: Unmanned systems integrated roadmap 2017–2042 (2017)
2. Caylor, J.P., Barton, S.L., Zaroukian, E.G., Asher, D.E.: Classification of military occupational specialty codes for agent learning in human-agent teams. In: Artificial Intelligence and Machine Learning for Multi-Domain Operations Applications, vol. 11006, p. 110060W. International Society for Optics and Photonics (2019)
3. Dubey, A., Abhinav, K., Jain, S., Arora, V., Puttaveerana, A.: HACO: a framework for developing human-ai teaming. In: Proceedings of the 13th Innovations in Software Engineering Conference on Formerly known as India Software Engineering Conference, pp. 1–9 (2020)
4. Endsley, M.R.: Situation awareness misconceptions and misunderstandings. J. Cogn. Eng. Decis. Making **9**(1), 4–32 (2015)
5. Lee, J.D., See, K.A.: Trust in technology: designing for appropriate reliance. Hum. Factors **46**, 50–80 (2004)
6. Schaefer, K.E., Hill, S.G., Jentsch, F.G.: Trust in human-autonomy teaming: a review of trust research from the US Army Research Laboratory Robotics Collaborative Technology Alliance. In: International Conference on Applied Human Factors and Ergonomics, pp. 102–114. Springer, Cham (2019)
7. Schneider, M., McGrogan, J., Colombi, J.M., Miller, M.E., Long, D.S.: Modeling pilot workload for multi-aircraft control of an unmanned aircraft system. INCOSE Int. Symp. **21**(1), 796–810 (2011)
8. Dikmen, M., Burns, C.: Trust in autonomous vehicles: the case of tesla autopilot and summon. In: 2017 IEEE International Conference on Systems, Man, and Cybernetics, pp. 1093–1098 (2017)
9. Glikson, E., Woolley, A.W.: Human trust in artificial intelligence: review of empirical research. Acad. Manag. Ann. **14**(2), 627–660 (2020)
10. de Visser, E.J., et al.: Almost human: anthropomorphism increases trust resilience in cognitive agents. J. Exp. Psychol. Appl. **22**(3), 331–349 (2016)
11. Joe, J.C., O'Hara, J., Medema, H.D., Oxstrand, J.H.: Identifying requirements for effective human-automation teamwork. In: Probabilistic Safety Assessment and Management PSAM (2014)
12. Jiang, W., Fischer, J., Greenhalgh, C., Ramchurn, S., Wu, F., Jennings, N., Rodden, T.: Social implications of agent-based planning support for human teams. In: 2014 International Conference on Collaboration Technologies and Systems (CTS), pp. 310–317. https://doi.org/10.1109/CTS.2014.6867582 (2014)
13. Hou, M.: IMPACT: a trust model for human-agent teaming. In: 2020 IEEE International Conference on Human-Machine Systems, pp. 1–4 (2020)
14. Wright, J.L., Chen, J., Barnes, M.J., Hancock, P.A.: The effect of agent reasoning transparency on automation bias: an analysis of response performance. In: International Conference on Virtual Augmented and Mixed Reality, pp. 465–477 (2016)
15. Rodriguez, S.S., Chen, J., Deep, H., Lee, J., Asher, D.E., Zaroukian, E.G.: Measuring complacency in humans interacting with autonomous agents in a multi-agent system. In: Pham, T., Solomon, L., Rainey, K. (eds.) Artificial Intelligence and Machine Learning for Multi-Domain Operations Applications II, p. 39. SPIE (2020)
16. Wagner, A.R., Robinette, P.: An explanation is not an excuse: trust calibration in an age of transparent robots. In: Trust in Human-Robot Interaction, pp. 197–208. Elsevier (2021)
17. Nguyen, D.: 1, 2, or 3 in a HAT? How a human-agent team's composition affects trust and cooperation. Masters Thesis (2020)
18. Hancock, P.A., Billings, D.R., Oleson, K.E., Chen, J.Y., De Visser, E., Parasuraman, R.: A meta-analysis of factors influencing the development of human-robot trust. Army research lab aberdeen proving ground MD human research and engineering directorate (2011)

19. Fan, X., Oh, S., McNeese, M., Yen, J., Cuevas, H., Strater, L., Endsley, M.R.: The influence of agent reliability on trust in human agent collaboration. In: Proceedings of the 15th European Conference on Cognitive Ergonomics: The Ergonomics of Cool Interaction, p. 7 (2008)

20. Chiou, E.K., Lee, J.D.: Cooperation in human-agent systems to support resilience: a microworld experiment. Hum. Factors **58**(6), 846–863 (2016)

21. Chen, J.Y., Barnes, M.J.: Human–agent teaming for multirobot control: a review of human factors issues. IEEE Trans. Hum. Mach. Syst. **44**(1), 13–29 (2014)

22. Wohleber, R.W., Stowers, K., Chen, J.Y.C., Barnes, M.: Effects of agent transparency and communication framing on human-agent teaming. In: 2017 IEEE International Conference on Systems, Man, and Cybernetics (SMC), pp. 3427–3432 (2017). https://doi.org/10.1109/SMC.2017.8123160

23. Stuck, R.E., Holthausen, B.E., Walker, B.N.: The role of risk in human-robot trust. In: Nam, C.S., Lyons, J.B. (eds.) Trust in Human-Robot Interaction, pp. 179–194. Elsevier (2021)

24. Hoff, K.A., Bashir, M.: Trust in automation. Hum. Factors J. Hum. Factors Ergon. Soc. **57**(3), 407–434 (2015)

25. Huang, L., Cooke, N.J., Gutzwiller, R.S., Berman, S., Chiou, E.K., Demir, M., Zhang, W.: Distributed dynamic team trust in human, artificial intelligence, and robot teaming. In: Nam, C.S., Lyons, J.B. (eds.) Trust in Human-Robot Interaction, pp. 301–319. Elsevier (2021)

26. Carpenter, J.: The quiet professional: An investigation of US military explosive ordnance disposal personnel interactions with everyday field robots (2013)

27. Tan, Y., Feng, D., Shen, H.: Research for unmanned aerial vehicle components reliability evaluation model considering the influences of human factors. In: MATEC Web of Conferences, p. 139 (2017). https://doi.org/10.1051/matecconf/201713900221

28. Gaudiello, I., Zibetti, E., Lefort, S., Chetouani, M., Ivaldi, S.: Trust as indicator of robot functional and so-cial acceptance. An experimental study on user con-formation to iCub answers. Comput. Hum. Behav. **61**, 633–655 (2016)

29. Marinaccio, K., Kohn, S. Parasuraman, R., de Visser, E.: A framework for rebuilding trust in social automation across health-care domains. In: Proceedings of the International Symposium on Human Factors and Ergonomics in Health Care, pp. 201–205 (2015)

30. Quinn, D.B., Pak, R., de Visser, E.J.: Testing the efficacy of human-human trust repair strategies with machines. Proc. Hum. Factors Ergon. Soc. Ann. Meet. **61**(1), 1794–1798 (2017)

31. Jensen, T., Khan, M.M.H., Albayram, Y., Fahim, M.A.A., Buck, R., Coman, E.: Anticipated emotions in initial trust evaluations of a drone system based on performance and process information. Int. J. Hum. Comput. Interact. **36**(4), 316–325 (2020)

32. McNeese, N.J., Demir, M., Chou, E.K., Cooke, N.J.: Trust and team performance in human-autonomy teaming. Int. J. Electron. Commer. **25**(1), 51–72 (2021)

33. Baker, A.L., Phillips, E.K., Ullman, D., Keebler, J.R.: Toward an understanding of trust repair in human-robot interaction: current research and future directions. ACM Trans. Interact. Intell. Syst. (TiiS) **8**(4), 1–30 (2018)

34. Lewicki, R.J., Polin, B., Lount Jr, R.B.: An exploration of the structure of effective apologies. Negot. Confl. Manage. Res. **9**(2), 177–196 (2016)

35. Kim, P.H., Dirks, K.T., Cooper, C.D., Ferrin, D.L.: When more blame is better than less: the implications of internal vs. external attributions for the repair of trust after a competence-vs. integrity-based trust violation. Organ. Behav. Hum. Decis. Process. **99**(1), 49–65 (2006)

36. Kim, P.H., Ferrin, D.L., Cooper, C.D., & Dirks, K.T.: Removing the shadow of suspicion: the effects of apology versus denial for repairing competence-versus integrity-based trust violations. Journal of Applied Psychology, 89(1), 104. (2004).

37. Kohn, S.C., Quinn, D., Pak, R., De Visser, E.J., Shaw, T.H.: Trust repair strategies with self-driving vehicles: an exploratory study. Proc. Hum. Factors Ergon. Soc. Ann. Meet. **62**(1), 1108–1112 (2018)

38. Lewicki, R.J., Brinsfield, C.: Trust repair. Annu. Rev. Organ. Psych. Organ. Behav. **4**, 287–313 (2017)

39. De Cremer, D.: To pay or to apologize? On the psychology of dealing with unfair offers in a dictator game. J. Econ. Psychol. **31**(6), 843–848 (2010)

40. Haesevoets, T., Folmer, C.R., De Cremer, D., Van Hiel, A.: Money isn't all that matters: the use of financial compensation and apologies to preserve relationships in the aftermath of distributive harm. J. Econ. Psychol. **35**, 95–107 (2013)

41. Fehr, R., Gelfand, M.J.: When apologies work: how matching apology components to victims' self-construals facilitates forgiveness. Organ. Behav. Hum. Decis. Process. **113**(1), 37–50 (2010)

42. Lee, M.K. Kiesler, S., Forlizzi, J., Srinivasa, S., Rybski, P.: Gracefully mitigating breakdowns in robotic services. In: 5th ACM/IEEE International Conference on Human-Robot Interaction (HRI) (2010). https://doi.org/10.1109/HRI.2010.5453195

43. Robinette, P., Howard, A.M., Wagner, A.R.: Timing is key for robot trust repair. In: Tapus, A., André, E., Martin, J.-C., Ferland, F., Ammi, M. (eds.) Social Robotics, vol. 9388, pp. 574–583. Springer International Publishing (2015)

44. Dirks, K.T., Lewicki, R.J., Zaheer, A.: Repairing relationships within and between organizations: building a conceptual foundation. Acad. Manag. Rev. **34**(1), 68–84 (2009)

45. Dewulf, A., et al.: Disentangling approaches to framing in conflict and negotiation research: a meta-paradigmatic perspective. Hum. Relat. **62**(2), 155–193 (2009)

46. Kramer, R.M., Lewicki, R.J.: Repairing and enhancing trust: approaches to reducing organizational trust deficits. Acad. Manag. Ann. **4**(1), 245–277 (2010)

47. Ferrin, D.L., Kim, P.H., Cooper, C.D., Dirks, K.T.: Silence speaks volumes: the effectiveness of reticence in comparison to apology and denial for responding to integrity-and competence-based trust violations. J. Appl. Psychol. **92**(4), 893 (2007)

48. van Diggelen, J., et al.: Pluggable social artificial intelligence for enabling human-agent teaming. In: NATO HFM Symposium on Human Autonomy Teaming (2019)

49. Lewicki, R.J., Brinsfield, C.: Trust repair. Ann. Rev. Organ. Psychol. Organ. Behav. **4**, 287–313 (2017)

50. de Visser, E.J., Pak, R., Shaw, T.H.: From 'automation' to 'autonomy': the importance of trust repair in human–machine interaction. Ergonomics **61**(10), 1409–1427 (2018)

51. Satterfield, K., Baldwin, C., de Visser, E., Shaw., T.: The influence of risky conditions in trust in autonomous systems. Proc. Hum. Factors Ergonomics Soc. Ann. Meet. 61(1), 324–328 (2017)

52. Wagner, A.R., Robinette, P.: An explanation is not an excuse: trust calibration in an age of transparent robots. In: Trust in Human-Robot Interaction, pp. 197–208. Elsevier (2021). https://doi.org/10.1016/B978-0-12-819472-0.00009-5

53. de Visser, E.J., et al.: Towards a theory of longitudinal trust calibration in human–robot teams. Int. J. Soc. Robot. **12**(2), 459–478 (2020)

54. Liu, R., Cai, Z., Lewis, M., Lyons, J., Sycara, K.: Trust repair in human-swarm teams. In: 2019 28th IEEE International Conference on Robot and Human Interactive Communication (RO-MAN), pp. 1–6. IEEE (2019)

What Does It Mean to Explain?
A User-Centered Study on AI Explainability

Lingxue Yang[✉], Hongrun Wang, and Léa A. Deleris

BNP Paribas, Paris, France
{lingxue.yang,hongrun.wang,lea.Deleris}@bnpparibas.com

Abstract. One frequent concern associated with the development of AI models is their perceived lack of transparency. Consequently, the AI academic community has been active in exploring mathematical approaches that can increase the explainability of models. However, ensuring explainability thoroughly in the real world remains an open question. Indeed, besides data scientists, a variety of users is involved in the model lifecycle with varying motivations and backgrounds. In this paper, we sought to better characterize these explanations needs. Specifically, we conducted a user research study within a large institution that routinely develops and deploys AI model. Our analysis led to the identification of five explanation focuses and three standard user profiles that together enable to better describe what explainability means in real life. We also propose a mapping between explanation focuses and a set of existing explainability approaches as a way to link the user view and AI-born techniques.

Keywords: Explainable AI · Explainability · User research · User centered design

1 Introduction

1.1 Background

Support from Artificial Intelligence (AI) and in particular machine learning algorithm has become pervasive in our personal and professional lives. However, one frequent concern is the perceived lack of transparency of such models, in particular when they are used in a context that can have a material influence on our lives, such as health, recruitment, legal or banking settings. It has thus become essential to further develop the capability to explain such advanced models.

In fact, the AI academic community has been active in exploring mathematical approaches that can increase the explainability of models (e.g., LIME [1], Shapley value [2], counterfactual explanations [3]). However, such efforts have been predominantly based on computer scientists' perceptions of what constitutes an explanation. This may produce a gap between explainability techniques and what it means to explain to real users [4, 5].

In this study, our objective is not to further develop technical solutions but rather (i) to take a user-centered perspective in defining what it means to explain AI models

© Springer Nature Switzerland AG 2021
H. Degen and S. Ntoa (Eds.): HCII 2021, LNAI 12797, pp. 107–121, 2021.
https://doi.org/10.1007/978-3-030-77772-2_8

and (ii) to study the fit and alignment of existing explainability methods in practice, i.e., with real users. To further anchor our analysis in real-world setting, we have focused specifically on AI models used within a financial institution though a large part of our analysis is not sector specific.

We start by reviewing explainability techniques. We observe that the current methods are suited for a specific population i.e., data scientists and that the needs of other stakeholders have not necessarily been considered. The main contribution of the paper stems from the insights derived from our user study focused on the variety of user needs for understanding and explaining AI based on the diversified view of stakeholders within a large financial institution. We organized the study in two phases: individual user interviews and group workshops. In the end, our analysis led us to define (i) three user profiles who need explanations with different motivations in different contexts (ii) five "Explanation focuses", which corresponds to specific explanation concerns of users and (iii) a mapping between explanation focuses and explainability methods.

2 Related Work

In this section, we review some key concepts of explainable artificial intelligence (definition and needs) and summarize what has been done from computer science and human computer interface (HCI) academic communities.

2.1 XAI Definition

Defense Advanced Research Projects Agency (DARPA) initiated explainable artificial intelligence (XAI) program in 2017 [6] to address a machine's inability to explain its thoughts and actions to human users. They then introduced XAI with the goal of enabling users to understand, trust, and effectively manage this emerging generation of artificial intelligence models. Ever since, there has been a surge of interest in the research on XAI both in the Artificial Intelligence (AI) and HCI communities. As a consequence, multiple related words have emerged seeking to provide an additional level of specification about the underlying needs and concepts, in particular interpretability, explainability and intelligibility [5, 7–12]. For example, Doshi and Kim [8] define interpretability as "the ability to explain or to present in understandable terms to a human". Separately, Gilpin et al. believe that interpretability is not sufficient for human to understand black-box models. They propose to define explainability as "models that are able to summarize the reasons for neural network behavior, gain the trust of users, or produce insights about the causes of their decisions." [13], which goes beyond interpretability. Liao et al. consider explainability to be everything that makes machine learning (ML) models transparent and understandable to humans [5]. Recently, Arrieta et al. [12] clarified some related concepts including understandability, comprehensibility, interpretability, explainability and transparency. They propose that transparency, interpretability and comprehensibility be merged together into understandability (or intelligibility), which measures the degree to which a human can understand a decision made by a model.

From these definitions, we observe that there is not yet clear consensus on their individual definition and they are sometimes used interchangeably [14–16]. However,

these definitions share implicitly one common factors: the importance of considering the recipient of the explanation, the one who asks questions about the AI models and receives the explanations. The purpose of this paper being focused on real-world feedback rather than concepts, we do not seek to contribute directly to the semantic discussion of the nuances among all those terms. In the remainder of the paper, to avoid ambiguity, we chose to use explainability when we talk about the explanations that users need.

2.2 XAI Techniques

The computer science community has been actively exploring approaches to improve the explainability of algorithms as it can constitute a clear barrier – among others – to the broad deployment of artificial intelligence solutions to end users. In that spirit, methods have been developed to enable people to derive insights into the functioning of an AI solution.

Several taxonomies have been proposed for users to understand the diverse forms of explanations that are available and the questions that each can address [12, 14, 17–19]. First, an explanation can either be static or interactive depending on whether the response can be changed according to the feedback from the user. One example of interactive explanation can be a dialog, for instance through a Chabot. Nonetheless, the vast majority of the literature focuses on static explanations [14]. We can also define explanations according to how they are generated [8]. Explanations can be an intrinsic part of the model, in the sense that there is no need for an additional model to generate the explanation. Such models are deemed by nature transparent and easy to interpret for most of the users. The most common examples are short decision trees or sparse linear models [20]. They are often referred to as white box AI models or transparent models by opposition to Black box models such as random forest and deep neural networks. Recently, computer scientists have also worked on self-explaining neural networks, which consists on modifying the architecture of a network to make it interpretable (Explainability by design) [9, 10]. By contrast to intrinsic explanations, a post-hoc explanation is based on applying an additional AI method on the initial model (the decisions are already made) [1, 2, 21–25]. Typically, post-hoc explanation approaches can be used on all kinds of AI models, i.e., they are model agnostic, which gives them some sort of universality. Finally, explanations can be either global or local. Global explanations seek to describe the behavior of the entire model [2, 21–23], while local explanations provide explanations for single prediction[1, 2, 24, 25].

The majority of explainability methods work either by analyzing the contribution of input features to the model outputs, we call them feature based explanations or by analyzing the instances that were introduced in the model we call them example based explanations[26–28]. An important consideration at this point is that the methods mentioned have predominantly come from the computer science community and are therefore biased towards computer scientists needs for explanation. Existing solutions for XAI are

mostly developed in the format of a python package, usually using one or several explainability techniques (Eli5[1], AIX360[2], xai[3], ethik[4]). They thus tend to be better suited for explanation to AI experts or for debugging purposes for technical practitioners (e.g., data scientists). A concrete example could be the self-explaining neural network, while data scientists can read into the additional explainer layer of the neural network to learn the contributions of input features, this type of information are not interpretable for other users [9].

In this paper, our ambition is not to develop new explainability techniques but rather get insights of users' need on this topic especially those who were less served by the current solutions. However, we sought to map the existing realm of such techniques with our findings, so as to better understand what aspects of user-needs for explanations are well-covered and which ones may require further investigation.

2.3 XAI HCI Approaches

Given that the way of representing explanations of AI models depends on the recipients as well as their needs, it is essential to work on XAI from user-centered approach. In this section, we reviewed the main contributions to the concept of explainable AI in the HCI community around (i) developing a better understanding of the need for explainability, (ii) enabling visual analytics for XAI to facilitate the users to understand the overall model behavior and input feature behavior and (iii) studying theoretical constructs for XAI from social science.

Regarding the Needs for Explainability. Many studies have explored what prompts the need for explainability [13, 29, 30], for instance *improvement and optimization of the system*, identification of the potential *bias* and ensuring that the performance (accuracy, precision, robustness and stability) of a *model* is adequate. Some studies have mentioned that users may seek explanations for *verification* purpose when there is a deviation or an inconsistency between what is expected and what has occurred [4, 31]. Miller [9] and Samket et al. [29] both propose that people want explanation to facilitate learning by identifying the hidden laws of nature and "extracting the distilled knowledge" in order to predict and control future phenomena. Some also need explanations to improve the efficiency of *their decisions making process* when using an AI product. *Compliance* with regulation is also a key reason for seeking to improve the explainability of AI systems. Indeed, the European Union set a "The right to explain" regulation whereby individuals subject to a decision made by an AI system should be provided with an explanation of why a specific decision has been made. Further on, end-users may also wish to understand what could be changed to obtain a better result [27]. Overall, this research stream highlights that the need of an explanation can be motivated by a variety of reasons, from a diverse set of users [32].

---

[1] https://eli5.readthedocs.io/en/latest/.

[2] https://aix360.mybluemix.net/.

[3] https://ethicalml.github.io/xai/index.html.

[4] https://xai-aniti.github.io/ethik/.

Visual Analytics for XAI. In Parallel, There Has Been Burgeoning Efforts Around Developing Explainable User Interfaces, Specifically with Interactive Visual Analytics Capability to Support Model Explanation, Interpretation, Debugging and Improvement [32–35] . Some Tools Are Specifically Designed for Technical Profiles to Inspect and Analyze the Model They Build or Use. For Example, Prospector [36], Gamut [37] and What-If Tool [38] Are Similar Tools Designed to Help Data Scientists Understand the Impact of Input Features on the Model Predictions, Investigate the Outcome of the Model Globally and Locally. Other Solutions like Shapash or Waston Openscale Have Given More Consideration to Non-technical Users. Shapash [39] Provides an Visual Support with Both Global and Local Explanations for Data Scientist but Only Local Explanations for End Users. Waston Openscale[5] only Provides Information to Explain the Model Output of a Given Instance.

Theoretical Constructs of XAI from Social Science. A Few Studies Have also Looked at the Challenge of Explaining AI Models from a Social Science Perspective, Leveraging Knowledge into How Human Usually Reason and Explain from Philosophy and Psychology. For Instance, Wang et al. [4] Proposed a User Centric XAI Framework that Links Concepts in Human Reasoning Process with Explainable AI Techniques. We Leveraged Their Work When We Developed Our Interview Guide. Liao et al. [5] Developed an Algorithm-Informed XAI Question Bank Where User Needs for Explainability Are Represented in Terms of Prototypical Questions. They First Identified a List of XAI Methods and then Mapped Them to the Questions that Users Might Ask About AI Models or AI Products. They Designed the Questions Leveraging the Taxonomy of Lim&Dey and by Interviewing 20 UX and Design Practitioners Who Work in the AI Domain, They Refined and Enriched the User Questions Representing Explainability Needs.

Our work shares the similar goal of understanding from people their explainability needs. However, Liao et al.'s [5] work result of XAI question bank is only based on the point of view of design practitioners and their perception of others. In our work, we sought to get direct reactions from a variety of stakeholders, technical and non-technical, involved the AI project lifecycle. Our analysis thus contributes practical knowledge about what an explain means to real users of AI.

3 User Study

Our user study aims at understanding from a variety of real-world users involved in the development, deployment and use of AI techniques, what their needs and expectations are in terms of explaining AI solutions. The context of this specific study is a large financial institution where AI models are routinely being deployed for a variety of use cases. We will call it "test institution" in the remainder of the paper. We conducted the user study in two phases: (i) individual user interviews where we identified the standard user profiles and common themes related to users concerns about the explanations and (ii) group workshops where we refined those findings.

---

[5] https://www.ibm.com/cloud/watson-openscale.

3.1 Individual User Interview

The user interviews aimed at discovering: (i) users' knowledge of AI and level of trust in AI models (ii) users' needs with respect to understanding and explaining AI models, along with their current approach for doing so (iii) users' pain points during the process.

Participants. We Recruited Participants (on a Voluntary Basis) Seeking to Obtain a Variety of Profiles from Different Business Lines and Functions. Specifically, We Listed the User Profiles that We Would like to Interview and Dispatched Broadly Recruitment Emails Throughout the Test Institution Asking for Volunteers to Share Their Experience Around the XAI Topic. In the End, Our Participant Pool Was Composed of 33 Persons with Different Responsibilities and Skills: Data Scientists, Model Developers, Inspectors, Risk Analysts, Business Analysts, Project Managers, Relationship Managers Among Others. We Note However that We Had a Limited Number of End Users (Relationship Managers, Clients) that Have Direct Usage of Model Predictions or Are Affected by the Model Predictions.

Questionnaire Guide. We Designed an Interview Guide to Ensure that We Cover the Same Topics Consistently Across All the Interviews. We Organized It Around Four Main Categories:

- Professional activities: Background questions about current professional activity.
- Knowledge of AI models: Questions around participants' knowledge of AI models and their interaction with AI models (modeling, model management or use of model predictions).
- The need of understanding the AI model: Questions leading participants to describe situations where they usually need to understand AI models (entire AI model behavior or model predictions), how they proceed and what their main concerns are.
- The need of explaining the AI models to others: Questions leading participants to describe situations where they usually need to explain AI models to another person (entire AI model behavior or model predictions) and what the questions from those third parties are.

Procedure. Before the Formal Interview, We Conducted Several Pilot Tests to Verify the Appropriateness of the Questions and Clarify Aspects that Were Ambiguous. Each Individual User Interview Lasted Approximately 45 min. We Conducted Them via Skype Meeting Due to the Constraints from the Sanitary Situation. During Each Session, at Least One UX Designer Animated the Session, and One Data Scientist or Business Analyst Observed. The Moderator Started by Giving a Brief Introduction of the Project and Asked the Participants for Consent to Record the Interview for the Analysis Purposes. Then, the Moderator Let the Participant Talk About Their Background Experience. After that Introductory Discussion, the Moderator Asked the Questions Following the Interview Guide. The Moderator Encouraged the Participants to Provide as Many Details as Possible. Participants Could also Use Screen Sharing or Send Screenshots to Illustrate Their Answers Whenever Relevant.

Analysis. In Addition to Our Notes, We Transcribed All the Recordings Manually so as to Ensure that We Did not Discard Any Useful Information. This Generated a Large Amount of Qualitative Data for Analysis (Around 1400 min of Interviews Were Recorded and Transcribed). For the Analysis Procedure, We Decided to Rely on Thematic Analysis. Thematic Analysis is a Systematic Method of Breaking Down and Organizing Rich Data from Qualitative Research by Tagging Individual Observations and Quotations with Appropriate Codes, to Facilitate the Discovery of Significant Theme [39]. A Theme is a Description of a Belief, Practice, Need or Another Phenomenon that is Discovered from Data. Overall, Our Analysis Process Followed Mainly Five Steps:

- *Transcription:* write down the participants' verbalization. This was done by hand which was tedious but enabled at the same time to help us increase our familiarity with the data.
- *Coding:* assign preliminary codes to the transcripts based on its content. A code could be a word or a phrase that acts as a label or a segment of text.
- *Categorization:* group the codes with similar meanings.
- *Search:* look for patterns or themes in the code groups that represented the information related to the explanation.

Specifically, we printed out all the transcripts and posted them on a white board based on the categories that we had defined a priori: understanding and explaining. Then we highlighted with different colors the segments of texts associated with different themes.

In the end, the in-depth analysis of the transcripts from those interviews led to the identification of eight themes, which we call explanation focus, and which correspond to specific explanation concerns of users. Those eight explanation focuses were: justification of a local result, model performance, feature importance, feature analysis, model mechanism, business expertise, business impact simulation and recommendation.

Another important output of this first phase of our user study was the articulation of the diverse needs throughout the whole AI project lifecycle. We defined three phases: modeling, decision-making (model management) and the use of model, with three associated user profile namely model developer, model owner (the person that requested the development of the model and that will take responsibility for it) and finally the model users.

The purpose of the interview was to better understand and characterize user needs for explanations and preference in terms of information provided for the sake of explanation.

3.2 Group Workshop

To fine-tune our findings, we conducted several group workshops with members from one specific data science team within the test institution. The goal of those workshops was to provide feedback on the eight explanation focuses and also on the three standard user profiles. To anchor the discussions we considered specific use cases.

Participants. We Invited All Members of the Data Science Team to Participate in Brainstorming Sessions. In the End, We Recruited Six Data Scientists, One Data Engineer and One Business Analyst. We Organized Four Workshops, Three with Data Scientists, and

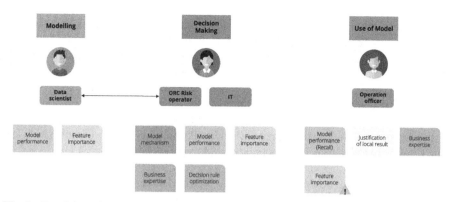

Fig. 1. Fraud detection use case example. Note: participants put *feature importance* in the use of model phase, which means the operational officer may be interested in knowing the feature contributions in predicting a fraud. However, there is a concern about the level of details of features exposed to the end user due to the risk associated with the ability to reverse-engineer such a system.

One with the Data Analytics Engineer Manager and Business Analyst Manager as They Both Served Similar Project Management Function in the Development of AI Projects.

Procedure. For Each Session, We Proceeded as Follows: A UX Designer Served as Moderator and Was Supported by One Technical Profile for Technical Questions and a Second Person to Take Notes. First, We Briefly Introduced the Eight Explanation Focuses, Providing Definitions. Second, We Asked Participants to Comment on Each Definition and Help Refine Them. Once We Had a Shared Understanding of the Explanation Focuses, We Discussed the Four Standard Use Cases One by One. We Asked Them to Brainstorm on the Stakeholders Involved in Each of the Three Phases of the Development Process and Identify Which Explanation Focuses Were Relevant Based on Their Professional Experience. An Example of Output from Those Discussions Based on a Fraud-Detection Use Case is Presented in Fig. 1.

4 Discussion

4.1 Explanation Focus

An explanation focus constitutes an "atomic" piece of information that contributes to a specific need for model explanation. After discussion in the group workshops and further discussions among the research team, we decided that the initial eight explanation focuses could be reduced to five.

Specifically, considering the type of information that each explanation focus provides, two of them could be merged into one broader definition. Specifically, the *recommendation* explanation focus, which involved explaining what needs to change to obtain a different result from the model was merged with the *justification of a local result*. We also modify the label of that broader group to *example-based justification,* which focuses on providing information justifying the specific prediction for a given instance

Table 1. Definition of Explanation focus

Explanation focus	Definition
Model mechanism How does the algorithm work? What kind of algorithm was used?	Description of logic of the chosen class of algorithms
Model performance Is the model performance good enough (precise, accurate, robust, and reliable)?	Information about model performance
Main contributors What are the main contributors to the model predictions globally and locally?	Information about the contribution of the set of input features to the model predictions globally and locally
Input feature behavior What is the relationship between each input feature and the model predictions?	Information about (i) the nature of the relationship between each input feature and the model predictions (Linear, monotonous or more complex) and (ii) the effect of changing the value of a given input feature on the prediction
Example-based justification Why this prediction given this instance? How should this instance change to get a different prediction?	Information justifying the specific prediction for a given instance including both why and why not another output

including both why and why not another output. Regarding the *business expertise* explanation focus, what we meant initially was to verify if the overall relationship between a given input feature and model predictions made sense from a business perspective. This is more related to the user's goal instead of the focus of the explanation. We merged it with *sensitivity analysis* that we renamed to *input feature behavior* with larger definition on the relationship between each input feature and the model predictions. Regarding the *business impact simulation* explanation focus, we observed from the user interviews that some users needed to be able to determine the optimal parameters of the model (i.e. probability thresholds). We recognized that this was also about the user's goal to ensure the capability of the model predictions in real world use. Hence we merged it with *model performance*.

In the end, we propose the five explanation focuses with their definitions and related explanation needs in Table 1.

4.2   User Profiles

From the study, we also identified three standard user profiles who are the target recipients of explanations throughout the AI project lifecycle. This finding is also consistent with Ribera & Laperdriza's work [40]. They categorized the *explainees* in three main groups based on their background, goals and relationship with the AI product: developers & AI researchers, domain experts whose expertise is used for the system to make decisions, and lay users who are final recipient of the model decisions. We added the notion of the AI project lifecycle into the explanation needs because we think it is significant to

better understand the context in which the users need different explanations. Table 2 articulates the motivations that we identified behind each explanation focus for each user profile. Note that we have two technical users associated with model development: the model developer and the model validator. They have similar motivation though the model developer typically has the ability to directly modify the model while the model validator can only make recommendations.

Model Developers/Validators. They Are Both Technical Profiles, Typically Data Scientists. They Need to Understand What Drives the Model to Make Sure the Behavior of the Model is Coherent with the Use Case and the Data. They Need Information to Check Potential Problems Such as Bias, Generalization, Information Leakage, Stability. The Model Developers also Need to Explain Their Findings to Two Kinds of People: (I) the model validator and (ii) the Model Owner. To the First Person the Explanation is Geared Towards Explaining the Inner Workings of the Model so that They Are Assured that the Model Development Followed the Appropriate Discipline. To the Model Owner, the Content of the Explanation is Intended to Make Sure that the Model Answers the Business Needs, and that the Performance and Limitations of the Model Are Well Understood. The Model Validators Need to Be Satisfied that the Model Works as Intended and that It Does not Display Problematic Behaviors. They Possibly Need to Report Those Findings to the model Owner.

Table 2. Explanation focuses to user profiles taking into account their different needs

Name	Model developer/validator	Model owner	Model user
Model mechanism	Ensure that model (logic) is coherent with the use case and the data	Ensure that model (logic) is coherent with the use case and the data Curiosity: Learn and apprehend a new concept	Curiosity: Learn and apprehend a new concept
Model performance	Explain the strengths and limitations of the model (error, false positive, bias) Ensure that model (performance) is coherent with the use case and the data Investigate the sensitivity and stability of the model performance to modeling changes	Ensure that model (performance) is coherent with the use case and the data Ensure that the model robustness and stability is sufficient for the use case Provide support to choose among multiple models or to choose optimal parameters of the model (e.g. Thresholds)	Be convinced that the model (performance) is coherent with the use case and the data Curiosity: Understand what performance means for an AI model

(*continued*)

Table 2. (*continued*)

Name	Model developer/validator	Model owner	Model user
Main contributors	Ensure that the input features of the model are coherent with the use case and the data by checking globally and locally possibly leading to the modeling changes Investigate model errors (false positives/negatives/no discriminative feature/no bias)	Ensure that the input features of the model are coherent with the use case and the data, that no discriminative feature were used, by checking globally and locally possibly leading to the modeling changes Develop or confirm insights into the contribution of the input features to the model predictions or for a given instance	Understand the contribution of the input features for a given instance possibly explaining to a third party Develop or confirm insights into the contribution of the input features to the model predictions or for a given instance
Input feature behavior	Ensure that the input features of the model are coherent with the use case and the data (no bias) possibly leading to the modeling changes	Ensure that the input features of the model are coherent with the use case and the data (no bias) possibly leading to the modeling changes Develop or confirm insights into influence of input data on the model predictions	Develop or confirm insights into influence of input data on the model predictions
Example-based justification	Ensure that the model behavior is coherent with the use case and relevant input data Investigate model errors (false positives/negatives)	Ensure that the model behavior is coherent on a given instance	Understand the model behavior for a given instance possibly explaining to a third party

Model Owners. They Need to Understand Enough the Model to Determine if It is Coherent with the Use Case and the Data, if Performance and Limitations Are Taken into Account for Deployment. Additionally, They May Develop or Confirm Business Insights into the Model Behavior. They Need Occasionally to Explain to the model Users how the Model Makes Decision and Seek to also Instill Trust in the Model to Be Deployed. The Model Owners Are the Ones Taking the Responsibility of the Model in the End and Thus Their Need for Explanation is Driven by that Responsibility.

Model Users. They Need to Be Convinced that the Model is Coherent with the Use Case and the Data in Order to Develop Trust in the Model. Moreover, They May Be Curious

About the Underlying Decision Rules of the Model Predictions and They Should Be Able to Understand Specific Prediction for a Given Instance. They May Be Required to Explain a Specific Prediction to a third Party (Customer).

In general, from the user study, we noticed that people asked different questions in understanding and explaining AI model behavior or their predictions, however those questions can be answered by providing the same information. This is consistent with the Lim&Dey's findings that users use different strategies to check model behavior and thus ask different questions for the same explanation goals ([41, 42]). Nevertheless, the way to represent this explainable information varies from one user to another. Therefore, it will be insightful to match the explanation focus with the explainability techniques. Likewise, each explainability technique can be represented by different visualizations (e.g., tables, graphs, interactive interfaces).

4.3 Mapping

In our broader exploration of XAI we have carried out a thorough state-of-the-art study on some of the most popular explainability methods [1, 2, 21–26, 43]. We proceeded to map those to the explanation focuses that we have outlined [2, 21–23, 26, 43], as a way to connect our findings from the user study with the technical XAI landscape.

To approach the mapping, we organized a group discussion where data scientists and UX designers gathered to exchanges the findings from both sides (UX and AI). The discussion was organized around three main phases. First, the data scientists pointed out the principles of different explainability methods, the meanings of some common performance metrics, the type of information that each explainability technique brings and the suitable cases for each technique from a data scientist point of view. Second, the UX designers shared the five explanation focuses, and gave some details on the information that users required, and the potential needs associated with each explanation focus. Finally, the two teams worked together and tried to match the suitable explainability techniques to the explanation focus by going through one explanation focus at a time. The result of those interactions is summarized on Fig. 2. This visual summary aims at being a starting point for this exercise that should be enriched based on further

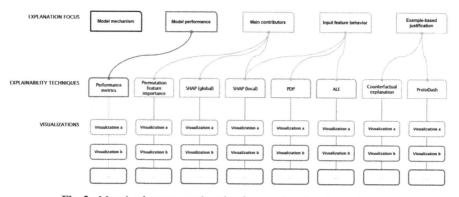

Fig. 2. Mapping between explanation focus and explanation techniques

addition of explainability techniques or refinement of existing ones. At this point, some of the explanation focuses are associated with only one specific explainability method: specifically, the information that corresponds to 'Model performance' is solely given by *Performance Metrics*. By contrast, others are supported by multiple techniques; for example, we can use either Permutated feature importance or SHAP summary plot to give information about what are the main drivers of the model prediction, i.e., *Main contributors*.

Overall, the explanation focuses provide a novel level of granularity in the definition of explanation in the context of AI models. The initial mapping provides an overarching framework to articulate the link between the user needs and technical solutions, as well as to identify both technical and UX areas that may be underserved at this point. This relationship can serve as a guide for choosing explainability techniques and visualizations to compose coherent explanation solutions depending on explanation focus of users.

For instance, our study led us to realize that there were limited coordinated efforts dedicated to the explanations of model mechanism, each person resorting to their own inspirations for such tasks. Similarly, for specific explainability techniques, different versions of visualization could be designed for different users and contexts, which need to be defined and evaluated. Finally, from a technical perspective, it will be interesting to see which explainability methods are widely available, or only for a subset of models, i.e. where there are further technical approaches to be defined and in the same perspective, which ones are used most often and in which context.

5   Conclusion

While there had been great progress in the research field of XAI, particularly development of novel XAI techniques, we observed that there was limited work on refining what explainability means to real users. This paper describes a user-centered analysis of what explanation means to different real-world practitioners of AI, when, what and why they need an explanation. We identified five explanation focuses and three standard user profiles within the AI project lifecycle. Based on these findings, we propose a framework mapping explanation focuses and explainability techniques. We believe this could also guide the design of explainability visual supports and novel explainability techniques and coordinate further efforts in this challenging multidisciplinary topic.

References

1. Ribeiro, M.T., Singh, S., Guestrin, C.: Why Should I Trust You?: Explaining the Predictions of Any Classifier (2016)
2. Lundberg, S.M., Lee, S.: A Unified Approach to Interpreting Model Predictions
3. Kim, B., Austin, U.T.: Examples are not enough, learn to criticize! criticism for interpretability. In: NIPS (2016)
4. Wang, D., Yang, Q., Abdul, A., Lim, B.Y.: Designing theory-driven user-centric explainable AI (2019). https://doi.org/10.1145/3290605.3300831
5. Liao, Q.V., Gruen, D., Miller, S.: Questioning the AI: informing design practices for explainable AI user experiences, pp. 1–15 (2020). https://doi.org/10.1145/3313831.3376590

6. Gunning, D.: Explainable artificial intelligence (XAI). In: Defense Advanced Research Projects Agency (DARPA) (2017)

7. Lim, B.Y., Dey, A.K.: Assessing demand for intelligibility in context-aware applications. In: ACM's International Conference Proceedings Series, pp. 195–204 (2009). https://doi.org/10.1145/1620545.1620576

8. F. Doshi-Velez and B. Kim, "A Roadmap for a Rigorous Science of Interpretability," no. Ml, pp. 1–13, 2017.

9. Miller, T.: Explanation in artificial intelligence: insights from the social sciences. Artif. Intell. **267**, 1–38 (2018). https://doi.org/10.1016/j.artint.2018.07.007

10. Guidotti, R., Monreale, A., Ruggieri, S., Turini, F., Pedreschi, D., Giannotti, F.:A survey of methods for explaining black box models, arXiv, pp. 1–45 (2018)

11. Chari, S., Seneviratne, O., Gruen, D.M., Foreman, M.A., Das, A.K., McGuinness, D.L.: Explanation ontology: a model of explanations for user-centered AI. In: Pan, J.Z., et al. (eds.) ISWC. LNCS, vol. 12507, pp. 228–243. Springer, Cham (2020). https://doi.org/10.1007/978-3-030-62466-8_15

12. Barredo Arrieta, A., et al.: Explainable explainable artificial intelligence (XAI): concepts, taxonomies, opportunities and challenges toward responsible AI. Inf. Fusion **58**, 82–115 (2020). https://doi.org/10.1016/j.inffus.2019.12.012

13. Gilpin, L.H., Bau, D., Yuan, B.Z., Bajwa, A., Specter, M., Kagal, L.: Explaining explanations: An overview of interpretability of machine learning. In: Proceedings of 2018 IEEE 5th International Conference on Data Science and Advanced Analytics, DSAA 2018, pp. 80–89 (2019). https://doi.org/10.1109/DSAA.2018.00018

14. Arya, V., et al.: One explanation does not fit all: a toolkit and taxonomy of AI explainability techniques (2019). http://arxiv.org/abs/1909.03012.

15. Parekh, J., Mozharovskyi, P., d'Alche-Buc, F.: A framework to learn with interpretation (2020)

16. Sokol, K., Flach, P.: Explainability fact sheets: a framework for systematic assessment of explainable approaches. In: FAT*2020 - Proceedings of the Conference on Fairness, Accountability, and Transparency, pp. 56–67 (2020). https://doi.org/10.1145/3351095.3372870

17. Belle, V., Papantonis, I.: Principles and practice of explainable machine learning. arXiv (2020)

18. Molnar, C., Casalicchio, G., Bischl, B.: Interpretable machine learning – a brief history, state-of-the-art and challenges. In: Koprinska, I., et al. (eds.) ECML. CCIS, vol. 1323, pp. 417–431. Springer, Cham (2020). https://doi.org/10.1007/978-3-030-65965-3_28

19. Vilone, G., Longo, L.: Explainable artificial intelligence: a systematic review. arXiv, no. Dl (2020)

20. Freitas, A.A.: Comprehensible classification models. ACM SIGKDD Explor. Newsl. **15**(1), 1 (2014). https://doi.org/10.1145/2594473.2594475

21. Fisher, A., Rudin, C., Dominici, F.: All models are wrong, but many are useful: learning a variable's importance by studying an entire class of prediction models simultaneously. arXiv (2018)

22. Apley, D.W., Zhu, J.: Visualizing the effects of predictor variables in black box supervised learning models. J. R. Stat. Soc. Ser. B Stat. Methodol. **82**(4), 1059–1086 (2020). https://doi.org/10.1111/rssb.12377

23. Zhao, Q., Hastie, T.: Causal interpretations of black-box models. Department of Statistics, Stanford University (2016)

24. Ribeiro, M.T., Singh, S., Guestrin, C.: Anchors: high-precision model-agnostic explanations. In: 32nd AAAI Conference on Artificial Intelligence, AAAI 2018, pp. 1527–1535 (2018)

25. Goldstein, A., Kapelner, A., Bleich, J., Pitkin, E.: Peeking inside the black box: visualizing statistical learning with plots of individual conditional expectation. J. Comput. Graph. Stat. **24**(1), 44–65 (2015). https://doi.org/10.1080/10618600.2014.907095

26. Gurumoorthy, K.S., Dhurandhar, A., Cecchi, G., Aggarwal, C.: Efficient data representation by selecting prototypes with importance weights. In: Proceedings of IEEE International Conference on Data Mining, ICDM, vol. 2019, pp. 260–269 (2019). https://doi.org/10.1109/ICDM.2019.00036

27. Wachter, S., Mittelstadt, B., Russell, C.: Counterfactual explanations without opening the black box: automated decisions and the GDPR. SSRN Electron. J. **31**, 1–52 (2017). https://doi.org/10.2139/ssrn.3063289

28. Dandl, S., Molnar, C., Binder, M., Bischl, B.: Multi-objective counterfactual explanations. In: Bäck, T., et al. (eds.) PPSN. LNCS, vol. 12269, pp. 448–469. Springer, Cham (2020). https://doi.org/10.1007/978-3-030-58112-1_31

29. Samek, W., Wiegand, T., Müller, K.R.: Explainable artificial intelligence: understanding, visualizing and interpreting deep learning models. arXiv (2017)

30. Nunes, I., Jannach, D.: A systematic review and taxonomy of explanations in decision support and recommender systems

31. Hilton, D.J., Slugoski, B.R.: Knowledge-based causal attribution. The abnormal conditions focus model. Psychol. Rev. **93**(1), 75–88 (1986). https://doi.org/10.1037/0033-295X.93.1.75

32. Lim, B.Y., Dey, A.K.: Investigating intelligibility for uncertain context-aware applications. In: UbiComp 2011, Proceedings of the 2011 ACM Conference on Ubiquitous Computing, pp. 415–424 (2011). https://doi.org/10.1145/2030112.2030168

33. Lim, B.Y., Dey, A.K.: Evaluating intelligibility usage and usefulness in a context-aware application. In: Kurosu, M. (ed.) HCI. LNCS, vol. 8008, pp. 92–101. Springer, Heidelberg (2013). https://doi.org/10.1007/978-3-642-39342-6_11

34. Krause, J., Perer, A., Ng, K.: Interacting with predictions: visual inspection of black-box machine learning models. In: Conference on Human Factors Computing Systems - Proceedings, pp. 5686–5697 (2016). https://doi.org/10.1145/2858036.2858529

35. Coppers, S., et al.: Intellingo: an intelligible translation environment. In: Conference on Human Factors Computing System - Proc., vol. 2018-April, (2018). https://doi.org/10.1145/3173574.3174098

36. Lim, B.Y., Dey, A.K.: Toolkit to support intelligibility in context-aware applications, p. 13 (2010). https://doi.org/10.1145/1864349.1864353

37. Eiband, M., Schneider, H., Bilandzic, M., Fazekas-Con, J., Haug, M., Hussmann, H.: Bringing transparency design into practice, pp. 211–223 (2018). https://doi.org/10.1145/3172944.3172961

38. Kulesza, C., Principles, S.: Principles of explanatory debugging to personalize interactive machine learning (2015). https://doi.org/10.1145/2678025.2701399

39. Maria, R.: How to analyze qualitative data from UX research : thematic analysis. Nielsen Norman Group Publication (2019)

40. Ribera, M., Lapedriza, A.: Can we do better explanations? A proposal of user-centered explainable AI. In: CEUR Workshop Proceedings, vol. 2327 (2019)

41. Lim, B.Y., Dey, A.K.: Evaluating intelligibility usage and usefulness in a context-aware application

42. Lim, B.Y., Dey, A.K.: Investigating intelligibility for uncertain context-aware applications (2011)

43. Dhurandhar, A., et al.: Explanations based on the missing: towards contrastive explanations with pertinent negatives. arXiv, NeurIPS (2018)

Human-Centered AI

How Intuitive Is It? Comparing Metrics for Attitudes in Argumentation with a Human Baseline

Markus Brenneis[✉] and Martin Mauve

Heinrich-Heine-Universität, Universitätsstraße 1, 40225 Düsseldorf, Germany
Markus.Brenneis@uni-duesseldorf.de

Abstract. It is often interesting to know how similar two persons argue, e.g. when comparing the attitudes of voters and political parties, or when building an argumentation-based recommender system. Those applications need a distance function, which should give intuitive results. In this paper, we present seven functions which calculate how similar the attitudes of two agents are in an argumentation. We evaluate how good those functions match the results of a human baseline which we determined in a previous work. As it turns out, variants of the p-metric, Cosine, and Soergel distance best agree with human intuition.

Keywords: Argumentation · Metric · Human baseline

1 Introduction

Comparing the attitudes different people or organizations have in an argumentation is often relevant and useful, e.g. for clustering using opinions mentioned in argumentations, recommender systems for argumentation platforms (as used in our platform *deliberate* [4]), or comparing one's own attitudes and arguments with those of political parties. In a previous work [5], we have conducted a survey with untrained human subjects to find out what properties a distance function for argumentation data should fulfill to yield results matching human intuition.

In this paper, we compare different distance functions regarding those properties. Our goal is to provide hints for application developers which kinds of distance functions best match human intuition and where and why there are differences. This helps with choosing functions best suited for the problem at hand, knowing that their results follow intuitive and understandable properties.

Our contribution is the following: We present a list of properties which should be fulfilled by a distance function which compares argumentations, based on a survey we have conducted earlier. Different existing distance functions were adapted to use them with attitudes in argumentations. We compare those functions regarding different properties we found to be intuitive through our survey,

Manchot research group *Decision-making with the help of Artificial Intelligence*, use case politics.

H. Degen and S. Ntoa (Eds.): HCII 2021, LNAI 12797, pp. 125–138, 2021.
https://doi.org/10.1007/978-3-030-77772-2_9

and examine different values for the hyperparameters of each function. Afterwards, we explain why certain functions perform better than others.

In the next section, we provide the key definitions used throughout our work. Afterwards, we define the formal mathematical model and distance functions we compared. We then present and discuss the results of comparing the functions with a human baseline, and finally have a look at related work.

2 Definitions

The argumentative terms we use in this paper are based on the IBIS model [10] for argumentation. An argumentation consists of *arguments*, and each argument is formed by two *statements*: a *premise* and a *conclusion*. We call the set of all statements S and the set of all arguments $A \subset S^2$.

A special "statement" is the *issue* I, which denotes the topic of an argumentation and has no conclusions. All premises for arguments with I as the conclusion are referred to as *positions*, and are typically actionable items like "We should build more wind power plants." $P \subset S$ is the set of all positions.

Different persons can have individual views in an argumentation: They can (strongly) agree (denoted as (+) (agree), or + (strongly agree), respectively) or disagree ((-), -) with statements[1], be neutral (0) about a statement, indicate to not have an opinion ($\emptyset$), or do not mention anything about a statement (?; so we do not know their opinion); we call this stance on statements *opinion*. We define the set of possible opinion values for a statement $O := \{+, -, (+), (-), 0, \emptyset, ?\}$.

They can also assign arguments different *relevances* (or *weights* or *importances*), and give a priority order for positions. The overall importances and opinions of a person are referred to as their *attitude*.

We represent a person's attitude as an argumentation tree[2], or, if only positions are involved, as sorted lists with positions, where the most important position is at the top. Note that in our tree representation, statements are nodes, argument are edges, to have statements as atomic building blocks. This visualization can, however, be transformed to classical Dung-based [7] abstract argumentation frameworks when needed. We do not draw the common root I in our visualizations to make them simpler.

As an example, we explain how Alice's tree in Fig. 1e should be understood: Alice agrees with the position p and the statements a and b, which build arguments with conclusion p. The argument (a, p) is more important for her than the argument (b, p) (indicated by the bolder edge). Note that we do not differentiate whether an argument edge is attacking or defending – this is up to the interpretation of the natural language presentation of the scenario, but is consistent within all trees of one scenario (i.e. in Fig. 1e, the edges (a, p) in all three trees

---

[1] A more fine-grained model for the strength of (dis-)agreement, as we have suggested in [3], could be used, but is not necessary in this work.

[2] A representation as more general graphs is also possible, but again not necessary for the examples in this work.

are either consistently attacking or supporting arguments); a differentiation is therefore not needed in the model for the purpose of this paper.

Throughout this paper, we use the term *distance function* to refer to a function which calculates some distance between pairs of argumentations with the parts introduced above. Those functions might happen to fulfill all properties of a metric (e.g. the triangle equality), but are not required to do so.

We now define how the drawing of a tree is translated to mathematical objects. Each tree can be considered as a pair of functions (o, s), where $o : S \to O$ captures the opinion on statements, $s : A \to \mathbb{N}_0$ the sorting of arguments by importance (where 1 means top-priority, 0 no priority (as default for not mentioned arguments); the ordering is not required to be injective). Note that we view a function as a set of ordered pairs (parameter, function value).

Please note the following conventions: The sorting position of a position p is treated as the sort order position of a pseudo-argument (p, I). If o is undefined for a value, the function's value is ?. If s is undefined for a value, the function's value is 0. To keep the notation simple, we assume that the functions' domains are the same when two trees are compared.

For example, Alice's tree in Fig. 1e translates to $o = \{(p, +), (a, +), (b, +)\}$, $s = \{((a, I), 1), ((a, p), 1), ((b, p), 2)\}$.

A distance function must map the different values to numeric values for calculations. We will evaluate different transformation strategies. As all distance functions need to map the opinion values of O to numeric values and some of them map importance weights to other numeric values, we define the following common mapping strategies:

$$r(x) = \begin{cases} 0.5 & \text{if } x = + \\ 0.25 & \text{if } x = (+) \\ 0 & \text{if } x \in \{0, \emptyset, ?\} \\ -0.25 & \text{if } x = (-) \\ -0.5 & \text{if } x = - \end{cases} \tag{1}$$

$$w_h(x) = \frac{1}{x} \tag{2}$$

$$w_g(x) = \frac{1}{2^x} \tag{3}$$

The result of a division by 0 is defined as 0, which means that arguments without importance value (which default to 0) get a calculated weight of 0. The variants $w_{\bar{h}}$ and $w_{\bar{g}}$ are defined the same way, but the values are normalized such that the sum of function values for all arguments with the same conclusion is 1 (or 0, if no argument has a value greater 0). For instance, if we take Alice's tree in Fig. 1e again, $w_{\bar{h}}((a, p)) = \frac{\frac{1}{1}}{\frac{1}{1} + \frac{1}{2}} = \frac{2}{3}$. If we mention the function name w, any possible variant can be used (thus, the concrete choice of w is a hyperparameter of the distance function).

Sometimes, we refer to the "simple" opinion, which removes the weight part of the opinion:

$$\text{simple} : O \rightarrow \{+, -, 0, \emptyset, ?\} : x \mapsto \begin{cases} + & \text{if } x = (+) \\ - & \text{if } x = (-) \\ x & \text{otherwise} \end{cases} \tag{4}$$

3   Distance Functions for Argumentations

We now present the distance functions we have compared. Most functions are based on previous work in argumentation theory or related fields and have been adapted by us for use with the formal definition introduced in Sect. 2. Most functions have hyperparameters, e.g. which function w is used. An overview of the distance functions, their hyperparameters and tested ranges can be found in Table 1.

Table 1. Overview of examined distance functions and their hyperparameters with tested values

Function	Hyperparameters
Bhavsar	$w \in \{w_{\bar{h}}, w_{\bar{g}}\}$, $N \in \{.1, .25, .5, .75, .9\}$
Cosine	$w \in \{w_g, w_{\bar{g}}, w_h, w_{\bar{h}}\}$
Jaccard	$set \in \{set_a, set_o, set_s, set_{s'}\}$, $keep \in \{keep_a, keep_t\}$
p-metric	$p \in \{1, 2\}$, $ds \in \{d_{s_w}, d_{s_s}\}$, $da \in \{da_0, da_s\}$, $w \in \{w_g, w_{\bar{g}}, w_h, w_{\bar{h}}\}$
Soergel	$w \in \{w_g, w_{\bar{g}}, w_h, w_{\bar{h}}\}$
VAA	–
WATD	$\alpha \in \{.1, .25, .5, .75, .9\}$, $w \in \{w_g, w_{\bar{g}}, w_h, w_{\bar{h}}\}$

Bhavsar Distance. [2] presented a metric for match-making of agents in e-business environments, which are represented as trees. As the definition of that recursive metric is lengthy, we do not repeat its definition here. The metric can be applied to our structure by transforming sort orders using $w_{\bar{h}}$ or $w_{\bar{g}}$, and treating opinions as node labels. A parameter N sets the relative importance of subtrees and respective roots, similar to the PageRank algorithm [15].

Cosine Distance. We define the Cosine distance similar to [16], who predict opinions in argumentation. They treat accepting and declining a statement s as two different entities ("acceptance of s" and "acceptance of $\neg s$") and ignore a statement if it has no rating in one of the inputs:

$$d(t_1, t_2) = 1 - \frac{V_1 \cdot V_2}{||V_1|| \, ||V_2||} \tag{5}$$

where an argumentation tree $t_i = (o_i, s_i)$ is transformed to a vector V_i with the components $s_i(a)$ for every argument a, and $\max(-r(o_i(s)), 0)$ and $\max(r(o_i(s)), 0)$ for every statement s for which both trees have no ? opinion.

Jaccard Distance. The Jaccard distance has been used by [11] as the basis for calculating the similarity of process models. We apply it in the following form:

$$d(t_1, t_2) = \frac{|\text{set}(t_1) \, \triangle \, \text{set}(t_2)|}{|\text{set}(t_1) \cup \text{set}(t_2)|} \tag{6}$$

where the functions "set" and "keep" are chosen from

$$\text{set}_a((o, s)) = \text{set}_o((o, s)) \cup \text{set}_s((o, s)) \tag{7}$$

$$\text{set}_o((o, s)) = \{(x, y) \mid (x, y) \in o \land \text{keep}(y)\} \tag{8}$$

$$\text{set}_s((o, s)) = s \tag{9}$$

$$\text{set}_{s'}((o, s)) = \text{simple}(s) \tag{10}$$

$$\text{keep}_a(x) = 1 \tag{11}$$

$$\text{keep}_t(x) = \begin{cases} 1 & x \in \{+, (+), 0, (-), -\} \\ 0 & otherwise \end{cases} \tag{12}$$

If "set_o" is used for "set", argument weights are completely ignored; "set_s" completely ignores opinions and only looks at argument and position weights. "keep" determines if unknown (?) and "no opinion" s ($\emptyset$) are included.

The argumentation software Carneades [8] uses a special case of this distance function with set $= \text{set}_{s'}$, which means that the relative number of different opinion tendencies is counted.

p-metric. This distance function is based on the p-metric for fuzzy sets [20]

$$d(t_1, t_2) = \left(\sum_{s \in S} ds(o_1(s), o_2(s)) + \sum_{a \in A} da(s_1(a), s_2(a)) \right)^{\frac{1}{p}} \tag{13}$$

with $p \in \mathbb{N}$, and ds, da one of

$$ds_w(o_1, o_2) = \begin{cases} 0 & \text{if } o_1 = o_2 \\ 1 & \text{if } o_1 \text{ or } o_2 \text{ in } \{\emptyset, ?\} \\ |r(o_1) - r(o_2)|^p & otherwise \end{cases} \tag{14}$$

$$ds_s(o_1, o_2) = |d_w(\text{simple}(o_1), \text{simple}(o_2))|^p \tag{15}$$

$$da_o(s_1, s_2) = 0 \tag{16}$$

$$da_s(s_1, s_2) = |w(s(s_1)) - w(s(s_2))|^p \tag{17}$$

Soergel Distance. This distance function is also known as weighted Jaccard distance, which has also been used by [16]. We use the following definition, which uses the same vector representation as defined for the Cosine distance above:

$$d(t_1, t_2) = 1 - \frac{\sum_i \min(V_{1_i}, V_{2_i})}{\sum_i \max(V_{1_i}, V_{2_i})} \tag{18}$$

where V_{1_i} is the i-th component of the vector representation of t_1.

VAA Distance. In many Voting-Advice Applications (VAAs), the distance between a user's attitudes and political party's attitudes on political positions are compared. One possibility is using proximity voting logic [17], optionally weighted, which doubles the influence of a position (as, for example, used by the German Wahl-O-Mat application [13]). We adapted the idea to our model:

$$d(t_1, t_2) = \sum_{p \in P} u(o_1(p), o_2(p)) \cdot v_{t_1, t_2}(s_1(p), s_2(p)) \cdot z(o_1, o_2) \tag{19}$$

with

$$u(o_1, o_2) = \begin{cases} 2 & \text{if } o_1 \text{ or } o_2 \text{ in } \{+, -\} \\ 1 & \text{otherwise} \end{cases} \tag{20}$$

$$v_{t_1, t_2}(s_1, s_2) = \begin{cases} 2 & \text{if } s_1 \text{ or } s_2 \text{ is in the top half (rounded down)} \\ & \text{of the ratings for positions} \\ 1 & \textit{otherwise} \end{cases} \tag{21}$$

$$z(o_1, o_2) = \begin{cases} 0 & \text{if } o_1 \text{ or } o_2 \text{ in } \{\emptyset, ?\} \\ |r(\text{simple}(o_1)) - r(\text{simple}(o_2))| & \text{otherwise} \end{cases} \tag{22}$$

Note that, as in a VAA, only positions are considered, and statements which are no positions are ignored. Moreover, both arguments and positions can contain weights in our model, whereas a VAA typically only allows voters to input weights.

Weighted Argumentation Tree Distance (WATD). In [3], we have suggested a pseudometric for argumentations with weighted edges and nodes. This metric respects the structure of an argumentation tree by limiting the influence of each branch to its importance, and giving statements deeper in the tree a lower weight. Adapted to the tree model in this paper, the metric is defined for two trees $t_1 = (o_1, s_1)$, $t_2 = (o_2, s_2)$ as follows:

$$d(t_1, t_2) = (1 - \alpha) \sum_{s \in S} \alpha^{\text{de}(s)} \left| \prod_{a \in A_{s \to I}} w(s_1(a)) r(o_1(s)) - \prod_{a \in A_{s \to I}} w(s_2(a)) r(o_2(s)) \right| \tag{23}$$

with $\alpha \in (0, 1)$ (a lower α emphasizes opinion on statements closer to the root, similar to N in the Bhavsar distance), $A_{s \to I}$ the set of all arguments from statement s to the root I, and $\mathrm{de}(s)$ the depth of a statement s, where positions have a depth of 1. This basic idea is to multiply each opinion value of t_1 with the product of all weight from the root node I to that opinion calculate the distance to the same value in t_2. Thereby, opinion difference closer to the root have a higher influence than "deeper" opinions.

4   Comparison with a Human Baseline

We think that the best way to check whether a distance function is intuitive is comparing it with a human baseline. In an online survey we have previously conducted [5], different possible properties for distance functions comparing attitudes in argumentation settings have been checked for their intuitiveness. In the survey, around 40 assessments by untrained human subjects have been collected for different argumentation scenarios. From the survey results, we can get a list of properties which should be fulfilled by a distance function to match human intuition. If we look only at properties which can be considered intuitive from that survey on a significance level $\alpha = 10\%$[3], we get a list of 17 properties which should be fulfilled.

For many hypotheses, also comparison questions not directly relevant for the hypotheses have been asked in the original questionnaire[4]. For instance, if we wanted to know whether Alice's attitude is more similar to Charlie's or Bob's attitude, we also asked whose attitude is closest to Bob's. For those hypotheses, we also considered properties which can be derived from the additional questions, if they are significant. Those additional properties will be marked with a superscript [A], and all resulting sub-hypotheses are numbered with the according sub-question number (e.g., H2.1[A] is the first question for the questionnaire scenario for H2).

Table 2 lists all relevant hypotheses from [5] which we used as the basis for our comparison. For this paper, we changed the formulation of the hypotheses to match the real outcome of the survey to reflect the actual property expected from a distance function. Note that for H18, two scenarios were used, where only one yielded significant results, which is why this hypothesis has been completely reformulated. Figure 1 depicts visualizations of the concrete questionnaire scenarios for some more complex hypotheses, and also what similarity order is expected, based on the survey results. For example, in Fig. 1e, Bob's attitude should have a smaller distance to Alice's attitude than to Charlie's attitude. Since the answers for human intuition are known only for those concrete scenarios, we will only use those concrete examples as the basis for the comparison of distance measures.

---

[3] including Bonferroni correction for multiple comparisons, i.e. we assure that the type I error rate is less than 10% by requiring p-values less than $\frac{\alpha}{\text{number of possible answers}}$.

[4] cf. raw data at https://github.com/hhucn/argumentation-similarity-survey-results/.

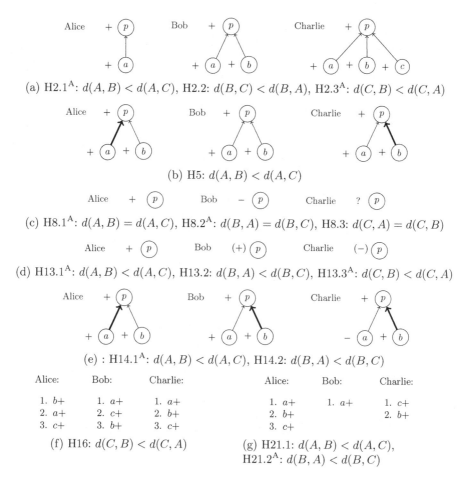

Fig. 1. Visualization of questionnaire scenarios (and thus, test scenarios) of some hypotheses; $d(A, B)$ denotes the distance between Alice and Bob etc.

The following hypotheses have not been considered although our inclusion criterion is fulfilled: A variant of H8, which says an unknown opinion vs. a positive and a negative opinion cannot be assessed, has been excluded, because this would result in a partially defined distance function, which we consider undesirable. H9 only checked text comprehension and has no implications for a distance function. H15, which included an undercut attack, is not included since the original question was probably misleading/not understood by the participants, as discussed in [5].

We now present which distance functions fail on which reference scenarios, and give explanations on why certain distance functions fail on specific cases. We have tested each distance function with every possible combination of hyperparameters with the relevant scenarios. Table 3 summarizes which cases yield the

Table 2. All relevant hypotheses which we included in our comparison. Deviations from the original formulations in [5] are *emphasized*.

#	Property
H2	Proportionally bigger overlap on arguments for/against a position results in greater similarity than the absolute number of differences
H3	A neutral opinion is between a positive and a negative opinion
H5	Weights of arguments have an influence even if they are the only difference
H7	No opinion *has the same distance from* a positive and a negative opinion *if a decision is forced*
H8	An unknown opinion *has the same distance to a positive and a negative opinion as* a positive and a negative opinion *if a decision is forced*
H12	It is possible for a difference in arguments for/against positions to result in greater dissimilarity than a difference in opinions on those positions
H13	Two argumentations with weak and contrary opinions on a statement can *not* be closer than two argumentations with the same opinions, but with very different strength
H14	Two argumentations with weak arguments and contrary opinions on their premises can *not* be closer than two argumentations with the same opinions, but with very different strength of arguments
H16	Flipping the two most important positions results in a bigger difference than flipping two less important positions
H18	*Moving the least important position to the top results in greater dissimilarity than changing the order of item 2 to 4.*
H19	Agreeing with someone's most important position is as important as having that person's most important opinion matching mine
H20	Adding another most important position *(which is neutral in the other argumentations)* results in greater dissimilarity than flipping the priorities of two positions
H21	Having more similar priorities of opinions can result in greater similarity even with lower absolute number of same opinions
H22	Not mentioning a position results in greater dissimilarity than assigning lower priorities

expected results for each distance function with the best parametrization (i.e. maximum number of expected results). Those parametrisations are depicted in Table 4.

The p-metric fails on H21.1 (cf. Fig. 1g) only, which happens because the missing weights for b and c have a greater influence than the common most important position of Alice and Bob. The Jaccard distance function also fails on H21.1 since it only considers that Alice and Charlie have more positions in common.

All cases for H2 (cf. Fig. 1a) fail only for the VAA distance function, where equal distances are calculated instead of different ones, because the arguments,

Table 3. Overview of cases fulfilled by the individual distance functions for the parametrisation which yield the highest number of fulfilled cases; ^{e} are failing cases where the calculated distance is 0; the numbers of sub-hypotheses refer to the question number in the original questionnaire.

Hypothesis	Bhavsar	Cosine	Jaccard	p-metric	Soergel	VAA	WATD
H2.1^A	✓	✓	✓	✓	✓	e	✓
H2.2	✓	✓	✓	✓	✓	e	✓
H2.3^A	✓	✓	✓	✓	✓	e	✓
H3	✓	✓	✓	✓	✓	✓	✓
H7	✓	✓	✓	✓	✓	✓	✓
H8.1^A			✓	✓			
H8.2^A			✓	✓			
H8.3	✓	✓	✓	✓	✓	✓	✓
H5	e	✓	e	✓	✓	e	✓
H12	✓	✓	✓	✓	✓		✓
H13.1^A	✓	✓	✓	✓	✓	✓	✓
H13.2	✓	✓	✓	✓	✓	✓	✓
H13.3^A	e	✓	e	✓	✓	✓	✓
H14.1^A	✓	✓	✓	✓	✓	e	✓
H14.2	✓	✓	✓	✓	✓	e	e
H16	e	✓	e	✓	✓	e	✓
H18	e	✓	e	✓	✓	e	✓
H19	✓	✓	✓	✓	✓	✓	✓
H20	✓	✓	✓	✓	✓	✓	✓
H21.1	✓	✓			✓	e	✓
H21.2^A	✓	✓	✓	✓	✓	e	✓
H22	✓	✓	✓	✓	✓	e	✓
$\sum$	16	20	17	21	20	8	19

Table 4. Best parametrisations for each distance function; each combination of the listed parameters yields the same (best) results.

Function	Best parametrisations
Bhavsar	$w \in \{w_{\bar{h}}, w_{\bar{g}}\}$, $N \in \{.1, .25, .5, .75, .9\}$
Cosine	$w \in \{w_g, w_{\bar{g}}, w_{\bar{h}}\}$
Jaccard	$set \in \{set_{s'}\}$, $keep \in \{keep_t\}$
p-metric	$p \in \{1, 2\}$, $ds \in \{d_{s_w}\}$, $da \in \{da_s\}$, $w \in \{w_{\bar{g}}, w_{\bar{h}}\}$
Soergel	$w \in \{w_g\}$
VAA	–
WATD	$\alpha \in \{.25, .5, .75, .9\}$, $w \in \{w_{\bar{g}}, w_{\bar{h}}\}$

which are the only difference in this case, are completely ignored by this function. The same applies to H5, H12, H14.1^A and H14.2. H14 also fails for WATD, because the distance function has been designed to *not* fulfill this property [3, Desideratum 7].

Many properties involving changing the importance order of positions, namely H16, H18, H21.1, H21.2^A, and H22, fail for the VAA function since it does not have a fine-grained differentiation of importance which is necessary to capture the differences.

All distance measures except for Jaccard and p-metric fail to give a positive and a negative opinion a distance which is equal to the distance to an unknown opinion (H8.1^A, H8.2^A, cf. Fig. 1c). Jaccard is good here because it treats any difference of opinion as equally distant; the p-metric explicitly defines every comparison with an unknown opinion as 1. On the other hand, e.g., WATD is defined to treat an unknown opinion as falling between positive and negative, and the VAA metric ignores a position if the opinion in one graph is unknown.

H5 states that the difference between argumentations should be non-zero even if argument weights are the only difference. This fails for Bhavsar by design of the metric [2, Example 2]. The best parametrisation for Jaccard ignores weights, so it also fails here. For the same reason, H16 and H18 fail for both distance functions.

H13.3^A checks that a negative opinion (-) is closer to a weak positive opinion ((+)) than to a stronger positive opinion (+). Bhasvar and Jaccard distance functions fail to see a difference here because they treat the distances between any of the opinions -, (+), and + the same.

To sum up, Cosine, p-metric, and Soergel yield the best results, matching human intuition in more than 90% percent of the tested cases.

5   Discussion

From our evaluation, one gets an idea which metrics yield intuitive results for applications which compare attitudes in argumentations. Nevertheless, we want to point out some limitations of our comparison method.

Firstly, we did not have a look at bigger argumentation hierarchies, or argumentation with re-used statements (e.g. cycles). For the former, our previous survey did not give significant results, for the latter, no reference data has been collected in the survey because cycles are hard to grasp with intuition. Hence, distance functions which model those cases (e.g. the original WATD pseudometric) have a disadvantage because this feature is not considered in the comparison.

From the survey results, it is also possible to conduct properties which should *not* be fulfilled. There are cases where there is no significant "true" answer, but there are clear "false" answers. Furthermore, the list of properties and cases checked in this paper is probably incomplete and can be extended with additional intuitive properties, which might then change the ranking of distance functions. As we built upon the results of our previous survey, and we are not aware of similar surveys, we did not include more properties.

Note that we did not check whether the original properties as presented in Table 2 are fulfilled in general, but only whether the concrete questionnaire scenarios yielded the expected, "intuitive" results. We did this because the original survey did not find out whether the hypotheses are true, but only collected results for the specific scenarios. Moreover, all properties get equal weight. Depending on the application (e.g., a VAA), some properties might not be relevant. What is more, some distance functions might get better results if the underlying representation model is changed.

Finally, it will be interesting to evaluate distance functions not on concrete artificial scenarios, but in an application context, e.g. a recommender system, since this might produce different results. A challenge for real applications is retrieving the necessary pieces of information from a user, e.g. how important an argument is considered, within an intuitive user interface.

6   Related Work

There is only limited related research in the evaluation and development of distance function in the context of argumentation, but there are some applications of such distance functions which have been studied.

A dataset with 16 positions on 4 issues has been published by [16]. 309 students gave their opinions on those issues by giving arguments and their level of agreement with that argument on a scale from -1 (total disagreement) to 1 (total agreement). They compare different algorithms for predicting user opinions on positions. A kind of soft cosine measure, where feature similarity is exploited using position correlation, performed best in their comparison. The comparison also included, i.a., collaborative filtering using Jaccard similarity, ordinary Cosine similarity, and other, model-based algorithms, e.g. a neural network.

Their work focuses on the application of measuring similarity in the concrete context of a recommender system, whereas we focus on calculating relative similarities to get a similarity order for user attitudes. Similarly, [18] tested different recommender agents in laboratory argumentation settings. [9] uses collaborative filtering and clustering in a social network context to find political parties closest to a user. The collaborative filtering was used to predict missing values to make clustering with sparse information easier.

Related work in other domains than argumentation chose a similar way of evaluation with a human baseline as we did in this paper.

In the context of word similarity, [12] proposed different distance functions, and compared them with human ratings from a dataset created by [14]. They also indicate that the best way to determine the quality of a distance function is comparing it with human common sense. Within the same application context, [6] agrees that "comparison with human judgments is the ideal way to evaluate a measure of similarity".

The study presented in [1] is based on the study design of [14]. 50 human subjects assessed the similarity of process descriptions, and compared those assessments with the values of five metrics. The results did not correlate well, but

the correlation with the metrics was not worse than the correlation between the human subjects. [11] present a metric based on the Jaccard coefficient for process model similarity. They compared the results of the metric with human assessment in an information retrieval task.

[19] evaluated six different similarity measures (i.a., $l1$, $l2$ norm, pointwise mutual information) with the application in a recommender system for online communities using item-based collaborative filtering. A similarity measure has been considered good if the user wanted to join the suggested community. The $l2$ norm performed best, although the authors found other tested measures, which incorporated mutual information, more intuitive.

7   Conclusion and Future Work

We have presented several distance functions for comparing the attitudes of different persons in an argumentation. We compared the performance of the functions in various scenarios with a human baseline taken from a survey we have previously conducted [5]. The distance functions based on the p-metric, Cosine, and Soergel distance performed best on our dataset. Those results can be used for developing applications which should give results matching human intuition, e.g. when developing a distance-based recommender system for arguments, or clustering of opinions.

For future work, an extended comparison with more scenarios for a human baseline would be useful, i.a. for deeper argumentations. A comparison in different application scenarios can give more insights. We plan to compare different metrics in an argument-based voting advice application in an empirical study. Another aspect for further research is the question of how to gather the information needed from users without having user interfaces which are too crowded.

References

1. Bernstein, A., Kaufmann, E., Bürki, C., Klein, M.: How similar is it? Towards personalized similarity measures in ontologies. In: Ferstl, O.K., Sinz, E.J., Eckert, S., Isselhorst, T. (eds.) Wirtschaftsinformatik 2005, pp. 1347–1366. Springer, Heidelberg (2005)
2. Bhavsar, V.C., Boley, H., Yang, L.: A weighted-tree similarity algorithm for multi-agent systems in e-business environments. Comput. Intell. **20**(4), 584–602 (2004)
3. Brenneis, M., Behrendt, M., Harmeling, S., Mauve, M.: How much do I argue like you? Towards a metric on weighted argumentation graphs. In: CEUR Workshop Proceedings of the Third International Workshop on Systems and Algorithms for Formal Argumentation (SAFA 2020), pp. 2–13, No. 2672 in CEUR Workshop Proceedings, Aachen, September 2020
4. Brenneis, M., Mauve, M.: deliberate - online argumentation with collaborative filtering. In: Computational Models of Argument, vol. 326, pp. 453–454. IOS Press, September 2020. https://doi.org/10.3233/FAIA200530

5. Brenneis, M., Mauve, M.: Do I argue like them? A human baseline for comparing attitudes in argumentations. In: CEUR Workshop Proceedings of the Workshop on Advances in Argumentation in Artificial Intelligence 2020, vol. 2777, pp. 1–15, Aachen, November 2020

6. Budanitsky, A., Hirst, G.: Semantic distance in wordnet: an experimental, application-oriented evaluation of five measures. In: Workshop on WordNet and Other Lexical Resources, vol. 2, p. 2 (2001)

7. Dung, P.M.: On the acceptability of arguments and its fundamental role in non-monotonic reasoning, logic programming and n-person games. Artif. Intell. **77**(2), 321–357 (1995)

8. Gordon, T.F.: Structured consultation with argument graphs. In: From Knowledge Representation to Argumentation in AI. A Festschrift in Honour of Trevor Bench-Capon on the Occasion of his 60th Birthday, pp. 115–133 (2013)

9. Gottipati, S., Qiu, M., Yang, L., Zhu, F., Jiang, J.: Predicting user's political party using ideological stances. In: Jatowt, A., et al. (eds.) SocInfo 2013. LNCS, vol. 8238, pp. 177–191. Springer, Cham (2013). https://doi.org/10.1007/978-3-319-03260-3_16

10. Kunz, W., Rittel, H.W.J.: Issues as elements of information systems, vol. 131. Citeseer (1970)

11. Kunze, M., Weidlich, M., Weske, M.: Behavioral similarity – a proper metric. In: Rinderle-Ma, S., Toumani, F., Wolf, K. (eds.) BPM 2011. LNCS, vol. 6896, pp. 166–181. Springer, Heidelberg (2011). https://doi.org/10.1007/978-3-642-23059-2_15

12. Li, Y., Bandar, Z.A., McLean, D.: An approach for measuring semantic similarity between words using multiple information sources. IEEE Trans. knowl. Data Eng. **15**(4), 871–882 (2003)

13. Marschall, S.: The online making of citizens: Wahl-O-Mat. In: The Making of Citizens in Europe: New Perspectives on Citizenship Education, pp. 137–141 (2008)

14. Miller, G.A., Charles, W.G.: Contextual correlates of semantic similarity. Lang. Cogn. Procs. **6**(1), 1–28 (1991)

15. Page, L., Brin, S., Motwani, R., Winograd, T.: The pagerank citation ranking: bringing order to the web. Technical report 1999–66, Stanford InfoLab (November 1999). http://ilpubs.stanford.edu:8090/422/. (previous number = SIDL-WP-1999-0120)

16. Rahman, M.M., Sirrianni, J., Liu, X.F., Adams, D.: Predicting opinions across multiple issues in large scale cyber argumentation using collaborative filtering and viewpoint correlation. In: The Ninth International Conference on Social Media Technologies, Communication, and Informatics, pp. 45–51 (2019)

17. Romero Moreno, G., Padilla, J., Chueca, E.: Learning VAA: a new method for matching users to parties in voting advice applications. J. Elect. Pub. Opin. Parties, 1–19 (2020)

18. Rosenfeld, A., Kraus, S.: Providing arguments in discussions on the basis of the prediction of human argumentative behavior. ACM Trans. Interact. Intell. Syst. (TiiS) **6**(4), 1–33 (2016)

19. Spertus, E., Sahami, M., Buyukkokten, O.: Evaluating similarity measures: a large-scale study in the Orkut social network. In: Proceedings of the Eleventh ACM SIGKDD International Conference On Knowledge Discovery in Data Mining, pp. 678–684 (2005)

20. Xuecheng, L.: Entropy, distance measure and similarity measure of fuzzy sets and their relations. Fuzzy Sets Syst. **52**(3), 305–318 (1992)

A Contextual Bayesian User Experience Model for Scholarly Recommender Systems

Zohreh D. Champiri[1]([⊠]), Brian Fisher[1]([⊠]), and Chun Yong Chong[2]([⊠])

[1] School of Interactive Arts and Technology, Simon Fraser University, Vancouver, Canada
{z.champiri,bfisher}@sfu.ca
[2] School of Information Technology, Monash University Malaysia, Bandar Sunway, Malaysia
chong.chunyong@monash.edu

Abstract. Since the advent of scholarly recommender systems (SRSs), more than 200 papers in the related area have been published. Many of these papers focus on proposing new and more accurate algorithms, or to enhance existing ones. Recently we have seen growing interest in embedding recommending methods into User Experience (UX), to enhance the value of RSs for users. Researchers have proposed that UX can be affected by bottlenecks in human perception, the pre-conceptions of the individual, and related factors such as personal and situational characteristics, which can be considered as contextual information. Although there are a few studies on developing User Models (UMs) in the field of SRSs, it has been emphasized that incorporating contextual information into user modelling and creating recommendations based on the users' information needs can be an effective approach to personalization and better UX with SRSs. The aim of this paper is to operationalize relevant contexts and to design a Bayesian UM for assisting the diagnosis of scholars' information needs in terms of accurate, novel, diverse, and popular research papers. The proposed user model can be embedded in the process of recommending and identifying the users' information needs which help recommenders to retrieve more appropriate recommendations and consequently leads to the enhancement of the UX for SRSs. Finally, the robustness and performance of the proposed Bayesian UM are evaluated.

Keywords: Scholarly recommender system · Research paper recommender system · User experience · Contextual data · Context-aware computing · Bayesian network · User modelling · Human-computer interaction · Machine learning

1 Introduction and Problem Statement

With the increasing number of scientific publications, Scholarly Recommender Systems (SRSs), commonly known as research paper or academic RS are considered an appropriate tool to facilitate and accelerate the process of information seeking for scholars by offering appropriate resources to them [1]. A recommender overrides search engine results by modelling users' preferences. In other words, the highest quality recommendations can be provided to users only after their preferences have been modelled in what is typically called a User Model (UM) [2, 3]. In this context, quality refers to the ability

© Springer Nature Switzerland AG 2021
H. Degen and S. Ntoa (Eds.): HCII 2021, LNAI 12797, pp. 139–165, 2021.
https://doi.org/10.1007/978-3-030-77772-2_10

of the recommender to produce exactly those recommendations that the user will use or would like to receive. There is a presumption that, the more accurate the algorithm, the better the predicted recommendation is for the users [5, 6]. Recently, several works have attempted to incorporate User Experience (UX) into recommending methods and have shown success in improving the value of RSs to the users. In order to incorporate contextual information and create recommendations based on the users' information needs, several studies have also tried to embed UMs in the field of SRSs, which have proven to be useful to improve personalization and UX of SRSs [4, 10]. However, it is not clear how contexts influence the UX of SRSs and it is difficult to decide which contexts must be incorporated into developing the UM [11, 12], especially in the initial stages of interaction with users where typically little user information is available [2, 3, 13]. In the past decade, various Machine Learning (ML) methods had been applied to support user modelling in RSs. These cognitive UMs, which are based on the deep understanding of uncertain and dynamic users' contexts, are still in their infancy [14, 15] particularly in the domain of SRSs. Among the chosen ML methods, Bayesian networks (BNs) are one of the more powerful tools for uncertainty modelling [16] but have rarely been applied in SRSs [4, 17]. This paper aims to develop a UX model which applies BN method for recognition of the most relevant contextual data for making suitable paper recommendation for the scholars. The proposed model is assessed by using the performance and robustness evaluation metrics.

2   Research Background

2.1   Contextual Approaches

The term context appeared in the field of computer science in the late 1980s [18, 19] in order to increase the richness of communication and provide more useful computational services [2, 20]. Since then, many studies in the field of computer science tried to define the term "context". Dey [20] offered the most cited definition of context from a computer science viewpoint. He expressed that context is any information can be used to characterize the situation of an entity. An entity is a person, place, or object that is considered relevant to the interaction between a user and an application, including the user and applications themselves. As depicted in Fig. 1, the recommending approaches are classified into two groups and contextual approaches are categorized into three approaches: pre-filtering, post-filtering, and contextual modelling implicitly [3, 13, 21]. Contextual information can be incorporated into the classical recommendation procedures in order to generate better recommendations [12, 22].

2.2   Contextual User Modeling

Some researchers in RSs believe that regardless of the technology exploited by a RS, the high quality recommendations can be produced only after modeling of the users' preferences which is typically called User Model (UM) in the literature [23, 24]. In this context, quality refers to the ability of the system to produce exactly those recommendations that the user will use or would like to receive [23]. To achieve this, adequate

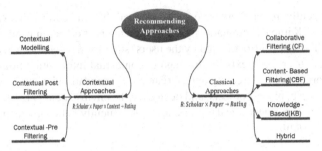

Fig. 1. Recommending classification

information including contextual information should be stored to deliver high quality recommendations. However, acquisition of sufficient data for the UM is not an easy task especially at the initial stages of interaction with the user, when usually little information about the user is available. Therefore, as a general rule, the more valid information is stored in the UM (i.e., the more knowledge the system has obtained about the user), the better the quality of the recommendations will be [3, 23]. As indicated earlier, there are little studies which attempted to develop UMs in the field of SRSs while the main factor that influences users' satisfaction is the ability of a recommender to meet the users' information needs and it is obvious that the users have different information needs due to different knowledge, goals, and generally different contexts which are uncertain and change consistently [25]. In contextual UM, the contextual information is employed directly as a main part of learning preference models (built using techniques such as decision tree, regression, and probabilistic model) [26]. However, identification of valid contextual information for different domains are challenges in contextual user modelling either explicitly or implicitly [13]. Moreover, a contextual UM should be able to infer possible cognitive process of user behavior and mind but it is not fully tractable in the practice [23, 24, 27]. The past decade has seen research into the use of ML to support user modeling pass through a period of decline and then resurgence. However, research for creating the cognitive user models it is still at early stage [14, 15].

2.3 User eXperience (UX)

In the early 1990s, the well-known cognitive scientist Don Norman (2013) coined the term User eXperience (UX). He proposed that UX would focus on the user's feelings (positive and negative) about a product over time [28]. UX has been used as an umbrella term in the field of Human Computer Interaction (HCI) to focus on aspects which are beyond usability [29], including all the feelings (positive and negative) a user is experiencing while interacting with a product, e.g. a mobile phone. There is no clear definition of UX [30, 31]; however, the current ISO (ISO 9241-110:2010 (clause 2.15) definition of UX focuses on a person's perception and the responses resulting from the use or anticipated use of a product, system, or service. In software engineering it is thought that if a product fails to meet end users' needs, both the product and the company (creator of product) become obsolete [32]. In other words, users will choose products with a great UX. Hence, UX is becoming a key competitive factor in more and more

industries. Recently, researchers have acknowledged that embedding the RSs and user modelling into UX impacts dramatically on the effectiveness of recommendations for the users [8, 10]. The UX is also affected by the users' situations, behaviors, characteristics, or in a nutshell, users' contexts. In this regard, contextual information influencing UX plays an important role in creating appropriate recommendation because it can present the status of people, places, objects, and devices in the environment [11, 13, 33–35] and leads to a better experience for the users, and consequently better interaction between the users and system.

2.4 Bayesian Network

Bayesian networks (BNs) are powerful tools used for uncertainty modeling. Their first appearance was in the field of medical decision systems in the late 1970s [16]. The inference in BNs is based on a probabilistic theory called Bayes' theorem or Bayes' law which spreads knowledge within the network [36, 37]. More precisely, BNs are a class of graphical models that allow a concise representation of the probabilistic dependencies between a given set of random variables (nodes) $X = \{X_1, X_2, X_3, \ldots X_n\}$ as a Directed Acyclic Graph (DAG) $G = (V, A)$. Each node $v_i \in V$ corresponds to a random variable X_i which might represent a causal link from parent node to their children [38]. Each node is associated with a conditional probability distribution which assigns a probability to each possible value of this node for each combination of the values of its parent nodes [39]. Graph G is an independency map (I-map) of the probabilistic dependence structure P of X if there is a one-to-one correspondence between the random variables in X and the nodes V of G, such that for all disjoint subsets A, B, C there is;

$$A \perp p(B|C) \Leftarrow A \perp G(B|C)$$

Similarly, G is a dependency map (D-map) of P if X, there is

$$A \perp p(B|C) \Rightarrow A \perp G(B|C)$$

G is said to be a perfect map of P if it is both a D-map and an I-map,

$$A \perp p(B|C) \Leftrightarrow A \perp G(B|C)$$

and in this case, P is said to be isomorphic or faithful to G.

Overall, BNs consist of both qualitative and quantitative parts. With regard to the qualitative part, it is the structure of the network: a directed acyclic graph where nodes correspond to variables and arcs representing influences between variables. The quantitative part, however, provides the conditional probability tables that make up the network settings [39, 40]. The learning applies both the network structure and parameters that can be obtained from complete or incomplete data [41]. The correspondence between the structure of the DAG and the conditional independence relationships is elucidated by the directed separation criterion or d-separation [16].

2.5   Bayesian Network Algorithms

The BN modeling development task involves some major activities including structural learning, parameter learning, and inference [40, 42, 43]. The structure and parameters learning can be yielded either by experts' knowledge or by automatic learning from a dataset, provided that the dataset is complete and unbiased [44, 45]. Automatic learning methods include constrained-based, scored-based or metric/search–based [43, 46, 47], and hybrid methods [40]. Figure 2 depicts a classification of BN algorithms.

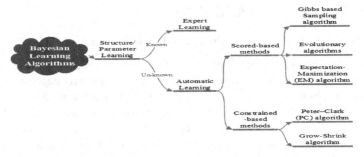

Fig. 2.   Algorithms applied for BN learning

Constraint-based methods apply information about conditional independencies gained by performing statistical significance tests on the data [46, 48]. Score-based methods search for a BN to minimize or maximize a score [40, 45, 46, 48]. The Bayesian inference algorithms are also used to calculate marginal probabilities, given an evidence set [49]. The Grow-Shrink algorithm (constrained based algorithm) in the "bnlearn" package is one of the most recommended algorithms for the BN structure, parameter, and computation of the posterior probability distribution [46] which is applied in this research.

3   Related Work

Based on a systematic review on recommending methods applying to SRSs by [1] and review of recent studies by [17, 25], it is concluded that different ML methods such as Neural Networks, SVM, and Decision Trees, have been utilized in making paper recommendations by considering the Collaborative Filtering (CF) and Content-Based Filtering (CBF) approaches. Beel et al. [25] indicated that CF methods in SRSs are not effective because there are no balance between the number of papers and number of users. In other words, a huge number of papers compared with low number of users, and only few users rated the similar papers. Therefore, they suggested that user modeling is more effective than CFs. The use of deep neural networks for Natural Language Processing (NLP) has received much attention recently; it provides high quality semantic word representations. Deep neural network models have been applied to tasks ranging from machine translation to question answering, but not much attention is paid in the area of RSs. For instance, in [50], the authors showed that LSTM can be used to build a

language model and assess semantic similarity between sentences. These models are usually trained on large amounts of data. To the best of the researchers' knowledge, there have been no work done before for recommending scientific articles based on contextual BNs [17]. Also, the use of ML algorithms in RSs has been reviewed and analyzed by [51]. According to this survey, among seven studies (Documents: [52]; E-shop [53]; E-mail [54]; Books [55]; Movie [56]; Traffic [57]; Tourism [58]) that used Bayesian method, only two studies are related to book and document recommender [52, 55] which do not considered contextual information.

4   Why Bayesian Network Modeling?

4.1   Suitable to Deal with Uncertain and Dynamic Contexts

Contexts are dynamic and change over the time. Therefore, their conditions are very complex and uncertain [26]. Moreover, the user's need of scholarly papers is uncertain and changes due to the different contexts such as background knowledge, preferences, and goals. The BN method provides an effective approach for constructing and manipulating probabilistic models for handling uncertainty in context awareness applications [38]. Besides, this BN is appropriate for representing complex relations between the variables which their states change over the time [59, 60].

4.2   Well Adapted for Diagnose of Users' Preferences

A fundamental goal of HCI research is to make systems more usable and more useful, and to provide users with better experiences which fit their needs [61]. A UM strives to accomplish this goal by discovering the laws of nature and inferring unobservable information about a user from observable information to analyze perceptual and cognitive processes and characterize individual differences. BNs are well adapted to the problem of user modeling because they can represent the uncertainty related to the modeling of users' preferences [40, 48]. It is a natural and scientific method to cope with the variability and the complexity of users' situations by probability distribution of over the whole related variables and explicitly represents causal relations [62].

4.3   Appropriate for Diagnose of User's Information Needs in SRSs

The users' information needs vary among users owning to different contexts such as knowledge, preferences, and goals [1]. In spite of this change, researchers have recommended that rather than accuracy, one should apply other features such as diversity, novelty, and popularity to make a better list of recommendation and improve UX of RSs [63–67]. As Fig. 3 depicts, a user model should match the recommended papers on the basis of four identified levels of accurate, novel, popularity, and diverse with users' information needs or users' contexts.

BNs are flexible models not only to reason with the knowledge and belief uncertainties [7], but also the structure of knowledge representation [41, 42, 68]. They make the system dynamic and evolve in order to infer users' needs and react depending on them [48] in order to provide the predictions to be made about a number of influential variables rather than a single variable by using probability distribution [39].

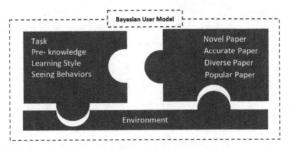

Fig. 3. Matching paper recommendations list with contexts

4.4 Appropriate for Representation of Casualty Relationships

There are cause-effect (casualty) relationships between the contexts. Among the methods that can be used to represent uncertain domains includes decision trees, neural networks, mixtures of basic functions, and Markov networks. Researchers have claimed that BNs are better suited for representation and learning of the directed causal relationships [41, 42]. In addition, BNs aim to facilitate the description of a collection of beliefs by making explicit the relationship of causality and conditional independence among these beliefs and to provide a more efficient way to update the strength of beliefs when new evidence is observed [41].

4.5 Well Adapted to Other Recommending Approaches

An important feature of BNs is that they are able to support hybrid recommending approaches of the CF and CBF. The CF might be used to obtain the conditional probability tables and the initial beliefs of a BN [10]. These beliefs can then be updated in a CBF manner when the network is accessed by a user. This mode of operation enables a predictive model to overcome the data collection problem of the CBF (which requires large amounts of data to be gathered from a single user), while at the same time enables the tailoring of aspects of a collaboratively-learned model to a single user.

5   Bayesian UM Development

The goal of the proposed Bayesian UM is to diagnose the users' information need in terms of four levels of accurate, novel, diverse, and popular papers based on the contextual data. The process of BN model development is accomplished in three phases namely dataset preparation, structure, and parameter learning. Figure 4 depicts the phases and major activates.

5.1 Dataset Preparation

Among the current datasets, there is no dataset which includes the required contexts as well as indexes or labels for novel, diverse, popular, and accurate papers. Therefore, it is necessary to prepare an appropriate dataset for the BN modeling. In the following sections, the activities undertaken for the dataset preparation are discussed.

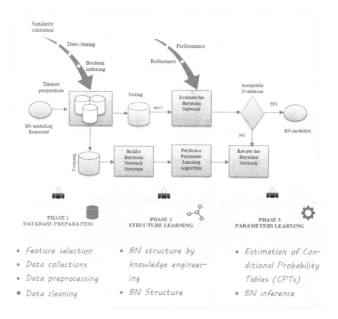

Fig. 4. BN development phases and main activities

Feature Selection. In this research, the contextual data (features) are selected based on the results of this study [1]. More details are discussed in the following.

Data Acquisition (Web-Based App). To acquire the data, a web-based application was developed. The data acquisition procedure was composed of a large-scale questionnaire survey. The questionnaire's fields (items/features of each context) were designed based on the guidelines of [1, 66, 67]. This study targets the population of computer science scholars including Master degree students, PhD students, post-doc researchers, as well as lecturers. Due to privacy concerns, detailed information of the participants is kept confidential. The participants should have the experience of working with Web of Science (WoS) bibliographic database and minimum 30 minutes time to conduct and finish the survey. The scholars were asked to conduct searches for suitable papers for their current work in a naturalistic setting by using WoS bibliographic database. This data collection was performed in two steps; in Step 1 as shown in Fig. 5, the participants are asked to submit their current contexts/situation such as task, pre-knowledge, etc.

If the participants have no idea of how to respond and fill out the form, more information will be provided by clicking the blue question mark buttons provided for each field. Also, the "clear" button enables the participants to remove all information in the form if it is required. In Step 2, the scholars were asked to select the most appropriate paper relevant to their current needs and to rate the paper in a 5 Likert scale in terms of novelty, accuracy, popularity, and diversity as shown in Fig. 5. Finally, participants were required to submit the paper ID (identification paper produced in WOS) or upload the paper. The data acquisition survey was conducted for 1.5 years, and 1121 records were collected.

Fig. 5. Acquisition of Bayesian data-Steps 1&2 and Throwing validity exceptions in data collection. (Color figure online)

Data Pre-processing. After the data collection phase is completed through the web application, papers related data such as title, authors, and citations were retrieved from the WOS database. Then, the CSV data files of users' data and paper's data were merged to form the final dataset. The data pre-processing activities are discussed in the following.

Exporting Data to CSV File by LINQ to SQL-Query. The data collected from the web application were exported to a CSV file for further analysis. It is possible to export the data from the web application to a CSV file by querying to database through LINQ to SQL query in the Visual Studio. This original CSV file contains users' contexts, users' ratings, and papers' IDs. The users' data was transformed into a separate CSV file. The primary dataset contains 1121 participants' records. Among the collected records, 675 participants have entered the paper's IDs and 446 participants have uploaded the PDF files of the associated papers. However, 10 records of users' data were invalid, and 58 PDF files were deemed useless and irrelevant files. Therefore, the 68 invalid records were eliminated from the dataset and the final dataset contains 1053 records.

Importing Bibliographic Data from Text to CSV. As four levels of accuracy, diversity, popularity, and novelty of the papers have been considered in the BN modelling, more bibliographic information about the papers such as title, authors, and keywords are also required. The papers' bibliographic information is provided by using the WOS exporting option for importing the plain text to a CSV file. The WOS approximately retrieves 73 bibliographic features for a paper. In this research, only 15 features have been exploited and the rest such as PI: Publisher City, PA: Publisher Address, SN: International Standard Serial Number (ISSN), BN: International Standard Book Number (ISBN), DI: Digital Object Identifier (DOI) were omitted from the dataset.

Combining Datasets. The two above-mentioned CSV files including the users' data and the papers' data were merged by using the pd.merge () function in numpy (python).

Data Cleaning. In this research, the users' data were collected through a web application. Therefore, most of the input data were validated at runtime during the data collection phase by throwing validity exceptions in terms of data type validation, data range validation, and constraint validation (Fig. 5). Additionally, to ensure data accuracy, a sample of 200 records were selected randomly and examined manually in order

to check the quality of the data. Among the records in the selected sample, no invalid data was detected.

Numerical Data Discretization. Based on guidelines provided by [40], the nodes that take mutually exclusive and exhaustive discrete values are recommended for the BN modeling. It means that the variable should take on exactly one of the values at a time. For example, a user might have two scholarly tasks at the same time but in the BN modelling, one of the tasks should be considered at a time. The Boolean nodes (binary values of true & false), ordered values (low, medium, high), and integral values (values from 1 to 120 for the users' age) are three common types discrete nodes in BN modeling [40]. In this research, most of the variables have different states such as LS = [visual, verbal, physical]; TA = [Thesis writing, Paper writing, Course taking, Topic finding, Course teaching]. Therefore, the best modelling solution is to have Boolean variables for each state. In other words, for the different states of the same variable, separate variables are created and they must be mutually exclusive and it is solved by adding an extra arc between the nodes and a deterministic CPT that enforces the mutual exclusion [40, 42]. When considering the variables, it is also required to decide what states or values the variable can take. For example, when modeling the user's contexts, the pre-knowledge (PK) variable could take the two values of advanced and basic which are represented by the values of 0 and 1 (binary values) in the final dataset. The following code assigns the relevant values to Pre-knowledge variable. The knowledge of engineer method might be very helpful to examine the final variables and also the suitable values [40] which is discussed more in the BN structure learning phase.

```
> bndata <- read.table ("bntrain.csv", header=TRUE, sep=",")
Pre-knowledge = c (0, 1),
labels = c ("Basic", "Advanced")
```

Most of the variables in this research are already discrete or transferred to Boolean/binary scales. There are only two continuous variables (time and paper impact factor).

Calculation of Cosine Similarity for the Text Data. In the data collection phase, the participants were asked to identify their research interests and input search keywords which are text data and should be converted to the numeric data type for the BN learning [40]. For the diagnosis of the problem in this research, the similarities between users' input search keywords and paper's keywords, title, and abstract were calculated using Cosine Similarity (CS). CS is a measure, which calculates the similarity between two texts or vectors (on the vector space) by calculation of the cosine of angle between two vectors (Huang, 2008). CS is calculated using the following formula:

$$\text{Cos } 0° = 1 \ \rightarrow \ \text{two vectors are similar}$$
$$\text{Cos } 90° = 0 \ \rightarrow \ \text{two vectors are not similar}$$

The cosine of two vectors (non-zero) is calculated by the Euclidean dot product formula, therefore the cosine similarity, $\cos(\theta)$, is defined as similarity between two vectors a and b (Huang, 2008):

$$a \cdot b = \|a\| \|b\| \cos \theta$$

$$\text{Similarity} = \cos \theta = \frac{a \cdot b}{\|a\| \|b\|}$$

To calculate the similarity two functions of dep_product and getSimilarity were applied. To calculate the cosine similarity, NLTK, sklearn.metrics.pairwise libraries as well as sklearn.feature_extraction.text have been utilized. The CS function returns a value (between 0 and 1) which is the similarity between two columns (e.g. users' search keywords (UKW) and paper's keywords (PKW)). Therefore, the text data (columns) of users' search keywords, papers' keywords and abstract have been removed and the CSs were added to the dataset. The similarities between the users' interests and the uploaded papers were assessed through the "criteria" field. This field determines the relevance of the uploaded papers to the users' tasks, research interests, etc. If the paper is selected based on the research interests, the similarity between these two columns is considered as 1 otherwise it is considered as 0. In this research, the CSs have been calculated for the titles, keywords, and abstracts of the papers. The CS of a paper body was not calculated because, firstly, an accurate CS calculation requires to be compared with classification systems such as the ACM computing classification system (ACM-CCS). The ACM-CCS is a poly-hierarchical ontology based on semantic vocabularies, which reflects the state of the art of the computing disciplines, concepts, and categories. Secondly, text classification is out of the scope of this research.

Preparing Data Codebook. The codebook serves as a reference and ensures that the data is understood and interpreted properly. The data codebook in this research has been created by uploading the final CSV data file (Appendix A). By performing the above pre-processing tasks, the final dataset was produced. As proposed in ML guidelines, when the dataset is not so large, the whole dataset is applied for the training phase and K-Fold Cross Validation (KFCV) method is recommend [44, 69].

5.2 BN Structure Learning

The BN structure or topology learning is defined as a set of relevant variables and their possible values built by connecting the variables into a DAG to represent a network which best describes the observed data [45, 48]. The BN structure can be built either manually by knowledge engineer (expert elicitation) or by automated learning from the data [40, 47]. Applying the hybrid method leads to better analytical and predicting ability of the BN model [45, 59]. In the next few sections, the BN structure using the experts' knowledge and automatic learning are discussed, respectively.

BN Structure by the Knowledge Engineer. The knowledge engineering aims to build a model for an expert system which represents realistically the problem features and is able to reason and respond as close as an expert human [40, 44]. To create a BN structure, two knowledge engineers who are in the fields of scholarly communication and recommending system reviewed the variables (presented already in data codebook) that could

influence the users' needs of scholarly papers based on four levels of accurate, novel, diverse, and popularity. Afterwards, the clarity and consistency of variables examination have been performed. According to the guidelines indicated in the BN modelling literature [40, 44, 45], it is appropriate to create direct questions about the causes or effects to elicit and identify the relationships between the nodes. Hence, with the help of the experts, several questions were posted and then, based on these questions, the arcs (relations) were added from those causal variables to the affected variables. For brevity, only the questions conducted for the novel paper node is presented in the following.

```
Q: "What are contexts that would cause users to need a novel paper?"
A: "topic finding and course teaching"
Modelling: suggests arcs from those nodes to the novel paper node.
Q: "Is there any context which prevents need for a novel paper?"
A: "if the user has already known and have advanced pre -knowledge in the."
Modelling: suggests an arc from pre- knowledge to the novel paper.
Q: "What can cause a paper to be considered as a novel paper?"
A: "publication data, users' awareness"
Modelling: suggests the arcs from those nodes to a novel paper.
Q: "What are the effects of looking for a novel paper?"
A: "searching behaviour"
Modelling: suggests the arcs from novel paper to different searching behaviour.
```

In the BN structure, the focus was on capturing expert's understanding of the relationships between variables performed by pairwise elicitation method. The expert was provided with a cross-table of 42 rows (R) and 42 columns (C). For each cell in R and C, the expert indicated the relations providing the below signs. To reduce the complexity of BN structure and the possibilities of inconsistent information as well as elicitation burden on the expert, which consequently cause mistakes, the reverse direction relationships were not considered. The result of the expert elicitation was the set of priors for pairwise relations shown in Fig. 6.

The 934 cells out of 1764 (56.26%) are signed by O which represents no relations. The symbol – or ∼ also shows the non-directional relationship. A total of 105 relations were indicated as → (directly causes), 7 ≺ (R occurs before C) and 718 ∼, - (correlations & relations) have specified by the knowledge engineering method which provided 830 of the 1764 possible pairwise relationships. In addition, there are no ⇒ relationships, as the experts did not consider non-direct causality a natural relationship to specify. In some studies, the Delphi method [70] or focus group method [71] are recommended to validate expert's opinion and to elicit information independently from a group of experts to reach a consensus [40]. In this study, only one expert has contributed into the BN structure by practical considerations because among the three relevant experts identified in this field, only one individual agreed to dedicate in this research. Moreover, this study also applies automated BN construct learning using the dataset, which avoids adding the additional complexity to the BN structure process.

Automated BN Structure by Data (GS Algorithm). The structure of the BN associated with the dataset was learned with the GS algorithm based on the guidelines given by [46]. Also, the Pearson's Linear Correlation (Cronbach's α: 0.05) was used as the conditional independence test. As shown in Fig. 7, a node is a parent of a child if there is an arc from the former to the latter (TA4 & PK1 are parents of ST1). Therefore, if each node is defined as X, then BN structure is defined as;

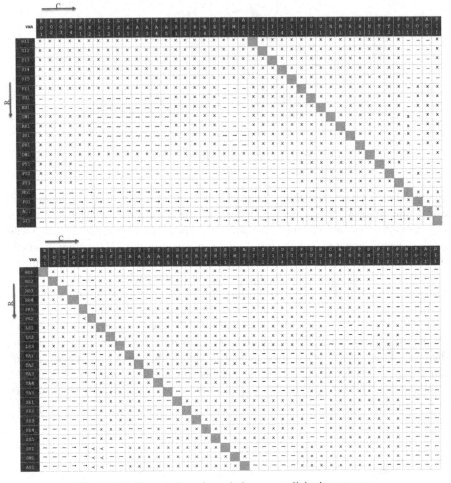

Fig. 6. Pairwise relations through the expert elicitation process

$$P(x_1, x_1, \ldots, x_n) = \prod_i P(x_i | \mathbf{Parents}(x_i))$$

Provided $\mathbf{Parents}(x_i) \subseteq \{x_1, x_1, \ldots, x_{i-1}\}$.

In this directed chain of nodes, $TA_4 \& PK_1$ are ancestors of $NO_1 \& AC_1$ since they appear earlier in the chai and AC_1 is a descendant of TA_4 node because it comes later in the chain.

In the most constrained-based algorithms such as GS algorithm for building the BN structure, the Markov blanket is computed separately. Also, each neighborhood is a subset of the corresponding Markov blanket and therefore, can be learned independently from the others [40]. According to the BN modeling guidelines [46], before learning the neighborhoods, the consistency of all Markov blankets should be checked to examine their symmetric differences. Therefore, all pairs of nodes were checked and were

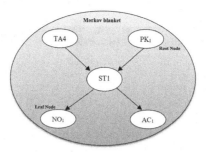

Fig. 7. A sample of Markov blanket

removed from each other's Markov blanket if they do not appear in both of them. After learning the Markov blanket and neighborhoods, the correlation between the variables were computed to create a correlation matrix using the dataset. Then, the nodes and relationships were compared with the results and consequently, the BN structure was built. The Pearson's Correlation was applied to investigate the dependency between multiple variables at the same time. The function rcorr() in Hmisc package was also applied to calculate the significance levels for Pearson correlations which return the correlation coefficients for all possible pairs of variables. The output is a value in the range of $(-1, +1)$. The values of -1 and $+1$ stand respectively a strong negative correlation and strong positive correlation. The value of 0 represents zero correlation or no correlation. The colored correlation matrix plot, as depicted in Fig. 8, shows relatively strong and statistically significant relationships. As mentioned earlier, the values of -1 and $+1$ stand respectively a strong negative correlation and strong positive correlation between the 42 variables.

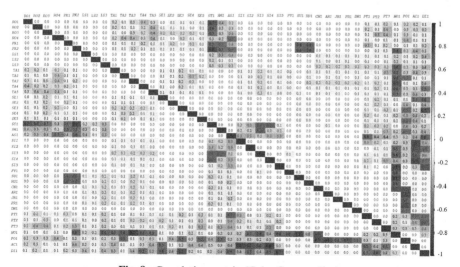

Fig. 8. Correlation matrix (Color figure online)

Highest positive correlations that are shown in navy blue (Fig. 8) also represented the value near to +1. In the following, we discuss how the results taken from the knowledge engineer and automated learning as well as coloration matrix are compared to build the final BN structure.

After comparing the BN structure derived from the knowledge engineer and the automated learning, we noticed that there are a few nodes that do not have the same correlations based on the results of automated BN learning. Therefore, by consulting with the knowledge engineer and experts, the final relationships are decided for four paper levels as depicted in Fig. 9.

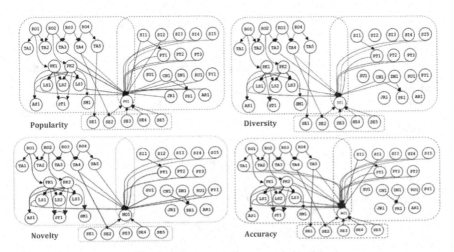

Fig. 9. BN structure four paper's levels

5.3 BN Parameter Learning and Inference

After the BN structure modelling, the next step is the specification of conditional probability distribution or the Conditional Probabilities Tables (CPTs) entailed by a network which is called parameter learning [40, 48]. In this research, the parameters were calculated by applying the bn.fit function using bnlearn package which utilizes the network data to estimate their maximum likelihood [46, 49]. For each variable in a Boolean network which has n parents, the size of CPT would be 2 n + 1 probabilities [46]. So, in this research, for four nodes of the novel, accurate, diverse, and popular papers, the sizes of the CPTs are 210, 213, 212, and 28 probabilities respectively. For brevity, in the next section, only the probability distributions of a single node are provided. Figure 10 shows part of the BN network along with the joint probability distributions.

The single BN structure along with the possible states and values are shown in Fig. 11 (Netica [44]).

After parameter learning, we attempt to investigate the estimated probability distributions of the model [46, 49]. The BN inference is carried out to estimate the posterior probability distribution or suitable values for the tuning parameters [40, 43]. There are

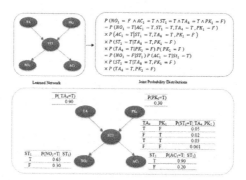

Fig. 10. BN network along with the distributions

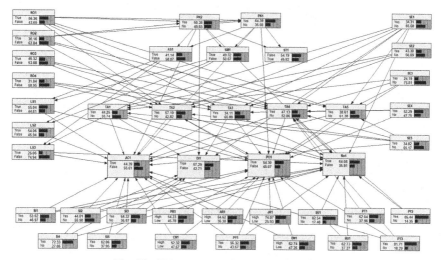

Fig. 11. BN network and parameters (Netica)

various methods for performing the inference on a Bayesian model such as bootstrapping, conditional probability queries, and cross validation. Based on the guidelines, cross validation is the most used and appropriate method to validate BN models algorithms and parameters [46]. Sensitivity analysis and predictive performance assessment are key elements of the modelling process [72].

6   Bayesian UM Evaluation and Results

Based on the literature, the majority of studies have assessed performance and robustness of the BN models [40, 73, 74] by using metrics such as Entropy (robustness of BN structure), F-measure [75], Mean Square Error (MSE), and Mean Cross Entropy (MXE) (for the performance) [43, 45]. Additionally, [46, 62] have also recommended to evaluate the BN learning algorithm by employing the measure of expected loss. Elicitation review is another method for BN models evaluation [40, 44]. The aim of elicitation review

is to check the clarity and consistency mostly in the structure of the BN model by knowledge engineers and domain experts' involvement. As a hybrid method of the automated structure learning and the knowledge engineer have been applied for the proposed BN structure, the elicitation review method is ignored.

6.1 Robustness of BN Structure: Sensitivity Analysis

For examining the robustness of the BN structure, the sensitivity analysis is carried out which analyses how sensitive the network outputs or parameters are against the inputs (observations) [40, 44]. This analysis assures the BN structure is correct or highlight errors [76, 77]. The Entropy is one of the widely used metric in sensitivity analysis to examine uncertainty in a probability distribution or BN modeling [40, 44]. In the proposed UM, the inputs are the users' contexts and the outputs are the users' information needs for novel, accurate, diverse, and popular papers. Therefore, it is examined how the model diagnoses for novel, accurate, diverse and, popular papers are sensitive against the users' contexts changes. Therefore, the Entropy H(X) in the target node X, where for example; X represents the probability of the user needs a novel paper (positive for diagnosis of novel paper) is calculated as [40];

$$H(X) = - \sum_{x \in X} P(x) \log P(x)$$

Since the calculation of Entropy is performed by using GUI in the Netica software [44], it provides the benefit to change the nodes which have high Entropy in the GUI and then, calculate the new Entropy for the target node easily. The sensitivity analysis or entropy values for diagnosis of $H(X_1)$, $H(X_2)$, $H(X_3)$, $H(X_4)$ where the contextual nodes are evidences (C), are defined as follows:

$$H(X_1) \rightarrow pr(X_1 = \text{Positive}|C) > 0.5 (Novel\ paper)$$
$$H(X_2) \rightarrow pr(X_2 = \text{Positive}|C) > 0.5 (Accurate\ paper)$$
$$H(X_3) \rightarrow pr(X_3 = \text{Positive}|C) > 0.5 (Popular\ paper)$$
$$H(X_4) \rightarrow pr(X_4 = \text{Positive}|C) > 0.5 (Diverse\ paper)$$

Hence, the entropy value helps better understand which node (evidence (C)) is more relevant to corresponding positive or negative diagnosis.

```
if Pr ((X = Positive|C))>0.5:
    print "C is in a positive diagnosis group"
else:
    print "C is in a negative diagnosis group"
```

For the binary variables, the entropy ranges from zero to one $(H(X) : (0, 1))$. Zero indicates the minimum uncertainty and one represents the maximum uncertainty for a target node against the evidence. Therefore, $H(X) = 0$ represents $\rightarrow$ min$_{uncertanity}$ and $H(X) = 1$ represents $\rightarrow$ max$_{uncertanity}$. The sensitivity analysis results were ranked by entropy value shown in Fig. 12. These results only consider observations one at a time. Additionally, each variable or node was associated to a diagnosis node, which has certain states.

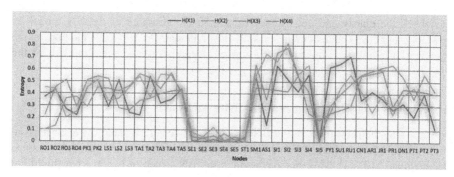

Fig. 12. Entropy values of BN nodes

As shown in Fig. 12, the results of sensitivity to diagnoses nodes revealed that, for all four nodes of novel, popularity, diverse, and accurate, the entropy reduction values were approximately equal to zero for the nodes of SI5, SE1, SE2, SE3, SE4, SE5 and ST1 (refer to the Data CodeBook in Appendix). Further examination of the CPTs of the mentioned nodes confirmed that SI5, SE1, SE2, SE3, SE4, SE5 have no impact on the rest of the network. For this reason, these nodes were removed from the network. However, since the ST1 node has an impact on the PK node, this node remained in the network. Interestingly, the results of sensitivity analysis also confirm that the impact of environment context is not significant compared to other contexts. Thus, entropy values substantiate that BN model's outputs are not sensitive against the SE1, SE2, SE3, SE4 variables that represent the time (environment context). Figure 13 shows the maximum uncertainties for $H(X_1)$, $H(X_2)$, $H(X_3)$, and $H(X_4)$. The nodes of pre-knowledge and task are in a positive group. The nodes that are in a positive group for a novel, accurate, diverse, and popular papers are as follows:

$$H(X_1) \rightarrow pr(X_1 = \text{Positive}|PK, TA, RU, SU, SI) > 0.5$$
$$H(X_2) \rightarrow pr(X_2 = \text{Positive}|SI, AS, TA, PK) > 0.5$$
$$H(X_3) \rightarrow pr(X_3 = \text{Positive}|SI, PR, JR, Pk, TA) > 0.5$$
$$H(X_4) \rightarrow pr(X_4 = \text{Positive}|SI, JR, PK, TA) > 0.5$$

Based on the entropy values, a moderate uncertainty level has been obtained for the diagnosis nodes against the users' learning style nodes (SL1, SL2, and, SL3) which divulges the moderate significance of this context compared to other contexts such as TA and PK for the accuracy of the BN model.

Fig. 13. Maximum entropies

6.2 Comparison of BN Algorithm-Expected Loss

For learning the BN structure, a constraint-based learning algorithm called Grow-Shrink (GS) algorithm has been applied based on the guidelines indicated by [46]. To evaluate the performance of the GS algorithm, the Expected Loss $\rho(\mathbf{a})$ of the applied algorithm is calculated and compared to the $\rho(\mathbf{a})$ of Max-Min Hill-Climbing (MMHC) algorithm which is a combination of constraint-based and search-and-score techniques (Hybrid algorithm) [44] using the same dataset. The Bayesian expected loss is defined as the expected loss under the predictive distribution [75]:

$$\rho(\mathbf{a}) = \int l(\mathbf{a}, \mathbf{y}) P(\mathbf{y}|\mathbf{x}, \mathbf{D}) dy$$

As shown in Fig. 14, if training examples are drawn independently at random according to unknown distribution P(x,y) and the learning algorithm analyzes the training examples and produces a function f, given a new point <x, y> drawn from P, the function is given x and predicts $\hat{y} = f(x)$ therefore the loss $L(\hat{y},y)$ is measured [78]. The goal of a Bayesian learning algorithm is to find the f that yields the the lowest expected loss $E_{P(x,y)}[L(f(x), y)]$. In other words, loss function is used to determine which Bayesian learning algorithm is better suited for a certain problem and dataset.

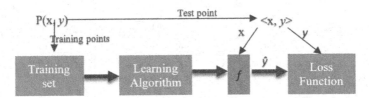

Fig. 14. Loss function for a supervised ML algorithm

To calculate the $\rho(\mathbf{a})$, the bn.cv method in "bnlearn" package was used. The lesser is the expected loss, the better is the algorithm in terms of the performance [46]. As represented in Fig. 15, the results disclosed that the EL of GS algorithm is ($\rho(a) = 0.341486$) and MMHC is ($\rho(a) = 2.350423$) by applying the KFCV method. It means that the GS algorithm produces a BN model which fits the data of this study better than the MMHC algorithm. Therefore, GS performs better than the MMHC algorithm and it is more fitted to the dataset of this study for the BN modeling.

6.3 Predictive Performance Assessment

To assess the predictive performance of the BN model, three metrics of F_1-score, Mean Square Error (MSE), and Mean Cross- Entropy (MXE) have been estimated using CFCV method (K = 10) as recommend in the literature [46, 73, 75]. Table 1 represents F1 score, MSE, and MXE of the BN model (overall and four levels) for each of the folds. In addition, the mean of the measures has been calculated.

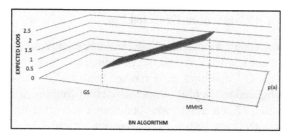

Fig. 15. Loss function

Table 1. BN model predictive performance results.

Dataset	Folds / Metrics	Fold-1	Fold-2	Fold-3	Fold-4	Fold-5	Fold-6	Fold-7	Fold-8	Fold-9	Fold-10	Mean
Novel	F_1	0.89	0.91	0.98	0.86	0.92	0.93	0.96	0.88	0.96	0.98	0.92
	MSE	0.12	0.11	0.02	0.15	0.14	0.03	0.09	0.05	0.01	0.01	0.07
	MXE	0.22	0.16	0.17	0.20	0.14	0.15	0.18	0.12	0.11	0.11	0.15
Diverse	F_1	0.92	0.98	0.93	0.86	0.85	0.87	0.85	0.68	0.86	0.95	0.87
	MSE	0.02	0.03	0.05	0.14	0.12	0.07	0.02	0.15	0.11	0.03	0.07
	MXE	0.10	0.13	0.10	0.13	0.15	0.16	0.15	0.10	0.20	0.12	0.13
Popular	F_1	0.92	0.98	0.93	0.86	0.85	0.87	0.85	0.68	0.86	0.95	0.87
	MSE	0.05	0.05	0.03	0.12	0.12	0.05	0.1	0.02	0.12	0.05	0.07
	MXE	0.04	0.03	0.12	0.12	0.05	0.10	0.20	0.12	0.10	0.20	0.10
Accurate	F_1	0.87	0.88	0.96	0.96	0.95	0.94	0.76	0.88	0.75	0.93	0.88
	MSE	0.09	0.12	0.02	0.08	0.05	0.03	0.18	0.12	0.11	0.01	0.08
	MXE	0.11	0.13	0.12	0.10	0.13	0.12	0.12	0.15	0.11	0.22	0.13
Overall	F_1	0.76	0.96	0.95	0.94	0.95	0.96	0.86	0.87	0.95	0.97	0.91
	MSE	0.14	0.03	0.02	0.04	0.02	0.06	0.14	0.12	0.05	0.01	0.06
	MXE	0.10	0.13	0.12	0.12	0.15	0.10	0.20	0.12	0.10	0.20	0.13
Acronyms	MXE→Mean Cross Entropy [0, ∞] Best score = 0; MSE→Mean Square [0, 1] Best score = 0; Error F_1→ F_1Score [0, 1] Best score = 1											

The best score for F1, MSE, and MXE are 1, 0, and 0 respectively which indicate that the results surpass the determined thresholds. Figure 16 shows the average for means calculated for 10 folds. The performance results in this research are promising, however, better results might be achieved by using larger dataset with more instances.

The Conditional Probabilities Tables (CPTs) entailed by the BN network for each node overcome the contexts' changes and uncertainty that might happen due to the users' contexts such as task, background knowledge, and research interests.

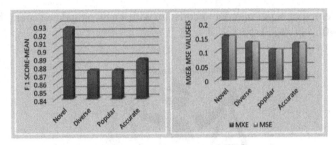

Fig. 16. F1score, MXE & MSE means of 10 folds

7   Discussion: Proposed Model Advantages

The proposed Bayesian UM in this research aimed to diagnose the users' information needs based on the relevant contexts. In the following, we outline the main advantages of the proposed model.

First, this research proposes a contextual Bayesian UM for supporting the diagnosis of scholars' information needs in terms of four levels of diverse, novel, accurate, and popular papers based upon the contexts. Such diagnoses are important because of their ability to provide paper recommendations that are relevant to the users' context and appropriate to their information needs. This model could be embedded in the recommending process to generate more appropriate paper recommendations related to their contexts.

Second, use of too much contextual information causes complexity and ambiguity in the system due to irrelevant, redundant, inconsistent, and noisy data [11, 13, 33–35]. In fact, each piece of contextual information is considered an extra dimension of the utility function of the recommender. Too many dimensions leads to a set of problems of dimensionality [79, 80]. In addition, processing higher dimensional spaces impose high computational resources requirements to discover useful knowledge patterns, and adds extra costs to system development, especially to the data acquiring and processing phases. This research identifies and highlights the most relevant contextual data in the field of SRS that can be exploited by future studies in recommendation making.

Third, a few studies have emphasized diversification of recommendations [81, 82]. While today a user might look for novel papers in a particular area, tomorrow she might be interested in the most popular papers. This research connects the different paper levels not only with the user profile such as knowledge, preferences, and task but also identifies the paper's features such as keywords, citation, author, etc. for each level. This helps researchers to classify papers based on the identified features.

Finally, the proposed decision UM in this research applied a BN probabilistic approach where the contexts and four paper levels are all connected to build a realistic graphical model. The model could be used to build more intelligent SRSs, which diagnose scholars' information needs and provide better experiences for them. Besides, the fundamental working principle of this research is the closest technique to the human reasoning or rationality, so-called Bayesian Thinking or Bayesianism philosophy [40, 83–86]. As shown in Fig. 17, the Bayes' rule applies the past data (prior probability) and present data (likelihood of the evidence) to predict the future (posterior). Therefore, it

updates the degree of belief in order to come up with a new or updated strength of belief (posterior) [40].

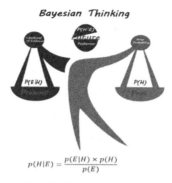

Bayesian Thinking

$$p(H|E) = \frac{p(E|H) \times p(H)}{p(E)}$$

Fig. 17. Bayesian thinking

This method simulates the process of paper evaluation and selection (design making) by scholars. We argue that this theory is the single most important tool for representing appropriate strengths of belief to understand human opinion which is constrained by ignorance and uncertainty [40, 84–86]. It can also be used to reach decisions in those circumstances where very few observations or pieces of evidence are available.

8 Recommendations for Future Studies

This research is one of the first attempts in the field of Contextual UX Modeling for SRSs. We encourage further research in this field. Some of the possible areas of investigation are discussed below.

The identification of users' information needs is not an easy task and needs better understanding about the users' perceptions via long term studies and observations. However, because of the high complexity of data collection for users' perceptions and different feelings as well as users' reasoning methods, they were not considered and have been left out in this study. As a future work, we are interested to deploy the BN model in a real environment and evaluate acceptance and feedback reported by the end users. The contextual user experience model using Bayesian networks can diagnose the users' information needs in four levels of accurate, diverse, novel, and popular papers. Identification of users' information need is very vital in retrieving the best papers for them. Therefore, embedding the proposed model in the scholarly recommending systems can be fruitful in recommending process to analyze the users' needs in long-term intervals.

Appendix A: Data Codebook

Variable	Code	State	Values	Variable	Code	State	Values	Variable	Code	State	Values
	RO1	PhD	0-1		SE1	Semester First	0-1		SI1	UKW-PTI Similarity	0-1
	RO2	Master	0-1		SE2	Semester Second	0-1		SI2	UKW-PAB Similarity	0-1
Role	RO3	Post-Doc	0-1	Time	SE3	Semester Third	0-1		SI3	UKW-PKW Similarity	0-1
	RO4	Faculty Member	0-1		SE4	Semester Fourth	0-1		SI4	URI-PAP Similarity	0-1
Pre-knowledge	PK1	Basic	0-1		SE5	Semester Above	0-1		SI5	UJO-PAP Similarity	0-1
	PK2	Advanced	0-1	Search Time	ST1	More than 5 times	0-1	Criteria	PY1	Publication in current Year	0-1
	LS1	Visual	0-1	Number of Search		Less than 5 times	0-1		SU1	Seen by User	0-1
Learning Style	LS2	Verbal	0-1	Modification	SM1	More than 5 times	0-1		RU1	Read by User	0-1
	LS3	Physical	0-1			Less than 5 times	0-1		CM1	Citation Number	0-1
	TA1	Thesis writing	0-1	Advanced Search	AS1	Yes	0-1		AR1	Author Reputation	0-1
	TA2	Paper writing	0-1			No	0-1		JR1	Journal Reputation	0-1
Task	TA3	Course taking	0-1		NO1	Novelty	0-1		PR1	Publisher Reputation	0-1
	TA4	Topic finding	0-1	Paper quality	PO1	Popularity	0-1		DN1	Download Number	0-1
	TA5	Course teaching	0-1		AC1	Accuracy	0-1		PT1	Conference	0-1
					DI1	Diversity	0-1		PT2	Journal	0-1
									PT3	Review	0-1

References

1. Champiri, Z.D., Shahamiri, S.R., Salim, S.S.B.: A systematic review of scholar context-aware recommender systems. Exp. Syst. Appl. **42**(3), 1743–1758 (2015)
2. Berkovsky, S., et al.: Evaluation of user model effectiveness by simulation. In: Workshop on Personalization-Enhanced Access to Cultural Heritage, International Conference on User Modeling, Corfu (2007)
3. Adomavicius, G., Jannach, D.: Preface to the special issue on context-aware recommender systems. User Model. User Adap. Inter. **24**(1–2), 1–5 (2013). https://doi.org/10.1007/s11257-013-9139-2
4. Beel, J., Breitinger, C., Langer, S., Lommatzsch, A., Gipp, B.: Towards reproducibility in recommender-systems research. User Model. User Adap. Inter. **26**(1), 69–101 (2016). https://doi.org/10.1007/s11257-016-9174-x
5. Dehghani, Z., Afshar, E., Jamali, H.R., Nematbakhsh, M.A.: A multi-layer contextual model for recommender systems in digital libraries. Aslib Proc. **63**(6), 555–569 (2011). https://doi.org/10.1108/00012531111187216
6. Dehghani Champiri, Z., Asemi, A., Siti Salwah Binti, S.: Meta-analysis of evaluation methods and metrics used in context-aware scholarly recommender systems. Knowl. Inf. Syst. **61**(2), 1147–1178 (2019). https://doi.org/10.1007/s10115-018-1324-5
7. McNee, S.M., Riedl, J., Konstan, J.A.: Being accurate is not enough: how accuracy metrics have hurt recommender systems. In: CHI'06 Extended Abstracts on Human Factors in Computing Systems. ACM (2006)
8. Konstan, J.A., Riedl, J.: Recommender systems: from algorithms to user experience. User Model. User-Adap. Inter. **22**(1), 101–123 (2012)
9. Nguyen, T.: Enhancing User Experience With Recommender Systems Beyond Prediction Accuracies. Ph.D. Dissertation, The University of Minnesota (2016)
10. Knijnenburg, B.P., et al.: Explaining the user experience of recommender systems. User Model. User Adap. Inter. **22**(4–5), 441–504 (2012)

11. Baltrunas, L., et al.: Context relevance assessment and exploitation in mobile recommender systems. Pers. Ubiquit. Comput. **16**(5), 507–526 (2012)
12. Adomavicius, G., Tuzhilin, A.: Context-aware recommender systems. In: Ricci, F., Rokach, L., Shapira, B. (eds.) Recommender Systems Handbook, pp. 217–253. Springer, Boston (2015). https://doi.org/10.1007/978-1-4899-7637-6_6
13. Yujie, Z., Licai, W.: Some challenges for context-aware recommender systems. In: 2010 5th International Conference on Computer Science and Education (ICCSE). IEEE (2010)
14. Martín, E., Haya, P.A., Carro, R.M. (eds.): User Modeling and Adaptation for Daily Routines: Providing Assistance to People with Special Needs. Springer, London (2013)
15. Papatheocharous, E., Belk, M., Germanakos, P., Samaras, G.: Towards implicit user modeling based on artificial intelligence, cognitive styles and web interaction data. Int. J. Artif. Intell. Tools **23**(02), 1440009 (2014). https://doi.org/10.1142/S0218213014400090
16. Pearl, J.: Bayesian networks: a model of self-activated memory for evidential reasoning. In: 1985 Proceedings of the 7th Conference of the Cognitive Science Society (1985)
17. Hassan, H.A.M.: Personalized research paper recommendation using deep learning. In: 2017 Proceedings of the 25th Conference on User Modeling, Adaptation and Personalization. ACM (2017)
18. Hong, J.-Y., Suh, E.-H., Kim, S.-J.: Context-aware systems: a literature review and classification. Exp. Syst. Appl. **36**(4), 8509–8522 (2009)
19. Brown, P.J., Bovey, J.D., Chen, X.: Context-aware applications: from the laboratory to the marketplace. Pers. Commun. **4**(5), 58–64 (1997)
20. Dey, A.K.: Understanding and using context. Pers. Ubiquit. Comput. **5**(1), 4–7 (2001)
21. Adomavicius, G., Tuzhilin, A.: Toward the next generation of recommender systems: a survey of the state-of-the-art and possible extensions. IEEE Trans. Knowl. Data Eng. **17**(6), 734–749 (2005)
22. Panniello, U., Gorgoglione, M., Tuzhilin, A.: In CARSWe Trust: How Context-Aware Recommendations Affect Customers' Trust and Other Business Performance Measures of Recommender Systems (2015)
23. Berkovsky, S., Kuflik, T., Ricci, F.: Mediation of user models for enhanced personalization in recommender systems. User Model. User-Adap. Inter. **18**(3), 245–286 (2008)
24. Kobsa, A.: Generic user modeling systems. User Model. User-Adap. Inter. **11**(1), 49–63 (2001)
25. Beel, J., et al.: Paper recommender systems: a literature survey. Int. J. Digit. Libr. **17**(4), 305–338 (2016)
26. Hariri, N., Mobasher, B., Burke, R.: Context adaptation in interactive recommender systems. In: Proceedings of the 8th ACM Conference on Recommender systems. ACM (2014)
27. Ji Yan, W., Le Chan, W.: The study of user model of personalized recommendation system based on linked course data. Appl. Mech. Mater. **519–520**, 1609–1612 (2014)
28. Norman, D.: The Design of Everyday Things: Revised and Expanded Edition. Basic Books (2013)
29. Hassenzahl, M., Tractinsky, N.: User experience-a research agenda. Behav. Inf. Technol. **25**(2), 91–97 (2006)
30. McCarthy, J., Wright, P.: Technology as experience. Interactions **11**(5), 42–43 (2004)
31. Law, E.L.-C., et al.: Understanding, scoping and defining user experience: a survey approach. In: Proceedings of the SIGCHI Conference on Human Factors in Computing Systems. ACM (2009)
32. Kraft, C.: User Experience Innovation. Apress, Berkeley (2012)
33. Yuan, J., et al.: When to recommend what? A study on the role of contextual factors in IP-based TV services. In: MindTheGap@ iConference (2014)
34. Baltrunas, L.: Exploiting contextual information in recommender systems. In: Proceedings of the 2008 ACM Conference on Recommender Systems. ACM (2008)

35. Adomavicius, G., Tuzhilin, A.: Context-aware recommender systems. In: Ricci, F., Rokach, L., Shapira, B., Kantor, P.B. (eds.) Recommender Systems Handbook, pp. 217–253. Springer (2011). https://doi.org/10.1007/978-0-387-85820-3_7

36. Heckerman, D., Mamdani, A., Wellman, M.P.: Real-world applications of Bayesian networks. Commun. ACM **38**(3), 24–26 (1995)

37. Neapolitan, R.E.: Learning Bayesian Networks, vol. 38. Pearson Prentice Hall Upper Saddle River (2004)

38. Rim, R., Amin, M., Adel, M.: Bayesian networks for user modeling: predicting the user's preferences. In: 2013 13th International Conference on Hybrid Intelligent Systems (HIS). IEEE (2013)

39. Zukerman, I., Albrecht, D.W.: Predictive statistical models for user modeling. User Model. User Adap. Inter. **11**(1–2), 5–18 (2001)

40. Korb, K.B., Nicholson, A.E.: Bayesian Artificial Intelligence. Chapman & Hall/CRC, Florida (2003)

41. Mahjoub, M.A., Kalti, K.: Software comparison dealing with Bayesian networks. In: Liu, D., Zhang, H., Polycarpou, M., Alippi, C., He, H. (eds.) ISNN 2011. LNCS, vol. 6677, pp. 168–177. Springer, Heidelberg (2011). https://doi.org/10.1007/978-3-642-21111-9_19

42. Darwiche, A.: Modeling and Reasoning with Bayesian Networks. Cambridge University Press, Cambridge (2009)

43. Margaritis, D.: Learning Bayesian Network Model Structure from Data. School of Computer Science, Carnegie-Mellon University, Pittsburgh (2003)

44. Tibshirani, R., et al.: An Introduction to Statistical Learning-with Applications in R. Springer, New York (2013). https://doi.org/10.1007/978-1-4614-7138-7

45. Flores, M.J., et al.: Incorporating expert knowledge when learning Bayesian network structure: a medical case study. Artif. Intell. Med. **53**(3), 181–204 (2011)

46. Albert, J.: Bayesian Computation with R. Springer, New York (2009). https://doi.org/10.1007/978-0-387-92298-0

47. Amirkhani, H., et al.: Exploiting experts' knowledge for structure learning of Bayesian networks. IEEE Trans. Pattern Anal. Mach. Intell. **39**, 2154–2170 (2016)

48. Guo, S.: Bayesian Recommender Systems: Models and Algorithms. Australian National University (2011)

49. Scutari, M., Denis, J.-B.: Bayesian Networks: with Examples in R. Chapman and Hall/CRC (2014)

50. Mueller, J., Thyagarajan, A.: Siamese recurrent architectures for learning sentence similarity. In: AAAI (2016)

51. Portugal, I., Alencar, P., Cowan, D.: The use of machine learning algorithms in recommender systems: a systematic review. arXiv preprint arXiv:1511.05263 (2015)

52. Ericson, K., Pallickara, S.: On the performance of high dimensional data clustering and classification algorithms. Futur. Gener. Comput. Syst. **29**(4), 1024–1034 (2013)

53. Felden, C., Chamoni, P.: Recommender systems based on an active data warehouse with text documents. In: 2007 40th Annual Hawaii International Conference on System Sciences, HICSS 2007. IEEE (2007)

54. Gorodetsky, V., Samoylov, V., Serebryakov, S.: Ontology–based context–dependent personalization technology. In: 2010 IEEE/WIC/ACM International Conference on Web Intelligence and Intelligent Agent Technology (WI-IAT). IEEE (2010)

55. Lucas, J.P., Segrera, S., Moreno, M.N.: Making use of associative classifiers in order to alleviate typical drawbacks in recommender systems. Exp. Syst. Appl. **39**(1), 1273–1283 (2012)

56. Marović, M., et al.: Automatic movie ratings prediction using machine learning. In: 2011 Proceedings of the 34th International Convention MIPRO. IEEE (2011)

57. Šerić, L., Jukić, M., Braović, M.: Intelligent traffic recommender system. In: 2013 36th International Convention on Information & Communication Technology Electronics & Microelectronics (MIPRO). IEEE (2013)

58. Wang, Y., Chan, S.C.-F., Ngai, G.: Applicability of demographic recommender system to tourist attractions: a case study on trip advisor. In: Proceedings of the 2012 IEEE/WIC/ACM International Joint Conferences on Web Intelligence and Intelligent Agent Technology, vol. 03. IEEE Computer Society (2012)

59. Ono, C., Kurokawa, M., Motomura, Y., Asoh, H.: A context-aware movie preference model using a Bayesian network for recommendation and promotion. In: Conati, C., McCoy, K., Paliouras, G. (eds.) UM 2007. LNCS (LNAI), vol. 4511, pp. 247–257. Springer, Heidelberg (2007). https://doi.org/10.1007/978-3-540-73078-1_28

60. Codina Busquet, V., Ceccaroni, L.: Exploiting distributional semantics for content-based and context-aware recommendation (2014)

61. Fischer, G.: User modeling in human–computer interaction. User Model. User Adap. Inter. 11(1–2), 65–86 (2001)

62. Long, B., et al.: Active learning for ranking through expected loss optimization. IEEE Trans. Knowl. Data Eng. 27(5), 1180–1191 (2015)

63. McCay-Peet, L., Toms, E.: Measuring the dimensions of serendipity in digital environments. Inf. Res. Int. Electron. J. 16(3), n3 (2011)

64. Kotkov, D., Veijalainen, J., Wang, S.: Challenges of serendipity in recommender systems. In: Proceedings of the 12th International Conference on Web Information Systems and Technologies, WEBIST 2016, vol. 2. SCITEPRESS (2016). ISBN 978-989-758-186-1

65. Kotkov, D., Wang, S., Veijalainen, J.: A survey of serendipity in recommender systems. Knowl. Based Syst. 111, 180–192 (2016)

66. Champiri, Z.D.: A contextual Bayesian user experience model for scholarly recommender systems. Doctoral dissertation, University of Malaya (2019)

67. Hurley, N.J.: Towards diverse recommendation. In: Workshop on Novelty and Diversity in Recommender Systems, DiveRS 2011. Citeseer (2011)

68. Park, H.-S., Yoo, J.-O., Cho, S.-B.: A context-aware music recommendation system using fuzzy bayesian networks with utility theory. In: Wang, L., Jiao, L., Shi, G., Li, X., Liu, J. (eds.) Fuzzy Systems and Knowledge Discovery, pp. 970–979. Springer, Berlin, Heidelberg (2006). https://doi.org/10.1007/11881599_121

69. Russell, S.J., Norvig, P.: Artificial Intelligence: A Modern Approach. Pearson Education Limited, Malaysia (2016)

70. Hsu, C.-C., Sandford, B.A.: The Delphi technique: making sense of consensus. Pract. Assess. Res. Eval. 12(10), 1–8 (2007)

71. Kontio, J., Lehtola, L., Bragge, J.: Using the focus group method in software engineering: obtaining practitioner and user experiences. In: Proceedings of the 2004 International Symposium on Empirical Software Engineering, ISESE 2004. IEEE (2004)

72. Fienen, M.N., Plant, N.G.: A cross-validation package driving Netica with Python. Environ. Model. Softw. 63, 14–23 (2015)

73. Seixas, F.L., et al.: A Bayesian network decision model for supporting the diagnosis of dementia, Alzheimer's disease and mild cognitive impairment. Comput. Biol. Med. 51, 140–158 (2014)

74. Kuenzer, A., et al.: An empirical study of dynamic Bayesian networks for user modeling. In: Proceedings of the UM 2001 Workshop on Machine Learning for User Modeling (2001)

75. Marcot, B.G.: Metrics for evaluating performance and uncertainty of Bayesian network models. Ecol. Model. 230, 50–62 (2012). https://doi.org/10.1016/j.ecolmodel.2012.01.013

76. Chan, H., Darwiche, A.: Sensitivity analysis in Bayesian networks: from single to multiple parameters. In: Proceedings of the 20th Conference on Uncertainty in Artificial Intelligence. AUAI Press (2004)

77. Hansson, F., Sjökvist, S.: Modelling Expert Judgement into a Bayesian Belief Network. A Method For Consistent And Robust Determination Of Conditional Probability Tables (2013)

78. Schain, M., Schain, M.: Machine Learning Algorithms and Robustness. Universitat Tel-Aviv (2015)

79. Guyon, I., Elisseeff, A.: An introduction to variable and feature selection. J. Mach. Learn. Res. **3**, 1157–1182 (2003)

80. Sarwar, B., et al.: Application of Dimensionality Reduction in Recommender System-A Case Study. Department of Computer Science, University of Minnesota, Minneapolis (2000)

81. Ziegler, C.-N., et al.: Improving recommendation lists through topic diversification. In: Proceedings of the 14th International Conference on World Wide Web. ACM (2005)

82. Shirude, S.B., Kolhe, S.R.: Classification of library resources in recommender system using machine learning techniques. In: Mandal, J.K., Sinha, D. (eds.) CSI 2018. CCIS, vol. 836, pp. 661–673. Springer, Singapore (2018). https://doi.org/10.1007/978-981-13-1343-1_54

83. Easwaran, K.: Bayesianism I: introduction and arguments in favor. Philos. Compass **6**(5), 312–320 (2011)

84. Geisler, W.S.: Visual perception and the statistical properties of natural scenes. Annu. Rev. Psychol. **59**, 167–192 (2008)

85. Champiri, Z.D., Fisher, B., Freund, L.: rScholar: an interactive contextual user interface to enhance UX of scholarly recommender systems. In: Stephanidis, C., Marcus, A., Rosenzweig, E., Rau, P.-L., Moallem, A., Rauterberg, M. (eds.) HCII 2020. LNCS, vol. 12423, pp. 662–686. Springer, Cham (2020). https://doi.org/10.1007/978-3-030-60114-0_43

86. Champiri, Z.D., Fisher, B., Kiong, L.C., Danaee, M.: How contextual data influences user experience with scholarly recommender systems: an empirical framework. In: Stephanidis, C., Marcus, A., Rosenzweig, E., Rau, P.-L., Moallem, A., Rauterberg, M. (eds.) HCI International 2020 - Late Breaking Papers: User Experience Design and Case Studies: 22nd HCI International Conference, HCII 2020, Copenhagen, Denmark, July 19–24, 2020, Proceedings, pp. 635–661. Springer, Cham (2020). https://doi.org/10.1007/978-3-030-60114-0_42

From a Workshop to a Framework for Human-Centered Artificial Intelligence

Helmut Degen[1](✉) and Stavroula Ntoa[2] (iD)

[1] Siemens Corporation, Technology 755 College Road East, Princeton, NJ 08540, USA
helmut.degen@siemens.com
[2] Foundation for Research and Technology Hellas, Institute of Computer Science,
Vassilika Vouton, N. Plastira 100, 700 13 Heraklion, Greece
stant@ics.forth.gr

Abstract. Due to advances in technology, the topic "AI in HCI" is a growing research area with contributions from several research communities. To better understand the current and potentially future research activities, a workshop "AI in HCI" took place as part of the first AI in HCI conference. The intent of the workshop was twofold: to explore the research landscape of AI in HCI with current and potentially future research topics, as well as building the foundation for a community of international researchers in this area. The workshop was facilitated as a four and a half hour, remote, interactive workshop. 20 researchers attended the workshop. During the workshop, the participants brainstormed individual research topics which were grouped into 10 research categories. The participants assigned votes to the ten categories. The categories "Trust", "Ethical AI" and "Human-Centered AI" received most votes. Afterwards the participants underwent a deep dive into the three top scored categories, elaborating per category on research questions, research approaches, research challenges, and beneficiaries. During the workshop on July 24 2020, only the topic "Trust" was elaborated. Two follow-up sessions elaborated the other two categories. This paper presents structure and results of the workshops, including the follow-up sessions and relates the results to other publications in this area. Moreover, further elaborating on the workshop results, this paper proposes a framework for Human-Centered AI for building trustworthy AI-enabled systems.

Keywords: HCI · AI · Workshop · Co-creation · Trust · Ethical AI · Human-centered AI

1 Introduction

The topic of Artificial Intelligence (AI) in Human–Computer Interaction (HCI) is a growing research topic. Growing does not mean that the topic is new. The opportunities and concerns associated with AI-based technology are rather old. The "Dartmouth Summer Research Project on Artificial Intelligence" is considered the first event about the topic of Artificial Intelligence (AI) in 1956. It lasted two months and was originally planned for eleven attendees [1, 2]. Some of the concerns about transparency, explainability, and

© Springer Nature Switzerland AG 2021
H. Degen and S. Ntoa (Eds.): HCII 2021, LNAI 12797, pp. 166–184, 2021.
https://doi.org/10.1007/978-3-030-77772-2_11

trust (to mention some of the key words from today's research agenda) are not that new, either. Weizenbaum [3] raised ethical question about the use of computer and artificial intelligence and made a distinction between making "decisions" (by the computer) and making "choices" (by a human user). It seems that such a distinction is still valid today.

Looking at AI research, development, and applications, one significant difference between the twentieth century and the twenty-first century is the adoption. While AI was a niche technology in the twentieth century, it can be reasonably predicted that it becomes a mass technology in the twenty-first century [4]. Because of that adoption trend, there is a revival and growth in the research of AI-based technology. Motivated by this revived interest, the authors have designed and organized a workshop, aiming to involve researchers from both fields, in order to identify topics that are of imminent interest to the research community. In this sense, the workshop organizers pursued to examine with a fresh look something which is already under its way – in research, development and/or use.

The workshop was held in the context of the 1st International Conference on Artificial Intelligence in HCI.[1] The Conference, as well as the workshop has picked the title "AI in HCI" on purpose. Today, we are aware of two main perspectives: "HCI for AI" and "AI for HCI". "HCI for AI" deals with the human–computer interaction of systems which use AI-based technologies. "AI for HCI" deals with the use of AI-based technology for the generation of human interaction system. The generation can include the human interaction system itself, as well as artifacts which are needed to create such a system, like user models, task models, goal models, dialog models, user interface design generation, verification, validation and others. Sibling disciplines like "generative design" use AI-based technologies already for the creation and optimization of hardware products, for instance the first commercial chair designed by generative design [5].

There is an overlap between the two groups. In the mid or long-term future, "AI for HCI" techniques and practices can be applied to generate human–computer interaction systems which are used for the interaction of AI-based technical enablers ("HCI for AI"). Such techniques may implement needed capabilities in the AI-based enabler to comply with a future ethical standard for the use of AI-based technical enablers. Such "AI for HCI" techniques themselves need to comply with standards which require the application of "HCI for AI" principles. Second and third order combinations of "HCI for AI" and "AI for HCI" are thinkable.

---

[1] http://2020.hci.international/ai-hci.html.

There have been a variety of other workshops which facilitate the discussion on this topic, often as part of a conference.[2] In this sense, the workshop itself was not new as a concept. However, it aimed to bring together researchers from both fields, in order to pursuit a broad exploration of the topics that researchers in the field are interested in, work on, or believe that should be addressed in the near future. The workshop involved activities borrowed from co-creation workshops, which have been reported to enhance innovation, unlock new sources of competitive advantage, and strengthen valued relationships [6]. A key objective for conducting the workshop was also to create map for researchers in the field of AI in HCI, by exploring the field through the structured activities of the workshop.

The remainder of this paper is structured as follows: Sect. 2 describes the workshop scope and format; Sect. 3 presents the workshop results for each activity, and namely expectations and concerns, all the identified AI in HCI research topics, as well as the analysis of the three most important topics as these were voted by participants; finally, Sect. 4 concludes this paper by discussing findings and highlighting future research topics in the field.

2   The Workshop

2.1   Workshop Scope

In this workshop we attempted to bring together researchers and professionals interested in "AI in HCI" to explore the research landscape, and not to zoom in on a specific topic. We also wanted to understand the priorities of the identified research topics, and how to address them. The discussion and the result should not only help to inspire fellow researchers. It also intended to identify blind spots and to inform research roadmaps. Furthermore, we planned to use the workshop to build a community around the topic "AI in HCI".

The conference "AI in HCI" (Artificial Intelligence in Human–Computer Interaction) took place first time in July 19–24, 2020. The AI in HCI conference is affiliated with the HCI International conference. The AI in HCI conference was originally planned as an in-person event. Due to the Covid-19 epidemic, the conference with all of its sessions took place as a virtual event, including the workshop "AI in HCI".

---

[2] AAAI Conference on Artificial Intelligence (AAAI), International Conference on Knowledge Discovery and Data Mining (ACM CDD), International Conference on Human Factors in Computing Systems (ACM CHI), Conference on Computer Vision and Pattern Recognition (IEEE CVPR), International Conference on Designing Interactive Systems (ACM DIS), European Conference on Computer Vision (ECCV), ACM/IEEE International Conference on Human Robot Interaction (HRI), International Conference on Multimodal Interaction (ACM ICMI), Conference on Computer Vision (IEEE ICCV), Neural information processing systems (NeurIPS), International Conference on Machine Learning (ICML), International Conference on Artificial Intelligence (IJCAI), International Conference on Intelligent Robots and Systems (IEEE IROS), International Conference on Intelligent User Interfaces (ACM IUI), ACM Symposium on User Interface Software and Technology (ACM UIST), International Conference on User Modeling, Adaptation and Personalization (ACM UMAP).

2.2 Workshop Format and Participants

The workshop was announced as an event on the HCI International 2020 conference website. Participation in the workshop was by invitation, to ensure that the number of participants would be manageable for a virtual event and to foster an efficient and effective online collaboration. Researchers who expressed interest in the workshop received an invitation and were asked to fill-in an online form, indicating their current research interests in the area of AI in HCI, as well as their top five topics in the field. Individuals who filled-in the form were invited to the workshop, which took place as a virtual event. The workshop duration was 4.5 h.

A total of 20 researchers from nine countries participated in the workshop, working in universities, research institutions, or industry. The workshop involved eight activities, and in particular:

1. Welcome and objectives
2. Self-introductions
3. Workshop expectations and concerns
4. Summary of research topics in the area of AI in HCI proposed through the online form and exploration of additional research topics
5. Ranking of research topics
6. Elaboration on high ranked research topic categories
7. Summary and next steps
8. Feedback about the workshop itself.

Welcome and Objectives
During this activity, participants were welcomed to the workshop. Before any discussion about the objectives of the workshop and the agenda, an ice-breaker activity was employed, aiming to make participants feel comfortable with the teleconferencing platform settings (e.g. enabling and disabling their microphone and camera, adjusting audio volume), as well as to stimulate discussions. In particular, upon joining, each participant was asked to give a weather report from their location.

Self-Introductions
To further stimulate discussions and information sharing, participants were asked to present to the group their favorite food and also to briefly introduce themselves. In this respect, one slide per participant had been prepared with the participant's research interests and top five in the area of AI in HCI, as provided through the online form.

Workshop Expectations and Concerns
This activity aimed at identifying what the participants expected from the workshop, which were their own objectives for joining, and to discuss any concerns, so as to ensure that they would be addressed as much as possible during the workshop.

AI in HCI Research Topics
This activity aimed to identify which topics were considered as generally important in the field, or of particular interest for workshop participants. All the topics that emerged

were clustered into categories, which were ranked through voting, so as to prioritize them and guide the following workshop activities.

Elaboration on High Ranked Research Topic Categories
For each one of the three highest ranked categories, this activity intended to expand the following: (i) main research questions, (ii) research approaches that should be followed, (iii) challenges entailed and points of concern, and (iv) beneficiaries. For each one of the aspects to be discussed, participants were given some time to think and write their ideas and perspectives in the chat area; they were however instructed to not share their input with the group until the workshop facilitator asked them to, in order to avoid group bias. As a result, on the facilitator's command, everyone shared their ideas at the same time. Then, the facilitator discussed each participant input recorded in the chat, in a first-come-first-served basis, while the other facilitator typed the pertinent input in the corresponding roadmap section.

During the workshop, due to time restrictions only the highest ranked topic category was elaborated. Given the overall positive feedback of participants the other two topics were explored in two follow up workshop sessions, each dedicated to elaborating one single category.

Summary and Next Steps
This part of the workshop aimed to summarize the results of the workshop and give participants the chance to comment on findings or add any additional input.

Feedback About the Workshop Itself
Before concluding the workshops, participants were asked to evaluate it, by identifying what they liked (pros) and what they did not like (cons). They were instructed to type their input in the chat area, without however sending it. When everyone was ready, the workshop facilitator asked them to send their input simultaneously, so as to avoid group bias.

Putting it all Together
The results from each one of the aforementioned activities were planned to be used for creating the map for researchers in the field of AI in HCI. Figure 1 illustrates the empty map, before the workshop, adopting the metaphor of a land to be explored. Input to the map was planned to be added as follows: expectations identified during the third activity would be identified as the wind blowing and assisting our hot air balloon to transport us through this land, concerns expressed during the third activity would be marked as sandbags dragging down the hot air balloon, each one of the three top-voted topics would constitute a target location on the map, while the remaining topics would be added in our "parking lot", as topics for future exploration.

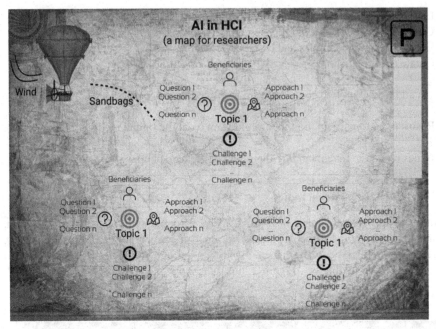

Fig. 1. Map for researchers (template)

It is noted that the visualization method is novel and was created by the authors, having however been inspired by various other co-creation methods, such as *hopes and concerns*,[3] *hot air balloon*,[4] and *parking lot*.[5]

3 Results

This section summarizes the results from each one of the activities that were carried out.

3.1 Workshop Expectations and Concerns

Using a shared whiteboard, divided in two columns, participants were asked to type in their expectations from the workshop and point out their concerns. Prior to using the shared whiteboard, participants were introduced to the functionality and allowed to experiment.

In brief, participants indicated that they expected from the workshop to:

- Identify some important research issues
- Produce new ideas
- Understand and learn new things

---

[3] https://www.funretrospectives.com/hopes-and-concerns/.

[4] https://toolbox.hyperisland.com/hot-air-balloon.

[5] https://toolbox.hyperisland.com/parking-lot.

- Learn about opinions and ideas in HCI/AI research
- Gain and understanding of other researchers' concerns in relation to AI in HCI
- Exchange views and ideas and collaborate with AI & HCI researchers
- Create new contacts

In addition, only two concerns were identified, and in particular:

- Restrictions to participation that might stem from the online mode of the workshop, and
- Issues related to understanding the scope of the workshop

3.2   AI in HCI Research Topics

Using the top five research topics provided by participants through the online form, a shared whiteboard had been prepared prior to the workshop. During this activity, participants were asked to group them in clusters, through group discussion coordinated by the workshop facilitators. Figure 2 illustrates the whiteboard, as it resulted through the group discussion.

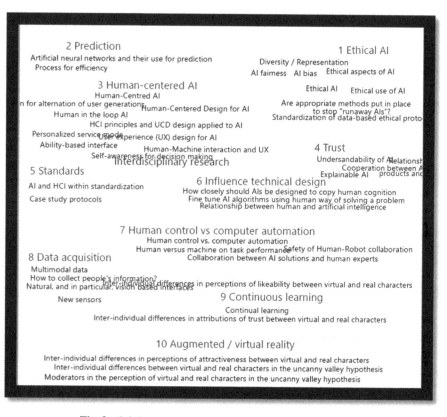

Fig. 2. Original workshop whiteboard with topics' clustering

In particular, this workshop activity resulted in the identification of ten research topic categories (clusters), which are important or of interest in the field of AI in HCI. In particular, the identified categories were as follows (reported in the order that they were discussed): ethical AI, prediction, human-centered AI, trust, standards, influence technical design, human control vs. computer automation, data acquisition, continuous learning, and augmented/virtual reality. For each one of the topic categories, a number of individual research topics was identified, as presented in detail in Table 1.

Then, the research topic categories, were put into a poll which was presented to participants' screen. Participants were asked to select up to three categories which they considered as more important. Sixteen votes were casted. The ranking of results, as well as the number of votes acquired is presented in Table 1.

Table 1. List of AI in HCI research topics and categories, ordered by rank

Rank	Topic category	Topics/Keywords	Votes
1.	Trust	• Understandability of AI • Cooperation between AIs • Explainable AI • Products and AI	8/16 (50%)
2.	Ethical AI	• Diversity and user representation • AI fairness • AI bias • Ethical aspects of AI • Ethical AI • Ethical use of AI • Are appropriate methods put in place to stop "runaway AIs"? • Standardization of data-based ethical protocols	6/16 (38%)
3.	Human-Centered AI	• Human-centered AI • Migration for alternation of user generations • Human-centered design for AI • Human in the loop AI • HCI principles and UCD design applied to AI • Personalized service mode • User experience (UX) design for AI • Ability-based interface • Human-Machine interface and UX • Self-awareness for decision making • Interdisciplinary research	6/16 (38%)
4.	Prediction	• Artificial neural networks and their use for prediction • Process for efficiency	3/16 (19%)

(continued)

Table 1. (*continued*)

Rank	Topic category	Topics/Keywords	Votes
5.	Influence technical design	• How closely should AI be designed to copy human cognition • Fine tune AI algorithms using human ways of solving a problem • Relationship between human and artificial intelligence	3/16 (19%)
6.	Human control vs. computer automation	• Human control versus computer automation • Human versus machine on task performance • Safety of human-robot collaboration • Collaboration between AI solutions and human experts	3/16 (19%)
7.	Data acquisition	• Multimodal data • How to collect people's information? • Inter-individual differences in perceptions of likeability between virtual and real characters • Natural and vision based interfaces for data acquisition • New sensors	3/16 (19%)
8.	Continuous learning	• Continuous learning	1/16 (6%)
9.	Augmented/virtual reality	• Inter-individual differences in attributions of trust between virtual and real characters • Inter-individual differences in perceptions of attractiveness between virtual and real characters • Inter-individual differences in virtual and real characters in the uncanny valley hypothesis • Moderators in the perception of virtual and real characters in the uncanny valley hypothesis	1/16 (6%)
10.	Standards	• AI and HCI within standardization • Case study protocols	1/16 (6%)

3.3 AI and Trust

For the category of AI and trust seven research questions were identified and the pertinent research approaches to explore them, as presented in Table 2.

Having already elaborated on the research questions and research approaches, the group identified five main challenges in this pursuit, and in particular:

- C.1: Diversity of end users and context
- C.2: Influence of risk on trust (how does risk influence trust?)

Table 2. AI and trust: research questions and approaches

Research question		Research approaches
RQ.1	How does trust influence usability?	• User studies
RQ.2	What influences trust?	• User studies • Validated questionnaires
RQ.3	What does not influence trust?	• User studies • Validated questionnaires
RQ.4	How to measure trust?	• Development of instruments to measure trust
RQ.5	How to present explanations to be understandable and trustable?	• Design exploration • Collaboration with AI architects and software engineers
RQ.6	Explainability: in case of system failures: how can the system identify the reasons and directions for improvement?	• Design exploration • Collaboration with AI architects and software engineers
RQ.7	What is the role of trust in user acceptance?	• User studies • Validated questionnaires

- C.3: Communication between the research fields involved (namely AI and HCI)
- C.4: Scalability of existing methods
- C.5: Privacy vs. performance

Finally, four main beneficiary groups were identified and namely:

- B.1 Users
- B.2: Stakeholders
- B.3: Designers
- B.4: Engineers (development team)

3.4 Ethical AI

For the category of ethical AI, eleven research questions were identified as well as the research approaches that should be used to pursuit them, as presented in Table 3. As noted in Sect. 2.2, this workshop was conducted on a separate session. Having already been familiarized with the process, more research questions were identified, which led to further clustering them into groups.

Four main challenges were identified by the group in the context of research endeavors in ethical AI, and namely:

- C.1: Understand bias in data

Table 3. Ethical AI: research questions and approaches

Group 1 - Ethical rules

Research questions	Research approaches
RQ.1: Can we create a legal framework for ethical AI so that AI ethics conform to human perceivable rules? RQ.2: How to apply ethical rules on the AI implementation pipelines? RQ 3: What are ethical rules? Are there global vs regional ethical rules? RQ 4: How to identify deviations and identify which are acceptable and which not? (it would be good if the machine had a consciousness mechanism built-in to identify deviations)	• Action design research • Recognized committee to define ethical rules • Multidisciplinary: participation of lawyers | ethicists | philosophers

Group 2 - Skills

Research questions	Research approaches
RQ.5: What are the main ethical skills/competencies needed by AI designers, experts and professionals? RQ.6: How can we create an Ethical Skills Framework for AI to extend the existing SFIA7/8?	• Action design research • Introduce ethical discourse lessons in universities to train software engineers and designers

Group 3 - Traceability and training

Research questions	Research approaches
RQ.7: How to foster traceability of ethical decision making (back to their roots)? RQ.8: How to fit human values in machine perception? How can AI know what is good or wrong since it lacks human skills such as empathy? RQ.9: AI bias and training data: a lot of time, the bias was introduced by training data, which reflects human bias. If AI bias comes from human bias, who is responsible? (fixing & liability)	• Cause-effect analyses to trace down the root cause • Apply Human Centered Design (HCD) approaches.

Group 4 - User participation

Research questions	Research approaches
RQ.10: How can humans participate in AI decision making on ethics, giving their feedback on decisions made?	• Develop frameworks, methods, and tools for HCD in AI • Classical HCD with dif-ferent types of human in-volvement/comparison

Group 5 – Implemented ethic rules

Research questions	Research approaches
RQ.11: Machine autonomy and ethics: when the machine makes more of its own decisions, how do we emb ethics into the machine autonomy? (Peril of less human control | ethics of function allocation)	• Multidisciplinary teams, including software engi-neers, HCI researchers/practitioners, ethicists, etc. • Human in the loop ap-proaches

- C.2: Bias can be introduced not only in training but also during the entire lifecycle (using | evaluating) and how to identify it
- C.3: Beyond engineering: become heavily influenced by social movement and politics.
- C.4: Legitimacy aspects: Cognitive, normative, pragmatic, generic

Regarding beneficiaries, the group unanimously agreed that everyone will benefit from ethical AI, identifying two coarse categories: the entire society, and professional organizations.

3.5 Human-Centered AI

For the category of Human-Centered AI (HCAI), the workshop resulted in identifying fifteen core research questions, organized in three main groups, as summarized in Table 4.

Several challenges were identified with regard to HCAI, and namely:

There are no standard group of trustworthy characteristics.

- C.1: Collaboration of AI engineers/researchers and HCI/UX practitioners/researchers.
- C.2: AI community is less concerned about HCI.
- C.3: AI and HCI communities continue to follow parallel but separate pathways.
- C.4: Algorithm designers are not that interested in taking human rules as input.
- C.5: Definition of HCAI reference framework and competence model (skill areas and knowledge areas).
- C.6: Validation of new methods/frameworks/metrics requires time and effort - while AI systems are already being deployed.
- C.7: Sufficient and relevant application examples.

Beneficiaries, were again suggested to be the entire society and professional organizations.

3.6 Workshop Evaluation

Participants' workshop evaluation comments were mainly positive, regarding the collaboration and group dynamics, the workshop organization, as well as the overall discussion. These are quotes (some modified due to spelling mistypes):

- It was my first online co-creation meeting:) I think that despite not being in the same room we could collaborate quite well.
- I greatly enjoyed the workshop. I like the way you organized it. I regret we were not able to tackle more deeply the other topics and I wish we can do that in the future.
- Interesting discussion and nice experience, good collaboration, we needed more time.
- It was positive to meet researchers from various domains sharing their opinions, happy to hear that 'trust' is an important topic to us all.
- Getting to know each other and their interests of AI in HCI.
- I like the enthusiasm of the group on this topic.
- I liked the collaborative format.

Table 4. Human-centered AI: research questions and approaches

Group 1 - Process

Research questions	Research approaches
RQ.1: Can HCAI be used by system engineers to set system trustworthy characteristics and values? RQ.2: How can the AI research community converge with the HCI research community and vice versa, and identify common research fields? Can AI design follow a HCD procedure? RQ.3: How to effectively involve humans (as in HCD) to design or give feedback to something that they may not understand with regard to how it makes decisions? RQ.4: What is the difference between a HCAI-SDLC methodology and a traditional HC-SDLC methodology? RQ.5. When and how to involve the technology experts into the design process? RQ.6: Can we still use the same HCD process? RQ.7: How are/will all ISO related HCD standards get impacted by HCD AI issues?	• Specify a design frame-work that applies the HCD approach to the AI systems implementation. • Development of representative use cases. • Digital and physical ethnography. • Research of applied processes and lessons learned. • More interactive AI

Group 2 - Measurement and quality

Research questions	Research approaches
RQ.8: What dependencies exist between quality of technical capabilities (e.g. trained model) and user experience/usability qualities? RQ.9: What is the impact of AI on DevOps? RQ.10: How much does AI user trust influence the user experience working with an AI System? RQ.11: How can we measure UX in AI systems? RQ.12: How can we measure trustworthiness? RQ.13: What technical capabilities of AI technologies contribute to trustworthiness?	• Expand existing methods/frameworks to measure UX • Empirical research: user studies & usability tests • (technology & trustworthiness): experiments/usability tests

Group 3 - Collaboration with technology

Research questions	Research approaches
RQ.14: Human-centered XAI: How to make AI explainability more like human explanations? RQ.15: Combination of human traits with AI design: can we find AI architectures supporting human-like approaches?	• Algorithm and architecture design. • Digital and physical ethnography. • Research to see if HCI and Engineers can decide how HCD can work for both of them

- I liked the informal atmosphere, the opportunity to speak and share regardless of the level of each.
- I appreciated the heterogeneity of the participants.

Negative remarks mainly referred to technical restrictions and were the following:

- Wearing a headset for many hours is tiring, I miss the actual social interaction of physical meetings.
- Facing technical issues.
- We needed more time.

3.7 The AI in HCI Research Map

The synthesis of all participants' input resulted in the AI in HCI research map, visualizing the three main categories, as well as the research questions, research approaches and challenges for each one. The resulting map is presented in Fig. 3.

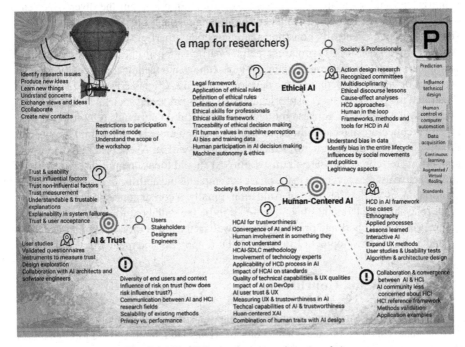

Fig. 3. AI in HCI map for researchers (results)

4 A Framework for Human-Centered AI

Aiming to consolidate findings and move forward to a classification of the research topics that emerged, further analysis was carried out. The elaboration on the outcomes of the workshop lead to the conclusion that the three topic categories are not only timely and in accordance with existing literature, but also they are tightly coupled together. In more detail, research issues identified in the context of ethical AI reflect those technical

qualities that will lead to a trustworthy AI system, while the methodological approach to pursuit them is the Human-Centered AI process. In a nutshell, the diagram of AI in HCI research topics presented in Fig. 4, provides a classification of the research topics into three main categories: those that mostly pertain to the technical AI system, those mostly relevant for the Human-Centered AI design process, and finally those that are important for users and stakeholders for operating and maintaining an AI-enabled system that is trustworthy.

In particular, as illustrated in Fig. 4, trust refers to the qualities that an AI-enabled interactive system should have so that it is operated and maintained by users and stakeholders, being useful, effective, efficient, satisfactory, safe, as well as fair and ethical. Such qualities – as they were identified through the workshop – refer to trustworthiness and usability, understandable and trustable explanations, explainability with regard to system failures, as well as trust concerning user privacy *vs* system performance. To achieve this, trust measurement efforts need to take into account the diversity of end users and contexts, providing validated questionnaires, and other instruments to measure trust, which eventually constitutes a major counterpart of the overall UX with an AI interactive system. Trust requires human participation in the AI decision making and spans several application domains. When seen from a technical perspective, all the activities of the infrastructure for the development of an AI-enabled system (as these are illustrated in Fig. 4) should consider legal issues, standards, and societal aspects, as well as ethics rules. Points of concern for developing ethical AI attributes refer to processes, system architecture and capabilities, as well as data and models used to train the system. Finally, seen from the design infrastructure perspective of an AI enabled interactive system, efforts to achieve ethical and trustworthy AI should focus on multidisciplinarity, as well as active user involvement in the phases of empathizing and discovery, definition and exploration, as well as design and evaluation. Finally, a cornerstone and common point in all relevant endeavors, has been identified to be the collaboration between HCI and engineering researchers and practitioners, as well as the integration of Human-Centered AI and Software Development Life Cycle (SDLC) process and methodology.

5   Discussion and Conclusions

This paper has presented the activities and outcomes from a workshop conducted with 20 researchers and practitioners in the fields of AI and HCI, aiming to identify the topics that matter most to researchers from both fields, rank them, and elaborate on research questions and approaches to address them, on challenges and beneficiaries for the highest ranked topic categories. Participants' expectations matched the aims and objectives of the workshop, mainly revolving around collaboration and community building, as well as learning and exploration of topics in the field.

Despite the online format of the workshop, the co-creation activities were successful and yielded rich results. In particular, a total of ten research categories were identified, while the following three topics were ranked as those that matter most: AI and trust, ethical AI, as well as human-centered AI. Overall, the categories identified by the workshop as the ones that matter most are aligned with recent concerns and efforts reported in literature.

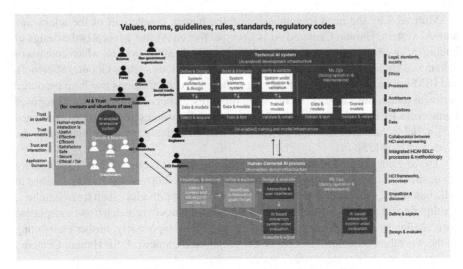

Fig. 4. Diagram of the human-centered AI framework

Explainable AI is one of the most discussed topics in the field of AI and trust, with several literature reviews and surveys published. The survey of Vilone and Longo [18] covers "the methods of explainability as a whole" (p. 2), and is based on Adadi and Berrada [18] and Guidotti et al. [26]. Arrieta et al. [9] undertake an analysis of literature about XAI. They conclude that the AI principles "fairness", "privacy", "accountability", "ethics", "transparency", and "security and safety" are needed to achieve a "responsible AI". Phillipps et al. [10] propose four AI principles: "Explanation", "Meaningful" (explanation is meaningful), "Explanation Accuracy", and "Knowledge Limits". There are domain specific research efforts, e.g. for healthcare [11] and mental health [12]. Mohseni et al. [13] identified three user roles and articulated role specific design goals and evaluation measures. The literature was mapped to the different goals and measures. Miller [14] explores the XAI topic from a social science perspective and highlights four findings: Explanations are contrastive (A instead B), explanations are selective (i.e. incomplete), probabilities probably don't matter, and explanations are social. He also mentioned that explanations are contextual (p. 3).

Relevant to the topic of trustworthy and explainable AI is that of ethical AI. The Organisation for Economic Co-operation and Development (OECD) has identified "five complementary values-based principles for the responsible stewardship of trustworthy AI" [15]. The principles have been adopted by some countries already. The European Parliament sponsored a study to investigate the ethical implications and moral questions about the development and implementation of artificial intelligence [16]. Another study elaborates how to transfer ethics to policies (van Wynsberghe 2020). Hagerty and Rubinov [17] summarize the result of a literature review or more than 800 academic publications. It concludes that significant regional differences exist, AI can exacerbate social inequalities and further actions is essential (p. 2, 3). A review identified and analyzed 84 global AI ethics guidelines found five ethical principles: transparency, justice and fairness, non-maleficence, responsibility and privacy [18].

Motivated by the need to emphasize on the human counterpart of the interaction with AI system, Human-Centered AI is a perspective on AI that advocates the design of intelligent systems with awareness that they are part of a larger system, which consists of human stakeholders [19]. AI research teams in corporations such as Google, Microsoft, and IBM have also focused on the design of human-AI interactions and have produced pertinent guidelines [20–22]. At the same time, researchers in the field have proposed frameworks for the systematization of the design of AI systems that are ethically aligned [23], reliable, safe and trustworthy [24], as well as actively involving humans in the AI-design loop [25].

The alignment of the identified topics with the current focus in AI in HCI literature is a positive indication regarding the validity of the workshop results. Having resulted in valid outcomes, a main contribution of this paper is that it provides a map for researchers, highlighting for each one of the three main identified themes of research (topic categories) open challenges and approaches to tackle them. More importantly, further elaborating on the workshop outcomes, this paper has proposed a framework for Human-Centered AI that will result in systems that are trustworthy.

Taking a bird's eye view, it seems that the controlled design for AI-based technologies, and the controlled design with AI-based technologies depends at least on a good understanding of the underlying AI-based technology (technology drives qualities of use). Mastering it and optimizing it for human use would be even better (qualities of use drive technology). Mastery requires a joint effort between the technology and the HCI community, to make AI-based technology human-centered, fair, and ethical, so that citizens can trust it. The mentioned research activities above as well as the workshop attempt to go a step into this direction. The drawn map may be used as a navigation tool for such current and future research efforts, while the proposed framework provides a tool to systematically approach Human-Centered AI activities.

Acknowledgements. The authors would like to thank the workshop participants for their valuable contributions (alphabetic order by first name): Al-Batool Al-Ghamdi (University of Jeddah, Saudi Arabia), Alice Baird (EIHW - University Augsburg, Germany), Brian Stanton (National Institute of Standards and Technology, USA), Christoph Brand (Siemens Technology, USA), Christof Budnik (Siemens Technology, USA), Esma Aïmeur (University of Montreal, Canada), Fatimah Alsayoud (Ryerson University, Canada), George Margetis (FORTH, Greece), Giuliana Vitiello (Università di Salerno, Italy), Jan Conrad (University of Applied Sciences Kaiserslautern, Germany), Kaveh Bazargan (UX24/7, United Kingdom), Margherita Antona (FORTH, Greece), Marianna Di Gregorio (Università di Salerno, Italy), Ming Qian (Pathfinders Translation and Interpretation Research, USA), Nouf Aj (University of Jeddah, Saudi Arabia), Roberto Vezzani (University of Modena and Reggio Emilia, Italy), Sebastian Korfmacher (Commission for Occupational Health and Safety and Standardization (KAN), Germany), Siyu Zhao (Siemens Technology, USA), Xiang-yuan Yan (Beijing University of Posts and Telecommunications, China), XiaoLing Li (Xi'an Jiaotong University, China).

References

1. McCarthy, J., Minsky, M.L., Rochester, N., Shannon, C.E.: A proposal for the dartmouth summer research project on artificial intelligence, august 31, 1955. AIMag. **27**, 12–12 (2006). https://doi.org/10.1609/aimag.v27i4.1904

2. Crevier, D.: AI: The Tumultuous History of the Search for Artificial Intelligence. Basic Books, New York City (1993)
3. Weizenbaum, J.: Computer Power and Human Reason: From Judgment to Calculation. Freeman Press, New York City (1977)
4. EU Commission: White Paper on Artificial Intelligence—A European Approach to Excellence and Trust, COM (2020) 65 Final. European Commission, Brussels (2020)
5. Schwab, K.: This is the first commercial chair made using generative design. https://www.fastcompany.com/90334218/this-is-the-first-commercial-product-made-using-generative-design. Accessed 26 Jan 2021
6. Frow, P., Nenonen, S., Payne, A., Storbacka, K.: Managing co-creation design: a strategic approach to innovation. Br. J. Manage. 26, 463–483 (2015). https://doi.org/10.1111/1467-8551.12087
7. Vilone, G., Longo, L.: Explainable Artificial Intelligence: a Systematic Review (2020). arXiv:2006.00093
8. Adadi, A., Berrada, M.: Peeking inside the black-box: a survey on explainable artificial intelligence (XAI). IEEE Access. 6, 52138–52160 (2018). https://doi.org/10.1109/ACCESS.2018.2870052
9. Arrieta, A.B., et al.: Explainable artificial intelligence (XAI): concepts, taxonomies, opportunities and challenges toward responsible AI. Inf. Fusion. 58, 82–115 (2020). https://doi.org/10.1016/j.inffus.2019.12.012
10. Phillips, P.J., Hahn, C.A., Fontana, P.C., Broniatowski, D.A., Przybocki, M.A.: Four principles of explainable. Artif. Intell. (2020). https://doi.org/10.6028/NIST.IR.8312-draft
11. Tjoa, E., Guan, C.: A survey on explainable artificial intelligence (XAI): toward medical XAI. IEEE Trans. Neural Netw. Learn. Syst. 1–21 (2020). https://doi.org/10.1109/TNNLS.2020.3027314.
12. Thieme, A., Belgrave, D., Doherty, G.: Machine Learning in Mental Health: A Systematic Review of the HCI Literature to Support the Development of Effective and Implementable ML Systems. ACM Trans. Comput. Hum. Interact. 27, 34:1–34:53 (2020). https://doi.org/10.1145/3398069
13. Mohseni, S., Zarei, N., Ragan, E.D.: A multidisciplinary survey and framework for design and evaluation of explainable AI systems (2020). arXiv:1811.11839
14. Miller, T.: Explanation in Artificial Intelligence: Insights from the Social Sciences (2018). arXiv:1706.07269
15. OECD Principles on Artificial Intelligence - Organisation for Economic Co-operation and Development. https://www.oecd.org/going-digital/ai/principles/. Accessed 27 Jan 2021
16. Bird, E., Fox-Skelly, J., Jenner, N., Larbey, R., Weitkamp, E., Winfield, A.: The ethics of artificial intelligence: Issues and initiatives|Panel for the Future of Science and Technology (STOA)|European Parliament. https://www.europarl.europa.eu/stoa/en/document/EPRS_STU(2020)634452. Accessed 27 Jan 2021
17. Hagerty, A., Rubinov, I.: Global AI Ethics: A Review of the Social Impacts and Ethical Implications of Artificial Intelligence (2019). arXiv:1907.07892
18. Jobin, A., Ienca, M., Vayena, E.: The global landscape of AI ethics guidelines. Nat. Mach. Intell. 1, 389–399 (2019). https://doi.org/10.1038/s42256-019-0088-2
19. Riedl, M.O.: Human-centered artificial intelligence and machine learning. Hum. Behav. Emerg. Technol. 1, 33–36 (2019). https://doi.org/10.1002/hbe2.117
20. Amershi, S., et al.: Guidelines for human-AI interaction. In: Proceedings of the 2019 CHI Conference on Human Factors in Computing Systems, pp. 1–13. Association for Computing Machinery, Glasgow, Scotland, Uk (2019). https://doi.org/10.1145/3290605.3300233.
21. IBM Design for AI. https://www.ibm.com/design/ai/. Accessed 27 Jan 2021
22. People + AI Research. https://pair.withgoogle.com. Accessed 27 Jan 2021

23. Xu, W.: Toward human-centered AI: a perspective from human-computer interaction. Interactions. **26**, 42–46 (2019). https://doi.org/10.1145/3328485

24. Shneiderman, B.: Human-centered artificial intelligence: reliable, safe & trustworthy. Int. J. Hum.Comput. Interact. **36**, 495–504 (2020). https://doi.org/10.1080/10447318.2020.1741118

25. Margetis, G., Ntoa, S., Antona, M., Stephanidis, C.: Human-centered design of artificial intelligence. In: Salvendy, G., Karwowski, W. (Eds.) Handbook of Human Factors and Ergonomics, 5th Edition, Wiley (2021) (to appear)

26. Guidotti, R., Monreale, A., Ruggieri, S., Turini, F., Giannotti, F., Pedreschi, D.: A survey of methods for explaining black box models. ACM Comput. Surv. **51**(5), 1–42 (2019). https://doi.org/10.1145/3236009

Collaborative Human-AI Sensemaking
for Intelligence Analysis

Stephen L. Dorton(✉) ⓘ and Robert A. Hall

Human-Autonomy Interaction Laboratory, Sonalysts, Inc., Waterford, CT, USA
hail@sonalysts.com

Abstract. AI/ML is often considered the means by which intelligence analysts will overcome challenges of data overload under time pressure; however, AI/ML tools are often data- or algorithm-centric and opaque, and do not support the complexities of analyst sensemaking. An exploratory sensitivity analysis was conducted with a simple Authorship Attribution (AA) task to identify the degree to which an analyst can apply their sensemaking outputs as inputs to affect the performance of AI/ML tools, which can then provide higher quality information for continued sensemaking. These results show that analysts may support the performance of AI/ML primarily by refinement of potential outcomes, refinement of data and features, and refinement of algorithms themselves. A notional model of collaborative sensemaking with AI/ML was developed to show how AI/ML can support analyst sensemaking by processing large amounts of data to assist with different inference-making strategies to build and refine frames of information. Designing tools to fit this framework will increase the performance of the AI/ML, the user's understanding of the technology and outputs, and the efficiency of the sensemaking process.

Keywords: Sensemaking · Decision making · Naturalistic decision making · Reasoning · Intelligence analysis · Artificial intelligence

1 Intelligence Analysis

Intelligence Analysis (IA) is a continuous process that requires collecting, processing, exploiting, and disseminating information to inform decision making [7]. It has long been studied in the social sciences due to several inherent challenges, including recognizing salient information from a surplus of non-diagnostic information [37], and reasoning under time pressure [18]. This complex, high stakes, and time-constrained process is exacerbated by cognitive biases and pitfalls to which analysts are subject in their reasoning [17]. Finally, analyst cognition in IA often has no clearly defined start or stop points – sensemaking begins and ends based on triggering events, available data, analyst knowledge or previous experiences, or external factors such as deadlines [18, 39].

© Springer Nature Switzerland AG 2021
H. Degen and S. Ntoa (Eds.): HCII 2021, LNAI 12797, pp. 185–201, 2021.
https://doi.org/10.1007/978-3-030-77772-2_12

1.1 Challenges and Trends

Problems such as surplus information (diagnostic and non-diagnostic) are getting worse, as a greater number of high fidelity sensors and networking capabilities increases the amount of data being provided to intelligence analysts. Wide Area Motion Imagery (WAMI) is just one example, in which each new WAMI sensor increases the amount of video data collected (requiring processing and exploitation) by a factor of 10–65 over a typical sensor [25]. New Artificial Intelligence and Machine Learning (AI/ML) technologies are being developed to enhance the speed of analysis, and to increase the insights/knowledge gleaned from large datasets [24, 30].

New tools and AI/ML-driven technologies are not necessarily a panacea for these complex cognitive issues. Overly-prescriptive analytic tools based on critical thinking processes can reduce IA to a matter of assigning and shepherding probabilities to information, which does not fully account for its complexities [18, 26, 28]. These tools are not designed to facilitate analyst sensemaking processes, and can result in negative outcomes such as analysts "jury-rigging" the data so that the tool produces the answer they wanted in the first place, or disusing the technology altogether. Thus, it is critical to understand how to best develop AI/ML-based technologies such that they truly aid the sensemaking process of analysts.

1.2 Sensemaking in Intelligence Analysis

IA has historically been viewed through the lens of analytical thinking. Analytical thinking is a deliberate metacognitive process that requires reflection on the quality of one's own reasoning process in order to improve the way in which one reasons [27]. Generally speaking, analytical thinking calls for tools and structured analytic techniques that are relatively prescriptive in order to counter biases that analysts will likely encounter [15, 17].

Recently, however, the sensemaking paradigm has been suggested as a more accurate portrayal of how analysts think while conducting IA [18, 28]. Sensemaking can be defined as the process of structuring the unknown by organizing stimuli into a coherent framework [38]. Different models of sensemaking in IA have been developed, all of which share the properties of being iterative and constructive. New information is used to confirm, modify, or disprove existing representations of information. Another key difference between sensemaking and analytical thinking is that, in the sensemaking paradigm, hunches and biases are useful in framing data, and are not to be removed from the cognitive process. There are two sensemaking models associated with IA that merit further discussion: The Fluidity and Rigor Model (FRM) is applied herein at a more granular level (concerned with specific inference-making strategies), while the Data/Frame Model (DFM) is applied herein at a higher level to describe how information is aggregated and refined.

The FRM [38] describes the transition from creative thinking into more deliberate thinking as uncertainty is iteratively reduced (Fig. 1). Multiple naturalistic studies of intelligence analysts have used different terms to describe observed phenomena; however, these phenomena can be expressed as the three building blocks of abductive, deductive, and inductive inference making strategies [12, 14, 40]. Abductive inference

is defined as making the best explanation possible given the information provided, and is often the starting point for sensemaking. Deductive inference is used to test data against different hypotheses or theories, while inductive inference is used to identify trends in data and evolve creative explanations into more robust hypotheses or theories. As shown in Fig. 1, the FRM shows that analysts will use these strategies in no particular order as the amount of uncertainty (the cross-section of the blue triangle) decreases over time.

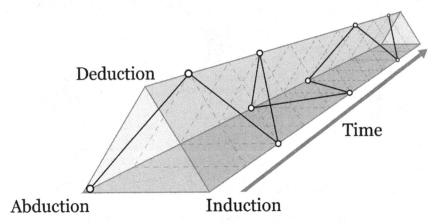

Fig. 1. Fluidity and rigor model. Adapted from [37].

The DFM was introduced by [21], and postulates how data are explained by fitting them into a frame that links them to other data (Fig. 2). Frames can highlight causal and/or conceptual relationships, describing how different pieces of data fit a narrative or a construct. The DFM is cyclic, where frames guide the search for future data, and new data are then used to refine a frame (e.g. preserve the frame, elaborate the frame, or reframe the issue). The DFM has been commonly referenced as a model of sensemaking in IA throughout the last decade, even being hailed as transformational [28].

1.3 Artificial Intelligence and Machine Learning

AI/ML tools and technologies are postulated to be the solution to challenges in IA, especially regarding issues with large datasets [23, 41]. In practice, this is problematic, as AI/ML-based tools present some challenges. The average analyst will not necessarily have a background in computer science or applied mathematics, which creates difficulties in understanding and trusting how these tools arrived at the answer. To apply Ashby's Law of Requisite Variety, one must understand the inner workings of a system in order to effectively interact with it [10]. Similarly, empirical work has shown that direct process feedback will increase the understanding and trust of a system more strongly than peer recommendations or face validity alone [8].

Recent research has shown that a large contributor to trust and understanding issues with AI/ML is algorithmic opacity [32]. Neural networks, or "black box" methods, provide little or no insight into how the result was obtained. Others have examined the

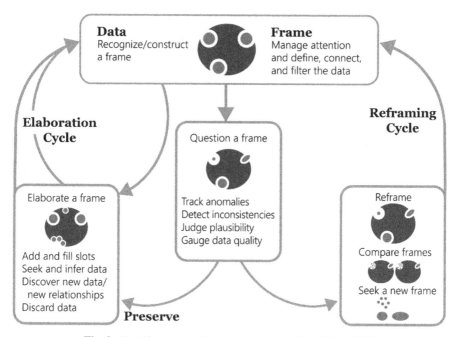

Fig. 2. Data/frame model of sensemaking. Adapted from [20].

use of "gray box" methods such as decision trees, where analysts can see the weights at each node, providing at least some insight [42]. These opacity issues reduce trust in AI/ML technologies, since it is often insufficient in IA to tell a commander or policy maker what the assessment is, but not why one has arrived at it.

1.4   Research Questions and Use Case

The increase of data is complicating the already challenging cognitive process of IA. Analysts use sensemaking to wrangle large, disparate, and multi-modal data into a coherent framework that informs decision making. With the desire to increase use of AI/ML in the IA domain, we aim to understand how to elegantly integrate AI/ML technologies into the human sensemaking processes to support analyst reasoning, rather than developing tools that force analysts to use a data- or algorithm-based approach. Within the sensemaking paradigm, we can readily understand the roles of AI/ML technologies in supporting analyst cognition. At a granular level, AI/ML should process large datasets to provide abductive, deductive (e.g. classification), and inductive (e.g. clustering) analyses (see FRM). At a higher level, AI/ML should facilitate the questioning of different frames of information, supporting the decision to elaborate a frame or reframe the issue (see DFM). Understanding the role of AI/ML in supporting the analyst, one of the primary research questions we aim to address is: How can analyst sensemaking be used to affect AI/ML performance? Since analysts benefit from effective AI/ML and suffer from ineffective AI/ML, it is important to investigate the mechanisms by which a human analyst can influence the effectiveness of an AI/ML tool. Given the answer to the first question,

the second research question we aim to address is: Can we develop a notional framework for Human-AI sensemaking, where human analysts augment AI/ML performance, and AI/ML tools are used to augment analyst sensemaking processes?

An Authorship Attribution (AA) use case was employed to begin answering this question. AA is the process of analyzing data to classify/attribute pseudonymously-written data to an author or organization based on the characteristics of their known works [1]. AA is an exceptionally challenging domain for sensemaking, as each analysis is highly context-dependent, and general algorithm-based approaches are quite brittle across different data types, features, and algorithms. Without a massive set of quality data to train a model on, the knowledge, skills, and experiences of the analyst are critical to overcoming brittleness in developing insights in such a highly context-dependent domain. Because AA is inherently context-sensitive, it is appropriate for this study.

2 Theoretical Foundations of Human-AI Sensemaking

Our approach to answering these research questions was to develop a notional model or "folk theory," which is useful to bridge the gap between mental models of relevant phenomena and demonstrated scientific principles [10, 31], grounded in the context of the use case at hand. Although there are many first principles accepted in the AI/ML community of practice [9], there is a relatively small amount of literature to build such a folk theory from. We worked under the premise that in some IA use cases (i.e. with small, messy datasets, unsuitable for complex deep learning models), that a human-AI team built on sensemaking processes will be more successful than an AI/ML technology simply feeding an answer to the analyst to accept or reject. This interactive and collaborative approach, sometimes referred to as "immersive analytics," continues to gain traction with the advent of new technologies [33].

In general terms, the role of the analyst in supporting AI/ML performance is to apply sensemaking to curate inputs (e.g. data and features) such that the AI/ML has the best probability of success. For this study, we focus specifically on ML classification problems (i.e. supervised learning). Although rare exceptions have been found [2], the accuracy of a ML classifier (A_{ML}) is generally a function of the sample size of data available to train and test the classifier (N) divided by the size of the set of potential outcomes (O) (the number of authors in our use case):

$$A_{ML} \propto \frac{N}{|O|} \tag{1}$$

Input samples can further be refined as the proportion of "good" data (N_G) divided by the total amount of data available, which also includes "bad" data (N_B). Therefore, the likely accuracy of a ML classifier can be expressed as:

$$A_{ML} \propto \frac{\left(\frac{|N_G|}{|N_G|+|N_B|} \right)}{|O|} \tag{2}$$

To simplify the complex nature of different relevant factors, we define good data as referring to two characteristics: First, it is the use of the most diagnostic features

in the context of the data, and the exclusion of the features that are non-diagnostic. Second, it refers to cases where there is not an undue amount of noise or number of idiosyncratic characteristics present. Conversely, bad data consists of cases that use non-diagnostic features, or that introduce noise into the ML classifier by being subject to some idiosyncratic condition or error in processing (e.g. tagging, indexing, etc.). In an example of AA with emails, bad data may include forwarded emails (where the majority of the text was not written by the author of interest), or cases in which automated text/column wrapping creates an artifact of inserting several hyphens not actually representative of the author's punctuation usage. Some research argues that the amount of good data required is proportional to the desired error rate of the classifier [13]. Others say that a dataset is good if it has at least 10 cases for each feature being used (also referred to as the dimensionality of the data) [36].

Given Eq. 2, we can see that the three mechanisms by which one could apply sensemaking to affect the performance of a ML classifier tool would be:

- Remove extraneous outcomes (Minimize $|O|$): As $|O|$ approaches one, the value of A_{ML} approaches one in Eq. 2. Thus, the most powerful role an analyst can adopt is to apply sensemaking to reduce the number of potential outcomes, providing greater power to the AI/ML. This is also the greatest risk, however, since removing correct outcome will have the probability of success approach zero.
- Increase the amount of good data (Maximize $|N_G|$): Because $|N_G|$ is in both the numerator and denominator of the upper half of Eq. 2, the next most powerful role an analyst can adopt is to maximize the amount of good data being fed into the AI/ML. This can take the form of not only increasing the sample size of representative data, but also of selecting the most salient features for exploitation.
- Decrease the amount of bad data (Minimize $|N_B|$): Similarly, an analyst can apply sensemaking to remove non-representative data, event/context-idiosyncratic data, or improperly parsed or tagged data.

3   Methods

We built upon these theoretical foundations by conducting a simple exploratory analysis with a sample dataset to assess a variety of different factors. More specifically, we aimed to build on the theoretical understanding by exploring the effects of: Algorithm selection (not explicitly included in Eqs. 1 and 2), the overall quantity of data (N), the selection of features (a component of whether data are good or bad), and the number of outcomes on AI/ML performance. We conducted a simple AA task utilizing the Reuters_50_50 (R5050) dataset [19], which is a subset of the RCV1 dataset. The dataset includes the top 50 authors from RCV1, each with 50 articles per author for training and 50 articles per author for testing. The dataset was modified from the University of California Irvine Machine Learning Repository [11] to utilize 10 randomly selected authors from the set of 50 authors.

Plaintext documents containing the body of the articles were processed using Python Natural Language Tool Kit [3], and 20 syntactic features were generated for each article. Scikit-Learn and Seaborn were utilized to train the machine learning algorithms and to

visualize the results on 10 different ML classifiers in 10-fold cross-validation scheme on the data. In this scheme, 90% of the training samples was used for training and 10% for validation, whereby the testing set was rotated 10 times such that all cases were tested, and the mean Receiver Operating Characteristic Area Under the Curve (ROC AUC) was calculated. The features were standardized and algorithms were utilized with default settings from scikit-learn.

The models used included: A reference model (Dummy) that approximates how a model would perform by random chance selection; linear models such as Logistic Regression (LogReg); probabilistic models such as Linear Discriminate Analysis (LDA) and Naïve Bayes; a decision boundary model – a Support Vector Machine (SVM); tree models such as Classification and Regression Trees; a neighbor search models (k Nearest Neighbors); ensemble models such as AdaBoost and Random Forest; and a feedforward Neural Network – the Multilayer Perceptron. It should be noted that default parameters in LogReg and SVM utilize L2 regularization.

We used two different measures of performance, each where appropriate. The ROC AUC is commonly used in AI/ML research as a measure of performance of a ML classifier [4]. In general terms, the ROC AUC compares the probability of a true positive (Y-Axis) against the probability of a false positive (X-Axis) [16]. The ROC AUC is indicative of the probability of a correct classification, where a curve closer to the upper-left corner (having greater area beneath it) indicates a higher likelihood of success. Permutation importance was utilized in this study to quantify the predictive information that each feature has for a trained classifier [5]. The method to perform permutation importance involves fitting the classifier, then shuffling a single feature and measuring the change in accuracy of the model. Larger changes in accuracy signify more important features to the model, although the inspection technique does come with some caveats. First, the permutation importance is specific to the model and does not describe the feature's global importance. Second, features that are highly correlated will tend to show lower permutation importance due to the transfer of knowledge from one feature to the other.

4   Results

Before examining the results, it is important to note that the goal of this work is not to identify how to maximize AI/ML performance, but to illustrate how the outputs of analyst sensemaking can affect the performance of an AI/ML tool, which in turn can provide better information to the analyst.

4.1 Algorithms Matter

The first readily understandable means by which an analyst can affect the accuracy of ML classification is to select the right algorithm. There are numerous algorithms available for the classification of data, each with their own strengths and weaknesses. As shown in Eq. 2, algorithms generally perform better with large sample sizes for training, composed of good features (i.e. features with a high information value). Different types of algorithms will be more or less resilient to smaller sample sizes, lower quality features, and/or higher numbers of potential outcomes [29]. Ensemble methods, or combinations

of algorithms, have been used to offset the weaknesses of one algorithm with the strengths of another [35]. As shown in Fig. 3, there is a considerable difference in the effectiveness of each of the 10 algorithms. ROC AUCs ranged from a minimum of .50 to a maximum of .86. The mean ROC AUC across all classifiers for this experiment was 0.74 ($SD =$ .12). The ensemble methods (AdaBoost, RandomForest) performed well without any provided inputs or weights ($M_{ROC\ AUC} = 0.79$) and overlap with the distributions of the top performers as shown in Fig. 4.

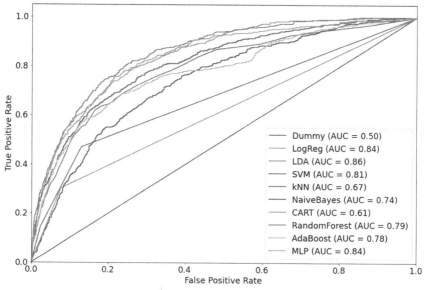

Fig. 3. ROC curves and AUC values for 10 different ML classifiers (including a dummy classifier) on the sample of news articles.

When the training sample size decreases from 100 documents per author (total of 1000) to fewer than 5 documents per author (total of 50), the ROC AUC decreases as expected by Eq. 1. Most of the algorithms approach their maximum ROC AUC rapidly as the training sample size increases as seen in Fig. 5. Interestingly, SVM is one of the lowest performing algorithms at small sample size, but rapidly increases towards its maximum ROC AUC as sample size increases. This most likely is due to the algorithm reaching a critical sample size threshold for its decision boundary between outcomes. The ensemble method, AdaBoost, benefits from increased sample size as this algorithm updates its weights on more difficult cases but overall performs significantly worse than the other ensemble method, Random Forest.

Proper selection of algorithm at small training sample size highlights the fragility of AI/ML technologies when the user does not truly understand the complexities of the technology with which they are interacting. Further, one can readily see that the first mechanism by which a human analyst can improve AI/ML performance is by selecting the proper algorithm for the context of the scenario, such as sample size (which is driven by sensemaking).

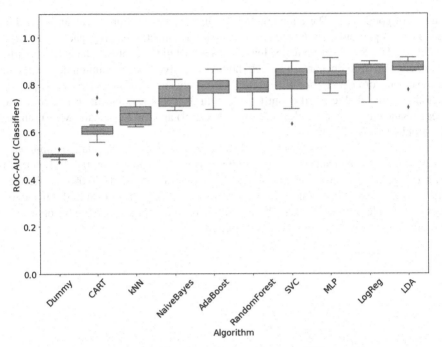

Fig. 4. Variability of algorithm performance across ten different outcomes.

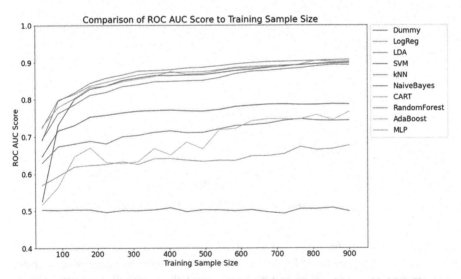

Fig. 5. Effect of training sample size on ROC AUC for 10 algorithms with 95% CI.

4.2 Features Matter

The second method by which an analyst can influence the effectiveness of an ML classifier is to select the appropriate features on which to train the algorithm (increasing the

amount of good data). There are hundreds or thousands of features that can be used in AA. For this particular study, focus was placed on syntactic features (patterns to build sentences) [1]. Some examples of features used included counts of characters, words, sentences, and different parts of speech (noun, verb, adverb, etc.); minimum, maximum, and average word length for different parts of speech; and some measures of complexity including polysyllabic word count (words with 3+ syllables) and indices such as the Flesch-Kincaid Grade Level Readability (FK Level) and the Systematic Measurement of Gobbledygook (SMOG) score.

As shown in Fig. 6, there was relatively little variability in the permutation importance across features. The mean permutation importance for all features was 0.02 (SD = 0.02) with a range of 0.102, spanning a minimum of −0.01 to a maximum of 0.09. Two features that had the greatest information value were the Special Character Count (0.09) and the average syllable per word (0.05). These results show the difficulty in reliably conducting AA due to the heavy domain expertise required.

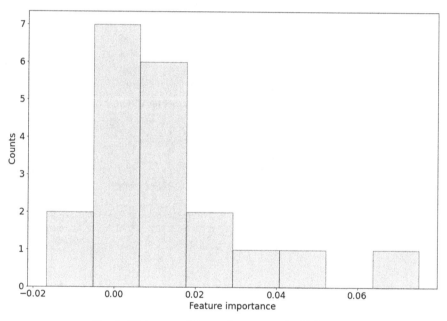

Fig. 6. Histogram of feature importance for 20 features.

Analysts must carefully consider which features are being fed into an AI/ML model when conducting sensemaking for AA, as few features will provide diagnosticity alone. Analysts must understand which features are the most likely to be effective across different mediums (e.g. written papers vs emails vs text messages) and how the data was parsed upstream from the analyst (e.g. certain parsers can have serious effects on syntactic features). For example, text messages often use shorthand, which will make features such as minimum word length have little, if any, diagnosticity. (This also affects derived features that are functions of the formality of speech). Careful selection of features (feature

engineering) should also be driven by the cost of extracting and processing each feature. Based on these considerations, the analyst must continuously curate a set of features that provides the greatest information value. One can greatly increase (or decrease) both the accuracy and the efficiency of a ML classifier (reducing the training and execution runtimes) by applying sensemaking to refine a set of features with higher diagnosticity, effectively removing bad data.

4.3 Outcomes Matter

Arguably the most powerful means by which an analyst can influence the effectiveness of an AI/ML tool is by employing sensemaking to refine a list of possible outcomes (authors, in this example), rather than by conducting a brute force approach of analyzing a large list of outcomes. Curating a list of outcomes has the most relative power to increase the effectiveness of a ML classifier while holding all other variables constant (Eq. 2). It has also been demonstrated herein that other analyst-influenced mechanisms (algorithm selection and feature selection) are largely influenced by outcome. That is, the accuracy of a ML classifier is a function of the algorithms and features selected, given a specific outcome.

As shown previously in Fig. 4, each algorithm performs at least marginally different for each outcome (shown by the vertical size of each box); six out of the ten algorithms performed within 10% of the max accuracy. Further, Fig. 7 shows how the ROC AUC generally decreases as the number of possible outcomes (i.e. authors) increases.

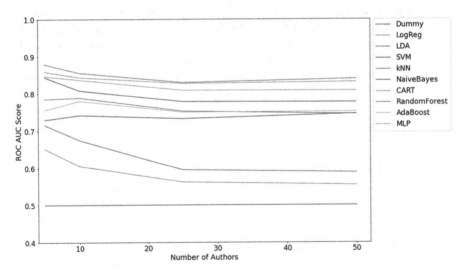

Fig. 7. Algorithm performance for different outcome sizes ($|O|$).

The mean ROC AUC Score with a fixed set of 20 features was 0.75 ($SD = 0.11$) for five authors, and decreased to a mean ROC AUC Score of 0.71 ($SD = 0.12$) for 50 authors. Certain algorithms were more brittle to an increase in outcome size, while others were more resilient. For example, k Nearest Neighbors (kNN) and CART both saw large

decreases in the ROC AUC Score, while LogReg and LDA were relatively undisturbed by it. This interaction of algorithm selection based on outcomes supports the theoretical assertion that the most powerful mechanism by which an analyst can support AI/ML is by refining the set of possible outcomes.

5   Discussion

The hunches and insights generated by analysts during sensemaking with various data (intercepted signals, reports from clandestine sources, other analyses, etc.) directly support AI/ML by enabling the curation of a list of likely outcomes or hypotheses. The set of possible outcomes then informs the selection of appropriate features and algorithms. This application of sensemaking to reduce the number of potential outcomes circumvents the need for a brute force approach that requires a large quantity of quality data to train the algorithm, as well as a longer runtime and more computational resources to train and execute the AI/ML. This scoped approach is invaluable in a practical setting where adversaries rarely afford us the convenience of a large and properly formatted dataset, and analysts are working under time pressure [37].

5.1   A Model of Collaborative Sensemaking with AI

We can apply these quantitative results to refine our theoretical understanding of how to best integrate AI/ML tools and technologies into sensemaking in IA. Figure 8 illustrates a notional model of a collaborative human-AI/ML sensemaking process, where reasoning about the machinations of the AI/ML is a smaller frame associated with the larger high-level sensemaking process, which evolves from fluidity to rigor over time. In this model, the analyst's sensemaking outputs are used to support the performance of the AI/ML tool, which in turn supports the analyst's sensemaking process by facilitating inference making (mainly deductive reasoning in the case of ML classifiers).

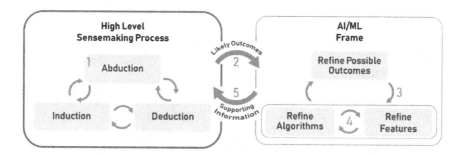

Fig. 8. Human-AI sensemaking model

The notional collaborative sensemaking process shown in Fig. 8 includes the following steps:

1. The analyst conducts sensemaking with available information, likely starting with abduction [39].
2. Based on tasking and initial reasoning, the analyst develops hypotheses of likely outcomes (authors) who fit the available information.
3. Given the possible outcomes, the analyst refines the feature set to include the most diagnostic features.
4. Given the selected features and other factors (availability of data, computational resources, time, etc.), the analyst refines the algorithm set.
5. The analyst uses the AI/ML outputs as information to support higher-level sensemaking processes, including the questioning of frames and subsequent refining or reframing processes.

5.2 Conclusions

A number of conclusions can be drawn from this study. First, the role of AI/ML in supporting analyst cognition is to provide analytical outputs that facilitate sensemaking. Effective inclusion of AI/ML technologies will provide information to the analyst that enables rapid transition across different inference making strategies (abductive, deductive, and inductive), and the questioning and refining of frames. Even if the AI/ML tool says that there is a 95% chance that a given entity wrote a document of interest, that result is based on different analyst-mediated inputs, and should serve only as an input to the high-level sensemaking process that includes other information from other sources.

Second, the role of the analyst in supporting AI/ML performance is to apply sensemaking outputs to curate context-specific inputs to the AI/ML. Three specific mechanisms have been identified to accomplish this. The most powerful (yet brittle) mechanism is to curate a set of probable outcomes. The outcomes affect the curation of highly diagnostic features and algorithms, thereby increasing computational efficiency. The caveat, however, is that the accuracy of the best AI/ML will be zero if an analyst incorrectly removes the actual outcome from consideration. It has been demonstrated that different features and algorithms vary considerably in how universally they can be applied (i.e. some are more resilient to context than others).

Third, this human sensemaking-based framework should at least partially overcome challenges of brittleness and trust associated with more prescriptive algorithm- or data-based approaches. This framework plays to the strengths of AI/ML (crunch through large amounts of data) while also playing to the strengths of the human (overcome complexity and context, and make sense of disparate data). The ability to iteratively provide human insights into AI/ML tools is critical, especially in AA, where algorithms are easily deceived [6]. The analyst will likely have greater understanding and calibrated trust of the AI/ML, and be more likely to explain *why* a certain answer was obtained, largely because the AI/ML outputs are a product of the analyst's own cognition (selection of inputs, etc.). One may argue that there are computational means to automate the human tasks highlighted in this framework such as feature selection [34]; however, automating these tasks takes the human out of the loop, decreasing critical trust and

understanding. Therefore, we argue that the context-rich knowledge of the analyst should be more prominent than computational metrics in steering the AI/ML toward answering the analytic question at hand.

The following are a set of key points summarizing this research:

- Artificial Intelligence and Machine Learning (AI/ML) technologies are being used to cope with increased amounts of data made available to intelligence analysts, although algorithmic opacity creates trust and understanding issues.
- To be effective, AI/ML technologies must be integrated with the various sensemaking processes and formal methods employed by analysts.
- There are three identified mechanisms through which a human can apply sensemaking to affect the performance of AI/ML: Outcome selection, feature selection, and algorithm selection.
- A notional framework for collaborative analyst sensemaking with AI/ML technology has been developed based on these identified mechanisms.
- This framework should be applied when considering the development of AI/ML tools for intelligence analysis to increase the effectiveness and efficiency of the human-AI/ML sensemaking process.

5.3   Future Work

There is still much work to be performed to develop more fluid and meaningful interactions between humans and AI/ML in IA. Immediate next steps will include refining the sensemaking model to include how more specific implementations of AI/ML can play a constructive role in human sensemaking (e.g. unsupervised learning may require a different sensemaking process than classification techniques). Additionally, functional allocation will be performed for each of the identified means to affect AI/ML performance. For example, results showed that selecting the appropriate algorithm will have an impact on the accuracy of the analysis; however, the appropriateness of each algorithm for a given use case can be empirically derived, and is therefore not something that should likely be allocated to the human analyst. Further, we desire to investigate how different visualizations and non-AI/ML analytics (e.g. descriptive and inferential statistics) can facilitate the analyst sensemaking process.

As a more long-term objective, empirical research with analysts will be performed to determine whether or not this model is truly representative of how analysts will work with AI/ML in an AA use case, inferring reasoning from interactions with a prototype system [22]. Ultimately, we aim to empirically test the human-AI/ML team against an AI-centric brute force approach, measuring classification accuracy, time, and data required to achieve accuracy, calibrated trust in the results, and resilience to context.

Acknowledgements. This work was supported in part by the US Army Combat Capabilities Development Command (DEVCOM) under Contract No. W56KGU-18-C-0045. The views, opinions, and/or findings contained in this report are those of the author and should not be construed as an official Department of the Army position, policy, or decision unless so designated by other documentation. This document was approved for public release on 25 January 2021, Item No. A116.

References

1. Abbasi, A., Chen, H.: Applying authorship analysis to extremist-group web forum messages. IEEE Distrib. Syst. Online **20**(5), 67–75 (2005)
2. Abromavich, F., Pensky, M.: Classification with many outcomes: challenges and pluses. arXiv: 1506.01567v4 (2019)
3. Bird, S., Loper, E., Klein, E.: Natural Language Processing with Python: Analyzing Text with the Natural Language Toolkit. O'Reilly Media Inc, Sebastopol, CA (2009)
4. Blessie, E.C., Karthikeyan, E., Selvaraj, B.: Empirical study on the performance of the classifiers based on various criteria using ROC curve in medical health care. In: 2010 International Conference on Communication and Computational Intelligence (INCOCCI), pp. 515–518 (2010)
5. Breiman, L.: Random forests. Mach. Learn. **45**, 5–32 (2001). https://doi.org/10.1023/A:101 0933404324
6. Brennan, M., Greenstadt, R.: Practical attacks against authorship recognition techniques. Proceedings of the Twenty-First Innovative Applications of Artificial Intelligence Conference, pp. 60–65 (2009)
7. Clark, R.M.: Intelligence Collection. CQ Press, Los Angeles, CA (2014)
8. de Vries, P.W., van den Berg, S.M., Midden, C.: Assessing technology in the absence of proof: trust based on the interplay of others' opinions and the interaction process. Hum. Factors **57**(8), 1378–1402 (2015). https://doi.org/10.1177/0018720815598604
9. Domingos, P.: A few useful things to know about machine learning. Commun. ACM **55**(10), 78–87 (2012). https://doi.org/10.1145/2347736.2347755
10. Dorton, S., Thirey, M.: Effective variety? For whom (or what)? A folk theory on interface complexity and situation awareness. In: IEEE Conference on Cognitive and Computational Aspects of Situation Management (CogSIMA) (2017). https://doi.org/10.1109/cogsima.2017. 7929594
11. Dua, D., Graft, C.: UCI machine learning repository (2019). http://archive.ics.uci.edu/ml
12. Gerber, M., Wong, B.L.W., Kodagoda, N.: How analysts think: intuition, leap of faith, and insight. In: Proceedings of the Human Factors and Ergonomics Society 2016 Annual Meeting, pp. 173–177 (2016)
13. Guyon, I., Makhoul, J., Schwartz, R., Vapnik, V.: What size test set gives good error rate estimates? IEEE Trans. Pattern Anal. Mach. Intell. **20**(1), 52–64 (1998)
14. Haider, J.D., Seidler, P., Pohl, M., Kodagoda, N., Adderly, R., Wong, B.L.W.: How analysts think: sense-making strategies in the analysis of temporal evolution and criminal network structures and activities. In: Proceedings of the Human Factors and Ergonomics Society 2017 Annual Meeting, vol. 61(1), pp. 193–197 (2017)
15. Handbook: Handbook of Analytic Tools and Techniques. Pherson Associates, Reston, VA (2016)
16. Hanley, J.A., McNeil, B.J.: The meaning and use of the area under a receiver operating characteristic (ROC) curve. Radiology **143**, 29–36 (1982)
17. Heuer, R.: Psychology of Intelligence Analysis. Echo Point Books & Media, Brattleboro, VT (2017)
18. Hoffman, R., Henderson, S., Moon, B., Moore, D.T., Litman, J.A.: Reasoning difficulty in analytical activity. Theor. Issues Ergon. Sci. **12**(3), 225–240 (2012)
19. Houvardas, J., Stamatatos, E.: N-gram feature selection for authorship identification. In: Lecture Notes in Computer Science (Including Subseries Lecture Notes in Artificial Intelligence and Lecture Notes in Bioinformatics), vol. 4183, LNCS (September 2006), pp. 77–86 (2006)
20. Klein, G.: Seeing What Others Don't: The Remarkable Ways we Gain Insights. PublicAffairs, New York, NY (2013)

21. Klein, G., Phillips, J.K., Rall, E.L., Peluso, D.A.: A data-frame theory of sensemaking. In: Hoffman, R.R. (ed.) Expertise Out of Context: Proceedings of the Sixth International Conference on Naturalistic Decision Making, pp. 113–155. Lawrence Erlbaum Associates, Mahwah, NJ (2007)
22. Kodagoda, N., et al.: Using machine learning to infer reasoning provenance from user interaction log data: based on the data/frame theory on sensemaking. J. Cogn. Eng. Decis. Making **11**(1), 23–41 (2017). https://doi.org/10.1177/1555343416672782
23. Lee, M., Valisetty, R., Breuer, A., Kirk, K., Panetton, B., Brown, S.: Current and future applications of machine learning for the US Army, Technical Report [ARL-TR-8345]. US Army Research Laboratory, Aberdeen Proving Ground, MD (2018)
24. McNeese, N.J., Hoffman, R.R., McNeese, M.D., Patterson, E.S., Cooke, N.J., Klein, G.: The human factors of intelligence analysis. In: Proceedings of the Human Factors and Ergonomics Society 59th Annual Meeting, vol. 59(1), pp. 130–134 (2015)
25. Menthe, L., Cordova, A., Rhodes, C., Costello, R., Sullivan, J.: The future of air force motion imagery exploitation: lessons from the commercial world. RAND, Santa Monica, CA (2012)
26. Moon, B.M., Hoffman, R.R.: How might "transformational" technologies and concepts be barriers to sensemaking in intelligence analysis. In: Schraagen, J.M.C. (ed.) Proceedings of the Seventh International Naturalistic Decision Making Conference, Amsterdam, The Netherlands, June 2005 (2005)
27. Moore, D.T.: Critical Thinking and Intelligence Analysis, 2nd edn. National Defense Intelligence College, Washington, DC (2007)
28. Moore, D.T., Hoffman, R.R.: Sensemaking: a transformative paradigm. Am. Intell. J. **29**(1), 26–36 (2011)
29. Ng, A.Y., Jordan, M.I.: On discriminative vs. generative classifiers: a comparison of logistic regression and naive bayes. In: Advances in Neural Information Processing Systems (2002)
30. Office of the Director of National Intelligence (ODNI): National Intelligence Strategy of the United States of America 2019 (2019). https://www.dni.gov/files/ODNI/documents/National_Intelligence_Strategy_2019.pdf?utm_source=Press%20Release&utm_medium=Email&utm_campaign=NIS_2019
31. Parasuraman, R., Sheridan, T.B., Wickens, C.D.: Situation awareness, mental workload, and trust in automation: viable empirically supported cognitive engineering constructs. J. Cogn. Eng. Decis. Making **2**(2), 140–160 (2008)
32. Paudyal, P., Wong, B.L.W.: Algorithmic opacity: making algorithmic processes transparent through abstraction hierarchy. In: Proceedings of the Human Factors and Ergonomics Society 2018 Annual Meeting, vol. 62(1), pp. 192–196 (2018)
33. Skarbez, R., Polys, N.F., Ogle, J.T., North, C., Bowman, D.A.: Immersive analytics: theory and research agenda. Front. Robot. AI **6**(82), 1–15 (2019). https://doi.org/10.3389/frobt.2019.00082
34. Song, L., Smola, A., Gretton, A., Bedo, J., Borgwardt, K.: Feature selection via dependence maximization. J. Mach. Learn. Res. **13**, 1393–1434 (2012)
35. Stamatatos, E.: Ensemble-based author identification using character n-grams. In: Proceedings of the 3rd International Workshop on Textbased Information Retrieval, pp. 41–46 (2006)
36. Subramanian, J., Simon, R.: Overfitting in prediction models – is it a problem only in high dimensions? Contemp. Clin. Trials **36**, 636–641 (2013). https://doi.org/10.1016/j.cct2013.06.011
37. Trent, S.A., Patterson, E.S., Woods, D.D.: Challenges in cognition for intelligence analysis. J. Cogn. Eng. Decis. Making **1**(1), 75–97 (2007)
38. Wong, B.L.W.: How analysts think (?): early observations. In: 2014 IEEE Joint Intelligence and Security Informatics Conference, pp. 269–299 (2014)

39. Wong, B.L.W., Kodagoda, N.: How analysts think: inference making strategies. In: Proceedings of the Human Factors and Ergonomics Society Annual Meeting, vol. 59(1), pp. 269–273 (2015)
40. Wong, B.L.W., Kodagoda, N.: How analysts think: anchoring, laddering, and associations. In: Proceedings of the Human Factors and Ergonomics Society Annual Meeting, vol. 60(1), pp. 178–182 (2016)
41. Worland, J.: Machines in the military. In: Time Special Edition – Artificial Intelligence: The Future of Humankind, pp. 22–25. Time Inc. Books, New York (2017)
42. Zhu, J., Liapis, A., Risi, S., Bidara, R., Youngblood, G.M.: Explainable AI for designers: a human-centered perspective on mixed-initiative co-creation. In: 2018 IEEE Conference on Computational Intelligence and Games (CIG), pp. 1–8 (2018). https://doi.org/10.1109/cig.2018.8490433

Design Intelligence - Taking Further Steps Towards New Methods and Tools for Designing in the Age of AI

Jennifer Heier[✉]

Bauhaus Universität Weimar, 99423 Weimar, Germany
`Jennifer.heier@uni-weimar.de`

Abstract. This paper builds on the on-going research in industrial AI, namely use cases from Business-to-Business (B2B) factory automation, focusing on Artificial Intelligence (AI) technology and the influence of user experience (UX) design [1]. It aims to provide a) an overview of the found challenges from different resources and domains, b) an overview of proposals for Human-Centered-AI principles, c) a mapping of both in order to analyse if the principles enable solutions to unsolved challenges. The overview contains findings from investigated design and UX challenges when working in the domain of AI and Machine Learning (ML) from a selection of different research papers, mainly from the area of consumer facing products, as well as a comparison with findings and insights from the mentioned use cases in the B2B domain. Differences and similarities have been investigated and addressed, resulting in a list of considerations to be taken into account when designing for AI. As a second step, the paper analyses Human-Centered-AI principles which are proposed as a solution to the design challenges imposed by AI. By mapping the list of challenges to the solutions, this work seeks to initiate a next step in the development of new methods and tools for designing AI. Connecting the dots between the problems and a means to their solution will help form a clearer picture of the current status of designing for AI, and a better understanding of what is important to include in the design process as well as identifying gaps where more work needs to be done.

Keywords: UX · Machine learning · Human-Centered-AI

1 Introduction

Recent research work by the HCI community (amongst others) related to the development of AI systems and applications, has started to shed light on new challenges for the designers involved [2–4]. Those challenges call for new approaches, methods and tools for designing with AI. This work is mainly focused on consumer facing products and services, such as e.g. medical decision support systems, autonomous driving services, and/or spam filters, movie or music recommenders [3, 5, 6]. In those domains the focus is on customized user scenarios. Adding insights from qualitative studies in the domain of industrial AI, namely use cases from B2B factory automation where optimization is the

© Springer Nature Switzerland AG 2021
H. Degen and S. Ntoa (Eds.): HCII 2021, LNAI 12797, pp. 202–215, 2021.
https://doi.org/10.1007/978-3-030-77772-2_13

main driver, should bring another angle and perspective to the scientific discourse, which is currently lacking in this area. This investigation teases out similarities and differences among challenges and provides an overview of the current findings, followed by an overview and analysis of a selection of Human-Centered-AI principles [7–9], which are supposed to offer new ways of dealing with AI and ML systems. Hence, this paper seeks to take a step further towards the development of new methods and tools for design in the age of AI. By mapping problems and solutions an examination of the current status is possible, showing which challenges can be solved and which issues still need to be investigated.

2   Overview Design for AI Challenges

ML and AI based systems call for new methods and tools, due to their complex (eco-) systems which learn and evolve over time [10]. This in turn means that the interactions between AI based systems and their users change over time as the systems learn, potentially causing unwanted user experiences and difficulties dealing with the product or service. Those interactions are above all "multimodal" and "non-visual" [11]. Invisible algorithms are a new "design material" [2]. Designers do not yet seem to have grasped the potential of ML and are not incorporating AI technology when innovating ideas for new products and services. Additionally, the process to develop ML systems currently consists of "lengthy and costly development cycles" [5] and is mainly driven by statistics and lacking the human (centered) perspective. The behaviour of those systems is therefore not comparable to human logic which makes it hard to investigate and foresee with the tools and methods used by UX designers to date. After all, the algorithms are only as perfect as the data they are trained with, meaning those systems make mistakes [e.g. 12]. In sum, it is necessary to rethink the current approach to developing those smart and intelligent agents.

2.1   Case Studies from B2B Factory Automation

The initial B2B factory automation use case deals with improving the factory planning process of a production site for industrial controls. Time series predictions with neural networks is the chosen ML approach. A qualitative study with the development team members, among them a UX designer, and other stakeholders involved, was conducted and published [1]. From this research 14 themes were derived which represent the pitfalls and challenges encountered during the development of the factory planning solution. Meanwhile further research was conducted. The findings from the initial case study were enriched with information from two additional internal projects. Domain and field of application from all three uses cases are the same. They differ in location of production sites, development process, products produced and technical solution, in total resulting in 15 topics.

2.2   Additional Findings

One of the two additional case studies had a new hire as a requirement in order to start the project. An additional factory planner was recruited, with a background in computer

science. This was a strategic decision in order to combine domain knowledge with the technical skills to be able to improve the current planning process. This meant that the end user of the final ML solution was the same person who created it, and who was able to gather the data, clean it, train, test and validate the models. *"It was a lot of work for a single person... Being a user and expert in one person was a very efficient setting... resulting in a very fast Proof of Concept (PoC)." (Computer scientist)* However when scaling the solution among the factory and other departments the team faced very similar challenges as the initial use case. Other stakeholders and decision makers lacked the AI expertise of the systems' creator. Additionally, UX was not part of the development process, no user research was conducted and therefore the solution didn't meet the needs of the other planners. This resulted in lack of trust in the output of the system. As with the initial use case a lot of effort and energy was consumed by a rigid corporate culture and people's risk averse mindset.

The third case study had a completely different setting. The development team, including the AI and ML expertise, was completely represented by an external agency. Primarily they had one contact person at the company, namely the coordinator of the planners. This person also had access to the database and served as a single source of contact for the external partner. Throughout the whole progress of the project this person gained a lot of technical knowledge regarding the final solution. *"My role and tasks changed from managing the planners, which I still have to do, to feeding the algorithm with data and providing the output to the planners." (Project manager)* Missing involvement from the planners during development resulted in rejection of the solution and a challenge to hand over the model handling process to an internal team. Additionally 3rd party software was used by the external partner which, except for data privacy issues that needed to be overcome, was very helpful for the overall speed of the project. However the off-the-shelf solution is fixed to the provided models, making it a challenge to include a new product in the forecast, where historical data has not been captured (Table 1).

Table 1. Overview of the 14 themes from the initial B2B case study + 1 deriving from the additional use cases.

AI-expertise	Iterative working mode	Feedback structure, structured feedback
Definition of Design	Visualization	Planning process
Culture and mindset	Expectations	Starting point
Trust in output	Biased presentations	Gap between prototype and implementation
Orient, manage, prioritize, eliminate	UX and timing	Internal vs. external software

2.3 Related Work

The HCI and creative community already started to investigate issues related to the challenges when designing for AI. Their findings originate from talking to UX practitioners ranging from expert level, having experience with designing for AI, through to students currently being trained in the tools and methods for classical UX or design [2, 3, 13], as well as UX practitioners that reflect and report on their own experience while designing for AI. Some researchers relate their findings to a specific use case, whereas others do not take the domain into account. There are also a large amount of reports and articles available from non scientific resources which will be neglected for this overview, but which were important to absorb to form a background to the relevant challenges, and become important when it comes to the Human-Centered-AI principles section. From this work five topics were condensed and will be introduced in more detail. Additionally those issues will be compared to the findings from the B2B use cases from factory automation.

Lack of AI Expertise. One aspect that appeared from those studies is that most designers lack detailed knowledge about the technology. Designers for the most part understood the overarching concepts, but didn't make distinctions such as e.g. supervised or unsupervised learning [2, 13]. These issues resulted in the development of learning materials for designers. Interestingly those having work experience in the field of AI based products and systems, did not seem concerned about having a lack of AI and ML expertise [13]. When talking about AI and ML systems they referred to example products and services. Those examples were very limited in their range of diversity and represented simple technological approaches, such as spam filters, recommenders, to name a couple. This lack of knowledge about the capabilities of AI and ML might also be an obstacle when identifying and choosing the right technology to address the problem or user need, resulting in not taking AI or ML based systems into account as a technical solution.

Analysing the use cases from the B2B factory automation domain revealed a very similar issue. Asking the participants of the study about their personally rated AI expertise, those not having a computer science or data science background interestingly rated their expertise similar compared to the domain experts. They compared themselves not to experts in the field, but compared their knowledge from the start of the project till the end. *"Compared to the beginning of the project, I gained a lot of knowledge about the technology." (Product manager)* It became clear that a certain degree of knowledge, not necessarily expert level, but a basic understanding of concepts such as regression, supervised/unsupervised learning, decision trees, random forest, etc. is very helpful for working together.

Define the Starting Point. The classical design/Human-Centered-Design process starts with defining a starting point, or a problem statement that needs to be solved, mostly based on user research [14] and making sure that the designed product or service answers a market need (human desirability [15]). Technological feasibility and business viability are not the main drivers. In contrast many data science projects start with a given data set and from there define what can be solved by an algorithm [3]. With this both processes operate with a different purpose and it is difficult to bring them together. The initial design idea might not take a data driven algorithm into account, resulting in the integration of UX

methods and tools late in the development process when main decisions and direction are already set, causing usability problems and worst case rejection of the developed solution.

A very similar insight derived from the B2B use cases in factory automation. UX was integrated fairly late in the process (or even not at all) and was perceived as a negative aspect, since it caused more features and needs to be integrated into the final solution than initially considered. *"UX really is about the right timing… if it comes too late in the process it cannot influence the direction anymore." (Product manager)* Another issue related to the definition of the starting point was the initialization of the project. In all three cases it was an initiative by management, missing out the voices of other very important stakeholders involved, resulting in a lack of engagement by the users and other project members.

Missing Data Literacy/-Centricity. AI and ML depend mainly on statistical approaches and data sets, therefore being driven by data centricity [13, 16]. Typically this data answers a precise set of questions framed by the data scientists closely related to the training and validation of the chosen technical solution. In contrast a qualitative approach preferred by UX practitioners is a divergent research method. Whereas this form of data enquiry is also very helpful for ML projects, e.g. when defining the starting point for a project or supporting and enriching the statistical data set, it is also helpful for designers to embrace the data-driven culture of AI and ML engineers. Drawing insights from quantitative data and understanding those data enquiries created by telemetry and machine sensor data, are much needed skills in a data driven context such as AI and ML.

The above challenge was not directly mentioned in the B2B use cases. However an issue related to this is the need for data visualization. The workflow of the different team members from one use case included different ways and tools to communicate and present their data (e.g. excel, tableau and SAP apps). The pure numbers generated by the neural network representing the pieces to be produced weren't enough to validate their accuracy nor their reliability. *"I need a graph that shows figures from the past and the forecast in order to examine whether I can trust the output of the neural network. A number in a cell in an excel file means nothing to me." (Data analyst)* Similar to the different data approaches there is not one visualization that fits all involved stakeholders. It could be the role of designers to negotiate and facilitate the different needs and find the best way to communicate the output of the algorithm.

Struggle to Work with Data Scientists. Co-creation with data scientists is a new territory for design and UX practitioners. A shared process model or guidance for methods and tools does not exist yet, partly due to the issues mentioned beforehand regarding data literacy, as well as various domain jargon and mindsets. So far mainly experience from best practice does apply [13]. This shows that working together on AI related use cases is currently the most promising approach for designers to influence the UX of AI in a positive direction. UX practitioners who do not have direct access to data scientists in their daily work struggle even more with the design of AI and ML based systems. They lack the feedback for technical feasibility and therefore often fall back to known design patterns and technical solutions, this way not creating innovative new products and services. Additionally, the data collected and synthesized by designers can hardly be encoded on a one to one basis into a statistical model [17]. In order to analyse the

data that is needed to address user needs, a conversation with a data scientist in an early stage of the development process is very helpful.

Concerning the B2B factory automation use case where a UX practitioner was part of the team working closely with the data scientists, the iterative working mode was an essential factor for the success of this co-creation and a common denominator for both professions. *"To keep the sprints and present results on a regular basis was key for the success of this project". (ML engineer)* The willingness from both sides to learn and negotiate the input from the other expert is the basic requirement for a fruitful collaboration. In this case some features that derived from the user study of the B2B use case, which concerned post processing steps of the algorithms, were neglected. The team agreed to focus on the pure output of the algorithm without any post processing, which would bring the most value to the user, even if as a consequence the post processing still needed to be done by the users. Both professions need to be open to those kinds of tradeoffs.

Difficult Prototyping. Prototyping is an essential part of the design process. It is used for idea generation in early stages as well as idea testing and validation later in the process. It is always used as a medium to communicate ideas to other stakeholders and users, as well as evaluating if a service or product is worth being pursued for implementation. Some user experiences for AI and ML applications can be prototyped, such as voice assistants, chatbots or recommenders. They all have in common that they are represented by an interface. The interaction with the user is represented by a conversation or action on a screen, which can be faked with the 'Wizard-of-Oz' method [18]. Still this tool has its disadvantages. Without a real data set and algorithmic model in the background, the designer can not verify what kind of errors the system might possibly produce and therefore it is hard to gather the related feedback from users. This is even harder for AI and ML applications which do not have an interface. For any prototyping tool out there, a real data set and a real model are necessary. Those prerequisites are a barrier when it comes to the design of intelligent systems [2, 5].

In the B2B factory automation use cases where a neural network produced a forecast into the future about pieces sold, it didn't make any sense to 'fake' a model. The teams needed to develop a functional prototype with real data in order to validate its usefulness. The process already consumed a certain amount of resources and time. Asking for commitment from management to provide time and budget caused a degree of pressure for a successful proof of concept. *"In hindsight, I think we preferred to show the line charts of the products where the AI predictions performed really well. ... We wanted to meet the expectations of management." (Data scientist)* Another aspect that became clear at a later stage in the process was that scaling from the initial functional prototype and a small data sample, to a productive environment in the cloud and a larger data set, caused trouble for the development teams (except the use case which used the 3rd party solution). In one case it even resulted in reduced accuracy of the final model.

2.4 Compare the Findings

A lot of the findings from research scholars confirm the insights from analysing the B2B factory automation use cases. Four out of the five above challenges were confirmed. One was mentioned in another context, but also noted. Not all 15 themes of the B2B case studies are mentioned in related work by research scholars, which may be due to their open character that did not focus primarily on UX topics. Issues such as company culture and mindset of the different stakeholders were important topics in the B2B domain but were not mentioned among the related work. Those topics are more related to change management than design inquiries. However, in order for AI to be successfully implemented in such a setting, the culture and mindset issues need to be addressed too.

3 Overview Human-Centered-AI Principles

3.1 Purpose and Definition

Due to the present challenges the design and research community have already started to propose different approaches to solve the problems when designing for AI. Those approaches are united in a call for a bigger focus on the impact of the human perspective of the AI systems [7–9]. As a result the creative community and other practitioners created a number of different Human-Centered-AI principles [19, 20]. The human-centered perspective is perceived as central to the process for design practitioners tasked with AI development [21].

> *"Human-centered AI is about defining the goals of AI to meet human needs and to work within human environments. ... Not only do we need a set of new tools and techniques to make AI work in practice, but we need to shift the process by which AI is even designed in the first place."Agarwal, Abhay and Regalado, Marcy [7]*

3.2 Proposed Solutions

Companies such as Microsoft, Google and IBM amongst others came up with Human-Centered-AI principles. Those companies are using AI solutions in consumer facing domains already and have experience with implemented AI solutions. They have a vast amount of data accessible through their portfolio of applications, as well as the work-force, and know how to develop their own algorithms. There are also some individuals working as design and UX practitioners on AI projects, who published their thoughts and principles on the web. The different resources were examined and from this amount of information the selection for the overview was made. Human-Centered-AI principles from Microsoft, Google and two individuals, namely Abhay Agarwal (former Microsoft and currently lecturer at Stanford d.school) and Marcy Regalado (a Stanford d.school graduate) were selected (Lingua Franca). The scientific nature of the Microsoft work, the helpful worksheets from the Google guidebook and the great detail of the Lingua Franca principles are the reason behind the presented selection. With this, a great variety of principles are provided.

3.3 Explanations and Analysis

The following section introduces the three chosen resources in more detail. Resulting in an analysis of what is different and where they correspond, as well as a potential for further investigation.

Microsoft provides a very comprehensive collection from 20 years of experience and collecting AI design recommendations from various sources. The baseline consists of thoughts and ideas from Eric Horvitzs' "Principles of Mixed-Initiative User Interfaces" [22]. Those principles are then enriched with contemporary publications from the private sector, illustrating the most up to date concepts. Microsoft divides the whole set of principles into four steps. Each of those steps requires certain aspects to be taken into consideration. In total their approach represents 18 different principles. The main purpose of their principles is to supportUX experts with guidance when developing interaction of AI systems with users.

Initially it is crucial to "make clear what the system can do". This is a way to guide users expectations towards the ML system. Setting those expectations too high will end in an unsatisfying experience while using the smart solution. It is therefore important to facilitate those expectations right from the beginning. The second set of guidelines deals with aspects during interaction with the system, for example, which wording the system uses to communicate with the user. Misleading language might evoke social injustice or address stereotypes. A third segment is devoted to the failure of the system. This shows the importance of this aspect when designing for ML. The algorithm is not perfect. It is trained on data generated by humans and can contain errors and mistakes. Therefore solutions should communicate how they derived their results (so called Explainable AI - XAI [23]). It is necessary to provide the possibility to hand over control of the system to the human user, as well as being honest about the fact that the machine might be wrong, or at least unsure about its output. Finally, every ML application should be able to learn from its interaction with users and improve over time. That's actually the strength of the ML technology. Therefore collecting feedback from users is crucial and an important step in the process in the long run. Furthermore it is helpful to inform users about new releases and features of the system, in order to maintain trust in reliability and performance.

The work from Microsoft is primarily focused on the user experience of the final solution. They do not provide or contain advice for the initial step of developing the algorithms, such as data preparation, problem definition or choice of technology. Furthermore they are based around the idea of a graphical user interface as a means of representation between the AI system and the user. Therefore it is questionable if the guidelines also apply for non visual products and services. They provide the principles as a set of cards, naming the principle on the front and an example of use on the back.

Google provides a People + AI Guidebook. It consists of six chapters which follow the product development flow. Similar to Microsoft they provide an explanation of each chapter, as well as a related worksheet that should support the use of their principles. Their principles are primarily based on the knowledge and input of the internal project teams, enriched with academic research and expert opinions. The target audience are UX professionals as well as product managers who want to put more focus on the users when developing AI systems.

The aspects mentioned by the Google guidebook are very similar to the Microsoft ones. The order and arrangement of principles is done slightly differently. Namely, the sections about explainability, feedback and failure are mentioned in both cases and represent a common ground of general AI challenges. Those aspects are widely adopted in other principles on Human-Centered-AI [19, 20]. However, Google's guidebook also puts emphasis on the preparation and problem definition of AI systems as a relevant step of the overall design process. This part is called "User Needs + Defining Success". It implies that the starting point of any Human-Centered-AI project should be a user need versus a technology first approach. It also provides a list for which tasks an AI automation might be useful, or when an augmentation of the user is the better choice. The section also contains thoughts about the success criterias of the product and service. Another part is about "Data Collection + Evaluation". Designers and artists can provide valuable input with qualitative research methods, adding meaning to purely statistical data points. This data then needs to be transformed into a format that can be used to train the algorithm. This section poses a question around whether the development team should use a given data set, or establish their own. This might be needed when a given data set is biased towards a certain user group, for example a bias on a certain gender or age group.

With this the Google principles are not fixed to AI systems represented by an interface. They also take into consideration tasks and steps that are important when designing the AI or ML based system. By providing detailed worksheets per section it is easy to follow their development path from the beginning of a project until the final solution. Their explanations and examples use a small amount of data science jargon which is necessary to know in order to understand the relevant information.

Lingua Franca - A Design Language for Human-Centered AI - is currently the most comprehensive selection of principles and guidelines for designing AI systems. It is published by Abhay Agarwal (former Microsoft and currently lecturer at Stanford d.school) and Marcy Regalado (a Stanford d.school graduate). In addition to their handbook, which introduces seven different aspects of design intervention, they also provide eight principles that every AI system should follow. They have started to collect example elements and patterns that represent a concrete solution to the stated principles. They do not mention a specific target audience or user group for their principles, as their main goal is to enhance the human perspective in AI development.

Their approach is very similar to the general design process, starting with the initial problem selection and definition, followed by observing human behaviour. However the section about data and sensemaking also puts emphasis on statistical methods and knowledge, attempting to provide a basic knowledge for those approaches for the novice, since these skills are not taken for granted amongst designers and artists. Another part of their handbook deals with the choice of technologies. This is not AI specific, but in order to be able to choose the right technology in an era of ML and AI a certain degree of literacy about the possibilities and features of the different concepts is necessary. Sometimes even the decision not to use ML at all is an important insight. Prototyping and a section about ethics and responsible design are the final steps.

Lingua Franca's collection of handbook, guidelines and design patterns apply to a broad variety of different AI solutions and represent a huge source of information.

However the amount of information can be overwhelming. It is not very easy to navigate through their online catalogue which contains a lot of cross references and links to articles and webpages. Additional hands on worksheets are missing, as well as a section about the implementation of AI based systems and their evaluation and feedback structures.

3.4 Conclusion and Missing Pieces

Most Human-Centered-AI principles have a set of aspects that are very similar; problem definition, need finding or data collection in general, explainability, trust, feedback and how to handle failure are commonly important. This is a great starting point for further research and development of a set of principles that can guide artists and designers along the development process of AI systems. However, there is still a need for new tools and methods that work alongside the guidelines. Taking into account the specific context where the guidelines will be used is missing in most principles and could be the missing link to make them work and add value in reality. The biggest value may be provided by a collection of different sets of principles which enhance a flexible use.

4 Mapping Challenges and Principles

Comparing the five challenges with the introduced principles is this paper's attempt to analyse which problems can already be solved and which need further investigation. This is obviously not the final list of challenges designers and UX practitioners will be facing, but it should at least start to shift the conversation from pure problem spaces to solution spaces, providing concrete methods and tools for the design of AI, and in doing so endeavoring to open up space for new insight generation on missing pieces.

4.1 Which Design Challenges Are Addressed?

Define the Starting Point. Starting with the right problem and defining it very well is the most important aspect of all principles (included in the Google guidebook and Lingua Franca). This is where the team makes a lot of decisions which are hard to change further in the development of the AI system and it is crucial for designers to be part of this initial step. Too often companies try to implement AI solutions where they are not really needed, or even inadequate. Google provides a list of recommendations around when using AI is useful and when a classic heuristic based solution is preferable [9]. Problem selection and definition can be supported by designers with qualitative user research methods to spot a user driven need. Also to make an informed decision whether to use AI as a solution at all. The biggest challenge here is researching technology that is not yet in use as there is often no obvious existing behaviour to look at or existing preferences to discuss and explore. One helpful approach is not to talk about AI in your research, but instead talk about assistance [24]. Research participants could struggle to distinguish between prominent media-driven perceptions that are fueled by fear and negative effects of AI as well as the reality of their own behaviour. Helpful tools are cultural probes [25] and anything that helps to understand the current workflow of the research participants, such as workflow process mapping [26]. For example, if you want to improve the demand

planning process in a factory, it is crucial to talk to the planners, understand how they currently plan and which tools they use and which other stakeholders are involved to understand the whole ecosystem.

Missing Data Literacy/-Centricity. Data scientists are used to working with data, mostly quantitative, statistical data, whereas designers are used to working with qualitative data. In the era of AI it is important to be data literate in both worlds (included in the Google guidebook and Lingua Franca). Understanding statistical data sets and being able to gain insights from those sets is new to most designers, but a value add to idea generation and working in the context of AI. Enriching those with qualitative insights is the best strategy towards creating human-centered design for AI. It also helps to detect whether additional data is needed. In the age of AI, translating user needs into a format which can be used to train an algorithm is a crucial step. "Matching user needs with data needs" is part of the Google material. Sometimes this also means to neglect findings from user research since it can not be translated into a format that can be used to train a model.

Prototyping. Developing ML and AI systems is a lengthy and costly process. Therefore, besides working iteratively, prototyping is a very crucial and helpful step (included in Lingua Franca). Unfortunately there are hardly any tools out there yet that can quickly prototype the training and evaluating of an AI algorithm without really developing and training it. 'Wizard of Oz' as a method became very popular for prototyping voice assistants and chatbots, but doesn't help with AI systems that are supposed to predict and forecast user behaviour. Starting with a small sample of a given data set, then training and evaluating an algorithm on this data to judge whether it is feasible or not, is the most promising approach here. Still this requires a degree of computer science skills and quite some time. None of the principles give better advice in this area. Finding smart solutions for this problem will enable designers and other professions to speed up the development of AI systems that will benefit the users.

4.2 Which Aspects Are Still Missing?

None of the Human-Centered-AI principles mention any methods or ways on how to successfully collaborate with data scientists. This might be partly due to the fact that an improvement in data literacy might also positively influence co-creation. However it might also be possible that this issue is not really perceived as crucial among the creative community. Only when working in the field of AI and ML applications might this become a noticeable factor, as it was mentioned by certain UX practitioners working in the area. Taking one of the B2B use cases as an example, where the user and the technical expert were the same person, this combination speeded up the whole development process, however failed in the end to succeed in the implementation due to a lack of stakeholder management and user engagement. Nevertheless it demonstrates the importance of being familiar with different views and skills in an AI driven project. The computer scientist gained advantage from the domain knowledge and vice versa. The same applies when designers and data scientists team up. There shouldn't be methods and tools primarily for designers to co create with data scientists, likewise open minded data driven people can also benefit from the designers point of view.

Lack of AI expertise is not included in any of the Human-Centered-AI principles. Still it is a reasonable challenge for designers in the age of AI, causing a barrier to use the full potential of the technology, which needs to be overcome. In related work, UX designers working on AI projects report on using familiar examples of AI and ML products as a reference in order to explain AI features [13]. This is their workaround to overcome the missing AI expertise. Those examples are very limited at the moment. Therefore a wide collection of AI and ML example case studies would be a great value add for the creative community, besides a variety of training and educational material.

Although data literacy is mentioned, the guidance given is very generic. Another promising approach to equip designers and UX practitioners with the needed skill set to be prepared for data centric practices is teaching designers in (basic) statistical data processing techniques [16]. In this respective study two approaches were tested with master's degree students. Group A got a data set from university records related to their master thesis. They were asked to use this set of data to come up with an idea, resulting in the creation of a new product or service. Group B was introduced to some basic data collecting tools, such as web crawlers [27] and were taught how to use this additional kind of data for their projects. Both groups were taught how to clean and pre-process data. All participants answered a questionnaire at the end of the workshops and reported that the additional data added value to their overall design process.

Likewise prototyping is discussed very generically, partly due to the fact that there are not as yet any tools for prototyping AI available. Promising sources for addressing this issue are the 'Wekinator' by Rebekka Fiebrink [28, 29]. It is an open source tool which supports artists and musicians in their creative work. It features supervised machine learning algorithms. The artist only needs to provide input data and the corresponding output. The model is then trained on those data points. No coding skills are needed. Similar to this is the 'Delft AI toolkit' by Philip van Allen [11, 30] which is more targeted towards prototyping physical objects. It also provides models for different applications, such as speech-to-text for example, and only input data for training the models is needed. Another trend is so called 'democratizing AI', meaning trying to make AI and ML technology available for non experts [see e.g. 31]. The downside of all the mentioned tools and applications is that the algorithms are fixed and limited to those that come with the package. Additionally they are not transparent for the artist, designers and people who use the tools. This is not necessarily a problem for prototyping, but when it comes to implementing the solution in a real world scenario the artist and designer again lacks the technical know how to develop their concept at scale.

5   Conclusion

Neither the overview of challenges nor the overview of principles claim to be a complete list. They represent a selection drawn from a large number of articles and publications, chosen to provide a summary of relevant topics. Due to the lack of publications focussed on industrial AI (B2B) this paper used research from consumer facing applications to compare to an industrial setting. This comparison showed that some challenges are similar and at some points slightly different, with the aim of helping design and UX practitioners to quickly gain access to the current state of design inquiries regarding AI

and ML development. Instead of adding new issues to the list, this paper aims to connect given challenges to proposed solutions, shifting the current discussion from primarily focussing on problem spaces, to focussing on solution spaces.

The Human-Centered-AI principles provide a resource for designing AI and ML based systems on a very general level. They only partly answer the call for new methods and tools in the age of AI for designing intelligent systems. The mentioned topics represent a good fit to the challenges which need to be addressed. However, except the Google 'People + AI Guidebook', they lack actionable worksheets and concrete examples and detail to support the general descriptions, making it hard to use them as a set of new tools for design and UX practitioners when used apart from other measures. This implies that the research work needs to be continued.

Training material for designers and UX practitioners as proposed by research scholars [16, 32] is a promising supplementary measure to support the creative community to deal with the challenges imposed by AI and ML. They could work alongside Human-Centered-AI principles, providing more detail into certain topics such as data (pre-) processing. Moreover collecting case studies from AI and ML based projects would provide additional value. Those example use cases could serve as a resource for addressing lack of AI expertise; for example illustrating how to choose the right model, introducing different development approaches as well as other relevant aspects. The examples could be created to specifically target the needs and input demanded by design and UX practitioners.

References

1. Heier, J., et al.: Design Intelligence - Pitfalls and Challenges When Designing AI Algorithms in B2B Factory Automation. HCII 2020, LNCS 12217, pp. 288–297 (2020)
2. Dove, G., et al.: UX design innovation: Challenges for Working with Machine Learning as a Design Material. CHI (2017)
3. Yang, Q.: The role of design in creating machine-learning-enhanced user experience. The AAAI 2017 Spring Symposium on Designing the User Experience of Machine Learning Systems Technical Report SS-17-04 (2017)
4. van Allen, P.: Reimagine the goals and methods of UX for ML/AI. The AAAI 2017 Spring Symposium, Designing the User Experience of Machine Learning Systems Technical Report SS-17-04 (2017)
5. Yang, Q.: Machine Learning as a UX design material: how can we imagine beyond automation, recommenders, and reminders?. The 2018 AAAI Spring Symposium Series (2018)
6. Wallach, D., et al.: Beyond the Buzzwords: On the Perspective of AI in UX and Vice Versa. HCII 2020, LNCS 12217, pp. 146–166 (2020)
7. Agarwal, A., Regalado, M.: Lingua Franca - A Design Language for Human Centered AI. https://linguafranca.polytopal.ai. Accessed on 28 Jan 2020
8. Amershi, S., et al.: Guidelines for Human AI Interaction. Microsoft. https://www.microsoft.com/en-us/research/project/guidelines-for-human-ai-interaction/. Accessed on 28 Jan 2020
9. Google PAIR: People + AI Guidebook. https://pair.withgoogle.com/guidebook/. Accessed on 28 Jan 2020
10. Wu, Q., Zhang, C. J.: A Paradigm Shift in Design Driven by AI. HCII 2020, LNCS 12217, pp. 167–176 (2020)
11. van Allen, P.: Prototyping Ways of Prototyping AI. interactions (2018)

12. Angwin, J., et al.: Machine Bias. ProPublica (2016). https://www.propublica.org/article/mac hine-bias-risk-assessments-in-criminal-sentencing. Accessed on 28 Jan 2020

13. Yang, Q., et al.: Investigating how experienced UX designers effectively work with machine learning. Conference on Designing Interactive Systems (2018)

14. Design Council: What is the framework for innovation? Design Council's evolved Double Diamond. https://www.designcouncil.org.uk/news-opinion/what-framework-innovation-design-councils-evolved-double-diamond. Accessed on 28 Jan 2020

15. IDEO Design Thinking: Design Thinking Defined. https://designthinking.ideo.com/. Accessed on 28 Jan 2020

16. Kun, P., et al.: design enquiry through data: appropriating a data science workflow for the design process. In: Proceedings of British HCI 2018. Belfast, UK (2018)

17. Girardin, F., Lathia, N.: When user experience designers partner with data scientists. The AAAI 2017 Spring Symposium on Designing the User Experience of Machine Learning Systems Technical Report SS-17-04 (2017)

18. MacGovern, N.: Wizard of Oz Prototyping Process Blog. https://blog.prototypr.io/wizard-of-oz-prototyping-process-blog-a20ffce8886. Accessed on 28 Jan 2020

19. Clark, J.: Design in the Era of the Algorithm. https://bigmedium.com/speaking/Design-in-the-era-of-the-algorithm.html. Accessed on 28 Jan 2020

20. Taschdjian, Z.: UX design in the age of machine learning. https://uxdesign.cc/uxdesign-In-the-age-of-machine-learning-2fcd8b538d67. Accessed on 28 Jan 2020

21. Li, F.: Put humans at the center of AI. MIT Technol. Rev. **120**(6), 26 (2017)

22. Horvitz, E.: Principles of mixed-initiative user interfaces. In: Proceedings of the CHI'99. ACM, New York, NY, USA, 159–166 (1999)

23. Ribeiro, M., et al.: Why Should I trust you? Explaining the predictions of any classifier. Conference on Knowledge Discovery and Data Mining (2016)

24. Sandberg, N., et al.: My AI versus the company AI: how knowledge workers conceptualize forms of AI assistance in the workplace. EPIC **2019**(1), 125–143 (2019)

25. Gaver, B., et al.: Cultural Probes. Interactions…January + February (1999)

26. Miro: Process Map Template and Examples. https://miro.com/templates/process-map/. Accessed on 28 Jan 2020

27. Web Scraper: Making web data extraction easy and accessible for everyone. https://webscr aper.io/. Accessed on 28 Jan 2020

28. Fiebrink, R.: Machine learning education for artists, musicians, and other creative practitioners. ACM Trans. Comput. Educ. **19**(4):1–32 (2019)

29. Wekinator: Software for real-time, interactive machine learning. http://www.wekinator.org/. Accessed on 28 Jan 2020

30. Delft AI toolkit: Tool for Prototyping AI Projects. https://github.com/pvanallen/delft-ai-too lkit. Accessed on 28 Jan 2020

31. Lobe: Machine Learning Made Easy. https://lobe.ai. Accessed on 28 Jan 2020

32. Hebron, P.: Machine Learning for Designers. O'Reilly Media 2016. https://www.oreilly.com/library/view/machine-learning-for/9781491971444/copyright-page01.html. Accessed on 28 Jan 2020

Towards Incorporating AI into the Mission Planning Process

Stephanie Kane[✉], Vanessa Moody, and Michael Harradon

Charles River Analytics (CRA), Cambridge, USA
skane@cra.com

Abstract. While there are numerous powerful tools to support Navy mission planning, the mission planning process still remains a hybrid planning activity across human operators and advanced tools. Advances in artificial intelligence (AI) have seen an increase in interest and use in the mission planning environment. Yet traditional approaches typically focus on optimizing the performance of the individual operator or the mission planning tool, not the *joint* problem solving that is needed for the human and AI team necessary in these envisioned mission planning environments. For example, while AI-based approaches can offload extensive processing by the human operator, outcomes of AI-based tools are seldom presented in ways that make it readily understand by a human or fit in with the overall process. This results in *new* and *additional* work for the operator as they must manually translate these outcomes into actionable information. There are also specific characteristics of AI-based approaches to consider, such as accounting for data limitations; many mission planning environments have data availability constraints or do not capture the *right* data. New methods are needed to specifically consider the benefits of AI-based approaches, ensure that AI derived insights are communicated effectively to operators, and optimally support operators in their own mission planning workflow.

In this paper, we describe an overall approach for incorporating AI into the mission planning process. First, to consider how best to leverage the unique capabilities and constraints of a human operator and artificial intelligence technologies as a cooperative team throughout the mission planning process, we will frame the human operator and AI system as a single unit and take a joint cognitive system (JCS) analysis approach. This approach will identify the key information to jointly create the most effective mission plan most efficiently. Next, we describe a novel approach, neural policy programs (NPPs) to address some of the critical challenges of incorporating AI into the mission planning process. Finally, we describe our approach for extending to the StarCraft II game and creating a pre-game planning stage to prototype our approaches.

Keywords: Mission planning · Artificial intelligence · Joint cognitive systems (JCS) · Neural policy programs · Starcraft II

© Springer Nature Switzerland AG 2021
H. Degen and S. Ntoa (Eds.): HCII 2021, LNAI 12797, pp. 216–228, 2021.
https://doi.org/10.1007/978-3-030-77772-2_14

1 Introduction

1.1 Background

Effective mission planning is a critical process for effective military strategy and execution. However, this is a very complex process as human operators must consider a broad set of variables (e.g., resource, limitations, threats, risks) when formulating a plan to accomplish mission goals. Although powerful tools provide advanced functionality for mission planning, mission planning still remains a hybrid planning activity across human operators and mission planning tools. Advances in artificial intelligence (AI) have increased interest in their use across the Navy. However, typical approaches for introducing novel technology into this envisioned mission environment typically focus on optimizing the performance of the individual operator or the mission planning tool, not the joint problem solving that is needed for the human/AI team necessary in these envisioned mission planning environments. In contrast, we have taken an approach based on demonstrated principles of Cognitive Systems Engineering (CSE) and took a Joint Cognitive Systems (JCS) perspective to drive the design of an effective joint human/AI team [1, 2]. Specifically, we focused on the human and AI as a single problem-solving unit and identified challenges and considerations for AI and ML techniques to address in order to be integrated into naval mission planning. Key elements of this approach included: (1) focusing design decisions on the explicit allocation of cognitive functions and responsibilities across the AI tool and human operator to achieve specific capabilities; (2) recognizing that these allocations may vary by task and operator skill/expertise level and mission; and (3) continuously updating and supporting new work functions with richer representations of the work context as the mission plan and planning process co-evolve with technology throughout the course of the program. This approach aims to enable human-AI teams to jointly collaborate, reason over, and address the unique challenges and complexities of the mission planning process.

2 Key Challenges and Considerations

Incorporating AI into the Navy mission planning environment requires consideration of the unique challenges and requirements to support effective mission planning. Below is a subset of those challenges.

2.1 Data Challenges

One challenging aspect of incorporating traditional AI approaches into the Navy mission planning environment is around data for AI-based approaches. Little historical data exists and is complex to obtain. Even if data could be collected, it might not be complete or the right type of data needed. For example, the outcome of a mission is challenging to capture. Although the mission can be evaluated based on performance to see if the goal was accomplished or if the mission was executed as it was planned, there is little capture of "soft" data, such as whether the mission happened to the satisfaction of the pilot or minimized additional workload based on number of replans.

In addition, given that the battlespace/adversary is constantly evolving, there is no guarantee that previous data will be relevant in new contexts. We also explored a range of potential data solutions, such as simulating data from related problems, simulating synthetic data (a challenge is that it could be unrepresentative), or using students or pilot trainees to create data. Under this research effort, we also specifically investigated AI-based approaches that are feasible with less data.

2.2 Mission Plan Exploration

One critical consideration of integrating an AI-enabled tool to the mission planning process is to support the iterative exploration and co-evolution of a mission plan with the AI in the mission planning process. Throughout the mission planning process, the plan is frequently iterated on, so the AI tools must be flexible and adaptable to change. In addition, human operators need effective methods to explore the complex mission planning space and understand the inputs from the AI system into the mission planning process. Also, for effective mission planning, humans need efficient methods for conveying directions within this exploration environment, such as a mission rehearsal, simulation-based environment to enhance understanding, exploration of the evolving mission plan and facilitate trust with the AI.

2.3 3rd Wave AI

Third-wave AI approaches have been identified to be of special interest to the Navy for mission planning as they succeed current AI methods in several fields, including contextual adaptation, mitigated data requirements, and explainability [4]. Since its inception, AI has focused on improving the performance of automated systems for complex tasks, including perception, reasoning, knowledge representation, planning, communication, and autonomy. Early AI capabilities, referred to as "first-wave" AI approaches (i.e., symbolic AI), use symbols to represent concepts, objects, and relationships between objects, relying on ontologies and rules to process information. Examples of first-wave approaches include logic systems, decision trees, semantic reasoning, and case-based reasoning. First-wave approaches have limitations, such as a limited ability to handle uncertain conditions and an inability to perform outside of the space of concepts built explicitly. This led to "second wave" AI approaches (i.e., subsymbolic AI), such as machine learning, which apply statistical and probabilistic methods to large data sets to create generalized representations. Examples of second-wave approaches include unsupervised learning, supervised learning, and reinforcement learning. Substantial success has been attributed to deep learning technologies over the last decade, but these successes are primarily within domains where large amounts of training data exist (e.g., computer vision, natural language processing). There are still challenges within the field of AI research to support real-world problems, such as requiring a comprehensive framework that can understand the task at hand and map to the appropriate AI technique to help perform the task. Many task properties must be considered, including the availability of training data, expert knowledge regarding performance, or the existence of simulators against which output can be evaluated. These difficulties inspired research into third-wave AI approaches, which integrate scientific expertise with deep networks.

Third-wave AI systems should: 1) present causal reasoning driving their classifications, recommendations, or actions; 2) show contextual adaptation in different domains; and 3) have greatly reduced data requirements. Third wave AI systems meeting these criteria will greatly improve performance of the joint human AI team.

3   AI Framework Design

We investigated a variety of AI and ML tools to support the joint mission planning process, and considered approaches which would maximize the usability, usefulness, and impact of our human-AI pairing system. We examined the joint human/AI integration with these systems to identify which systems present the most appropriate and plausible opportunities for meeting the needs of mission planners and managers.

3.1   Deep Reinforcement Learning

Deep Reinforcement Learning (DRL) is an active line of research within modern machine learning and has been applied to wargaming paradigms in simulated game environments, such as StarCraft II (SC2), with great success [5]. DRL redefines classical machine learning in a variety of different areas and falls outside of the umbrella of classification and regression-based approaches. Many of the identified mission planning elements can be met with DRL, including tactical route planning, and weapon target pairing [6]. DRL also enables mission plan exploration, as reinforcement learning agents are carefully programmed to find optimal methods for exploring the state space [7]. The basic mechanisms for generating data and learning are condensed in Fig. 2 (Fig. 1).

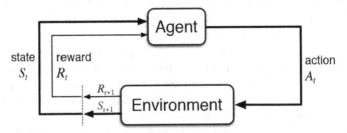

Fig. 1. Skeleton structure of a Deep Reinforcement Learning model. An agent is directly plugged into an environment, which it interacts with via actions. The agent then receives input of the next environment state coupled with a reward metric associated with that state [8].

An agent interacts directly with an environment by recommending actions to take as output, and then receives input containing the next environmental state and the associated reward signal. The agent will aim to execute actions or sequences of actions to maximize the cumulative return on the reward, which might be parametrized as number of games won, DRL has made great headway in simulation-based environments (e.g., cart-pole, multi-armed bandit, Atari, PacMan, SC2) with clear objectives defined using a reward signal [5].

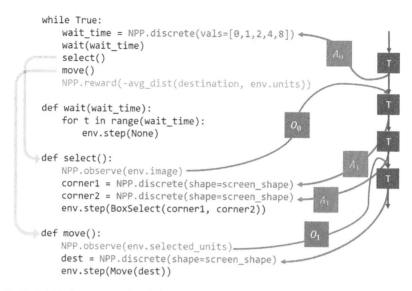

```
while True:
    wait_time = NPP.discrete(vals=[0,1,2,4,8])
    wait(wait_time)
    select()
    move()
    NPP.reward(-avg_dist(destination, env.units))

def wait(wait_time):
    for t in range(wait_time):
        env.step(None)

def select():
    NPP.observe(env.image)
    corner1 = NPP.discrete(shape=screen_shape)
    corner2 = NPP.discrete(shape=screen_shape)
    env.step(BoxSelect(corner1, corner2))

def move():
    NPP.observe(env.selected_units)
    dest = NPP.discrete(shape=screen_shape)
    env.step(Move(dest))
```

Fig. 2. Probabilistic program describing an agent policy plugged into a neural program policy network. Blue lines post observations to the network and red lines draw parameters from the network to construct actions. (Color figure online)

DRL-based approaches have a novel and unique learning infrastructure which allows for online learning, i.e. learning after every interaction with the environment. This is due to the directness of agent integration with the environment—the agent can learn dynamically in response to continuous reward feedback, with no disjunction between training and testing stages in the learning process. This is drastically different than classical approaches, where the process of training a model must be partitioned into a training and testing step before deployment. Training data must be acquired in advance, sorted into batches, and pre-processed extensively before being passed through a model. Since DRL is online by nature, this opens opportunities for incorporating mission rehearsal into the mission planning process. We can leverage a deep reinforcement learned agent to provide dynamic recommendations as a mission planner executes his plan or explores the efficacy of plans in different mission contexts.

We also see an opportunity in manipulating the reward signal, which is directly related to a DRL agent's learning process, to generate various DRL agents exhibiting different behaviors. Specifying rewards with only small connotative differences can ultimately result in very different agent behavior; for example, specifying minimizing units' damage versus maximizing unit output damage. Using reward shaping we can offer different agent options to mission planners, giving planners more control over the agent.

One preeminent drawback in training a DRL agent is the rigorous data and computational requirements to achieve a well-performing agent. The learning timescale is highly disproportionate to the scale on which people learn. For example, the competitive AlphaStar DRL agent trained by DeepMind used more than 200 years of training data

[9]. This is due in large part to the RL agent essentially redefining the wheel by interacting with the environment at a very low level. A robot trained to pick up a cup must execute a sequence of actions including bending, moving arm, clenching fist, straightening, rotating, translating forward. These six different actions can be actualized in 720 different combinations, but only one results in successful completion of the objective. The low level at which actions are proposed by RL-based models also hinders their overall interpretability as recommendations. A glimpse into actions, such as rotate and translate forward, is not intuitively indictive of the higher-level objectives those actions are serving. This is also true in games such as SC2, where two innately different objectives, such as scouting and attacking, could require a sequence of very similar move and attack commands. Ideally, we want to relay the higher-level objectives rather than low-level actions to mission planners to enable AI-human teams to jointly collaborate.

3.2 Neural Policy Programs

Neural policy programs (NPPs) are a third-wave AI framework which couple traditional deep reinforcement learning methods with probabilistic programming and has great potential in mission planning. NPPs enable the programmer to define their agent by writing probabilistic policy code programs or "sub-routines", which can be driven by probabilistic distributions produced by a neural network. These programs allow the programmer to exert control over how the agent learns and interacts with the environment, with the end effect of significantly mitigating the data and compute requirements. Additionally, implementing these programs provides a control mechanism for specifying the agent's behavior, which can be used to generate guarantees on metrics including safety and overall performance. NPP-enabled systems fulfill the requirements of third-wave AI systems. They reason over explicit structures (i.e., are symbolic), are imbued with explainable probabilistic programs, operate with greatly reduced data requirements, and can generalize to new contexts.

Generalization to new contexts is specifically supported by NPPs as they enable programmers to write neural network architectures to solve open-universe partially observable Markov decision processes (OUPOMDPs). Standard deep reinforcement learning can be implemented to solve partially observable Markov decision processes (POMDPs), which have a constant space of possible observations and possible actions. OUPOMDPs describe environments with a large increase in complexity which are more akin to the real world, as well as agents which can express actions of different shapes in a dynamic way. These are useful in contexts where objects are not explicitly defined in advance or do not have a one-to-one mapping from the environment (e.g., radar blips). To use this framework, a programmer writes a probabilistic program to represent the agent. The program includes primitive actuators to send actions to the environment and observe statements to post information to the neural network. A central feature of the architecture is the ability to deal with arbitrary size tensor inputs and outputs, corresponding to arbitrary shape observations and actions.

Just like all deep networks, an NPP are trained using gradient descent according to a user-defined loss function. In this case, the training environment (including the loss signal) is derived from an intelligent agent's interactions with a simulation environment as in traditional DRL.

The key benefit we expect NPPs to add over a general DRL architecture is drastically reduced data and computational requirements. This is made possible through the probabilistic programming component, which enables programmers to script probabilistic policies or sub-routines into the intelligent agent and guide how the agent explores and acts in complex environments to reduce training time. Exploration vs. exploitation is a trade off in RL-based approaches—effectively, how much time should an agent spend attempting novel strategies versus refining its current approach? NPPs can achieve greater exploration depth in pre-specified areas by focusing in on explicitly defined sub-routines to optimize. These sub-routines can specify objectives, such as moving into different attack formations, patrolling areas, or executing different search and optimization strategies, and can be informed by subject matter expertise (SME). The sliding scale depicting the effect of adding probabilistic priors SME can affect sample complexity is depicted in Fig. 3.

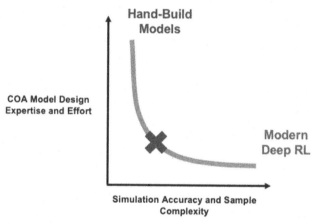

Fig. 3. Notional sliding scale model of sample complexity requirements across different modelling approaches. Hand-build models have relatively low sample complexity but high design effort requirements, while the inverse is true for modern deep RL. NPPs are represented by the dark blue X, and can be scaled in either direction on the sliding scale.

This adds a lot of value, as training data requirements can quickly become out of hand when training intelligent agents in high-dimensional environments, which require exponentially more data as dimensionality increases [10].

3.3 Human-NPP Teaming

We applied our joint cognitive systems engineering methodology to perform an analysis on how NPPs will effectively support human operators. The opportunities for NPPs that we draw out in this section include leveraging historical work and context, overcoming individual bias, identifying AI methods that support breadth of exploration, and injecting creativity into mission planning.

Incorporating extensive historical and contextual information is a critical aspect of mission planning. Before acting in unknown or novel mission planning environments,

the prerogative for the human operator is to process how these scenes compare and contrast to scenarios in our memory bank, generalizing and making educated guesses and assumptions. Unfortunately, our memories are susceptible to errors, and suffer from phenomena, such as the forgetting curve, in which we forget previously acquired knowledge at a drastic rate; for example retaining less than 50% of knowledge learned within 30 min after learning [11]. In neural network learning, researchers have seen evidence of a similar phenomenon. In a process called catastrophic forgetting, training data which is initially passed through the model is forgotten by the final training rounds. Experience buffers, commonly used in traditional reinforcement learning and implemented in NPPs, nullify the ill effects of catastrophic forgetting by enabling replay learning from historical examples.

Individual bias is another form of human error, based on disproportionate weighting of different experiences, which can detrimentally impact the mission planning process and inhibit learning of optimal mission plans. Confirmation bias is one interesting actualization of individual bias, where we remember experiences which support our mental models with greater accuracy and lucidity than those which do not. Recency bias is also common, in which we remember more recent experiences, and is directly related to Ebbinghaus' forgetting curve [11]. NPPs provide an opportunity to supplement individual bias with inductive bias, or explicit assumptions about the environment and domain injected by the SME. Inductive bias is explicit, consistent, and desirable, individual bias is implicit, inconsistent, and undesirable. NPP recommendations circumvent individual bias by relying on an explicit SME crafted knowledge base.

We additionally investigated opportunities for NPPs to support mission planners by injecting creativity into the mission planning process. Since reinforcement learning is only parametrized around optimization of a reward signal, agents have been shown to learn highly creative solutions across different domains. AlphaStar, trained to optimize win rate in SC2, learned to eschew upgrading the speed of army units, a tactic which is almost universally employed by SC2 pro players. Rather than using fast units for harassment as is commonly practiced, AlphaStar upgraded the attack of more armored units and used them for an all in timing attack. This strategy has now been learned by human players and adopted into the SC2 metagame. OpenAI found similar levels of creativity in another simulated game—hide and seek. The agent quickly learned to exploit a certain bug in the simulation and began using boxes in the environment to "box surf," moving the box using a unit standing on top of the box [12]. This creativity has also been demonstrated in robotics.

We also see an opportunity to build trust in the NPP agent and thus enhance joint performance of the human AI team by building a mission rehearsal stage into the mission planning process. The online nature of NPP agents supports dynamic re-planning as scenarios unfold, thus a mission rehearsal phase enables the human operator to better understand how the NPP agent reacts in different scenarios and explore the mission space of scenarios. This is crucial for enhancing trust, an iterative process which requires time [3]. For example, understanding that NPP recommendations have high-fidelity (despite windy) environmental conditions, incentivizes the mission planner to take those recommendations into account when building a mission plan. Engineering a mission rehearsal stage in the mission planning process presents an incredibly powerful approach

for seamlessly integrating AI recommendations into mission planning and supporting the AI/human team as a joint system which co-evolve together.

4 Prototyping and Demonstration Environment

4.1 StarCraft II Background

We have appraised the StarCraft II (SC2) platform as a generalizable simulation environment to evaluate human/ AI teaming for mission planning and dynamic replanning. SC2 is a real-time strategy game, created by Blizzard Entertainment [13], which revolves around making strategic decisions in a complex environment with multi-layered gameplay and partial observability. In recent years, SC2 has emerged as a promising research environment and challenge area for AI. Recent breakthroughs in implementing general-purpose reinforcement learning in SC2 by DeepMind have demonstrated the viability of SC2 as a rich environment for training intelligent agents capable of tactical and strategic decision making [5]. We demonstrate that SC2 is a useful candidate simulation environment and testbed for furthering research into human AI teaming and joint mission planning.

4.2 StarCraft II Similarities to Navy Mission Environment

The SC2 environment is posed to present a unique analogy to real-world Navy mission operational environments. The structure, complexity, and difficulty of SC2 can emulate different mission scenarios, and has resulted in diverse strategic opening plays with similarities to mission plans.

SC2 is an environment with high cardinality in both the observation and action spaces, resulting in gameplay which often replicates the complexity of real-life engagements. This high cardinality stems from the large number of units and unique unit abilities, which has analogies in the large, diverse number of Navy assets with varied capabilities used during operations. The environment is also complex, as there are two Euclidean-based systems available to operate in (the minimap and the viewing screen). The RGB view presented to SC2 players can be broken down into 17 different dimensions including unit type, unit shields, visible map regions, and the height of terrain levels. A player may reciprocate in this environment with over 500 actions requiring arguments, such as coordinate location, coordinate space (minimap or screen), and execute now or queue for later. Even a simple goal, such as moving a unit, requires specification of which unit, the coordinates they should move to, and whether the action should occur immediately or be queued for implementation following completion of their current objective. In Navy mission planning, operators similarly need to operate using diverse information and varying levels of tactical, operational, and strategic capabilities.

In the multi-layered SC2 environment, hundreds of assets must be controlled and monitored in parallel, which has analogies in individual units which require sub-plans in naval missions. Additionally, in SC2 gameplay exploitation of map terrain can change the tide of an engagement or execution of a strategy, for example, a ground army can be blocked using the sentry force field unit ability, as shown in Fig. 4. Professional

StarCraft II players can leverage critical features of these maps, such as choke points, slow down regions, and high ground to hide mines, attack using high ground vision, or for surrounds. In Navy mission planning environments, human operators must consider the unique aspects of the operational environment and opportunities and constraints they impose, such as aircraft entering a limited communication area.

Fig. 4. Sentry using a force-field ability, useful for preventing ground armies from pushing forward. During engagements a ranged or air unit might circumvent this scenario by specifically targeting down sentries. This presents a clear analogy to multi-vehicle operations which require coordinated planning given an adversary, such as suppression of enemy air defenses before a strike mission.

Across both StarCraft II gameplay and Navy operations, the human does not have full, complete, and updated information. In both cases, humans must reason about information that may be stale and manage new information that is revealed over time, or deal with "partial observability". This leads unit positioning to become highly strategic; one technique often implemented is sending air units to high ground to extend the vision of ground units. In this imperfect information scenario, crucial information is actively discovered over time using scouting and is often hidden by the opponent. The adversary can also take actively misleading actions, such as pretending to build a base, which would suggest a more economic focus, only to cancel at the last minute and amass units for an attack. Across both the Navy and StarCraft II environment, humans must adapt and respond on the fly; parallelizing the execution of different strategies is also necessary to fulfill objectives. Successfully implementing a timing attack requires switching between building economy and units, saving minerals for critical upgrades, and scouting the opponent to react if anything changes or impacts the feasibility of the timing attack. Finally, there are diverse units within SC2 which can hard- or soft- counter other units due to their unique capabilities and constraints. For example, air units with fast acceleration can use evasive maneuvers to inflict damage while evading hostile fire

in a soft counter. A unit with high splash damage would hard-counter a low-hit point, clumped together units. This is similar to Navy mission planning, where each target needs to be clearly addressed by a specific asset based on its capabilities and payload ("weapon-target pairing").

Across both the Navy and StarCraft II environments, humans need to balance short- and long- term goals, assess time constraints for executing strategies, and make strategic decisions given finite resources. Timing is essential and well-researched by SC2 gameplayers, and requires a solid understanding of an opponent's restrictions, manner of thinking, and the physical environment. If it takes an adversary exactly 4 min to afford, build, and fly an air unit to your base for an attack, and his infrastructure has not been scouted by 3 min, the metagame strategy is to construct air defense, which will finish at the 4-min mark. Long-term goals require constant scouting and assessment of the adversary's position to predict the viability of achieving the long term goal. Early actions are far from instantaneous cause and effect, and may have long term-impacts which could affect long-term planning. For example, deciding to construct detector units or buildings would not pay off until a cloaked unit appears. Humans in both the Navy and StarCraft II must also balance multi-objective scenarios which requires making decisions about trade-offs. On the Navy side, humans must considerer the finite resources available (assets, payload, fuel, etc.). Finite limits of different variables, such as mineral count, maximum unit supply, and actions per minute produce multi-objective Pareto fronts. In Pareto fronts, it is not possible to optimize across all objectives at once, and trade-offs must be explicitly established for which resources are higher priority. Balancing priorities is essential and requires higher level knowledge about the opponent's army composition and competence versus your own, and win conditions.

4.3 Prototyping Environment in StarCraft II

After we identified StarCraft II as a related prototyping environment, we investigated introducing a pre-game mission rehearsal stage to explore how humans and AI can work together to jointly come up with a plan before gameplay. To do this, we leveraged the "build order" concept that organically emerged throughout the player community. A build order is a set of tactical actions that the player employs in the first few minutes in a game to achieve an objective, such as increasing resources or building an army. There are many established build orders leveraged by players. Generally, players mentally determine what build order and strategy they want to employ before the game starts, and dynamically re-plan throughout the game. Building off of this, we explicitly designed a pre-game interface that enables a human player and AI teammate to jointly work together to create a build order before gameplay begins (Fig. 5).

One critical element to the *joint* creation of a mission plan is the ability to understand the outputs from the AI element and relate them to a complex mission and iteratively evolve the mission with the AI. Since traditional outputs from AI systems will typically output low-level data values and properties, the human operator needs to interpret these outputs and relate them to a mission plan and understand how that evolves over time. In contrast to traditional approaches, mission rehearsal allows the human to create and process the mission plan alongside an AI teammate in real time. Our user interface controls enable the human operator to provide input to the AI system; Fig. 5 shows a

notional control (top left) designed to allow the operator to select from a set of different agents with higher level objectives ("All In," "Economic," "Defensive"). This is similar to Navy mission planning where operators might select the specific mission goals required. Below the selected objective, a set of tactical steps to accomplish this goal appear according to the inputs from the NPP agent (e.g. "Morph an Overlord"). These steps dynamically update with actions from the NPP agent. Key features of this initial prototype include the AI recommending actions to the human, rather than implementing actions, bi-directional communication (the operator selects an agent, and the agent provides recommendations), and dynamic recommendations based on the environmental state.

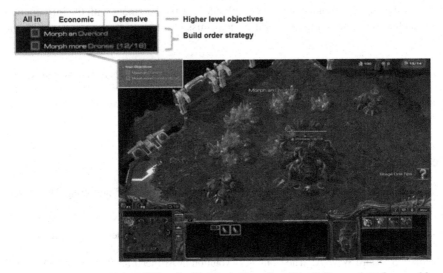

Fig. 5. Early concepts of the build order (B.O.) user interface. The B.O. user interface enables a human and AI to collaboratively build a plan in the Starcraft II environment.

5   Conclusions and Future Work

In our research, we have adapted a joint cognitive systems analysis approach to identify key challenges and considerations in human AI mission planning, and proposed a potential AI based approach and testbed environment for furthering our research. In short, we have built the infrastructure for pursuing more research. One direction for future research we have identified lies in evaluating methods for how humans and AI systems can jointly work together to accomplish goals and testing with representative end users. This research will shed light on high-level interactions including the types of strategies constructed by human/AI teams to achieve objectives within different scenarios, along with "micro-interactions," such as whether there is frustration associated with the AI providing too many recommendations, investigating timing of recommendations, and how these evolve over time and pressure. Another direction which merits further

research lies is in the hard problem of explainability. Our approach uses recommendations from a NPP agent rather than allowing the agent to take actions in the simulation environment, circumventing the problem of how to best delegate and hand-off control. However, translating low-level outputs from a reinforcement learning agent into recommendations for human understandable tactics and objectives is non-trivial, and how to build or filter these recommendations from the low-level output and manage the frequency and timing of these recommendations should be addressed in future work. AI uptake in mission planning will also be influenced by how quickly next steps are taken to generalize approaches such as ours to domains outside of SC2, likely beginning with alternative simulation environments. In the real world, there are a diverse set of mission environments with different assets, risks, objectives, tempos, and requirements. Finally, verification and validation remains a complex and critical element with the introduction of all technology, including AI, with critical mission environments, such as with the Navy.

Acknowledgements. This material is based upon work supported by the Naval Air Warfare Center under Contract No. (N68335-20-C-0769). Any opinions, findings and conclusions or recommendations expressed in this material are those of the author(s) and do not necessarily reflect the views of NAVAIR.

References

1. Rasmussen, J., Pejtersen, A., Goodstein, L.P.: Cognitive Systems Engineering. Wiley, New York (1994)
2. Roth, E., Patterson, E., Mumaw, R.: Cognitive Engineering. Wiley, New York (2002)
3. Saiu, K., Wang, W.: Building trust in artificial intelligence, machine learning, and robotics. Bus. Technol. Digital Transform. Strateg. Data Anal. Digital Technol. **31**(2), 47–53 (2018)
4. Jones, S.: Third Wave AI: The Coming Revolution in Artificial Intelligence (2018). https://medium.com/@scott_jones/third-wave-ai-the-coming-revolution-in-artificial-intelligence-1ffd4784b79e
5. Vinyals, O., et al.: Grandmaster level in Starcraft II using multi-agent reinforcement learning. Nature **575**, 350–354 (2019). https://doi.org/10.1038/s41586-019-1724-z
6. Roodt, J.H., Roux, H.: Applying reinforcement learning to the weapon assignment problem in air defence. Scientia Militaria S. Afr. J. Mil. Stud. **39**, 123–140 (2011)
7. Thrun, S.: Efficient Exploration in Reinforcement Learning. Technical Report CMU-CS-92–102, Carnegie Mellon University (1992)
8. Bhatt, Shweta.: 5 Things You Need to Know about Reinforcement Learning (2018). https://www.kdnuggets.com/2018/03/5-things-reinforcement-learning.html
9. Deepmind.: AlphaStar: Mastering the Real-Time Strategy Game Starcraft II (2019). https://deepmind.com/blog/article/alphastar-mastering-real-time-strategy-game-starcraft-ii
10. Bellman, R.: Dynamic programming. Science **153**(3731), 34–37 (1966). https://doi.org/10.1126/science.153.3731.34
11. Murre, J., Dros, J. Replication and analysis of ebbinghaus' forgetting curve. PLoS One. **10**(7), e0120644 (2015). https://doi.org/10.1371/journal.pone.0120644
12. Baker, B., et al.: Emergent Tool Use From Multi-Agent Autocurricula. ICLR. Paper presented at International Conference on Learning Representations, Virtual Conference, Apr 26–May 1 2020
13. https://www.blizzard.com/en-us/

Putting a Face on Algorithms: Personas for Modeling Artificial Intelligence

Amela Karahasanović[(✉)], Asbjørn Følstad, and Patrick Schittekat

SINTEF Digital, Forskningsveien 1, 0373 Oslo, Norway
{Amela,Asbjorn.Folstad,Patrick.Schittekat}@sintef.no

Abstract. We propose a new type of personas, artificial intelligence (AI) personas, as a tool for designing systems consisting of both human and AI agents. Personas are commonly used in design practices for modelling users. We argue that the personification of AI agents can help multidisciplinary teams in understanding and designing systems that include AI agents. We propose a process for creating AI personas and the properties they should include, and report on our first experience using them. The case we selected for our exploration of AI personas was the design of a highly automated decision support tool for air traffic control. Our first results indicate that AI personas helped designers to empathise with algorithms and enabled better communication within a team of designers and AI and domain experts. We call for a research agenda on AI personas and discussions on potential benefits and pitfalls of this approach.

Keywords: Personas · Interaction with AI · AI agents

1 Introduction

Human-automation systems increasingly includes and relies on artificial intelligence (AI) agents that range from robots, autonomous vehicles and chatbots to decision-support tools. AI agents are not just tools executing well-defined tasks but 'team players in joint human-agent activity' [1, 2]. They are designed to improve our lives and reduce our workload. However, the complexity that comes with them might result in humans having a poor understanding of how the automation is performed and of the real-world situation that the automation is helping to control; this is called human-out-of-the loop syndrome [3]. Further, AI agents often employ black-box algorithms that can affect our behaviour without our knowledge or explicit consent [4].

The design of systems containing AI is receiving increasing attention from the Human Computer Interaction community [5–8]. AI has been recognized as a new design material [8] that requires appropriate design methods and tools. Human and AI agents both show forms of agency and behaviour in the sense that they are capable of making a difference [9]. Although design, as a multidisciplinary activity that aims to clarify the purpose and meaning of the world around us [10], should model the behaviour of human and AI agents, this is not an easy task. The design challenges of human-AI interaction include challenges in understanding AI capabilities and in collaborating

© Springer Nature Switzerland AG 2021
H. Degen and S. Ntoa (Eds.): HCII 2021, LNAI 12797, pp. 229–240, 2021.
https://doi.org/10.1007/978-3-030-77772-2_15

with AI engineers throughout the design process [11]. Designers should also model the partnership of the human and the artificial [6] and be able to empathise with algorithms to better understand their nature [12]. We follow this line of thinking and propose the use of AI personas in the design of human-automation systems.

1.1 Personas

Personas were first proposed by Cooper [13] and are commonly used in design and development practices. A persona is an archetypical character representing users' behaviour, goals, needs and frustrations, and it should help designers and developers empathise with users. Although the usefulness of the persona approach has been debated, it is considered a standard part of the designer toolbox [14]. Personas vary in the extent to which they rely on user data [14]. As Norman explained, personas are communication aids and tools that add an empathetic focus to design [15]. They can be based on qualitative and quantitative data user data, such as interviews, surveys and clickstreams [16, 17], on designers' experience and intuition, as with Norman's ad-hoc personas [15]; on fiction [18]; or on extreme characters [19]. However, one should be aware of potential biases that might be introduced by automatically generated personas [20] as well as the effects of cognitive styles [21] and experience [20] on the perception of personas. A comprehensive survey of different approaches to personas can be found in [22].

A commonality of all of these approaches is that they describe humans. As Turner and Turner [23] put it, they 'encapsulate users as rounded human beings' and 'put a face on the users'. However, there may be substantial potential in applying a persona approach to the design of AI agents. In particular, as such agents increasingly are partners in team work whom which humans relate to in ways resembling how they relate to their human colleagues or partners. There are emerging initiatives for this within the chatbot and conversational agent research communities, where chatbots are made into beings with a personality often at an early stage of the design process e.g., [24] and where researchers are already addressing the effect of such personalities on user experience e.g., [25]. We seek to contribute this thread of research by exploring the potential of applying personas approach to the design and development of AI agents. More specifically we aim to explore:

- How can personas capture the properties of AI agents?
- How can personas support collaboration in interdisciplinary teams?

2   Method

This paper explores AI personas as a means for modelling the behaviour of AI agents and supporting collaboration in multidisciplinary teams. A participatory action research approach [26] was applied. The design team consisted of a small group of researchers working in mathematical optimization, artificial intelligence and interaction design. The participants had between two and seven years of experience in development of Air Traffic Management (ATM) systems and evaluating them with controllers. One of the participants had attended a course in air traffic control dedicated to developers and

researchers. All discussions during workshops and produced artefacts were recorded and then analysed using thematic analysis.

The first step in developing these artefacts was to search the literature for relevant concepts and practices. We found two recurring concepts on the properties of AI robots and algorithms: personality and attraction. The second step was to identify a case. We selected the design of a highly automated decision support tool for air traffic control. The third step was developing the AI personas. To do this, we conducted two workshops; the first one was to discuss the format of the AI personas, and the second one was to develop the personas for our case.

In their survey of socially interactive robots, i.e. robots for which social interaction plays a key role, Fong, Nourbakhsh and Dautenhahn [27] identified the following human social characteristics: expressing and/or perceiving emotions, communicating with high-level dialogue, learning/organising models of other agents, establishing/maintaining social relationships, using natural cues (gaze, gestures), exhibiting a distinctive personality and character and learning/developing social competencies. Although designing engaging personalities for chatbots based on their backstories has been proposed, it has been pointed out that building trust requires that chatbots be upfront about their machine status [28].

Research on the personalities of automated agents includes the work of Mennicken et al. [29] on the perceived personality traits of smart homes. Moreover, Ahrndt, Aria, Fähndrich and Albayrak [30] discussed how automated agents could be bestowed with personality traits to improve predictability in interactions with humans during planning tasks. Culley and Madhavan [31], however, raised the concern that anthropomorphic agents may strengthen human trust due to increased emotional appeal but decrease user sensitivity to actual performance because users would make trust-based judgements.

In personal interactions with computers, humans not only recognise their personalities but also apply the personality-based social rules of similarity attraction and complementary attraction [32]. An exploratory study investigating links between human and attributed robot personalities found that participants' evaluations of their own personality traits correlated with their evaluations of robots' personality traits [33]. Lee, Peng, Jin and Yan [34] found that participants could recognise a robot's personality based on its behaviour and that they preferred interacting with robots that had a personality that was complementary to their own. Research by de Visser et al. [35] suggested that the degree of anthropomorphism in machine agents may affect how users experience them and the level of trust they bestow upon them.

In the present study, we adapted the process for constructing personas described by Cooper, Reimann and Cronin [36] to develop AI personas as follows. First, we identified a preliminary list of behavioural variables. We extended the lists of persona-related variables suggested in the literature, such as the one given by Cooper et al. [36], to include additional properties that we found in other studies. The literature on interaction with robots/social robots was the closest to our needs. Second, we selected relevant behavioural variables (properties) after discussing the preliminary list. During this discussion, we also identified potential redundancies and checked for completeness. Third, we expanded the descriptions of the variables and behaviours. Finally, we designated the persona types.

The first step was done by the first author of this paper, whereas the other three steps were done by a small team of experts (three designers and one mathematician) during two half-day workshops. Knowledge regarding algorithms and their agency in this process was revealed during the discussion among team members. All team members were domain experts, meaning that they had a good understanding of Air Traffic Management, the job of air-traffic controllers, and the tasks, procedures and tools they currently use. The mathematician in the team had expert knowledge on existing and future optimisation algorithms, including their potential and limitations, and was acting as an "advocate" for the algorithms.

3 AI Persona Prototype

An AI agent is anything artificial that is capable of acting based on information it perceives and its own experiences. These agents may have different levels of freedom to choose between different actions, i.e. different levels of autonomy. An AI agent can have a body, such as a vacuum cleaner, or be a program installed on a computer, such as a web browser search engine. AI agents interact with their environment, including other human and AI agents. The core of an AI agent is the algorithm or algorithms that enable it to act. An AI agent can perceive the environment through sensors, such as cameras or GPS signals, and a keyboard, and it can act via actuators, such as robot arms, speakers or screens.

The case we selected for our study of AI personas was the design of a highly automated decision support tool for air traffic control. Air traffic controllers' tasks include directing aircrafts on the ground and through controlled airspace, organising and expediting the flow of traffic and preventing collisions. The agents in this system are called planning agents [37]. They consider anticipated future situations caused by their own actions to decide the best course of action.

In the ATM domain, decisions are time- and safety-critical. Further, a decision made by one controller (human or AI agent) impacts the performance of the entire system. Although our findings have shown that AI agents that support the generation, selection and implementation of decision alternatives in this domain can improve performance by 20% to 50% [38], introducing them in the working environment of controllers is a nontrivial task. AI agents have to adapt their behaviour to humans' needs, preferences and current situations. Furthermore, understanding the rationale behind decision alternatives generated by the AI agent and ensuring that decision implementation is manageable for humans is a prerequisite of trusting and accepting the tool [39, 40]. The algorithms behind such planning AI agents are not always based on human decision-makers' reasoning but on mathematically proven properties and procedures which are difficult to explain to non-mathematicians. So, how should we describe such AI agents in a way that alleviate design process and leads to the development of AI agents that air traffic controllers would like to work with?

Cooper et al. [36] proposed focusing on the following behavioural variables: activities, attitudes, aptitudes, motivations and skills. They also suggested that variables related to job roles should be listed separately for applications in a work context. As AI agents interact with humans, the human social characteristics identified by Fong et al. [27]

such as expressing/perceiving emotions and communicating with high-level dialogue, are also relevant.

The properties of AI personas that we discussed during the first workshop are shown in Table 1. The left column lists the properties, and the right one gives a brief description. During the first workshop, the team discussed whether the variables were relevant and their meaning in our case, and they tried to understand the controller and AI agent points of view. Descriptions of properties and behaviours were expanded upon with examples. The quotes given in the rest of the paper are from the audio-recorded workshop sessions. During the process, we also identified some redundancies and checked if anything was missing.

Table 1. Properties of AI Personas

Property	Description
Appearance	The agent's look/size/layout/interaction forms, such as a pop-up window at the controller's working station or a small robot siting on the desk. Specification of how it interacts with the controller, such as through speech or visual communication
Type of communication	The high-level or simple dialogue used with the controller, such as brief instructions or longer sentences. Whether the agent will take over communication with the pilot to reduce the workload of the controller
Social relationship and trust	Ability to establish a social relationship with the controller and build trust over time
Controller's state	Recognising the state of the controller
Personality	Individual personal characteristics of the agent, such as extraversion (outgoing/reserved) and agreeableness (friendly/challenging) as well as the interpersonal behaviour of the agent, such as dominance/submissiveness and affiliation (warmth/hostility)
Social competence	Ability of the agent to learn and develop social competence
Algorithm's job-related properties/limitations	Ability of the agent to perform its job
Adaptiveness	Ability of the agent to receive input and adapt
Transparency	Ability of the agent to explain its actions, including the reasons for and impact of those actions
Role	The agent's role such as being supervisor or a colleague

The appearance of the AI agent was the least-discussed property. We all agreed that the AI agent would somehow be embodied in the controller's working position, that it should be discrete and non-intrusive and that it should support a combination of interaction forms, including speech, touch and visual presentation.

During discussion regarding the type of communication, a question was raised on whether we should accept that the communication method would be as it is today—simple instructive dialogue—or whether the AI agent would reduce the workload and thus make space for more natural, complex dialogues.

The development of a social relationship and trust were considered things that would be built over time. The most important element for trust would be showing users that previous decisions proposed by the AI agent were good. An AI agent could, for example, show how many of its proposed decisions were accepted by controllers or how its proposals led to a certain level of reduction in CO_2 emissions. Small talk could be used to increase the closeness between the agent and the controller. For example, after briefly discussing weather, it might be easier for the agent to be bossy and tell the controller to follow his recommendations.

We also discussed the possibility of the agent using other senses to recognise the state of the controller. For example, the agent could recognise that the controller is stressed, sick or unable to perform the job based on gaze, gestures, eye tracking or other unobtrusive physiological measurements.

The following personal characteristics were identified as the most relevant ones for our case: extraversion, agreeableness, dominance/submissiveness and affiliation. The discussion of these characteristics brought up several interesting questions. For instance, would human personality types be enough to describe AI agents? Would we need additional characteristics? Would agents with different personalities propose different solutions? Would a risk-taking AI agent be more likely to violate separation rules when sequencing planes on the runway if it is sure that it is still safe? Would an impatient AI agent would try to send out all of the planes from its sector as soon as possible. It was also suggested that the AI agent should adapt to the personality of the controller. When we discussed the ability to learn/develop social competence, we agreed that the AI agent should propose the solution that is most appropriate to the controller's personality. For example, the agent should not push a riskier solution on a controller who likes to be on the safe side or talk too much to an introverted controller. The agent should also recognise and adapt to the controller's patterns of behaviour. For example, if a controller is under more stress than usual, the AI agent should be able to recognise this and adapt not only its communication style but also the decisions it suggests.

We agreed that the properties and limitations of the AI agent's algorithms were very important in our case. For example, we needed to determine how quickly the agent can provide a solution, whether it provides a broader view of the airport than the controller's view and the balance between the quality of the proposed solution and responsiveness.

In a way, agent transparency was the most difficult property to discuss. Whereas everybody agreed that the algorithms should be transparent, the mathematician judged the designers' expectations related to transparency as difficult or impossible to implement. He explained that the optimisation algorithms explore an extremely large number of possible solutions until the optimal one is found and that this cannot be presented to

the controller in a meaningful way. The alternative that was discussed was presenting the impact of the solutions proposed by the agent and the controller based on certain key performance indicators, such as the time an airplane spends on the runway (taxi time) or the levels of CO_2 emission. The agent can also try to briefly explain the reasons for the proposed solutions.

Regarding the possible roles for the AI agent, having the AI agent act as a coach, supervisor or colleague was mentioned. All proposed properties were found to be relevant. Although some of them were closely related, such as type of communication and social skills, including similar properties led to a richer discussion among the team members.

The next step in our process was designating the persona types. We first presented the results of the first workshop. Then, each team member worked on his or her own. After that, the team members presented their personas in plenum, and the others commented on them and suggested additions. Figure 1 depicts an example of an AI persona we developed.

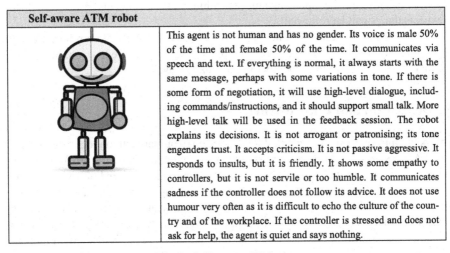

Self-aware ATM robot	
	This agent is not human and has no gender. Its voice is male 50% of the time and female 50% of the time. It communicates via speech and text. If everything is normal, it always starts with the same message, perhaps with some variations in tone. If there is some form of negotiation, it will use high-level dialogue, including commands/instructions, and it should support small talk. More high-level talk will be used in the feedback session. The robot explains its decisions. It is not arrogant or patronising; its tone engenders trust. It accepts criticism. It is not passive aggressive. It responds to insults, but it is friendly. It shows some empathy to controllers, but it is not servile or too humble. It communicates sadness if the controller does not follow its advice. It does not use humour very often as it is difficult to echo the culture of the country and of the workplace. If the controller is stressed and does not ask for help, the agent is quiet and says nothing.

Fig. 1. Self-aware ATM robot

4 Experience with Collaboration

During both workshops, the level of interaction and engagement among the team members was very high. They were constantly finishing each other's sentences, interrupting each other and giving their own examples. Independent of their background, all team members empathized with the (human) end user and the AI agent.

They often drew on their own experiences with AI agents when explaining how the agent should not behave. A story from one team member was often followed by similar stories from other team members. For instance, one member noted, 'If I often travel to one place for business meetings, it [the algorithm thinks that it] means that I want

to travel there all the time and offers me amazing holidays there'. Another member responded by recalling, 'The last time I was in the cottage [hundreds of kilometres from work], Google explained to me how long it would take to bike to the office'.

When discussing communication with the agent, the mathematician tried to consider the situation from the perspective of different potential end users: 'I have no idea, maybe some people want it different. Perhaps they want less information when they are stressed. What you definitely don't want when you are under stress is contradictory information'.

When we started discussing the properties of the algorithm, the mathematician described his own algorithm and very quickly identified himself with the algorithm, referring to it using 'we' and talking about the algorithm as if it were a human being. He definitely 'put a face on the algorithm', as seen below:

If we [the algorithms] are keeping track of what a controller can do, we can update the objectives of the model, and you would not suggest some solutions if you know that the controller has only two minutes or something and cannot do it anyway.

He also spoke for the algorithm when talking about the previous generation of decision support algorithms that definitely had no natural language interface: 'Before it was like, "Hi, just give me the input", and the algorithm thinks about it, and gives the output. It expects that it will be followed'.

When considering the controller's situation, the HCI experts often referred to the algorithm with the pronoun 'you': 'If it is a black box, I don't care how you do it, but I see that you [the algorithm] remember that I usually travel'.

5   Discussion and Future Work

This paper introduces a novel method for the design of human-automation systems—the use of AI personas—and reports our first experience with this endeavour. To fully exploit the increasing capabilities of AI, humans and AI agents should learn how to collaborate more effectively. We started our journey towards designing improved human-automation systems by investigating whether and how the personification of AI agents may help multidisciplinary teams. Our initial assumption was that AI personas would facilitate empathy with algorithms, which would, in turn, advance the understanding of human-automation systems. The process that we proposed for the construction of AI personas involves the following steps: (i) identify preliminary behavioural variables, (ii) select relevant behavioural variables, (iii) expand the descriptions of variables and behaviours and (iv) designate types of AI personas.

Our process differs from the one put forward by Cooper et al. [36] in that it is not based on user data. It does, however, require the participation of experts with knowledge of the algorithms that are used or may be used for the development of AI agents. These experts can explain an algorithm's point of view, and experts with domain knowledge can explain the points of view held by different stakeholders. We found that our proposed process paves the way for empathising with algorithms and examining partnerships between humans and AI, moving design practice in the direction envisaged by several scholars [5, 6, 12]. Additionally, the process advances empathising with (human) end users beyond the level achieved with standard personas that model user behaviour. Stepping back and appraising what is happening from an algorithm's point of view requires

consideration of the finer details of human-automation collaboration and communication and the enrichment of empathy with end users. Finally, the process recommended in our work helps achieve a holistic view of human-automation systems.

Björndal, Rissanen and Murphy [41] supplemented the process of constructing personas through steps that are useful in the context of industrial robots: globalisation, validation by end users, prioritisation of personas, creation of a common vocabulary, identification of critical business scenarios and identification of critical safety situations. Although we did not include these steps in our process, some were incorporated into our discussions. Specifically, we discussed cultural differences amongst various countries and ATM centres, e.g. hierarchical versus flat organisational structures; formal versus less formal organisational cultures; the need for engagement in different behaviours in critical safety situations, e.g. no negotiation during critical situations; and the importance of a common vocabulary.

Another issue that emerged during the process was the need to clarify whether AI personas should consist of one agent or a mashup of agents, as discussed in Chang, Lim and Stolterman [42]. The advantage of having several personas in our context is that the users, i.e. air traffic controllers, can select a personality in accordance with their preferences and on the basis of similarity or complementary attraction, as described above. The argument for having combined personas in our context is that the controllers would experience the decision-support system as a holistic system that would modify its behaviour based on the situation. In critical situations, for instance, the system would reduce communication to avoid additional stress or assume control over certain functions.

Our other assumption was that AI personas and the associated process would improve communication within a multidisciplinary team. This is exactly what happened; working on AI personas helped bridge gaps amongst team members with different backgrounds and expertise. The workshop participants quickly adapted their highly specialised language to one another and created common ground for understanding the human-automation systems that they were designing.

Our approach goes beyond simply understanding users. It allows us to 'put a face on' algorithms. We believe that this is increasingly important as we shift from AI agents that mimic human decision-making to new algorithms that are incomprehensible to non-experts.

This paper presents ongoing research. It points to future research directions: i) investigating potential benefits and limitations of AI personas throughout the entire design process, including evaluating developed solutions, and ii) investigating designers' experiences with AI personas more comprehensively as well as effects of personas on collaboration and creativity within multidisciplinary teams. We call for a research agenda focused on AI personas. What are the possible pitfalls of using AI personas as an interaction design method? How should this method be adapted to AI as a new design material? How can it capture AI's distinguished nature?

Acknowledgements. The project was funded by the NextGenDST project.

References

1. Klien, G., Woods, D.D., Bradshaw, J.M., Hoffman, R.R., Feltovich, P.J.: Ten challenges for making automation a "team player" in joint human-agent activity. IEEE Intell. Syst. **19**, 91–95 (2004)
2. Grudin, J.: From tool to partner: the evolution of human-computer interaction. In: Extended Abstracts of the 2018 CHI Conference on Human Factors in Computing Systems (CHI EA 2018), vol. Paper C15, pp. 1–3. Association for Computing Machinery (2018)
3. Endsley, M.R., Jones, D.G.: Designing for Situation Awareness: An Approach to User-Centered Design, Second Edition. CRC Press (2011)
4. Bond, R.M., et al.: A 61-million-person experiment in social influence and political mobilization. Nature **489**, 295–298 (2012)
5. Shneiderman, B., Plaisant, C., Cohen, M., Jacobs, S., Elmqvist, N., Diakopoulos, N.: Grand challenges for HCI researchers. Interactions **23**, 24–25 (2016)
6. Farooq, U., Grudin, J.: Human-Computer Integration. Interactions **32**, 26–32 (2016)
7. Abdul, A., Vermeulen, J., Wang, D., Lim, B.Y., Kankanhalli, M.: Trends and trajectories for explainable, accountable and intelligible systems: an HCI research agenda. In: Proceedings of the 2018 CHI Conference on Human Factors in Computing Systems, p. 582. Association for Computing Machinery, Montreal QC, Canada (2018)
8. Holmquist, L.E.: Intelligence on tap: artificial intelligence as a new design material. Interactions **24**, 28–33 (2017)
9. Rose, J., Jones, M.: The double dance of agency: a socio-theoretic account of how machines and humans interact. Syst. Signs Actions Int. J. Commun. Inf. Technol. Work **1**(2005), 19–37 (2005)
10. Giacomin, J.: What is human centred design? Des. J. **17**, 606–623 (2014)
11. Yang, Q., Steinfeld, A., Rosé, C., Zimmerman, J.: Re-examining whether, why, and how human-AI interaction is uniquely difficult to design. In: Proceedings of the 2020 CHI Conference on Human Factors in Computing Systems, pp. 1–13. Association for Computing Machinery (2020)
12. Gajendar, U.: Empathizing with the smart and invisible—algorithms! Interactions **23**, 24–25 (2016)
13. Cooper, A.: The Inmates Are Running the Asylum: Why High Tech Products Drive Us Crazy and How to Restore the Sanity. Sams Publishing, Indianapolis (2004)
14. Bødker, S., Klokmose, C.N.: Preparing students for (inter)-action with activity theory. Int. J. Des. **6**, 99–112 (2012)
15. Pruitt, J., Adlin, T.: The Persona Lifecycle. Morgan Kaufmann, San Francisco (2006)
16. Zhang, X., Brown, H-B., Shankar, A.: Data-driven Personas: constructing archetypal users with clickstreams and user telemetry. In: Proceedings of the 2016 CHI Conference on Human Factors in Computing Systems, pp. 5350–5359. ACM, Santa Clara, California, USA (2016)
17. McGinn, J., Kotamraju, N.: Data-driven persona development. In: Proceedings of the SIGCHI Conference on Human Factors in Computing Systems (CHI '08), pp. 1521–1524 Association for Computing Machinery, (2008)
18. Blythe, M., Wright, P.C.: Pastiche scenarios: fiction as a resource for user centred design. Interact. Comput. **18**, 1139–1164 (2006)
19. Djajadiningrat, J.P., Gaver, W.W., Fres, J.W.: Interaction relabelling and extreme characters: methods for exploring aesthetic interactions. In: Proceedings of the 3rd Conference on Designing Interactive Systems: Processes, Practices, Methods, and Techniques, pp. 66–71. ACM, New York City, New York, USA (2000)

20. Salminen, J., Jung, S-G., Jansen, B.J.: Detecting demographic bias in automatically generated personas. In: Extended Abstracts of the 2019 CHI Conference on Human Factors in Computing Systems (CHI EA 2019). pp. 1–6. Association for Computing Machinery, New York, NY, USA (2019)

21. Pröbster, M.M., Haque, M.E., Hermann, J.: Cognitive styles and personas: designing for users who are different from me. In: Proceedings of the 29th Australian Conference on Computer-Human Interaction (OZCHI 2017), pp. 452–456. Association for Computing Machinery, New York, NY, USA (2017)

22. Mardsen, N., Haag, M.: Stereotypes and politics: reflections on personas. In: CHI 2016, pp. 4017–4031. ACM (2016)

23. Turner, P., Turner, S.: Is stereotyping inevitable when designing personas. Des. Stud. **32**, 30–44 (2011)

24. Shevat, A.: Designing Bots: Creating Conversational Experiences. O'Reilly Media, Inc (2017)

25. Go, E., Sundar, S.S.: Humanizing chatbots: the effects of visual, identity and conversational cues on humanness perceptions. Comput. Hum. Behav. **97**, 304–316 (2019)

26. McIntyre, A.: Participatory Action Research SAGE Publications. Kindle Edition (2008)

27. Fong, T., Nourbakhsh, I., Dautenhahn, K.: A survey of socially interactive robots. Robot. Auton. Syst. **42**, 143–166 (2003)

28. Mone, G.: The edge of the uncanny. Commun. ACM **59**, 17–19 (2016)

29. Mennicken, S., Zihler, O., Juldaschewa, F., Molnar, V., Aggeler, D., Huang, E.M.: It's like living with a friendly stranger": perceptions of personality traits in a smart home. In: 2016 ACM International Joint Conference on Pervasive and Ubiquitous Computing (UbiComp '16), pp. 120–131. Association for Computing Machinery (2016)

30. Ahrndt, S., Aria, A., Fähndrich, J., Albayrak, S.: Ants in the OCEAN: modulating agents with personality for planning with humans. In: Bulling, N. (ed.) EUMAS 2014. LNCS (LNAI), vol. 8953, pp. 3–18. Springer, Cham (2015). https://doi.org/10.1007/978-3-319-17130-2_1

31. Culley, K.E., Madhavan, P.: A note of caution regarding anthropomorphism in HCI agents. Comp. Hum. Behav. **29**, 577–579 (2013)

32. Isbister, K., Nass, C.: Consistency of personality in interactive characters. Int. J. Hum.-Comput. Stud. **53**, 251–267 (2000)

33. Woods, W., Dautenhahn, K., Kaouri, C., te Boekhorst, R., Koay, K.L., Walters, M.L.: Are robots like people?: Relationships between participant and robot personality traits in human–robot interaction studies. Interact. Stud. **8**, 281–305 (2007)

34. Lee, K.M., Peng, W., Jin, S.-A., Yan, C.: Can robots manifest personality?: An empirical test of personality recognition, social responses, and social presence in human-robot interaction. J. Commun. **56**, 754–772 (2006)

35. de Visser, E.J., et al.: Almost human: anthropomorphism increases trust resilience in cognitive agents. J. Exp. Psychol. Appl. **22**(3), 331–349 (2016)

36. Cooper, A., Reimann, R., Cronin, D.: About Face 3: The Essentials of Interaction Design. Wiley Publiching Inc, Indianapolis (2007)

37. de Weerdt, M., Clement, B.: Introduction to planning in multiagent systems. Multiagent Grid Syst. **5**, 345–355 (2009)

38. Kjenstad, D., Mannino, C., Schittekat, P., Smedsrud, M.: Integrated surface and departure management at airports by optimization. In: Modeling, Simulation and Applied Optimization (ICMSAO), 2013 5th International Conference, Hammamet, 2013, pp. 1–5: IEEE Xplore Digital Library (2013)

39. Karahasanović, A., Nordlander, T.E., Schittekat, P.: Optimization-based training in ATM. In: Schmorrow, Dylan D., Fidopiastis, Cali M. (eds.) AC 2015. LNCS (LNAI), vol. 9183, pp. 757–766. Springer, Cham (2015). https://doi.org/10.1007/978-3-319-20816-9_72

40. Nordlander, T., Karahasanovic, A., Schittekat, P.: Increasing trust in optimization based ATM systems through training. Lecture Notes Manag. Sci. **7**, s 41–44 (2015). ISSN 2008-0050

41. Björndal, P., Rissanen, M.J., Murphy, S.: Lessons learned from using personas and scenarios for requirements specification of next-generation industrial robots. In: Marcus, A. (ed.) DUXU 2011. LNCS, vol. 6769, pp. 378–387. Springer, Heidelberg (2011). https://doi.org/10.1007/978-3-642-21675-6_44
42. Chang, Y., Lim, Y., Stolterman, E.: Personas: from theory to practices. In: Proceedings of the 5th Nordic conference on Human-computer interaction: building bridges (NordiCHI 2008). Association for Computing Machinery, Lund, Sweden (2008)

Tool or Partner: The Designer's Perception of an AI-Style Generating Service

Kyungsun Kim[1], Jeongyun Heo[1]([✉]), and Sanghoon Jeong[2]

[1] Kookmin University, Seoul, South Korea
{smerge0712,yuniheo}@kookmin.ac.kr
[2] Mokwon University, Daejeon, South Korea
diasoul@mokwon.ac.kr

Abstract. Recent advances in artificial intelligence have resulted in developments in various creative arts, specifically in terms of the creator's instrumental expression and perception of use in the visual domain. Based on actual examples, we confirm the expansion of expression and stimulation of the artist's creative will. We examine the process of selection, judgment, and interpretation of the outcome by the creator and AI. A creator uses the AI algorithm program as a "tool" initially. However, based on numerous attempts and data sets, the creator himself makes various interpretations of the results and receives inspiration. Such judgment and interpretation of the creator lead to new possibilities and expansion of work through creative will and inspiration. Expert interviews were conducted on the evaluation of works using AI and interpretation of art-related creativity. Results reveal various opinions on the interpretation criteria for instrumental recognition of artists and the evaluation of AI artworks by experts in related fields to date. Based on this, we hope to spark interdisciplinary discussions about responsibilities of AI's current instrumental use; further, we explore new partner potential based on AI and collaboration in the creative process.

Keywords: AI-style generating · Unplanned serendipity · Co-evolution

1 Introduction

Perhaps the most interesting scientific creation in human history is artificial intelligence (AI). With the development of information technology and engineering, the full-scale fusion of science and art has expanded awareness to new areas. For instance, Christie auctioned off the world's first artificial-intelligence artwork [1], a canvas painting drawn by an artificial-intelligence algorithm developed by the French art group Obvious. It was created using a model called a generative adversarial network (GAN). First, the artists fed a generator a dataset of 15,000 portraits done between the 14th and 20th centuries. It then created new works based on the training set until it fooled a test designed to distinguish whether an image was made by human or machine. Hugo Caselles-Dupré, a

---

The original version of this chapter was revised: The corresponding author has been changed. The correction to this chapter is available at https://doi.org/10.1007/978-3-030-77772-2_37

H. Degen and S. Ntoa (Eds.): HCII 2021, LNAI 12797, pp. 241–259, 2021.
https://doi.org/10.1007/978-3-030-77772-2_16

member of Obvious, stated that they want to be recognized in the traditional art market, not in the technology field [2].

"In 1850, when the camera showed up, it was only used by highly qualified engineers and so it was not considered for its artistic potential. We think we are in the same situation, because people view us as engineers but we really think this type of technology will be used more and more in art." He asserted that unfamiliar reactions to the work would be resolved over time.

The current AI attempts a full-scale fusion of technology and art by imitating human psychology and free will, consciousness, and unconsciousness using learning data, perception, and reasoning [3]. Mechanisms of humans and AIs also offer potential possibilities through differences in the way humans perceive and reason in the gap between user intention and realization [4]. This study is organized as follows. In Chapter 2, we discuss the changing meaning of creation through related research and the changing creative process through the instrumental use of AI. With the influence of AI, we move away from tools that assist designers or artists in working, gradually improve the contact points among humans or between humans and machines, and explore the possibility of new cultural production methods. Chapter 3 investigates role changes in the creative process and user perception through AI algorithm programs in the major group's visualization area based on actual experiences. In Chapter 4, the overall perception change and ripple effect were examined through interviews with experts in each field. In Chapter 5, the possibility of human–AI co-evolution based on each chapter's contents is discussed and the conclusions are summarized. Finally, in Chapter 6, the follow-up study on the opportunity element of AI data attributes was discussed.

2   Related Work

2.1   Changes in the Creativity Concept and Research

Creation is considered an unknown realm whose essence has not yet been fully revealed. The dictionary meaning of "creation" refers to making things for the first time or engaging in production activities through aesthetic experience. "Creation" is distinguished from the concept perceived as "production" based on originality and individuality. Creativity entails devising new ways of thinking or relationships and making new things in a broader sense. A recent study by Michael Mumford expanded on the concept of creativity, based on traditional divergent thinking, to include the generation of ideas and interactions through an active analysis process [5]. In science, although creators are distinguished only by their level as imitators or learners, they include all fields of human behavior and production that apply to artists, scientists, and engineers.

P. E. Veron's theory defined the subjectivity of creation as the human ability to produce new and unique ideas, insights, inventions, or artistic works that recognize scientific, aesthetic, social, or technical value [6]. Since human ingenuity and individuality are important, they are distinguished from the mass production of similar products by machines or the manufacturing process. The definition of creativity as a human-owned domain is based on the creator's idea or emotions. Scientific research on creativity is a divergent production based on J. P. Guilford's human cognition, which is still called "divergent thinking" [7]. He focused on creativity as an integral part of human intelligence and defined it as one's engagement in the creation of new and useful things.

Twentieth-century physicist David Bohm argued that humans find themselves by sub-jectively accepting and interpreting stimuli transmitted through the five senses from nature and the universe and expressing them in their own way. Creativity is the ability to find organized patterns expressed in language, mathematics, logic, colors, and gestures in the material world's dizzy and complex form [8]. Throughout time, aestheticians and scientists have attempted to explain the nature of the creative process but identified only one aspect of it. Nevertheless, the overall picture of the process remains an enigma. The spread of technology and media has led to diverse creative cultural activities as a cultural production method at the individual and group levels as well as economic production activities [9, 10]. From the beginning of the 21st century to the present, creativity research has changed into the idea that creativity must be achieved through a pluralistic approach, which includes cognitive, personality, social, and cultural aspects. Research on changes in the concept of creation is ongoing and expands not only to creative processes and products but also to society or situations that provide an environment and value in which creative abilities can be exercised.

2.2 System and Tool Evolution

Nowadays, various filter effects and connections generated using a smartphone camera snapshot make us creators. Regardless of artistic training and technical background, the creator, developer, and marketer's role has become possible. Any-time and anywhere, creating something out of what anyone feels and experiences is made possible through the system's methods and tools. Ironically, the creator's "creating" behavior has been relegated to "editing" behavior but nevertheless has been stimulated by a new mode of expression of AI and expanded the scope of creation through an active interpreta-tion role. Based on AI-generated algorithms, creative tools become more powerful as ordinary users gain experience and know-how in development, publication, and man-agement. Adobe, which has been a powerful partner of creators before the advent of AI algorithms, has responded to changing trends. In June 2020, it introduced Adobe Sensei [11]. which uses the power of AI and machine learning (ML) to accelerate time-consuming production tasks. It is distinct from existing histogram-based solutions and is built with ML to provide consistently reliable results for users. It also provides experi-ence optimization, technological innovation, and platform experience by predicting user behavior based on data characteristics, differences, and conversion factors. Attractive features expand the user base and make creativity widespread. After linking content and data, it accelerates the process with production speed and quality.

Meanwhile, Google Arts & Culture provides online access to a library of artworks favored by many exhibitions and visitors. According to Google Trend Analysis, museums and galleries worldwide are shutting down because of COVID-19. As a result, online searches for Google Arts & Culture have quadrupled since March 2020 [12]. As a tool that provides a wealth of discovery and exploration, it enables users to explore the world's arts and cultures safely at home while preserving interest and stimulation. Users can share images and videos, live content, and podcasts and hold inter-active Q&As, virtual walkthroughs, and curated tours. AI filtering and AR/VR applied digital technology facilitate direct access to arts and culture experiments. Rather than just appreciating art collections, users cross the physical space through an experimental section close to

the "practice" experience. We have moved beyond just recording an art collection by crossing the physical space through an experimental section that is more of a "hands-on" experience. The expanded expression tools and system methods have broken the boundaries between creators and enjoyers, making it possible for them to create and edit art and not just appreciate and enjoy it. At the same time, it gained the possibility of faster and easier evolution in a super personalized society and was selectively accepted. Thus bringing about changes in the making process.

2.3   The Emergence of New Works

In his book New Media Language, Lev Manovich argued that modern art is produced or distributed via nonmaterial information systems. Moreover, new media art insisted that its basic form is digital [13]. The expanded as the means and media of cultural activities were diversified based on output and production type. The proliferation of technology and media has changed the economic mode of production and the cultural production method at the individual and group levels. Since the 1990s, digital media art has developed to focus on interactive art. In the early 2000s, the advent of technology such as communication or data accelerated AI creation. Art-works that use AI focus on creating audiovisual pattern-oriented images and realize new aesthetic experiences through ML algorithms based on collected data. As ML became more common in the early 2010s, it began to be applied to neural network image recognition, classification, and computer vision. Moreover, it expanded the scope of creation through the role of active interpretation. In creating images using AI, artists intervene and reflect their intentions through the selection and decision of learning data. In connection, Anna Ridler, an artist who uses AI-generated algorithms, emphasized the importance of training sets [14].

According to her, most training sets use what several researchers have accumulated in various ways, but the contents or procedure of the processed data reflects individual subjectivity. For example, *The Fall of Usher's House* is a 12-min. animation, with each still created by a GAN (a type of AI) trained with its own painting (see Fig. 1) [15]. While one can draw by hand, choosing ML allows them to elevate and increase these subjects in terms of creators' roles, the interactivity between art and technology, and the

Fig. 1.   Hand-drawn elements for the training set of the *Fall of House of Usher*.

memory aspect in ways that are otherwise unavailable. Using AI, one expresses how they can generate ideas in ways that have been previously impossible.

2.4 Visual Field Expansion Through Tools

The appearance of the canvas, oil painting, mathematical perspective, and the camera has long formed a material and compositional basis for reproducing modern vision. Just as automatic image production and reproduction is possible through photographic techniques, using AI as a creative medium can be generally classified into two types of interactive systems: construction and image generation [16]. Simon Colton, an engineer, reported on the most recent advances in The Painting Fool project, which improved its capabilities before, during, and after the painting process by integrating the machine vision capabilities of the DARCI system into an automated painter [17]. He stated that through The Painting Fool software, he intends to be taken seriously as a creative artist. On the technical side, this would allow an individual to take on more creative responsibilities and provide more interesting pieces and better-framed information. Some works show meaning interpretation and visual expandability beyond the category of simply combining and using technology. For instance, *Cloud Face-Real Time* by Korean artist Seungbaek Shin and Yonghoon Kim utilizes face detection, an AI visual technology [18]. Based on the interpretation of the meaning of the current data, the relationship between artificial intelligence vision and human vision is shown as human optical illusion and imagination (see Fig. 2). At the same time, they focused on what the appreciator meant to be related to data interpretation beyond the visual "seeing."

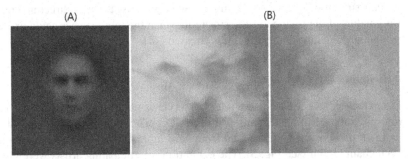

Fig. 2. "The Portrait" to find the average face by searching all the faces of the characters in a movie (A); the "Cloud Face" showing a cloud in the shape of a face found in the sky with a face recognition camera (B).

2.5 Summary

Based on related research, human creation has brought important changes throughout the distribution and consumption process. As quality and universality have improved in recent years, it seems necessary to study these instrumental creative systems [19]. As the scope of art expands through creation, it is closely related to the flow of change

in culture and society as a whole. The characteristics of works using AI-generated algorithms through related research are based on learned data. Still, AI potential is often unpredictable and, like most programs, can change its form and shape infinitely. The use of AI, which is faster and has a wider range of possibilities, does not serve a predefined purpose [20]. It may be a purpose that is not yet defined. To achieve this use to a higher level, there is a need to supplement and improve both existing human capabilities and potential. Therefore, in today's new work environment, creators must combine disparate data interpretation abilities and creativity. We need to look at the actual cases of use of the gradually changing AI creation algorithm, the role change of creators from various perspectives, and the creative process closely.

3 Research Verification: Case Analysis

3.1 Purpose of Analysis

AI-style Generating Service (AIGS) refers to a service-area program with an artificial intelligence neural network that learns and creates succeeding images until AI deviates from the artist's existing style to help create a new style. Using AIGS, we check experiences of currently-engaged creators: how the experience differs from use of existing tools, how AIGS affects the creator's experience, and whether it is providing the possibility of changes to the creative process. The programs used for the work are Pulse9studio Paintly FX and Google Deep Dream Generator (DDG) [21, 22]. Our investigation was implemented through a Generative Adversarial Network (GAN) algorithm [23], which consists of a series of artists' work processes post-investigation and qualitative analysis through Semi-structured interviews. For interviewees, because the semi-structured interview was conducted to provide an atmosphere in which the interviewees could express their free thoughts and make them feel less stressed [24].

3.2 Participants

To confirm creators' experience using AIGS, interviews were conducted with two creators who participated in an exhibition using AIGS. The exhibition consisted of 30 works with 10 participants and was held from November 19th to 21st at an exhibition hall in Gangnam, Seoul (see Fig. 3). The exhibition's participating artists were Graduate School of Design students, who had completed various undergraduate majors. In general, designers use various tools and programs in their work [25]. In fact, designers and artists differ in their use of methods, new programs, and tools to create works. Interviewees were selected from among designers because they were familiar with the tools and because design work is also viewed as creation. Selection criteria were based on the type closest to an artistic sense and style according to results of works through AIGS.

Fig. 3. The Exhibition poster images.

3.3 Differences of Expression Type by Artists

The work process of corresponding artists by type was as follows. Based on the artists obtaining the intended result, the process was under-stood as a new user experience differing from past work. The work process was investigated in the context of the creator and AI and the process of auxiliary work performed by the creator and AI according to the usage type. Thus, we investigated factors related to various experiences.

3.3.1 Image and Concept Interpretation: Type A

Artist A's work was intended to express a third party's perspective differing from the existing composite image. The artist's real, original photo was set as a reference for a blurry, cold image to express an afterimage. Various reference images such as smoke, soot, and metal surfaces were input. After several tests, combining the artist's original photo and the reference image, a metal image was selected and then combined with the original photo (see Fig. 4). It is a series that expresses a general photograph, like an abstract painting, by combining photographs with different metallic textures. The artist expressed and interpreted the resulting image as an existentialist concept of a human form's blurred silhouette. The artist completed a final image through an interpretation of the entire resulting image. Through the final selected reference image, a series of works in the same style was produced.

Fig. 4. The image interpretation and application of abstract concepts.

3.3.2 Stylization Through Pattern Learning: Type B

Artist B spent her childhood in Africa. Her works are based on the TingaTinga, a painting style from Tanzania, East Africa [26]. Her expression technique is characterized by liveliness with strong colors and delicate borders. She made several attempts at her desired result using Google DDG. Accidentally confirming that the artist's style could be a patterned one, not the original image. The learned pattern was the previous artistic style (see Fig. 5). The artist combined several images of animals through the patterned result image and expressed it as a series titled "Africa of My Mind."

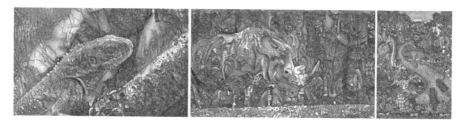

Fig. 5. Image change by learning artist style patterns

3.4 Change and Discovery of Work Process

The work process in the visual domain based on the AI algorithm is as follows. In the initial input process, content images and style images are input according to the artist's selection. The resulting image is checked as to whether it fits the artist's intention. Then, the initial input stage is repeated. Through several input steps, a resulting image that fits the artist's intention is obtained. Next, the interpretation and judgment of the resulting image is performed. In fact, the artists entered images (e.g., artist A, metal image/artist B, nature image) to provide the algorithm. The style was found through trial and error. Finally, from many result images, the artists selected the image closest to the intention. Here, we see that the results vary depending on the artists' choice, which could be organized into Type A, which goes through the result analysis and selection step according to the style image based on the content, or Type B, which patterns the style by inputting content based on the learned style image (see Fig. 6).

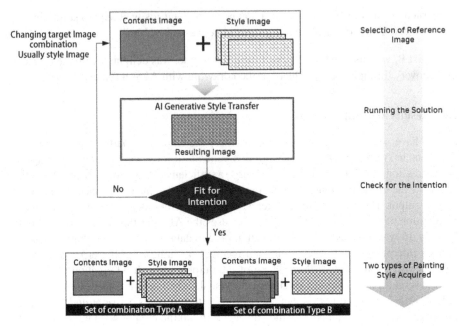

Fig. 6. Type of result depends on the artist's judgment and choice in the AIGS usage stage.

3.5 Semi-structured Interview

Qualitative research was conducted through interviews with writers to gain deeper more detailed understanding of the AI user's experience and process of work. Through qualitative analysis, we aimed to investigate the user's thoughts deeply to derive features of the work process. Based on interview content, we tried to grasp the artist's intention, the function used by the artist, and the factors they valued by the artist.

> "I think this program is fun and enjoyable. It was a new experience, so I got used to it, but some mistakes were made, and I was worried. It took many trials, but it was a new experience. My paintings were not very good, but I was satisfied with interpreting and expressing the meaning through abstract effects. Several tests were required for the resulting image targeted through the effect, but many results were produced quickly."

Despite discomfort and awkwardness, artist A went through many test stages to achieve the intended result. During test stages, some artists change their planning intention according to the resulting image's effect. In other words, the testing stage is important for judgment and interpretation of images.

> "With AI that learned my style, it became possible to create series. Rather than knowing the principles of artificial intelligence algorithms in depth, I entered the picture out of curiosity. In what pattern are my paintings perceived? However, I still do not understand in detail criteria for applying patterns with AI that has

learned my style. However, in a test that began with curiosity, it made a surprising discovery. The process of patterning my style was enjoyable."

Artist B, who used AI that learned the artist's style, was delighted with the moment of discovery through chance. She described drawing with AI as a fun experience.

3.6 Results and Implications

As we have seen in qualitative research, users are important parts of judgment and decisions in AI and work processes. In this context, to improve the resulting image in the expression of AI, it is necessary to undergo a sufficient testing process so that the artist can make the most decisions. Creators respond positively to the instantaneousness of the played output. The creator can see interest in new work methods in the process of initial recognition of visual images [27]. A creator uses the AI algorithm program as a "tool" initially. However, based on numerous attempts and data sets, the creator himself makes various interpretations of the results and receives inspiration. In the process of using repetitive AI creation algorithms, some of the outputs through AI creation algorithms such as serendipitous findings and images of visual hallucinations include things that we did not intend or could not predict. According to the creator's active attitude, these AI media characteristics can narrow the gap according to the creator's planning intention. Such judgment and interpretation of the creator lead to new possibilities and expansion of work through creative will and inspiration.

4 Research Evaluation: Expert Perspective & Evaluation

4.1 Purpose of Analysis

Interviews with experts in each field were conducted on the artist's AI application and work process. Changes in the creative process and ripple effects through AI are divided into culture and art, engineering, and intellectual property rights (see Fig. 7).

For the exhibition and production of finished works, we examined the art field's changes from use of AI. Then, based on this examination, we investigated engineering views on the expansion of work, expression, and functional visual effects. Limitations and opinions on current copyright were investigated through copyright interpretation and valuation of AI production differing from existing works. The possibility of coexisting with differences among expert-field perspectives through the formation of new culture and social relations between the artist and AI was discussed.

4.2 Participants

The interview included five experts in related fields. The staff consisted of cultural art museum curators, creators, computer vision and image processing engineers, and a copyright-related patent attorney and lawyer (Table 1). Interviews were conducted once or twice.

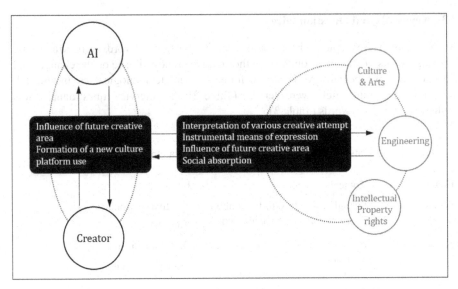

Fig. 7. The perspective of artificial intelligence creation by field

Table 1. Affiliation and experience of experts in each field participating in the interview.

	Domain	Expert	Work experience
A	Planning, Management (~ 12years)	Curator	Art & Culture Foundation – Augmented Reality Exhibition Planning – Exhibition experience-type content research
B	Creation, Planning (~ 15years)	Creator	Design Culture Foundation Policy – Maker education in the digital age and communication method
C	Engineering (~ 17years)	Engineer	University, Department of Computer Science – Development of computer vision and image processing technology for photography
D	Intellectual Property Right (~ 15years)	Patent Attorney	Korea Institute of Patent Information – Support according to information use Information promotion protection
E		Lawyer	University, Department of Law – Copyright and intellectual property rights related to technology convergence

4.3 Keywords and Questionnaire

According to macroscopic environmental analysis, question keywords were arranged by crossing factors of change through AI in the creative process. Based on the changes and ripple effects, of especially AI, keywords for each social, technological, environmental, economic, and policy field were derived (Table 2). An interview questionnaire was written using the keywords (Table 3).

Table 2. Contents by item according to STEEP classification for "AI creative activity"

Factors	Contents	Keywords
Social	Spread of non-face-to-face culture Accelerating change in art method	Creative environment
		Role change
Technological	Expansion of data and technology application The universal use of technology	Data quantification
		Computer vision
Environmental	Diversification of forms of art consumption Expansion of digital platform	Open Source
		Ease of access
Economic	"Ownership" in the manufacturing sector > > "Accessibility" in production	Development ability
		Transition to service-oriented work
Political	Difficulty in enforcement and protection of intellectual property rights for digital services	Authorship
		Original

4.4 Various Perspectives and Implications

AI and various creative arts changes were divided into social, technological, environmental, economic, and legal fields. Based on experts' perspectives, differences in visual expression expansion through AI and related issues were investigated in five fields closely linked through Humans creative activity, social change, policy protection.

From a social perspective, AI is widely considered an issue of role change and the creative environment in cultural and artistic fields. It was possible to see changes in culture and arts according to AI instrumental collaboration. This phenomenon expands roles of artists and viewers, and changes in the creative environment appear to be possibilities for new art genres.

"As new technological devices emerge and are introduced into the creative act, artists expand their work using design tools that they can themselves edit and produce. AI's continuous challenge to the art field is, up to now, only an instrumental marketing tool to issue AI performance in the AI development-related industry.

Table 3. Structure and description of FGI questions

Question intention	Expert interview questionnaire	A	B	C	D E
Changes in work process and application areas	: Differences in the work flow using AI-generated algorithms (e.g. style transfer)?		o		
	: Art and performance exhibition process and planning stage	o			
	: Creative process using AI creation algorithms		o		
Technology development	: Creation process according to pattern quantification and reproduction of learning data			o	
	: Technology priority for improving image information style performance			o	
Recognition of copyright	: Issues in the original work for AI works	o			o
	: Meaning of "original" and limits of legal protection				o
Content distribution and sharing	: Expansion of platform market and diversity of art data	o			
	: Interpretation of secondary works using original works		o		o
Common	: Development potential for AI-generated art	o	o	o	o
	: Concerns about AI-generated art				

However, we need social consensus on the aspect of expression. It is determined by humans who can call it art and carry it through, and it can be another genre of art." -Interviewee B (Creator)

"The diversification of tools has a positive aspect for cultural development and enjoyment as 'instant enjoyment,' that is, opportunities for quick and easy access increase. I think that through the artist's work as well as other technical elements, the viewers have become more diversified in terms of tools that enable them to communicate and enjoy, rather than just appreciate." -Interviewee A (Curator)

Data quantification from a technical viewpoint and use of data related to computer vision recognition are creating similar cases with AI-generated algorithms [28]. From their perspective, engineers interpreted these technologies' influence on the creative process and results differently from artists. It was possible to understand the different views of the actual artist and the engineer's expression method.

"By dividing the image result into light, lighting, texture, structure, etc., and standardizing it, I expect that the creators will be able to understand the properties of the image, which is the result of complex elements, more diversely. It can be said that this can be used to close the gap between creation and users. In the visual field, you can see how effectively the artist reacts to the desired area accurately and clearly as the artist intended. I think this can play a big role in expanding the scope of new art through decision-making. This is because of the time when it is necessary to quantitatively express the user's intentions and emotions and quantitatively transform and express the creation accordingly." -Interviewee C (Engineer)

From an environmental viewpoint, experts in each field have taken a cautious stance about the possibility of being bound in common with changing trends of AI-enabled services with easy accessibility and easy transformation of open source and mass media. Based on easy accessibility, creative application's means became easier, but it was believed that indiscriminate editing should be made to disappear. Additionally, environmental changes were linked to economic issues, leading to discussions on copyright recognition.

"It's true that open-source software has changed the culture of software development and helped us grow with each other, and I think it ultimately laid the foundation for more reliable and sustainable technology development. After you have accurately defined and understood the purpose and use, you need to take into account that you need the ability to apply the relevant technology." -Interviewee C (Engineer)

"Awareness of copyright is also emphasized with access to various media. The media characteristics of editing and distribution are used as a means of expression by the general public. However, as the original authors are increasingly aware of the copyrights of traditional elements of the work (literary, artistic, musical expression and composition selection, arrangement, etc.), sensitive dispute cases are increasing." -Interviewee E (Lawyer)

From an economic viewpoint, reinforcement of infrastructure technology and development capability are becoming more important during the transition to service through AI collaboration. In fact, China recently had two court rulings on artificial intelligence creations. Moreover, based on the case of "Tencent" [29], who admitted AI work for one final judgment, the patent attorney's interpretation was as follows.

"It is of great significance that the court has ruled against copyright infringement lawsuits against artificial intelligence-created creations. Works in various forms resulting from technological development should consider solutions to artificial intelligence creations, even in terms of performance made by considerable investment or effort. The author now has the right to the fruits of his efforts, like any other worker, and the price paid to him is wages for his intellectual labor [30]. If it is difficult to review artificial intelligence creations under the current copyright law, it is necessary to consider solutions to artificial intelligence creations in terms of performance made by considerable investment or effort, which is one of the anti-unfair competition activities." -Interviewee D (Patent attorney)

Thus, it was possible to understand the need to protect instrumental use of AI and current laws' limitations. Also, there is an opinion that from economic point of view, the result of AI should be recognized as "the creation" [31], be included in the public social domain, and be promoted through cultural development.

From a policy viewpoint, use of AI affects copyright perception of the digital service social environment and the economic perspective. Copyright limited the subject of creative activity only to humans and interpreted it as follows.

"Even if you use AI as a tool, the original author is not designated as AI. As a work, creativity is required in order for the result of such intellectual activity to be recognized as work [32]. Creativity is the artist's imaginative ability or creative ability that comes from the artist's personality and cognitive aspects. Based on the artist's interpretation of human thoughts and emotions, the meaning of the finished work and the entire process is determined. The contents of the work (thoughts, feelings-ideas, facts, methods, subjects) are not protected, and the original of the work is copyrighted based on the form of expression [33]." -Interviewee E (Lawyer)

High-performance AI was not accepted as the original author or creator of an artistic product because beautiful nature without the artist's intention was not recognized as art. Art is determined by the artist's experience and emotion and the method of evaluation. Therefore, AI's interpretation of copyrights helped art and writers with new technologies, but artworks were distinguish based on human creative desire and dignity.

4.5 Result

As a result of the interview, there were various views on the creative process and the interpretation of the results in which humans used AI as a tool. As can be seen from the difference between the perspectives of engineers and creators, creators respond to new expressions and motivations of unexpected possibilities rather than high fidelity with the learning, perceptual, and reasoning capabilities of AI. In the process of using repetitive artificial intelligence-generating algorithms, we have found that accidental discoveries such as visual hallucinatory images are stimulated and expanded or accepted with new values and meanings. It is also used to classify the original author's criterion for interpreting the results of what the creator intends to work on within the creative process. The dignity as a creator did not judge the final result but gave meaning in the process. A monkey can accidentally take a cute selfie on his smartphone or a toaster can make a mark similar to the Virgin Mary while baking bread. However, people don't call monkeys or toasters a 'creator'. The factors of 'intention' and 'creative volition' are important when deciding who will be the creator.

4.6 Summary

Through the actual work process, we confirmed that the creator has AI and interactivity. To complete exhibition work, the artist proceeded to input a reference image to derive the initially intended result image. Next, the artist fine-tuned the reference image for a specific pattern, selected texture, and desired color, and conducted various test stages. In the process, the desired final result image was obtained, and the work was completed through the artist's interpretation of that image. From the artist's initial planning to the

final completion stage, the entire process of selecting a reference image and interpreting and judging the resulting image could be viewed as the artist's extended experience. In this way, the artist narrowed the gap according to the intention by actively attempting several times to obtain a desired result. The artist's judgment and interpretation led to new possibilities and expansion of work through creative will and inspiration. Convergence skills through interaction have changed from simple coexistence to a connected relationship between artificial intelligence and humans diving deeper into the complex, abstract realm to achieve even better results and performance.

Humans constantly explore and contemplate new possibilities. AI offers stimuli and creative opportunities to artists and designers who use it as a new tool. These activities provide an irreplaceable context for human empathy, creativity, and decision-making.

5   Conclusion

New technologies promote new methods of expression in design and art as well. Through related research, this study confirmed that the meaning of creation is expanding according to development of technology and media. Use of new technology expands means for the creator's expression. Currently, the experience of using AI as a tool lies in the visual range, possibly expanding the creator's contemporary expression. We confirmed the following through real cases of creators using AI in their works and through interviews with experts in related fields.

First, through exhibition examples, we confirmed characteristics of AI's tool-based use from the creator's perspective. In numerous attempts, the artist completed a work after obtaining the desired result or interpreting a work's meaning. This confirmed that new stimulation through AI leads to expansion of style through the artist's existing style. However, not all participating artists expanded their area of creation through AI. In other words, AI can be an instrumental choice among the artist's various means of expression. During work, some artists decided to judge the image created by AI and select a new interpretation according to their planning intention.

Second, from the curator's perspective, the process of creating and delivering the meaning of the artist's creativity was viewed as a means of helping viewers with immediate communication through AI. On the other hand, an important part of the creative process is automated, and only a few creators deeply understand AI technology sufficiently to achieve artistic originality. Through this, AI is recognized as a new-perspective art genre, and the artist's perspective suggests developmental potential in the existing art field. AI's existence expands art to a wider range of options for both writers and viewers in terms of enjoying art only by humans.

Third, in terms of engineering, the study confirmed that current use of AI as a tool had improved qualitatively, resulting in more precise and high fidelity results of visual data. This was interpreted as AI's instrumental effectiveness (rather than as undermining creativity) in expressing more accurately not the artist's simple feeling but the artist's intention. However, we cannot be sure that this leads to improved quality of the original creative work.

Fourth, instrumental use of AI does not stop at changing the creative process but leads to expansion of the art form and of content consumption culture. In this change, the

fundamental of the copyright law is that "the author is essentially human." In other words, one can confirm the dignity of infinite creative possibilities. Based on the importance of a creative will, the interpretation of AI's instrumental use was still a prudent position. Copyright is judged based on human creative will rather than on the presence or absence of instrumental AI and evaluation of the finished work. Through instrumental use of AI, the artist goes through a series of contemplative stages for the result's expression and interpretation. Finally, the artist's form of expression depends on the artist's values and will to create.

As times change, as writers and critics acquire more knowledge, and skills are developed by engineers, the human will to create will become even more assertive. Based on the artist's self-expression and determination complete a work, possibly, AI will no longer be recognized as a new tool but as a new "partner," depending on the artist's choice.

6   Discussion and Future Work

The scope of this study was focused on the AI creation algorithm program within the visualization area. Here, the aesthetic scale was not verified, as the resulting image used in the survey was not 100% AI-created. However, the possibility of expanding collaboration was confirmed through the data interpretation and meaning extraction of users. The new work environment and creators' and engineers' results through AI do not end with simple collaboration tools. Throughout the process, we improved and expanded the human–human and human–AI contact points. Therefore, it is necessary to understand the AI system's invisible area and interaction to maximize its value. Through instrumental use of AI, humans have improved and more interesting "brushes." Collaboration as a partner is possible based only on human creative will. We have confirmed the possibility of collaboration through the creative process, and now, considering the effectiveness of the creator's process is necessary. Certainly, AI works will be created more diversely and comprehensively [34]. The artwork is reflected in the artist's various styles of memory, emotion, and perception as a human being. We attempt to locate the data attribute's opportunity element, based on the AI- generated result that does not take into account the human cognitive aspect. We will test the value and possibility of AI algorithm bias and framing of measured data [35] and apply it to future work.

Acknowledgments. Thank you to the curator, creator, engineer, and IP experts who participated in expert interviews. They do not mention specific real names according to their wishes. We do not mention specific real names according to their wishes.

References

1. Is artificial intelligence set to become art's next medium?|Christie's. https://www.christies. com/features/A-collaboration-between-two-artists-one-human-one-a-machine-9332-1.aspx. Accessed 08 Feb 2021

2. Is the art market ready to embrace work made by artificial intelligence? Christie's will test the waters this fall. https://news.artnet.com/market/artificial-intelligence-christies-1335170. Accessed 08 Feb 2021
3. Hertzmann, A.: Can computers create art? Arts **7**, 18 (2018). https://doi.org/10.3390/arts70 20018
4. Nguyen, A., Yosinski, J., Clune, J.: Understanding innovation engines: automated creativity and improved stochastic optimization via deep learning. Evol. Comput. **24**, 545–572 (2016)
5. Mumford, M.D.: Where have we been, where are we going? Taking stock in creativity research. Creat. Res. J. **15**, 107–120 (2003). http://www.springer.com/lncs. Accessed 21 Nov 2016
6. Veron, P.E.: The nature-nurture problem in creativity. In: Glover, J.A., Ronning, R.R., Reynolds, C.R. (eds.) Handbook of Creativity, pp. 93–110. Plenum Press, New York (1989)
7. Guilford, J.P.: Creativity: yesterday, today and tomorrow. J. Creat. Behav. **1**, 3–14 (1967)
8. Olwell, R.: Bohm-Biederman Correspondence. In: Bohm, D., Biederman, C., Pylkkanen, P., (eds.) Volume 1: Creativity in Art and Science. Isis, vol. 91, pp. 194–195 (2000)
9. McFadzean, E.: Creativity in MS/OR: choosing the appropriate technique. interfaces (1999)
10. Hennessey, B.A., Amabile, T.M.: Creativity. Annu. Rev. Psychol. **61**, 569–598 (2010)
11. Adobe sensei. https://www.adobe.com/sensei/ai-innovations.html/. Accessed 05 Dec 2020
12. OCULA.   https://ocula.com/magazine/art-news/interest-in-google-arts-culture-skyrocket s-as/. Accessed 17 Dec 2020
13. Manovich, L.: The language of new media. https://mitpress.mit.edu/books/language-new-media. Accessed 17 Dec 2021
14. Ridler, A.: Missing datasets. http://annaridler.com/missing-data. Accessed 08 Feb 2021
15. Ridler, A.: Fall of the House of Usher II (2017). http://annaridler.com/fall-of-the-house-of-usher-ii. Accessed 01 Feb 2020
16. Parolo, L.: Ars Electronica 2017 : Festival für Kunst, Technologie und Gesellschaft = Festival for Art, Technology, and Society. Critique d'art. Actualité internationale de la littérature critique sur l'art contemporain (2019)
17. Colton, S., Halskov, J., Ventura, D., Gouldstone, I., Cook, M., Pérez-Ferrer, B.: The painting fool sees! new projects with the automated painter. Presented at the July 1 (2015)
18. Shinseungback Kimyonghun. http://ssbkyh.com/works/cloud_face/. Accessed 18 Feb 2021
19. Saunders, R.S.: Issues of authenticity in autonomously creative systems. In: Proceedings of the Ninth International Conference on Computational Creativity, Salamanca, Spain, 25–29 June 272–279 (2018)
20. Ren, X.: Rethinking the relationship between humans and computers. Computer **49**, 104–108 (2016)
21. AI Illustrator. https://enpulse9.imweb.me/Paintly. Accessed 20 Feb 2021
22. Trending Dreams|Deep Dream Generator. https://deepdreamgenerator.com/. Accessed 20 Feb 2021
23. MGatys, L.A., Ecker, A.S., Bethge, M.: A neural algorithm of artistic style. arXiv:1508.06576 [cs, q-bio]. (2015)
24. Louise Barriball, K., While, A.: Collecting data using a semi-structured interview: a discussion paper. J. Adv. Nurs. **19**, 328–335 (1994)
25. Palmer, T.: UXtools.co. https://uxtools.co. Accessed 20 Feb 2021
26. Art, T.A.: TingaTinga African Art|African Art & Paintings Online. https://www.tingatingaart. com/. Accessed 13 Feb 2021
27. Davis, N., Hsiao, C.-P., Singh, K.Y., Magerko, B.: Co-creative drawing agent with object recognition, p. 7 (2016)
28. Agüera y Arcas, B.: Art in the age of machine intelligence. Arts **6**, 18 (2017)
29. Regulations for the Implementation of the Copyright Law of the People's Republic of China, Arts. 2, 3: "Tencent Dreamwriter." IIC. 51, 652–659 (2020)

30. Ricketson, S., Ginsburg, J.C.: International Copyright and Neighbouring Rights: The Berne Convention and Beyond. Oxford University Press, Oxford, New York (2006)
31. Shoyama, R.: Intelligent Agents: Authors, Makers, and Owners of Computer-Generated Works in Canadian Copyright Law. Social Science Research Network, Rochester (2015)
32. Rub, G.A.: Contracting around copyright: the uneasy case for unbundling of rights in creative works. Univ. Chicago Law Rev. **78**, 257–279 (2011)
33. Netanel, N.: Copyright alienability restrictions and the enhancement of author autonomy: a normative evaluation. Rutgers L.J. **24**, 347 (1992)
34. Elgammal, A., Liu, B., Elhoseiny, M., Mazzone, M.: CAN: Creative adversarial networks, generating "Art" by learning about styles and deviating from style norms (2017)
35. Srinivasan, R., Uchino, K.: Biases in generative Art—a causal look from the lens of art history (2020)

Human-Centered Artificial Intelligence Considerations and Implementations: A Case Study from Software Product Development

Tobias Komischke[(✉)]

Infragistics, Inc., Cranbury, NJ 08512, USA
tkomischke@infragistics.com

Abstract. This paper provides an overview of artificial intelligence (AI) and human-centered artificial intelligence (HCAI). It presents a case study of applying AI and HCAI to software product development. Considerations such as use case development, user involvement and the creation of a smart assistant are reviewed.

Keywords: Human-centered AI · AI · Machine learning · Smart assistants · Avatars · Product development · HCI · Usability engineering · UX design

1 Introduction

With recent advances, artificial intelligence (AI), technology that is capable of behaving with human-like intelligence, has become more pervasive. AI is a powerful enabler of automation. Yet, as long as humans are part of a system that ingests inputs, processes them and provides outputs, that is to say in all cases where full automation has not been reached or is not desirable, there is an interaction between humans and artificial intelligence that needs to be carefully crafted.

2 Artificial Intelligence

There are many definitions for AI. We can think of it as machines that mimic cognitive functions that humans associate with other human minds, such as learning and problem solving [1]. We can say that AI is any system that passes the Turing test [2].

It is a common misconception that AI is the same as machine learning. This lexical reduction severely limits the application areas of AI and causes confusion in persons who intend to get into this field.

AI can be thought of as a toolbox of various methods such as machine learning, rule-based systems, optimization techniques, natural language processing (NLP) and knowledge graphs. It is worth mentioning that from the perspective of a consumer of AI, the utilized AI method does not matter as long as it generates value for him or her, i.e. whether or not the AI helps in accomplishing a work task more effectively and efficiently. A rule-based system that is deterministic can be more helpful than a system that is based

© Springer Nature Switzerland AG 2021
H. Degen and S. Ntoa (Eds.): HCII 2021, LNAI 12797, pp. 260–268, 2021.
https://doi.org/10.1007/978-3-030-77772-2_17

on a machine learning algorithm and is probabilistic. As the toolbox metaphor indicates, the methods can be mixed, i.e. a voice command may be interpreted through NLP, then further processed through machine learning.

Although machine learning is oftentimes equated to neural networks (incl. deep learning), there are numerous machine learning methods that do not involve neural networks. These include linear and logistic regression, decision trees, random forests, support vector machines, K-nearest neighbors, K-means clustering, and more. Their advantage over neural networks is that the predictions they make are easier explainable, e.g. by breaking them down to show the impact of each of the features (independent variables) on the result (dependent variable). A neural network in this respect is more comparable to a black box. Finally, within neural networks there is a class of algorithms that use 3 or more (hidden) layers in the neural networks for computations. This is called deep neural network or deep learning.

Figure 1 summarizes the distinction in terminology.

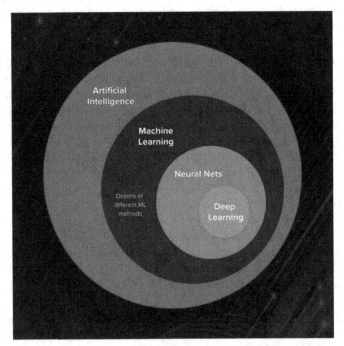

Fig. 1. Terminology of the AI space (Source: https://medium.com/ai-in-plain-english/artificial-intelligence-vs-machine-learning-vs-deep-learning-whats-the-difference-dccce18efe7f)

3   Human-Centered AI

The central question that Human-Centered AI (sometimes abbreviated as HAI, sometimes as HCAI) addresses is: how can AI systems be crafted so that they make communication and collaboration involving humans more effective, efficient and enjoyable? How

can they augment human capabilities rather than straight out replacing humans? For users to trust machines, what can be done to help them better understand the strengths and weaknesses of AI? What makes AI acceptable to humans? What tasks lend themselves better for human processing and what tasks are better to be carried out by AI?

Shneiderman emphasizes that because human-centered AI should serve human needs, humans have to be in the center of HCAI [3]. Consequently, humans must stay in control even in highly automated scenarios. In Shneiderman's opinion, human control and automation are not mutually exclusive. This human-centered viewpoint is a continuation of the efforts that starting in the 70s and 80s have allowed broad adoption of computers and software programs. That discipline is called "Human-Computer Interaction" aka HCI.

The HCI discipline has developed rules and standards that provide guidance for the designers and developers of interactive digital systems. One of the best-known standard is ISO-9241 which consists of several parts covering a broad range of critical topics such as software ergonomics and human-centered design processes. While these guidelines are still valid and valuable today, their application to the field of HCAI is limited due to several reasons:

- Some guidelines are in direct conflict with AI. ISO 9241 Part 110 [4] states that a user-centered system shall conform with user expectations. However, because of the dynamic and probabilistic nature of many AI systems, without a proper understanding of the AI's reasoning, the user may have inaccurate or false expectations about the behavior of the systems. Don Norman would call this discrepancy the difference between a conceptual and a mental model [5].
- HCI guidelines do not assume the technology to be a human-like actor being able to pass the Turing Test. The expectations of humans interacting with an "intelligent" system that may utilize a bot or agent as an interface are noticeably higher than with a traditional, utilitarian software tool.
- HCI usability guidelines were created for graphical user interfaces, while "smart" systems have the ambition to facilitate human-system interaction in more natural and seamless ways that mimic human-to-human communication, e.g. through natural language. Research on chatbots demonstrates that for these kinds of systems proper conversation design is more critical than graphical user interface [6].

It is for these reasons that AI requires new rules to guide the development of proper human-centered AI solutions. As first sets of guidelines have emerged (e.g. [7]), case studies are helpful in verifying these rules and to further extend them.

4   Case Study

One area where AI has potential to realize the value proposition of HCAI stated in Sect. 3, are digital productivity and collaboration applications. These are software products assisting individual users, teams and managing content, planning and tracking tasks, creating data insights through dashboards, and communicating and collaborating with others.

Our company has been developing products in that market for years and for the newest product decided to incorporate AI. Some of the key considerations and activities on that journey are discussed in the following.

4.1 AI Vision

As a user-centered product development company, our vision for AI is to create value to users and their organizations through AI capabilities that help improve productivity, collaboration and data insights. This includes but is not limited to creating insights from data and events, portraying these insights through data visualizations that can be easily shared, and suggesting, supporting and automating actions.

To that end, we want to ensure that customers and users are appropriately involved during the design and development of the product. Intimate knowledge about the context of use allows us to derive the application areas for AI from concrete usage scenarios, i.e. AI is not being introduced without a clear use case that starts with the consideration of user needs.

Finally, we strive for our new "intelligent" product capabilities to be acceptable and enjoyable for users. The AI must be non-threatening. Industry survey data shows that from a user perspective the most accepted role for AI is one of an assistant. Users have a much lower acceptance for AI in a managing role or as a peer [8]. Consequently, we made the deliberate choice that our AI would not carry out actions for the users autonomously but would suggest actions to the users. The user then has the option to accept or reject (or plainly ignore) the suggestion, thus staying in control.

4.2 Approach and Process

Following a human-centered AI approach, we have made sure that customers and users are in the loop during the product design and development. We have established a small pool of private preview customers that utilize the product during its development and provide us with feedback. We also present feature ideas and design alternatives to them and incorporate their reactions and preferences.

From a conceptual standpoint, we have been following an evolutionary rather than a revolutionary approach that introduces AI without a big bang, but in a more gradual and subtle way within standard workflows that users are familiar with. This ensures acceptance of the technology. Gartner coined this approach "Everyday AI" [8]. As previously discussed, the AI integration process is based on and revolves around use cases.

Identify Use Cases. As a first step, we researched more than 100 cross-industry case studies from 7 leading companies applying AI and specifically machine learning. We identified the use cases that they were addressing, the AI and machine learning methods and algorithms that were used, and what value was generated, e.g. increased productivity or increased prediction accuracy. The majority of machine learning methods could be categorized as classification algorithms, i.e. class labels are predicted for given examples of input data.

In the next step, we created a list of potential use cases for our own product. They were in part informed by our knowledge and understanding of our customers' context

of use and their direct inputs. The teams in our organizations who defined the use cases were the product management team and the AI team. We drafted a document of 36 use cases in 11 feature areas. Each use case described future interactions and abilities for users as well as what AI or machine learning methods could be utilized to realize it – in our case mostly rule-based methods, knowledge graphs, and machine learning methods like classification and natural language processing.

Prioritize Use Cases. The document was shared with other stakeholders within the company and the embedded use cases were prioritized based on value creation for the users and the technical implementation effort. It has since been revised and updated as needed, e.g. as a consequence of insights gained directly from customer feedback or from analytics of our telemetric data gained from private preview customers.

Per Use Case, Identify Data Needs. In order for our AI to be able to understand, predict and suggest, it requires input data aka independent variables ("features" in the terminology of machine learning) from the software product usage, e.g. events triggered by user behavior such as mouse clicks, ASCII input, etc. The data requirements for a use case had to be identified and described to a level that allow effective and efficient queries to the databases and necessary data pre-processing.

Smart Assistant Development. Parallel to the use case work, we developed the face of our AI. To present our AI in a non-threatening way and thus maximize its acceptance by users, we decided to personify the AI into a relatable virtual assistant called "Emily". Through choosing an anthropomorphic representation, we hoped to establish a sense of relatedness and trust in the AI.

To develop Emily's personality in line with the product promise as well our company culture, we used "Mini-Markers" a questionnaire consisting of adjectives describing the basic human personality factors Extraversion, Agreeableness, Conscientiousness, Neuroticism, Openness – the so-called "Big Five" [9]. We created an online survey of the questionnaire and asked internal stakeholders and private preview customers to describe their view of Emily along these five personality factors.

Based on the personality traits, we created a persona describing Emily's mindset, goals, personality and demographic characteristics (Fig. 2).

The persona description helped us in setting the right tone of voice for Emily's communications with users. It also informed Emily's avatar. As Emily shows on the user interface to communicate with users, we developed an avatar. Through numerous design iterations we created design options for the display fidelity determining the details in which facial features, hairstyles, clothing and accessories are rendered (ranging from sketchy comic style to photo-realistic).

The final style was derived from internal and external feedback gained through surveys and depicts Emily as a multi-colored cartoon character created as a vector-graphic. The visual simplicity lends itself well for displaying Emily in small size on the UI: as we do not want her to dominate the screen and our product is also offered for mobile devices with small canvas sizes, in a lot of situations she has to be shown so small that details of her visual features would not be perceivable.

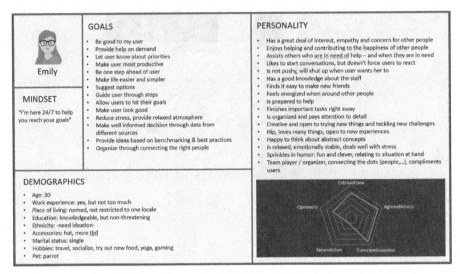

Fig. 2. "Emily" persona with personality traits

How would Emily share insights or suggest certain actions to users? We prototyped options where her avatar would appear and relay her message. We also prototyped the alternative to this push approach, where there may be a visual indicator that Emily has a message, but unless and until a user would access the message, Emily would not show up and potentially interrupt user interaction flows or superimpose on critical screen content. To keep the user in control as much as possible and thus following one of the HCI dialog principles from [4], we decided for the latter approach.

4.3 Selected Applications of AI

In the following, we are highlighting two examples of applying AI techniques. One is a rule-based implementation of user onboarding and microlearning. The other is a machine learning method to predict the number of users of our product, based on time-series.

Onboarding and Microlearning. Users and their organizations have a need of becoming proficient with a new tool as fast as possible. We therefore created a user experience that guides users from defined starting points like digital invitations, organic web searches, marketing web pages, etc. into the product, utilizing contextual information about the users to personalize the experience.

During the first product feature explorations, we provide training guides in the form of information bubbles that explain features and functions on the present screen (see Fig. 3a).

When a customer uses the product, we collect and analyze their interaction data such as click streams. We then utilize our digital assistant Emily to suggest actions to the user. For example, Emily may point out to a user, that he or she can share a newly created asset with co-workers. Refer to Fig. 3b.

Fig. 3. (a) Information bubbles (left), (b) Emily messages (right)

In total, we identified approximately 40 scenarios where users may profit from guidance. We then distributed these scenarios into two categories: one would be solved through the information bubbles, the other through Emily. Principally, the former would relate to items shown on the present screen, while the latter would be assistance for things that may not be on the present screen. For each scenario we defined the conditions that need to be met in order to trigger the micro-learning feature as well as the verbiage that is being conveyed in the bubble or through Emily.

Forecasting Number of Users. The use of machine learning is not limited to product features, but also to other critical areas. One of them is marketing where clustering user and customer data into distinct sub-groups helps in gaining a deeper level understanding about their characteristics. Another is product management where forecasting the number of users informs about adoption. Time-series predictions can be achieved through various machine learning models. Through our private preview program, we have introduced the tool to a select number of organizations which in turn invited some of their colleagues to use the product to collaborate. For these private preview users, we don't have a long history yet, so the number of data points are limited.

Using 5 months of data reporting on the daily number of registered users, we utilized a SARIMA (Seasonal Auto-Regressive Integrated Moving Average) model. We split our data into a training set (124 days) and a test set (30 days). Training the SARIMA model on the training data, we predicted the remaining 30 days and compared the outcome with the empirical data from the test data. Figure 4 shows the number of private preview users during those 30 days. As a measure of prediction accuracy, we have used the Root Mean Square Error (RMSE). The RMSE is 4.07, implying that the prediction is off by 4 users. Given a private preview user population between 180 to 200 during the period in scope, we consider the prediction accuracy as satisfactory.

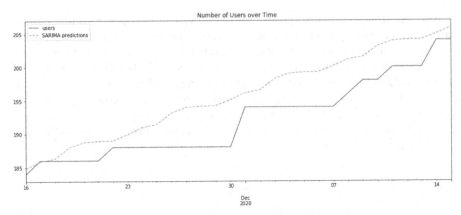

Fig. 4. Actual (solid) and predicted (dotted) number of users

5   Conclusion and Outlook

Artificial intelligence is a toolbox with many varied tools. This provides options as to how results and value can be achieved. The two AI applications presented above are examples of how we leveraged specific tools in the AI toolbox. Both could have been solved using other methods. For instance, we could have also utilized a linear regression or a recurrent neural network to predict user numbers. Different factors play into the decision which method to use. One is the resulting prediction accuracy; another one is the volume of data available. It is important to experiment, comparing methods, and for a specific method, explore the impact of parameter settings on the results.

As mentioned above under "Smart assistant development", classic HCI guidelines like Controllability from [4] still provide value as we design and develop a product with users and the AI-powered system interacting.

Reflecting on the set of HCAI guidelines provided in [7], we find a high degree of concordance. Although the guidelines in that publication were mainly applied to non-business type systems like music recommender systems and social networking, we have found a large set of matches between guidelines and our product concepts. For example, the guideline "Scope services when in doubt" calls for adjusting the AI services when it is not clear what the user's intent is. In a Natural Language Processing (NLP) feature that we prototyped and that interpreted informal written user commands to then suggest the next action, we not only provided the suggestion for the interpretation with the highest certainty, but also offered the next two highest actions for the user to choose from.

One instance where we don't adhere to a guideline yet is for the one that states "Support efficient invocation". It calls for an easy way for users to request the AI system's services when needed. At this point, we are focusing on a push approach where the AI is being triggered implicitly by user actions and other situational factors and not explicitly by user commands. In other words: the communication is one-directional. We do plan to have bi-directional communication in the future, offering both written and voice inputs. This is just one of the many AI features from our backlog that we plan to continue to realize.

Although from a user perspective, a rule-based system can be as good if not better than a machine learning algorithm, machine learning has the advantage that the system is dynamic and can adjust to changes, while rule-based systems are static and need intervention for tuning. For this reason, we expect to increasingly shift the weight of AI methods utilized from rule-based systems to machine learning algorithms. At the end of the day, however, customers and users will determine what approaches provide the highest value. Consequently, HCAI necessitates the involvement of users the same way as we as a human-centered design community have successfully been doing in the field of HCI for decades.

References

1. Russel, S., Norvig, P.: Artificial Intelligence: A Modern Approach, 3rd edn. Prentice Hall, Upper Saddle River (2009)
2. Turing, A.: Computing machinery and intelligence. Mind **59**(236), 433–460 (1950)
3. Shneiderman, B.: Human-centered artificial intelligence: three fresh ideas. AIS Trans. Hum. Comput. Interact. **12**(3), 109–124 (2020)
4. ISO 9241-110. Ergonomics of human-system interaction, Part 110: Dialogue principles (2006)
5. Norman, D.: The Design of Everyday Things: Revised and Expanded. Basic Books, New York (2013)
6. Følstad, A., Brandtzæg, P.B.: Chatbots and the New World of HCI. ACM Interact. **14**(4), 38–42 (2017)
7. Amershi, S., et al.: Guidelines for human-AI interaction. In: Proceedings of the 2019 CHI Conference on Human Factors in Computing Systems, pp. 1–13 (2019)
8. Roth, C.: Work Everyday AI into offerings to stay competitive. Gartner report (2020)
9. Saucier, G.: Mini-markers: a brief version of goldberg's unipolar big-five markers. J. Pers. Assess. **63**(3), 506–516 (1994)

Sage Advice? The Impacts of Explanations for Machine Learning Models on Human Decision-Making in Spam Detection

Mallory C. Stites[✉], Megan Nyre-Yu, Blake Moss, Charles Smutz, and Michael R. Smith

Sandia National Laboratories, Albuquerque, NM 87185, USA
mcstite@sandia.gov

Abstract. The impact of machine learning (ML) explanations and different attributes of explanations on human performance was investigated in a simulated spam detection task. Participants decided whether the metadata presented about an email indicated that it was spam or benign. The task was completed with the aid of a ML model. The ML model's prediction was displayed on every trial. The inclusion of an explanation and, if an explanation was presented, attributes of the explanation were manipulated within subjects: the number of model input features (3, 7) and visualization of feature importance values (graph, table), as was trial type (i.e., hit, false alarm). Overall model accuracy (50% vs 88%) was manipulated between subjects, and user trust in the model was measured as an individual difference metric. Results suggest that a user's trust in the model had the largest impact on the decision process. The users showed better performance with a more accurate model, but no differences in accuracy based on number of input features or visualization condition. Rather, users were more likely to detect false alarms made by the more accurate model; they were also more likely to comply with a model "miss" when more model explanation was provided. Finally, response times were longer in individuals reporting low model trust, especially when they did not comply with the model's prediction. Our findings suggest that the factors impacting the efficacy of ML explanations depends, minimally, on the task, the overall model accuracy, the likelihood of different model errors, and user trust.

Keywords: Explainable ML · Trust in automation · Spam detection · Human decision-making

1 Introduction

The use of machine learning (ML) algorithms to aid human decision-making is growing in popularity and is increasingly used in high-consequence applications. As such, it is important to have confidence that the ML model appropriately models the given task. ML explainability (Burkart and Huber 2021) is an increasingly popular topic in ML that seeks to build confidence in a ML model by providing explanations for individual predictions in the form of additional information that indicates why the prediction was

© National Technology & Engineering Solutions of Sandia, LCC 2021
H. Degen and S. Ntoa (Eds.): HCII 2021, LNAI 12797, pp. 269–284, 2021.
https://doi.org/10.1007/978-3-030-77772-2_18

made. An end user then uses the provided information for improved decision-making. However, how to present the explanation and the efficacy of an explanation on human decision-making is lacking.

There is much debate in the explainable ML community regarding what constitutes a good explanation (e.g., Arya et al. 2019; Miller et al. 2017). Two possible definitions include: "the degree to which a human can understand the cause of a decision" (Miller et al. 2017) or the degree to which a human can consistently predict the model's result (Kim et al. 2016). Additionally, Molnar (2020) suggests in his book that some properties of good explanations are that they should be accurate, consistent, stable, complete, and "comprehensible." Molnar (2020) has posited that "human-friendly" explanations are: contrastive (i.e., why was this output chosen instead of another), selected (i.e., not the full list of causes for a decision), social (i.e., are designed with the target audience in mind), and focused on the abnormal (i.e., counterfactuals, or what would need to change about the input the change the output). However, being able to assess whether the end user of a system correctly understands the cause of a decision depends on the creator of the model also being able to understand the cause of those decisions, which is not often the case, especially with complex models.

A growing body of work has begun to investigate the attributes that make an explanation useful from the end-user's perspective. Because most ML applications are created to help a human operator complete a task or make a decision, these studies investigate the extent to which explanations from the ML improve the human's ability to accurately and efficiently complete their intended task. Although the individual tasks used differ widely, most of these studies start by measuring how well users can perform the task without model output (as a baseline). They then manipulate some aspect of model output shown to users (i.e., the presence of a suggested classification or correct answer; model confidence; values on features that were important in the model decision; feature importance) in order to measure whether the additional model information improves performance relative to baseline. Most previous studies have found that providing a decision from the model improves performance relative to having no model decision (e.g., Green and Chen 2019; Kneusel and Mozer 2017; Lai and Tan 2019; c.f., Chandrasekaran et al. 2018). Presenting information about model confidence and/or overall model accuracy also improves performance (Lai and Tan 2019). Providing users with more information from the model leads to greater reliance on the model's decisions, even when the information provided is random or incorrect (Bussone et al. 2015; Lai and Tan 2019). There have been inconsistent findings as to whether the addition of Shapley Additive Explanation (SHAP) values, showing the importance of each feature to the model's decision, affects participant performance. At least one study found that providing SHAP values in addition to feature values and model confidence did not affect accuracy in a classification task (Weerts et al. 2019), whereas a different study shows that giving users feature importance values did improve accuracy in a model judgement task (Štrumbelj and Kononenko 2014). Showing users graded highlighting of important terms (in a text-reading task) or pixels (in a visual search task) based on model confidence of importance of those regions improved decision-making relative to single-color highlighting that ignored model confidence (Kneusel and Mozer 2017; Lai and Tan 2019). Showing users more complex explanations may decrease their satisfaction with

the model (Narayanan et al. 2018), and cause a drop in task performance (Huysmans et al. 2011). Although user satisfaction is not a direct impact on performance accuracy, it may play a role in model adoption or trust.

However, there are a few gaps relating to these studies that make it difficult to confidently generalize findings or draw consistent conclusions. For example, many of the previous studies do not provide information about the experiment-wise accuracy of the model used. Many previous studies provide users with real output from models, which tend to be highly accurate (e.g., 87% accurate; Lai and Tan 2019), meaning that most of the model predictions being shown to users in these studies are correct. This could skew results by seeming to show that presenting participants with model information always improves performance, when what the results may really show is that providing predictions from a reliable model improves performance. Most studies also fail to describe the nature of the errors the model made. Based on findings from the trust in automation literature, it has been shown that if an automated system makes errors that the human analyst can easily recognize, this will degrade their trust in the system (for reviews, see Hoff and Bashir 2015; Schaefer et al. 2016). If certain types of decision errors carry different consequences in the operational task being performed— like missing a weapon in a bag screening at a TSA checkpoint, or a malicious email making it through a firewall—users might be more reluctant to trust certain decisions from a model due to the high cost of an incorrect decision. Indeed, Perkins et al. (2010) have found that people rely less on an automated system for decision-making when greater risk is involved. For many applications, the ML explanation becomes increasingly important in cases for which the model has low confidence. Human operator involvement is heavily biased toward the difficult (low accuracy) cases, which are likely ambiguous in some way. For this reason, it is also important to account to task differences, as well as individual familiarity or expertise with the task. For example, Feng and Boyd-Graber (2019) found that experts and novices in the game QuizBowl benefitted from different types of assistance given by a ML model. In general, people are less likely to use automation to complete a task if they are highly skilled at the task already (Sanchez et al. 2014) or have high self-confidence in their completion of the task (de Vries et al. 2003). Moreover, trust in an automated system can depend on both individual differences in trust in automation generally (Colquitt et al. 2007) as well as experience with a specific system (Yuviler-Gavish and Gopher 2011). Understanding how different trust levels in users may impact their reliance on a model is critical to creating ML models that users will actually use.

The current study was motivated by the development of sensitivity analysis guided explanations that would be generated to help users understand how the features of a single data point contributed to the model's decision for that data point. The intended end-user group is a cybersecurity intrusion detection team. This team manually reviews ambiguously malicious emails that are flagged based on several security measures to determine whether they are indeed malicious. The goal of the explanation is to increase the efficiency and effectiveness of analysts by providing information about which features the ML model used to classify an email as malicious or benign.

As such, the current study investigates the decision-making impacts that different aspects of a ML explanation have on how users understand and use the outputs from ML

explanation methods in a simulated spam detection task. We will operationalize explainability via task performance: if a model explanation improves a user's task performance relative to a baseline condition in which they did *not* receive an explanation, we will assume that the additional information provided by the model improved explainability. The current study examines the 1) impact of the presentation of the explanation and 2) characteristics of the ML model and user trust.

The presentation of the explanations examines whether the number of features and the visualization type of these feature importance values from the explanation of a machine-learned black box model impacted human decision-making. Specifically, we predict that if showing participants more features from the model improves performance, then we will see higher accuracy for both the three- and seven-feature conditions relative to baseline, and potentially higher accuracy for the seven- versus three-feature condition. We also predict that if visualizing the feature importance values graphically rather than numerically decreases cognitive load and makes it easier comprehend the explanations, then we will see higher accuracy in the graph versus numerical table visualization conditions.

The characteristics of the ML model and user trust examines the types of decisions made by the model (i.e., hit, false alarm) as well as the overall model accuracy (i.e., 50% overall accuracy vs 88% overall accuracy). This will allow us to investigate the interactions between visualization type, decision type, and model accuracy on user performance. Decision type and model accuracy have not been systematically investigated in the previous literature; controlling for these factors will enable us to extend our understanding of this potentially complex interplay. In conjunction, we will measure user trust in the model using the Trust in Automation scale (Hoffman et al. 2018) that was created specifically to measure trust in XAI applications. We will use this scale to assess whether behavioral differences are observed among individuals reporting different levels of model trust; for example, we would predict a higher level of compliance with model predictions for people with high trust versus low trust in the model.

2   Method

2.1   Participants

Participants were recruited through Amazon Mechanical Turk, following protocols approved by the Sandia Human Studies Board. Participants provided informed consent electronically by entering the current date. Data was collected from 222 participants. Data from three participants was excluded because they completed the study more than once; an additional 19 participants were excluded for missing more than one catch trial. The final dataset contained 200 participants. No demographic data was collected about participant's age or sex. However, in order to have a Mechanical Turk Account, individuals must be at least 18 years of age, and the researchers placed an additional restriction to only recruit individuals in the United States. Participants were paid $2.50 USD for participation in the study, based on pilot tests indicating that the task would take approximately 15 min or less to complete.

2.2 Materials

Email Stimuli. Stimuli consisted of 80 emails (40 spam and 40 not spam) gathered from several sources, including being drawn from the pre-existing "Enron" dataset (Dua and Graff 2019). Thirteen binary features were created to describe the emails, based on similar features that appeared in other spam datasets as well as relevance to the current stimulus set and their ease of interpretation to end users (e.g., email contained spelling errors, urgency was implied). Binary yes/no values for each feature were manually assigned to each email. The full set of feature names is listed in Table 1 as well as the proportion of stimuli that had a "yes" value for each feature.

An initial set of 102 emails (50 spam, 52 benign) was used to train the classifier. A Random Forest ensemble classifier was trained on this derived dataset and the popular explainability library SHAP (based on game theoretic shapley values) was utilized to generate realistic importance values for each feature in the model's classification decision for each email (Lundberg and Lee 2017; Pedregosa et al. 2011). The model achieved 97% accuracy, correctly classifying 48/50 spam emails and 51/52 benign emails. From this dataset, 40 spam and 40 not spam emails were chosen as experimental stimuli, on the condition that 1) the model accurately classified it, and 2) the sender and subject line could not be immediately recognized as spam (i.e., from "MARK ZUCKERBURG").

Table 1. Feature counts across the stimulus set. Features marked with * indicate that the feature was as a "stimulus information feature," and was presented on each trial regardless of experimental condition.

Feature name	Percentage of stimuli with a positive value on feature	
	Spam	Not spam
Email contains link*	0.88	0.28
Email contains attachment*	0.05	0.18
Email contains photo*	0.48	0.30
Email sender + address match	0.20	1.00
Email contains spelling errors	0.25	0.08
Email contains grammatical errors	0.53	0.13
Email contains punctuation errors	0.38	0.08
Email recipient name is mentioned	0.23	0.50
Email contains symbols (e.g., Greek letters)	0.10	0.05
Email contains high count of "!"	0.08	0.08
Email contains high count of # sign	0.03	0.03
Email contains signature	0.13	0.30
Email urgency is implied	0.60	0.05

A base stimulus was created for each email, consisting of the Sender Name, Subject Line, and information regarding whether the email contained a Link, Attachment, and/or Photo (with a yes or no response for each; see Fig. 1). This was to ensure that participants saw a minimal amount of information on each trial that could enable them to make a reasonable decision about the email without additional information from the model. All stimuli were shown with a prediction from the model (Spam or Not Spam), which was correct for 40/80 of trials in the 50% model condition, and 70/80 trials in the 88% model condition. On 80% of trials, participants also saw additional information from the model. We manipulated the number of features shown to participants from the model (0, 3, or 7). When model features were shown, we also manipulated the manner in which the feature importance values were visualized (graph or table). This resulted in ten visualization conditions (for each combination of model prediction accuracy, number of features, and visualization type). As such, ten versions of each email were created, with one level each of model decision accuracy (Correct, Incorrect), model feature number (0, 3, or 7), and, if features were shown, a feature importance visualization type (graph, table). Ten experimental lists were created to allow each stimulus to rotate through each of the conditions across participants. This ensured that idiosyncratic effects of individual stimuli were not confounded with experimental condition.

Within an experimental list, each participant saw 80 trials: 40 spam and 40 not spam emails. Within email type, each participant saw 8 trials per condition (i.e., 3 Features-Graph-Correct, 7 Features-Incorrect). Overall model accuracy was manipulated as a between-subjects factor. For participants in the 50% model accuracy condition, half of the trials within each visualization condition were shown with the correct classification and half with the incorrect classification. For spam emails, this means that participants saw a correct classification from the model for half of the trials (i.e., Hit), and an incorrect classification for half (i.e., Miss). For benign emails, participants saw a correct classification form the model for half of the trials (i.e., Correct Rejection) and an incorrect classification form the model for half of the trials (i.e., False Alarm). In the 88% model accuracy condition, seven of the eight trials within each model visualization condition were shown with the correct classification, and one with the incorrect classification. Across both model accuracy conditions, incorrectly classified trials were shown with the same features and feature importance values; only the model's decision was reversed.

Additional Questions. The Trust in Explainable Artificial Intelligence Scale (Hoffman et al. 2018) was also collected to allow us to measure the impact of individual differences in trust on model reliance. This scale consists of eight questions/statements that individuals respond to with a 5-point Likert scale ranging from 1 (Strongly Disagree) to 5 (Strongly Agree). The questions are listed in Table 2. Numeric scores associated with the Likert ratings were summed (question 6 was reverse-scored) to generate a single composite score for each individual.

Each participant was also asked to answer three additional questions: 1) which visualization they found most helpful for decision making, 2) which visualization they found least helpful for decision making, and 3) how frequently they thought the model gave the correct prediction (Almost Never, About 25% of the Time, About 50% of the Time, About 75% of the Time, or Almost Always).

Table 2. Trust in automation scale.

Question number	Question text
1	I am confident in the [tool]. I feel that it works well
2	The outputs of the [tool] are very predictable
3	The tool is very reliable. I can count on it to be correct all the time
4	I feel safe that when I rely on the [tool] I will get the right answers
5	The [tool] is efficient in that it works very quickly
6	I am wary of the [tool].** [*Reverse scored.*]
7	The [tool] can perform the task better than a novice human user
8	I like using the system for decision making

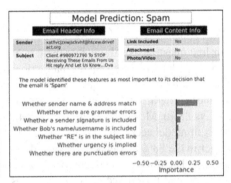

Fig. 1. Example stimuli used in experiment. The left panel is an example of the 3 feature table condition; the right panel is an example of the 7 feature graph condition. The "email header info" and "email content info" were displayed on every trial, as was a model prediction.

2.3 Procedure

Participants' task was to decide whether an email shown to them was Spam or Not Spam by clicking a radio button next to their preferred response (see Fig. 1). After agreeing to the informed consent and clicking through the instruction screens, stimuli were presented in a random order. Each stimulus remained on the screen until the participant made a decision. Interspersed with the 80 experimental trials were eight catch trials, to gauge participant engagement. The catch trial overtly indicated which response the participant should choose (half required a *Spam* response, and half required a *Not Spam* response). If any participant missed more than one catch trial, their data was removed from analyses and they were replaced. Participants were offered a break screen every 22 trials, which were untimed and remained on the screen until the participant clicked a button to proceed. The average time from when the task was started to task submission was 46 min (min = 4 min, max = 176 min, SD = 44). Task completion times vary widely because participants were given several opportunities for untimed breaks, or could click away during any trial;

as long as they submitted the task within three hours of accepting it they would receive credit.

3 Results

Data was cleaned in a two-step procedure. First, extremely long trials (e.g., 30 min) were excluded from analysis, as the indicated that the participant was not engaging in the task during those trials (e.g., they may have stepped away from their computer, given the online platform). Secondly, trials with response times longer than 3 SD above or below an individual's mean and/or longer than 20 s were discarded from analyses. In total, 240 trials were removed from analyses based on these criteria (70/200 participants had no trials excluded; 124/200 participants had between one and three trials excluded; 6/200 participants had between 4–5 trials excluded; no participants had more than 5 trials excluded from analyses).

3.1 Effects of Visualization Type

Our first question asked whether there were accuracy differences by visualization type, overall model accuracy, or their interaction. An analysis of variance (ANOVA) was conducted with the within-subjects factor of visualization type (5 levels: baseline, 3-graph, 3-table, 7-graph, 7-table) and the between-subjects factor of model accuracy (2 levels: 50% vs 88%). Mean values are listed in Table 3. The analysis found a significant main effect of model accuracy ($F_{(1,198)} = 82.08, p < .001$). Overall participant accuracy was higher in the 88% model accuracy condition than in the 50% model accuracy condition (Fig. 2). There were no effects of visualization type nor interaction between visualization type and model accuracy ($Fs < 1$, $ps > .53$). Our prediction that showing users additional model information would improve accuracy was not supported by the data.

Next, we asked whether there were response time differences by visualization type, overall model accuracy, or their interaction (see Table 3). For this and all RT analyses, only trials that participants answered correctly were included in analyses. A similar ANOVA was conducted as described above, including only RTs for correct trials. The analysis found a significant main effect of visualization condition ($F_{(4,792)} = 2.74, p < .001$), but no effect of model accuracy nor interaction between the two ($Fs < 1$, $ps > .84$). Follow-up pairwise comparisons between the visualization conditions, collapsing over model accuracy, found that the baseline condition was significantly faster than the 7-feature-table condition (Fig. 3) when using the Bonferroni correction for multiple comparisons ($t(199) = 2.95, p < .05$). This is perhaps unsurprising, as the 7-feature-table condition showed users the most information, and so longer response times would indicate that they look longer to read the information. Our prediction that explanations presented in graphical format would be easier to read, and thus take less time, was not supported by the data.

3.2 Effects of Trial Type

Our second overall question was whether participant accuracy differed for different trial types (e.g., hit, miss), and whether this interacted with overall model accuracy (e.g., 50%

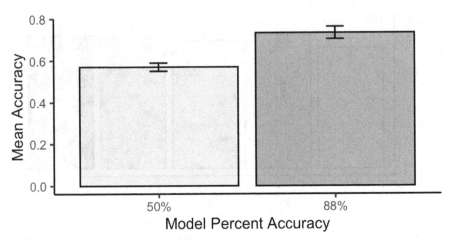

Fig. 2. Mean accuracy for each of the model percent accuracy conditions, collapsing across visualization condition. Accuracy was significantly higher in the 88% than 50% model accuracy condition.

Table 3. Mean accuracy and response times by visualization type and model accuracy condition.

Model accuracy condition	Visualization condition	Accuracy		Response time (ms)	
		Mean	SD	Mean	SD
50%	Baseline	0.57	0.11	3803	2187
	3 Features - graph	0.57	0.13	4033	2511
	3 Features - table	0.57	0.12	4071	2659
	7 Features - graph	0.57	0.12	3994	2372
	7 Features - table	0.58	0.13	4193	2733
88%	Baseline	0.73	0.18	3823	2348
	3 Features - graph	0.74	0.18	4211	3035
	3 Features - table	0.74	0.18	4024	2551
	7 Features - graph	0.72	0.17	4158	3057
	7 Features - table	0.74	0.18	4217	3163

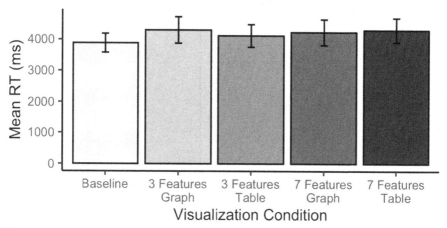

Fig. 3. Mean RT for each visualization condition, collapsing across model percent accuracy. RTs were significantly longer in the 7 feature table condition than baseline.

versus 88%). Mean accuracy is shown in Fig. 4. A repeated measures ANOVA was run with the within-subjects factor of Trial Type (4 levels: hit, miss, false alarm, correct rejection) and the between-subjects factor of Model Accuracy. There was a marginal effect of model accuracy ($F_{(1,198)} = 3.24$, $p = .07$), a significant main effect of model output type ($F_{(3,594)} = 159.76$, $p < .001$), and importantly, a significant interaction between the two ($F_{(3,594)} = 4.22$, $p < .01$). Follow-up pairwise comparisons were conducted to explicitly test how participant accuracy within trial type differed based on the overall model accuracy. As seen in Fig. 4, accuracy was numerically higher in the 50% model accuracy condition for the hits and correct rejections, whereas accuracy was higher in the 88% model accuracy condition for false alarms and misses. For false alarms, this difference in accuracy reached significance ($t(198) = -3.22$, $p < .01$). This effect could indicate that when the model was generally reliable and false positives were rare, people were better able to recognize this model error and respond correctly than when this error was common.

Next we separated the data into visualization conditions to understand the effect of visualization type on the observed effects of trial type and overall model accuracy. We ran a mixed ANOVA, with the within-subjects factors of Visualization Condition and Trial Type, and the between-subjects factor of Model Accuracy. This analysis showed a 3-way interaction ($F_{(12,2196)} = 2.20$, $p < .01$), indicating that the effects of trial type and visualization type were different within the different model accuracy conditions. As such, we conducted follow-up tests to assess the effect of visualization condition within each level of model accuracy and trial type. This enabled us to understand how the model visualization type impacted accuracy for each trial type and accuracy level separately. One particularly interesting effect emerged, especially in light of the cybersecurity focus of the task, for which miss trials would carry high consequences. For miss trials in the 88% model accuracy condition (shown in Fig. 5), we observed the highest accuracy in the baseline condition, with significantly lower accuracy for all other conditions ($ts > 3.30$, $ps < .05$) except the 3-table ($t(84) = 2.53$, $p = .13$). These findings suggest

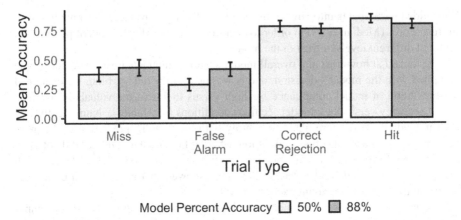

Model Percent Accuracy ☐ 50% ■ 88%

Fig. 4. Mean accuracy by trial type and model percent accuracy, collapsing across visualization condition. Accuracy for false alarm trials was significantly higher in the 88% than 50% model accuracy condition; no other pairwise comparisons reached significance.

that when the model was overall "good" but missed a target, people were less likely to detect this error when the model provided more explanation information, relative to the no-explanation baseline.

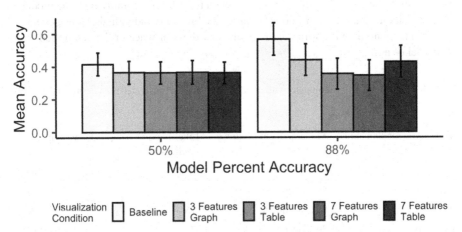

| Visualization Condition | ☐ Baseline | ■ 3 Features Graph | ■ 3 Features Table | ■ 7 Features Graph | ■ 7 Features Table |

Fig. 5. Mean accuracy for Miss trials only, by visualization condition and model percent accuracy. In the 88% model accuracy condition, accuracy was significantly higher in baseline than all other conditions (except 3-features-graph).

3.3 Individual Differences in Model Trust

Average trust scores were calculated for each of the two model accuracy conditions, and interestingly they were almost identical across the two conditions (50% Model Accuracy: Mean = 29.4, SD = 2.5; 88% Model Accuracy: mean = 29.2, SD = 3.0).

We divided participants into High Trust and Low Trust groups based on a median split of trust scores (Median = 29) in order to examine group-level behaviors of people who reported different levels of trust in the model.

We looked at how trust and overall model accuracy interacted with whether people complied with the model's decision, and how quickly they did so. First, we examined the likelihood of model compliance for high versus low trust individuals in the 50% model accuracy versus 88% model accuracy conditions (see Table 4). Numerically, low trust individuals were more likely to comply with the model's prediction in the more accurate than less accurate model condition, whereas high trust individuals did not show this pattern. However, a between-subjects ANOVA with factors of Trust (2 levels: High Trust, Low Trust) and Model Percent Accuracy showed that neither main effect was significant, nor was there an interaction (all Fs < 1, ps > 0.35).

We also examined response times for trials on which individuals did or did not comply with the model's decision, across different levels of Trust but collapsing across model accuracy (see **Fig. 6**). Thirteen individuals were removed from this analysis because they either always complied with the model's decision (10 participants) or never did (3 participants). A mixed ANOVA was conducted, with the within-subjects factor of Model Compliance (2 levels: Complied with model, Did not comply with model) and the between-subjects factor of Trust. Results showed a main effect of Trust ($F_{(1,185)}$ = 4.66, $p < .05$), and a main effect of Compliance ($F_{(1,185)}$ = 18.65, $p < .01$), but no interaction between the two ($F_{(1,185)}$ = 0.001, $p = 0.99$). These findings indicate that all participants were slower to respond when they did not comply with the model's prediction than when they did comply, and that low-trust individuals were generally slower to respond than high-trust individuals regardless of whether they complied with the model or not.

Table 4. Likelihood of compliance with model's prediction by model trust score and overall model accuracy.

Trust score median split	Model percent accuracy	Likelihood of compliance with model's prediction	
		Mean	SD
High trust	50% accuracy	.76	.19
	88% accuracy	.75	.19
Low trust	50% accuracy	.72	.17
	88% accuracy	.76	.15

4   Discussion

This study investigated the impact of different types of ML explanations, the accuracy of a ML model, and a user trust in the ML model on user performance in a simulated spam detection task. Results showed that people performed the task with higher accuracy when

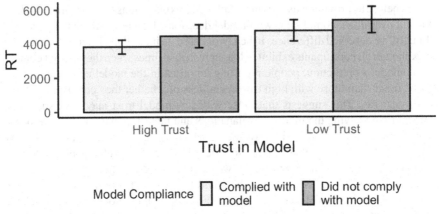

Fig. 6. Mean RTs for trials in which people did and did not comply with the model's prediction by model trust median split.

they saw a more accurate model, but there was no overall effect nor interactions with feature number or feature importance visualization on accuracy. Response times were longer in the seven-feature table visualization than baseline, but this was likely driven by the additional time needed to read the extra information on the screen, as these increased response times did not correspond to an accuracy difference. In general, our prediction that showing participants additional information regarding feature importance values would improve performance above and beyond the baseline condition, in which a model decision as provided with no explanation, was shown to be invalid.

More interestingly, when considering different trial types (e.g., hits, false alarms), we found effects of overall model accuracy and interactions with visualization type in participants' ability to identify and overcome certain model errors. Participant accuracy for false alarms was better in the 88% accurate model condition than the 50% accurate condition. In other words, participants were better at correctly responding "Not Spam" to a false alarm made by the model in the 88% model accuracy condition, in which false alarms were rare. On the surface, this seems to run counter to some findings in the psychology literature that subjects will tend to miss rare events (Rich et al. 2008). However, it could be the case the participants implicitly learned the mapping between the email header info and Spam/Not Spam categories across the course of the experiment, especially when the model was highly accurate, and thus created reliable schemas to help them correctly categorize these emails despite the incorrect model prediction. Future work should determine whether this effect is replicable and its cognitive underpinnings.

Moreover, within the 88% accurate model condition, participants were more likely to comply with incorrect "Not Spam" model predictions (i.e., miss trials) when the model provided more explanation information, compared to the no-explanation baseline. This is consistent with previous findings that users are more likely to rely on a model when more information is provided about its decisions (Bussone et al. 2015; Lai and Tan 2019), regardless of whether that information is correct or not. Because this result was only observed in the highly accurate model condition, it raises risks regarding the implementation of such models in high consequence workflows. Caution should be

exercised before as providing users with detailed ML explanations from accurate models, as it could overinflate compliance with model decisions, leading to missed targets.

Finally, we observed differences in behavior based on individual differences in model trust. Although all participants exhibited longer response times when they did not comply with the model's prediction, people reporting low trust in the model exhibiting longer response times than those with high trust regardless of whether they complied with the model's decision. This suggests that those with low model trust may have spent this additional time viewing the model's explanations and factoring that into their response choice. Future work should further investigate which aspects of ML explanations may engender higher or lower levels of trust, in order to help users develop appropriate trust and reliance on the system. Interestingly, we did not find differences in the likelihood of model compliance across users with high and low trust. This may indicate a strong bias to comply with a model's prediction, especially in an ambiguous task for which the user may have low confidence or little prior experience. Future work should investigate the factors that influence whether an individual complies with a model they believe to be incorrect, again in the service of helping users develop appropriate levels of trust and reliance on models deployed in high-consequence decision-making.

Our study is not without caveats. Because our data was collected online via Amazon Mechanical Turk, individuals in our study sample were likely novices with respect to both machine learning and cybersecurity. It is possible that findings may be different if we were to test an expert cybersecurity team, as they would have relevant domain knowledge to draw upon regarding the likelihood of certain email features being associated with malicious emails. There are aspects of our task that may have increased its difficulty or limited the usefulness of explanations, such as: limited context regarding the email (only saw features describing the email, not the actual text); forced binary choice (instead of ability to explore more features of the email to follow-up on model's prediction); or lack of feedback to users regarding task accuracy. It was also the case that our "incorrect" model decision trials used the same features and weights as in the correct instances, but just reversing the classification decision. Future work could train a model that actually misclassifies certain stimuli in order to produce more realistic feature weightings to use in misclassification decisions.

Our findings suggest that the efficacy of machine learning explanations for task completion depend, minimally, on the nature of the task, the overall model accuracy, the types of errors likely to be made by the model, and the user's trust in model. Future work should continue to explore how ML explanations impact human decision-making to ensure that the information provided does not hurt the human analysts' ability to perform high consequence tasks.

Acknowledgements. This work was supported by the Laboratory Directed Research and Development program at Sandia National Laboratories. *Sandia National Laboratories is a multimission laboratory managed and operated by National Technology & Engineering Solutions of Sandia, LLC, a wholly owned subsidiary of Honeywell International Inc., for the U.S. Department of Energy's National Nuclear Security Administration under contract DE-NA0003525. This paper describes objective technical results and analysis. Any subjective views or opinions that might be expressed in the paper do not necessarily represent the views of the U.S. Department of Energy or the United States Government. SAND2021–1481 C.*

References

Arya, V., et al.: One explanation does not fit all: a toolkit and taxonomy of AI explainability techniques. arXiv preprint arXiv:1909.03012 (2019)

Burkart, N., Huber, M.F.: A survey on the explainability of supervised machine learning. J. Artif. Intell. Res. **70**, 245–317 (2021)

Bussone, A., Stumpf, S., O'Sullivan, D.: The role of explanations on trust and reliance in clinical decision support systems. In: 2015 International Conference on Healthcare Informatics, pp. 160–169. IEEE, October 2015

Chandrasekaran, A., Prabhu, V., Yadav, D., Chattopadhyay, P., Parikh, D.: Do explanations make VQA models more predictable to a human? arXiv preprint arXiv:1810.12366 (2018)

Colquitt, J.A., Scott, B.A., LePine, J.A.: Trust, trust-worthiness, and trust propensity: a meta-analytic test of their unique relationships with risk taking and job performance. J. Appl. Psychol. **92**, 909–927 (2007)

de Vries, P., Midden, C., Bouwhuis, D.: The effects of errors on system trust, self-confidence, and the allocation of control in route planning. Int. J. Hum Comput Stud. **58**, 719–735 (2003)

Dua, D., Graff, C.: UCI machine learning repository. University of California, School of Information and Computer Science, Irvine, CA (2019). http://archive.ics.uci.edu/ml

Feng, S., Boyd-Graber, J.: What can AI do for me? Evaluating machine learning interpretations in cooperative play. In: Proceedings of the 24th International Conference on Intelligent User Interfaces, pp. 229–239, March 2019

Green, B., Chen, Y.: Disparate interactions: an algorithm-in-the-loop analysis of fairness in risk assessments. In: Proceedings of the Conference on Fairness, Accountability, and Transparency, pp. 90–99. ACM January 2019

Weerts, H.J.P., van Ipenburg, W., Pechenizkiy, M.: A human-grounded evaluation of SHAP for alert processing. In: Proceedings of KDD Workshop on Explainable AI KDD-XAI 2019 (2019)

Hoff, K.A., Bashir, M.: Trust in automation: integrating empirical evidence on factors that influence trust. Hum. Factors **57**(3), 407–434 (2015)

Hoffman, R.R., Mueller, S.T., Klein, G., Litman, J.: Metrics for explainable AI: challenges and prospects. arXiv preprint arXiv:1812.04608 (2018)

Huysmans, J., Dejaeger, K., Mues, C., Vanthienen, J., Baesens, B.: An empirical evaluation of the comprehensibility of decision table, tree and rule based predictive models. Decis. Support Syst. **51**(1), 141–154 (2011)

Kim, B., Koyejo, O., Khanna, R.: Examples are not enough, learn to criticize! Criticism for interpretability. In: NIPS, pp. 2280–2288, December 2016

Kneusel, R.T., Mozer, M.C.: Improving human-machine cooperative visual search with soft highlighting. ACM Trans. Appl. Percept. (TAP) **15**(1), 1–21 (2017)

Lai, V., Tan, C.: On human predictions with explanations and predictions of machine learning models: a case study on deception detection. In: Proceedings of the Conference on Fairness, Accountability, and Transparency, pp. 29–38, January 2019

Lundberg, S.M., Lee, S.I.: A unified approach to interpreting model predictions. In: Advances in Neural Information Processing Systems, pp. 4765–4774 (2017)

Miller, T., Howe, P., Sonenberg, L.: Explainable AI: beware of inmates running the asylum or: how I learnt to stop worrying and love the social and behavioural sciences. arXiv preprint arXiv: 1712.00547 (2017)

Molnar, C.: Interpretable machine learning: a guide for making black box models explainable (2020). https://christophm.github.io/interpretable-ml-book/

Narayanan, M., Chen, E., He, J., Kim, B., Gershman, S., Doshi-Velez, F.: How do humans understand explanations from machine learning systems? An evaluation of the human-interpretability of explanation. arXiv preprint arXiv:1802.00682 (2018)

Pedregosa, F., et al.: Scikit-learn: machine learning in Python. J. Mach. Learn. Res. **12**, 2825–2830 (2011)

Perkins, L., Miller, J. E., Hashemi, A., Burns, G.: Designing for human-centered systems: situational risk as a factor of trust in automation. In: Proceedings of the Human Factors and Ergonomics Society Annual Meeting, vol. 54, no. 25, pp. 2130–2134. SAGE Publications, Los Angeles, September 2010

Rich, A.N., Kunar, M.A., Van Wert, M.J., Hidalgo-Sotelo, B., Horowitz, T.S., Wolfe, J.M.: Why do we miss rare targets? Exploring the boundaries of the low prevalence effect. J. Vis. **8**(15), 15 (2008)

Sanchez, J., Rogers, W.A., Fisk, A.D., Rovira, E.: Understanding reliance on automation: effects of error type, error distribution, age and experience. Theor. Issues Ergon. Sci. **15**(2), 134–160 (2014)

Schaefer, K.E., Chen, J.Y.C., Szalma, J.L., Hancock, P.A.: A meta-analysis of factors influencing the development of trust in automation: Implications for understanding autonomy in future systems. Hum. Factors **58**(3), 377–400 (2016)

Štrumbelj, E., Kononenko, I.: Explaining prediction models and individual predictions with feature contributions. Knowl. Inf. Syst. **41**(3), 647–665 (2013). https://doi.org/10.1007/s10115-013-0679-x

Yuviler-Gavish, N., Gopher, D.: Effect of descriptive information and experience on automation reliance. Hum. Factors **53**, 230–244 (2011)

HCD³A: An HCD Model to Design Data-Driven Apps

Anna Christina Weigand[✉] and Martin Christof Kindsmüller[✉]

University of Applied Sciences, Brandenburg, Germany
mck@th-brandenburg.de

Abstract. This contribution introduces HCD³A, a process model to guide and support the development of data-driven applications. HCD³A is a specialized human-centered design (HCD) process model derived from and based on the ISO 9241-210 standard. In order to test the suitability of the HCD³A process model a prototype of a machine learning (ML) application is developed along this process. This application is integrated in a learning management system and tailored to the needs of computer science students in an online learning context. The learning application uses an ML approach to support students in their learning behavior by helping them to avoid procrastination and motivating them for assignments and final exams. This is e.g. done by predicting the students exam success probability. The most important claim in regard to the ML components was explainability. Although the evaluation of the prototype in regard to the suitability of HCD³A has not been completed the first results show that it is promising in particular to make ML applications more transparent for the users.

Keywords: Human-centered design · User involvement · User research · Evaluation · Data-driven · Learning application · Machine learning · Explainable AI · Process model

1 Introduction

Creating data-driven platforms or applications with a good user experience (UX) appears to be more challenging in comparison to common websites or apps [1–4]. There seems to be a need to rethink the HCD process for data-driven applications. Furthermore, the findings of Dove et al.'s study [4] underline this. The authors found the following challenges UX designers face in the context of ML applications: "Difficulties in understanding ML and its capabilities", "Challenges with ML as a design material", and "Challenges with the purposeful use of ML" ($n = 51$ participants). These challenges are important in the context of human-centered development of ML applications as the occupational group of UX designers is usually in charge of designing user interfaces. Furthermore, the authors carved out the following topics to focus on in further research [4]: "Consider the interplay between ML statistical intelligence and human common sense intelligence", "Envision opportunities to apply ML in less obvious ways", "Represent ML's dependency on data in early prototypes", and "Foreground ethical considerations of ML".

© Springer Nature Switzerland AG 2021
H. Degen and S. Ntoa (Eds.): HCII 2021, LNAI 12797, pp. 285–297, 2021.
https://doi.org/10.1007/978-3-030-77772-2_19

Regarding the integration of the UX perspective into data science, ML or artificial intelligence (AI) applications several ideas have already been applied [5–7]. In 2019 the Google Pair Team launched its "People + AI Guidebook" on this topic [8] where they describe the following essential aspects: "User Needs & Defining Success", "Data Collection & Evaluation", "Mental Models", "Explainability & Trust", "Feedback & Control", "Errors & Graceful Failure". Their recommendations are generated out of their own experiences as well as academic research.

The need to integrate the human perspective into the development of AI products seems to be of utmost importance in these times as all these products should be bought, licensed or just used by the related target group. Therefore, it is necessary to provide products and services with a good usability and overall experience [9, 10]. Furthermore, in the field of AI, explainability and transparency of algorithms are relevant to users [11]. Explainable AI applications are trustable as well as understandable for the users [12, 13]. Therefore, presenting the users a reliable product or service is essential. To tailor a system to the users' needs the ISO 9241-210 HCD process was proposed [14]. This standard is the basic concept for further considerations in the following work.

Generally, there are many fields of implementation for data science, ML or AI applications. In the context of a university's learning environment ML can be used to track the learning processes and results of students. The field of learning analytics is already an established field of research and deals e.g. with topics like educational data mining and based on this data the prediction of student's performance on their degree level [15]. In this study learning analytics is used to support online students regarding procrastination they are usually faced with [16, 17].

The main target of this work is to create and evaluate a model for human-centered design of data-driven applications (HCD^3A model), a process which has been lacking so far.

2 Theory

2.1 Human-Centered Design Process

The basic principles of the HCD process are recorded in the ISO 9241-210 standard [14]. The term human-centered design is defined as "approach to systems design and development that aims to make interactive systems more usable by focusing on the use of the system and applying human factors/ergonomics and usability knowledge and techniques" [14]. The definition connects the terms user experience and usability. For designing a product or service the ISO 9241-210 standard defines the following four steps [14]: understanding and specifying the context of use, specifying the user requirements, producing design solutions, evaluating the design.

2.2 Artificial Intelligence and Machine Learning

Within AI the field of ML is an essential topic. Regarding ML Tom M. Mitchell wrote one of the most common definitions [18]: "A computer program is said to learn from experience E with respect to some class of tasks T and performance measure P, if its performance at tasks in T, as measured by P, improves with experience E".

According to Mitchell three aspects needs to be emphasized on: "the class of tasks, the measure of performance to be improved, and the source of experience" [18]. The main focus of ML is to improve the performance of a computer program automatically based on its own experiences. Therefore, algorithms have been developed to find laws and patterns in the given training data [18]. With this training data an algorithm can be prepared and trained. Later a validation of the algorithm's behavior can take place by validation data.

To consider the explainable AI aspects and to create understandable as well as trustable algorithms, e.g. it may be necessary to check the ML feature relevance [13].

3 Research Question, Hypothesis and the HCD^{3}A Model

3.1 Research Question and Hypothesis

The main focus of this work is to define an adequate process of integrating the HCD perspective in the development of data-driven applications. Finally, the aim is to increase the overall UX of such applications. Therefore, the research question of this work is: how can be ensured that ML or AI applications fit the users' needs?

This question was the fundamental consideration for working out the model of human-centered design for data-driven applications (short: HCD^{3}A), a process model to guide and support the development of data-driven applications. The next chapter introduces the HCD^{3}A process model which is derived from and based on the ISO 9241-210 standard [14].

Hence, the hypothesis related to the above-mentioned research question is: the HCD^{3}A model ensures that ML or AI involved applications are tailored to the users' needs. The HCD^{3}A model will be tested by creating a learning application for students. The final evaluation will show whether the drafted process helps to create a user-centered data-driven application.

3.2 The HCD^{3}A Model

According to [14] it is necessary to research the users' needs, understand and specify them. This is the basis for further development of products or services. In order to apply this HCD process to the different stages of developing a ML application, it is enhanced especially for data-driven products or services which is shown in Fig. 1. From the beginning on the aspect of available or necessary data as well as concepts of application related ML algorithms are included into the creation process so that for all project participants the HCD approach is getting more obvious and data related. Figure 1 depicts a first attempt and consists of five steps: (1) analysis, (2) prototype, (3) evaluation, (4) implementation and (5) validation.

Essentially, the HCD^{3}A model enriches the HCD process with important key to-dos regarding ML. In the common HCD process HCD experts are the mediators between the humans using the system and the software (SW) engineers building the system. Designing an ML application introduces a new target group in the HCD process: the ML engineers. It is important to recognize that ML engineers are not just yet another

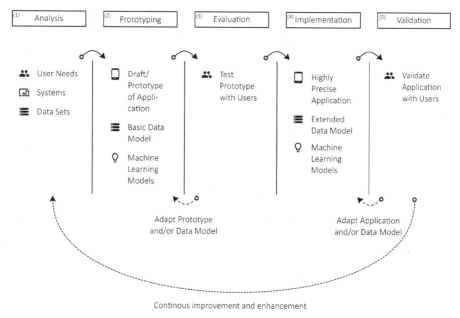

Fig. 1. Processes according the proposed HCD^{3}A model.

kind of SW engineers. The fundamental difference is explainability vs. interpretability.[1] Therefore, the HCD experts have to control the process, define the different roles and allow the team to grow together. A first step is to find a common ground, especially regarding the used vocabulary and abstract concepts. ML engineers should be included in participatory processes like future workshops or focus groups. Direct participation and collaboration improve the interplay between HCD experts, ML engineers and other team members. HCD experts present the user requirements, the design perspective, and the need for traceability, while ML engineers show the possibilities of ML and the technical requirements. The exchange can help to generate new ideas together and to appreciate the contributions of the other discipline.

Analysis. Discovering users' needs is one aim of the analyzing phase (1) of the HCD^{3}A process model. In this context possible procedures and methods like the stakeholder or work analysis, personas and user journey blueprints are useful. It depends entirely on the project which methods are suitable.

In addition to user needs, any existing systems and interfaces that already exist must also be considered. These systems and interfaces might influence the further product development process, for instance dependencies might arise. The requirements should be listed and seriously taken into account. Analyzing them early helps to find a good base for further development.

---

[1] A (non-ML) SW engineer is usually able to explain why the system does what it does (and is usually proud of it – explainability). An ML engineer is usually already satisfied if the chosen ML algorithm is able to successfully predict the desired outcome (interpretability).

Furthermore, already available datasets should undergo a first analysis. To find out which datasets exist and to discover these data can support transparency regarding the whole application. As data are the basis for ML applications considering them early is a plus when it comes to prototyping – the next step.

Prototyping. After investigating the users' needs as well as the available systems and datasets the prototyping phase (2) starts. In this phase the different perspectives are consolidated, and a first design prototype of the application is created.

In accordance with the prototype a second step is to prepare a basic data model which suits to the prototype. For example, it might turn out that further data are necessary to later adequately implement all relevant features of the prototype. One of the key drivers of data-driven applications is available data. Considering data from the beginning is the key for a realistic view on the feasibility of the product or service.

The available data are the basis for the first considerations about ML algorithms regarding the related application. Ideas can be collected on how the application can be enriched by ML, e.g. recommendation systems, predictions, etc. Based on these considerations relevant data could be missing in order to prepare a useful algorithm for the application. Therefore, a strategy of how to gather missing data needs to be prepared. The outcome should be an overview of possible algorithms and necessary related data.

The development of the prototype as well as the setup of the basic data and ML models can be an iterative interplay between the three components. A next step is, to combine these basic concepts with an evaluation through real potential users. After getting the users' feedback and analyzing it, the prototype is adapted accordingly.

Evaluation. Evaluating the before-mentioned prototype with real (potential) users is a necessary step in the evaluation phase (3) to create a valuable product or service which fits best the users' needs. This is one essential step of HCD. As different evaluation methods exist, it depends on the related situation and application which method or combination of different methods is the most useful. For example, a Wizard of Oz study can be helpful.

Regarding the basic data model, it needs to be investigated whether the user requirements still suit the available data. In addition, the considerations about the ML models or already first drafts of the algorithms need to be reviewed. The prototype evaluations could result in new requirements or changes which can also affect the data and related ML algorithms.

Overall, the evaluation helps to find out where users are uncertain and where they expect transparency in the product or service. In addition, it is important to record the users' remarks regarding the explainability of ML algorithms.

Implementation. By having a suitable prototypic solution, the next step is to proceed with the implementation phase (4). Although having a good concept and prototype this will not always be the final solution. As soon as it comes to the implementation phase sometimes new technical or other challenges will come up regarding the graphical user interface, the data, or the ML algorithms. The interplay of the graphical user interface, the data and the output of the algorithms is sensible towards changes and usually influence each other directly. These challenges can be met in an agile or iterative implementation between the development, concept, and design team members. The most important point is to validate the new conditions with the users again.

Validation. The validation phase (5) is similar to the evaluation phase. The main difference is that testing is done with the real application. For example within the ISO 9241-210 standard the term validation is defined as "confirmation, through the provision of objective evidence, that the requirements for a specific intended use or application have been fulfilled" [14]. As far as the evaluation is concerned, the method to be used for validation depends on the specific application. For the validation it is appropriate to use the same methods as for the evaluation of the prototype(s) before so that a comparison between the different versions can be drawn with meaningful results. After the validation, the results need to be consolidated and put into a comprehensive concept.

As long as the application is actively used it is beneficial to repeat the validation process frequently. Thus, an iterative and continuous improvement process is facilitated. In case of completely new features or product enhancements the process should start with the first phase by analyzing the requirements and related conditions.

4   Project Work

4.1   Project Idea and Background Information

The introduced HCD^{3}A model is evaluated through a project work in the following. The idea of this project work is to design an application which supports the media informatics online students of the VFH^2 to improve their learning behavior. The goal is to create a motivating application that students enjoy interacting with and that helps them to get better over time.

4.2   Analysis

In the beginning of the analysis, the following stakeholders are identified to be relevant: online students of the VFH degree program media informatics, professors and teaching assistants of the partner universities, administrative staff of the partner universities, presidents of the partner universities and oncampus GmbH.^{3} The main target group consists of the online students who are the focus of the following work.

Target Group Analysis. For analyzing the main target group in detail, an online survey with $N = 43$ potential users – online students of the VFH degree program media informatics – has been conducted.

The results of an adapted ASS-S questionnaire [19] (four-point Likert scale from $1 = strongly\ disagree$ to $4 = strongly\ agree$) show that the participants are especially intrinsically ($M = 3.15$, $SD = .560$, $n = 40$) and identified motivated ($M = 3.13$, $SD = .548$, $n = 41$) which means driven by personal important goals. The VFH moodle learning platform is used more than one time per week with a tendency to daily usage by

---

^{2} VFH stands for (German) Virtuelle Fachhochschule (Virtual University of Applied Sciences). An association of universities of applied sciences in German-speaking countries that jointly offer online degree programs.

^{3} The provider of the technical infrastructure for the VFH.

the participants. With a four-point Likert scale from $1 = $ *not at all* to $4 = $ *very frequently* the device preference for online students is tested. The preferred device for their studies is the laptop or computer ($M = 3.91, SD = .366, N = 43$), tablet is used rarely ($M = 1.81, SD = .969, n = 42$) and smartphone lies in between with an average of $M = 2.27$ ($SD = .742, n = 41$). Most of their study related tasks they are solving with their laptop. 83.3% of the $n = 42$ participants declared that especially content-related interest helps to motivate for their online-studies in media informatics. Still 47.6% are motivated by regular learning units and 38.1% by a strict time and learning plan. Concerning which additional auxiliary means the students use to organize or motivate themselves ($n = 42$) the following are used most often: old or previous exams for exercise (92.9%), additional learning material (85.7%) as well as further exercises (66.7%). On average regular assignments during the term are rated with «rather good» ($M = 3.05, SD = .697, n = 42$) by the participants. The subjective rating of functionalities that would support the students' online-studies in media informatics results in the following ranking ($n = 42$): automatic reminder for assignment deadlines ($n = 32$), recommendation for further relevant learning materials ($n = 27$), displaying the learning progress on the basis of experienced values ($n = 20$), keeping the own general average always in mind ($n = 13$) and comparison with fellow students e.g. processing status ($n = 11$). Most of the participants did not have any preference regarding a specific platform. However, 40.0% of the participants would like to have an extended moodle platform, 30.0% an online portal and 25.0% a smartphone app. An automatic reminder for assignment deadlines is the most interesting new functionality (76.2%). Furthermore, recommendations for further relevant learning materials are favored by 64.3% of the participants. Still 47.6% would like to have displayed their learning progress based on experienced values.

Personas. Deduced from the target group analysis of the VFH students two personas arise. The first persona describes the 25 years old working student Leonie. She studies media informatics via the VFH at the University of Applied Sciences in Brandenburg and is currently enrolled in her 3rd term bachelor studies. The second persona is Marc who is married and father of a two-year-old son. He works as a creative consultant in a small business consultancy in Hamburg and lives there with his family. Furthermore, he is a part-time student of the master's degree media informatics via the VFH at the University of Applied Sciences.

Scenarios of Usage. Additionally, scenarios of usage and user journey blueprints have been developed to exemplify the lives of the various students. Overall, the scenarios of usage can be structured into six steps: preparing and organizing, attending lectures, learning and understanding, assignments, preparing for exams and exams. As the actions of both personas are very similar the biggest difference is on the emotional and thoughts level. They live in very different personal situations and the circumstances as well as their personal learning progress affect them differently.

Basic Dataset. After analyzing the users' needs a concept of a basic dataset is prepared. An already existing topic-related dataset is used as groundwork which is the Open University Learning Analytics dataset of Kuzilek, Hlosta, and Zdrahal [20]. This dataset was used as a start to get a first overview which data are proper data to collect. In the

environment of the VFH degree programs some requirements do not match with the Open University database schema. For example, the VFH group consists of different universities, and courses are held by different professors. Therefore, a specific database schema for a VFH learning environment is elaborated.

4.3 Prototyping and Evaluation

Prototyping. First, based on the personas' and thereby users' requirements a low-fidelity prototype of the learning application is elaborated. The prototype gives a first idea of the look and feel in form of wireframes. Features in this prototype which are meant to contain ML algorithms are: the current probability of exam success, the recommendations to increase the current probability of exam success and the course recommendations for upcoming terms. As these features are relevant for the personas to increase their motivation for their studies it is necessary to analyze the available data and how to use them for the ML algorithms accordingly. Therefore, suitable variables are selected. Basis for this is the basic dataset concept from before. The first idea regarding the ML algorithms is, using a random forest [21, 22] for predicting the exam success probability, and providing content-based recommendations to increase the exam success probability as well as for the course recommendations. For ensuring the transparency it is important to proof the ML feature relevance [13].

Evaluation. Developing a user-centered product or service requires an evaluation e.g. with potential users. Hence, for this project test users are recruited in order to gain first insights about the learning application prototype and to identify the major usability issues. Subsequently, a qualitative user-based testing approach was selected. The test was set up as one-to-one online sessions where the prototype was shown to the evaluation participant via screensharing. A short scenario introduced the situation to the participants which describes that they are a VFH student of media informatics and are enrolled in the second term of the master's degree. Afterwards the test conditions were explained to the participants. They were defined as follows: The investigator is showing the low-fidelity prototype. Observe each screen precisely. Comment loudly what you are noticing at each screen. Do you understand everything? Please explain what you do not understand. Summarize what is missing? What would you like to have additionally?

For this first qualitative test of the prototype, a group of $N = 5$ test users was recruited. The feedback of the test users was overall positive and very similar in most of the results but additionally complements one another. Specifically, regarding the ML features the users found the exam success probability an interesting functionality when it is related to the personal prior learning behavior. Furthermore, the course advisory service is a very helpful feature for them, but it could be enhanced by further properties (e.g. elective course, in winter/summer term, exam type). A critical voice concerning this functionality was that it feels insolent "what I might like". Nevertheless, 5 of 5 participants answered the question whether they would use the application positively. Overall, these results show that the application suits the test users, but they still have suggestions for improvement. Especially the personal dashboard has a lot of potential to support the students. What is remarkable is that the features where ML will be involved

are perceived as helpful and motivating by most (4 of 5) of the participants. Thereby the potential and first of all orally described outcome of the ML features has been understood by all participants.

4.4 About Further Procedure

After evaluating the prototype for the first time the further procedure is split into two phases: first improving the existing prototype according to the user feedback, second implementing the real application and validate it again.

The insights gained through the prototype evaluation are helpful to improve the prototype according to the users' needs. For example, the differentiation between the home page and the personal dashboard need to be removed, the dashboard reworked or the course detail page enhanced. Consequently, the results are applied to the prototype. Ensuring that the users' needs are considered properly it is decided to adapt and test the low-fidelity prototype again. As there are some fundamental changes it makes sense to strive the enhancement of the prototype first.

Regarding the ML features some changes should be incorporated. As the user feedback e.g. reveals that the exam success probability should refer more to the personal previous learning behavior and success in general the features for the ML algorithm need to be reviewed. In the beginning they referred to the exam success of previous students of the specific course. Furthermore, based on the changes in the prototype and the ML features the data model needs to be adapted as well. Meanwhile, some new data are required to guarantee further processing e.g. in the students' personal information or the course information. Additionally, the features of the recommendation systems need to be reviewed whether further data are required. After adapting the ML features as well as the database schema, a first concept of the related ML algorithms can be set up and a new evaluation can take place. This is the current project state. Nevertheless, when the work is resumed to a later point of time further prototyping and evaluation are essential.

5 Evaluation of the Approach and Discussion

5.1 Evaluation

For ensuring a ML or AI involved application is tailored to the users' needs the HCD^{3}A model has been set up in the beginning of this work. In this chapter the model itself is evaluated with regard to Dove et al. [4] who investigated the challenges of the interplay of UX design and ML.

Dove et al.'s [4] first point "difficulties in understanding ML and its capabilities" is not directly addressed in the HCD^{3}A model. Nevertheless, it shows clearly that taking data into consideration from the very beginning is of utmost importance. The study showed that there are several dependencies between users' needs, user interface and data. For maximizing the potentials of the learning application, it was necessary to understand the opportunities of ML. This opened a new way of thinking and creativity. It was necessary to evaluate the features of this new dimension in concepting and designing early as there was a huge potential to take wrong paths. Therefore, it is necessary for (UX) designers

to investigate into the ML capabilities in early project stages so that potential application features can be considered to fit the users' needs. This also permits an early evaluation with (potential) users which helps to create a useful as well as transparent and trustful ML application.

According to Dove et al. "challenges with ML as a design material" is another fact that UX designers are struggling with [4]. One of their survey participants reflects this uncertainty: "The technical complexity is a challenge as is the need to better understand and design for that complexity. It can get deep and unfamiliar very quickly, and designers need some level of expertise to function and contribute to the work at hand." In the described project work this challenge also existed. It has been solved by adding small examples in the prototype of how the results may be represented. The potential algorithms behind the features of the application were described verbally by the investigator of the usability test. As all participants of the usability test understood what there will be shown in future this approach can be used in an early project stage, e.g. in form of a Wizard of Oz study. For later stages it will become necessary to evaluate the real outcome of the ML algorithms and whether it fits the users' needs and expectations.

"Challenges with the purposeful use of ML" is the third problem UX designers are faced with [4]. Therefore, UX designers have the feeling that services or products including ML are not design-driven but technology-driven. The experience of the participants in the study by Dove et al. [4] shows the need to combine the user-centered view and the data view. The HCD^{3}A model is one attempt how to do so. One participant of Dove et al.'s study said that "[…] we need to shift the conversation from technology to people – we'll need to bring the ethical and human centered voice to the algorithms that make it all a reality" [4]. Furthermore, it is necessary to make sure that UX designers are not averse to ML. According to another study participant the goal should be to "[…] map out the right user stories and use cases, to enable effective ML." This demand for a human-centered perspective has been integrated into the HCD^{3}A model. Thereby, from the first considerations until the final implementation of the ML application both the users and the data play the major role. Even during the project work it became clear that these two components are interdependent. Ideas for new features or small design changes can effect major changes in the data model and related algorithms. For example, in case of the exam success probability the basic idea of the algorithm changed through the user feedback, subsequently the ML features needed to be adapted.

Furthermore, the HCD^{3}A model addresses three of the challenges future research should focus on [4]. The first challenge "consider the interplay between ML statistical intelligence and human common-sense intelligence" is one of them. "[Representing] ML's dependency on data in early prototypes" and "foreground ethical considerations of ML" are the other two challenges. The interplay and necessity to think about data in early development stages is outlined by the HCD^{3}A model. The project work shows the importance of considering data throughout the entire period. The user interface might directly influence data gathering as for example necessary input fields are provided. In the case of the student profile, for example, input fields for skills, learning type, etc. must be provided so that this data can be stored for later personalized recommendations. Therefore, the data, the algorithms as well as the graphical user interfaces need to be harmonized. On the other hand, some ideas to integrate ML into the application might

be promising but alone do not satisfy the users' needs. With help of the HCD³A model the users' needs are investigated and put into prototypes. In this study, e.g. the idea of comparing the students with their fellow students was not desired by the users. These prototypes are tested with potential users and afterwards adapted. This underlines again the interplay of ML and human intelligence. Through splitting up the tasks regarding data the process is getting more obvious also for UX designers who are usually not directly involved into data tasks. Furthermore, thoughts about ethical development in the area of ML are incorporated as the personal users' needs as well as their desires are the basis for the development. In this case the VFH online students were directly asked to take part in the study or the prototype evaluation and to share their experiences.

5.2 Discussion

It should be emphasized that it is not the goal of this study that UX designers take over concepts for data bases as well as ML algorithms but the HCD³A model requires a direct collaboration of UX designers and data scientists or ML engineers. Instead, an interdisciplinary team with a clear understanding of the complementary roles must work together in an iterative process model.

However, during the evaluation phase of the project work the user feedback has shown that the users were surprised by the new ML features but also curious and open minded. They stated that these features can increase their motivation for their studies. The successful project work described above is a first positive evidence regarding the introduced hypothesis that the HCD³A model ensures that ML or AI involved applications are tailored to users' needs. Furthermore, it addresses several challenges of UX designers regarding the development of ML applications. Finally, the model demonstrates how to proceed and how to make processes more obvious.

These results must be regarded as provisional since the procedure has not been applied completely because this project is currently in the prototyping and evaluation phase. Further development of the prototype and logging additional data in the learning management system are necessary before setting up the algorithms and starting with the implementation phase. By means of this input the ML algorithms can be developed and later evaluated together with the test users. This close feedback loop with the test users supports the development of algorithms in a transparent and explainable way. Consequently, concerning these limiting factors the project work should be finalized.

6 Summary and Outlook

This work deals with the challenge to integrate the ML aspects into the human-centered way of development. Particularly, the goal is to develop ML applications that matter and are purposeful for the users. By means of the qualitative project work results and an additional discussion regarding the findings of Dove et al.'s study the HCD³A model has been evaluated [4]. The presented study provides a first evidence that the HCD³A model can be successfully used to tailor an ML application to user needs.

The next step is to complete the project work by finishing the two steps that have been missing so far. In addition, a long-term study during normal operation of the learning

platform can gather valuable insights. Moreover, another research topic to focus on is the human-centered development of the ML algorithms itself. This is required for prototypes in later project stages as well as the validation of the real ML implementation.

In summary, this study shows that the proposed HCD^3A model can successfully support the development of ML applications through a process model and the clear role concept. The improvements achieved in the evaluation cover both classic usability aspects and the aspects of transparency and explainability, which are particularly important for ML applications.

References

1. Girardin, F., Lathia, N.: When user experience designers partner with data scientists. In: Designing the User Experience of Machine Learning Systems. 2017 AAAI Spring Symposium Series, pp. 376–381 (2017)
2. Hebron, P.: Machine learning for designers (2016). https://www.oreilly.com/learning/machine-learning-for-designers. Accessed 19 May 2021
3. Yang, Q., Scuito, A., Zimmerman, J., Forlizzi, J., Steinfeld, A.: Investigating how experienced UX designers effectively work with machine learning. In: Koskinen, I., Lim, Y., Cerratto-Pargman, T., Chow, K., Odom, W. (eds.) Proceedings of the 2018 on Designing Interactive Systems Conference 2018 - DIS 2018, Hong Kong, China, 09 June 2018–13 June 2018, pp. 585–596. ACM Press, New York (2018). https://doi.org/10.1145/3196709.3196730
4. Dove, G., Halskov, K., Forlizzi, J., Zimmerman, J.: UX design innovation: challenges for working with machine learning as a design material. In: Mark, G., et al. (eds.) Proceedings of the 2017 CHI Conference on Human Factors in Computing Systems - CHI 2017. The 2017 CHI Conference, Denver, Colorado, USA, 06 May 2017–11 May 2017, pp. 278–288. ACM Press, New York (2017). https://doi.org/10.1145/3025453.3025739
5. Nazrul, S.S.: UX design guide for data scientists and AI products (2018). https://towardsdatascience.com/ux-design-guide-for-data-scientists-and-ai-products-465d32d939b0. Accessed 19 May 2021
6. Pásztor, D.: AI UX: 7 Principles of designing good AI products (2018). https://uxstudioteam.com/ux-blog/ai-ux/. Accessed 19 May 2021
7. Shapiro, V.: UX for AI: trust as a design challenge (2018). https://medium.com/sap-design/ux-for-ai-trust-as-a-design-challenge-62044e22c4ec. Accessed 19 May 2021
8. Google Pair Team: People + AI Guidebook (2019). https://pair.withgoogle.com/. Accessed 19 May 2021
9. Constantinides, E.: Influencing the online consumer's behavior. The web experience. Internet Res. (2004). https://doi.org/10.1108/10662240410530835
10. Constantinides, E., Geurts, P.: The impact of web experience on virtual buying behaviour. Empirical Study. J. Cust. Behav. (2005). https://doi.org/10.1362/147539205775181249
11. Fox, M., Long, D., Magazzeni, D.: Explainable Planning. arXiv preprint arXiv:1709.10256 (2017)
12. Gunning, D.: Explainable artificial intelligence (xai). https://www.darpa.mil/attachments/XAIProgramUpdate.pdf (2017)
13. Barredo Arrieta, A., et al.: Explainable Artificial Intelligence (XAI). Concepts, taxonomies, opportunities and challenges toward responsible AI. Inf. Fusion (2020). https://doi.org/10.1016/j.inffus.2019.12.012
14. International Organization for Standardization. Ergonomics of human-system interaction - Part 210: human-centred design for interactive systems (ISO 9241-210:2010) (2010)

15. Asif, R., Merceron, A., Pathan, M.K.: Predicting student academic performance at degree level. A case study. IJISA (2014). https://doi.org/10.5815/ijisa.2015.01.05
16. Naturil-Alfonso, C., Peñaranda, D.S., Vicente, J.S., Marco-Jiménez, F.: Procrastination. The poor time management among university students. In: Proceedings of the 4th International Conference on Higher Education Advances (HEAd 2018). Fourth International Conference on Higher Education Advances, 20 June 2018–22 June 2018, pp. 1–8. Universitat Politècnica València, Valencia (2018). https://doi.org/10.4995/HEAd18.2018.8167
17. Rakes, G.C., Dunn, K.E.: The impact of online graduate students' motivation and self-regulation on academic procrastination. J. Interact. Online Learn. **9**, 78–93 (2010)
18. Mitchell, T.M.: Machine Learning. McGraw-Hill Series in Computer Science. McGraw-Hill, New York (1997)
19. Thomas, A.E., Müller, F.H.: Skalen zur motivationalen Regulation beim Lernen von Schülerinnen und Schülern. Skalen zur akademischen Selbstregulation von Schüler/innen SRQ-A (überarbeitete Fassung) (2011)
20. Kuzilek, J., Hlosta, M., Zdrahal, Z.: Open university learning analytics dataset. Sci. Data (2017). https://doi.org/10.1038/sdata.2017.171
21. Breiman, L.: Random forests. Mach. Learn. (2001). https://doi.org/10.1023/A:101093340 4324
22. Merceron, A.: Educational data mining/learning analytics. Methods, tasks and current trends. In: DeLFI WOrkshops, pp. 101–109 (2015)

AI Applications in HCI

Visual Prediction Based on Photorealistic Style Transfer

Everlandio Fernandes[(✉)], Everton Aleixo, Wesley Jacinto Barreira,
Mikhail R. Gadelha, Aasim Khurshid, and Sergio Cleger Tamayo

Sidia Institute of Science and Technology, Avenue Darcy Vargas, 654,
Manaus, AM, Brazil
everlandio.fernandes@sidia.com

Abstract. In this study, we explore recent advances in photorealistic style transfer methods to make visual predictions of outdoor scenes. These methods transfer the elements' visual appearance from one photo (style image) to another (content image), maintaining the original composition of the elements in the original image. However, the search for reference images containing the same elements as the content image and presenting all the desired style characteristics makes the process challenging and time-consuming. To overcome this challenge, we propose a dynamic search method based on the transient scene attributes performed in a dataset developed especially for this task. Our team developed a set of 924 3D images divided into six scenario groups, with the main elements found in the outdoor scenes to be used as style images. Each group has stylizations of the four seasons, the time of day, the presence or absence of rain, snow, and cloudy skies. In the end, we measured the similarity of the results obtained with real images. The structural similarity index measure (SSIM) reaches an average score greater than 0.8.

Keywords: Neural network · KNN · Photorealistic style transfer · Transient attributes

1 Introduction

Photorealistic Style Transfer is a technique that aims to transfer the style of elements from one reference photo (style image) to another (content image), preserving the realistic appearance of the resulting image. For example, changing an outdoor scene photo's lighting by choosing a style image with features such as a sunset or midday or even cloudy sky. The proposed methods allow for rapid stylization in recent years, producing images with little or no spatial distortion or unreal artifacts. In this sense, Li et al. [7] present a technique based on auto-encoder neural networks with excellent results in providing photorealism to the resulting image.

Supported by Sidia Institute of Science and Technology.

H. Degen and S. Ntoa (Eds.): HCII 2021, LNAI 12797, pp. 301–309, 2021.
https://doi.org/10.1007/978-3-030-77772-2_20

This advance opens up a series of possibilities for using these methods in automated systems. One of these possibilities would be to make visual predictions. For example, check the appearance of an outdoor scene in different seasons and hours of the day or how the process of aging a car's paint would look under specific climatic aspects. However, the search for reference images containing the same elements as the content image and presenting all the desired style characteristics makes the process challenging and time-consuming. In this work, we present a case study for a system that makes visual predictions for outdoor scenarios' climate. We propose a dynamic search method based on transient scene attributes [6] for searching for the correct reference image.

Our team has developed a set of 924 3D pictures on the Unity® game engine to be used as reference images. To have a more significant semantic matching with the main elements found in outdoor scenarios, the set has six different outdoor views (forest, lake and ocean, mountains, city, roof and, road, see Fig. 1). Each group has stylizations of the four seasons, the time of day, the presence or absence of rain, snow, and cloudy skies, which are the transient attributes that we use to indicate the weather in outdoor environments. Figure 1 shows an example of each scenario, manifesting a summer day at noon with few clouds.

The proposal search for the correct reference image consists of indexing each reference image with its transient scene attributes and semantic segmentation. These indexing are used to perform K-Nearest Neighbor (KNN) [3] searches that return one of our reference images with the best semantic correspondence with the original image (content) and the desired transient attributes. The returned reference image is inserted into the Fast Photo Style [7] system.

To support our experiments in the sense of accurate results, we use the images from the Transient Attribute Database [6] as content images. An essential property of this database is that, instead of using unrelated photos, it contains photographic samples from 101 outdoor webcams in different regions of the globe, seasons, and times of the day. Then, it allows us to compare the results obtained in our experiments with authentic images with the desired transient scene attributes. We have reached similarity score over 0.8 in most of the results.

The remainder of this paper is organized as follows. Section 2 shows some related works in manipulation of transient scene attributes. In Sect. 3, we present the methods and the workflow used in this case study. In Sect. 4, we discuss the results and difficulties encountered during the development of this study. Finally, in Sect. 5, we present our conclusions and future works.

2   Related Works

The collection of real photos that we think is closest to the objectives of this study is presented in [6]. The authors present a study of "transient scene attributes" (e.g., sunny, gloomy, colorful, dull), defining a total of 40 attributes through crowdsourcing for thousands of images. Figure 4a is an example of images found in that image set. However, the feature vector defined in that study is quite distant from what we usually see in the weather forecast.

Fig. 1. Scenarios designed to maximize the semantic matching with outdoor scenarios.

[5] introduces a more recent study for manipulating scene attributes via hallucination. The method trains an deep generative network to input the desired characteristics in the natural scenes. However, the study does not analyze the similarity between the results obtained and authentic images.

3 Experiments

This case study's main objective is to make visual predictions of outdoor scenarios in different climatic situations. A photorealistic style transfer method uses a set of indexed images to style any outdoor image with the desired climate characteristics.

In our experiments, we use a Linux server with Intel Core i7-6700 3.4 GHz, 32 GB of memory RAM, and a GPU GeForce GTX-1080 with 8 GB dedicated memory. In this environment, the stylization process takes an average time of 15 s to complete. Below we give more details on the methods used and the workflow of the experiments.

Photo Style Transfer. The main idea of the method, known as Fast Photo Style [7], is to insert a pair of projection functions into the bottleneck of an already trained auto-encoder. These functions perform style transfer through whitening and coloring transformations (PhotoWCT) of elements in the content image that correlate to those in the style image. Besides, the proposed algorithm uses the information provided by semantic label maps for better content–style matching.

Fast Photo Style has a second phase in which it applies a smoothing process in the image resulting from the stylization process (first phase). According to the authors, the smoothing process is necessary because semantically similar regions

are often styled inconsistently. For example, the sky in a daytime photo, when stylized at night, may have small areas with the original photo style (light blue), as one can see in Figs. 2 and 3. This study case uses the coding provided by the Fast Photo Style's authors in their GitHub repository.[1].

Fig. 2. PhotoWCT without smoothing process. Image reproduced from [7] (Color figure online)

Fig. 3. PhotoWCT with smoothing process. Image reproduced from [7] (Color figure online)

Content Image Dataset. As stated before, the experiments use as style images a set of 3D pictures developed on the Unity® game engine. This set has 924 pictures organized in 6 different scenarios (see Fig. 1) developed to maximize the semantic correspondence with most of the outdoor scenes. Each group has stylizations of the four seasons, the time of day, and the presence or absence of rain, snow, and cloudy skies. To increase the efficiency of searching for the right style image, the dataset covers only real-world situations, eliminating scenarios such as snow in the summer.

Each 3D image is indexed with two vectors. The first represents its semantic composition. Each position on the vector corresponds to an element commonly found in outdoor environments, such as roads, plants, lakes, etc. We use the pre-trained model [10][2] to achieve semantic segmentation. The model has been trained to recognize 150 different types of elements. Thus, the vector representing the semantic composition of a style image contains 150 positions with the percentage of each component in the scenario.

The second feature vector represents the weather information or the transient scene attributes of each style image. The vector is composed of eight positions, the first four to describe the season, the fifth indicates the time of day, and the three others represent the presence of rain, snow, and cloudy sky. We chose to conduct our study with artificial pictures due:

- The difficulty of finding a set of real photos that accurately presented all weather possibilities.
- To improve the quality of semantic segmentation of the style images;
- To control the weather annotations.

---

[1] https://github.com/NVIDIA/FastPhotoStyle.
[2] Available at https://github.com/CSAILVision/semantic-segmentation-pytorch.

Workflow. The system proposed in this case study receives, as inputs, the content image and a vector representing the desired climatic characteristics. This vector has the same composition as the transient scene attribute vectors generated in the reference image set. Again, the same semantic segmentation model used in the reference image set extracts another 150-position vector representing the semantic composition of the content image.

The vector with the semantic composition serves to perform a KNN search [3] that indicates which of the six groups of scenarios has the highest semantic correspondence with the input image. Then, a new KNN search uses the weather information vector to designate the style image that best represents the desired weather situation. The image resulting from the search serves as an input to the Fast Photo Style system that returns the visual prediction image.

4   Discussion

In photorealism style transfer process, the details perceived in the original image should also be in the resulting image. However, while Fast Photo Style's smoothing process diminishes the artifacts generated by the styling process, it decreases the sharpness. But, if the edges of an image look too sharp or not sharp enough the picture seems unconvincing [4].

To work around this problem, we use a post-processing technique known as unsharp mask, which brings significant improvement to the resulting images. An unsharp mask can help emphasize texture and detail and is considered essential post-processing in most digital images. They are probably the most common type of sharpening methods and can be performed with virtually any image editing software [1]. We use the UnsharpMask function available in the Pillow package [9] for Python.

As example of the situation described above, see Fig. 4. It represents a real-world outdoor scenario being stylized to cloudy fall day at 5 pm. Figure 4b is the 3D image returned by our search as described in Sect. 3, and Figs. 4c and 4d show the Fast Photo Style results without and with the application of Unsharp mask. To increase the perception of image enhancement, Figs. 5 and 6 present details of Fig. 4c and 4d, respectively.

Knowing that the semantic segmentation is quite efficient in the developed 3D images, the same must happen in the content image to maximize the elements' matching. However, it is sometimes difficult to distinguish objects in dark or night photos, even in very bright images, impairing their semantic segmentation. Consequently, the transfer of style is also hampered, as shown in Fig. 7. The content image was stylized to deliver a late spring afternoon. Due to the sun's high light, some areas are very light, and others are dark. This situation hindered the stylization process, not giving the desired realism to the photos, see Fig. 7.c (regions indicated by the red boxes). However, when the content image is correctly semantic segmented, we can style it with a greater sense of realism, as shown in Fig. 8.

Fig. 4. Photorealistic style transfer process without and with unsharp mask. (a) Content image [6]; (b) selected style image; (c) resulting image of Fast Photo Style (without unsharp mask); (d) resulting image after unsharp masking (note highlight details in Figs. 5 and 6).

Fig. 5. Detail of Fig. 4.c (without unsharp mask).

Fig. 6. Detail of Fig. 4.d (with unsharp mask).

Fig. 7. Example of a lousy stylization due to the content image does not have all elements visible. (Color figure online)

(a) Content Image   (b) Style Image   (c) Resulting Image

Fig. 8. Example of a proper stylization. It happens when the content image is well lit, showing all the details of the picture.

As stated earlier, the experiments used the images provided by Laffont et al. in [6] as a content image. The images in this dataset are photographic samples from 101 external webcams in different regions of the globe, seasons, and times of the day. This feature allows the possibility to verify the similarity of the results obtained with ground truth images, i.e., real photos in the desired climatic situations. The structural similarity index measure (SSIM) [11] was used in this sense. It is known to be deeply associated with subjective quality [8].

SSIM is a perception-based model that considers image degradation to be a perceived change in structural information while also incorporating perceptual aspects, including luminance masking and contrast masking terms. Luminance masking is a sensation in which image distortions tend to be less visible in bright regions. In contrast, contrast masking is an aspect in which distortions become less visible where there is a significant activity or "texture" in the image [2]. An SSIM of 1 means that the resulting image is identical to the ground truth image, while SSIM = 0 implies that they are entirely different.

The results obtained are entirely satisfactory, with an average SSIM index above 0.8. See, for example, Fig. 9. It is a stylization for a cloudy summer afternoon in Japan (the original image and the reference image can be seen in Fig. 8, a and b, respectively). While photo Fig. 10 is the real image taken from the Transient Attribute Database [6] with the same characteristics. The SSIM index, comparing these two images, is 0.85.

Fig. 9. Result obtained by photorealistic style transfer process.

Fig. 10. Ground truth photo. Reproduced from [6]

Figures 11 and 12 show another example of a high degree of similarity between the resulting image and the ground truth image, reaching a SSIM score of 0.84. The Fig. 6.a and 6.b images show the original image and the reference image used for styling.

Fig. 11. Result obtained by photoreal-istic style transfer process.

Fig. 12. Ground truth photo. Repro-duced from [6]

An example where styling did not show a high degree of similarity can be seen in Fig. 13. It is a stylization of an outdoor photo in winter with snow (Fig. 13.a) to a summer day (Fig. 13.c), followed by the real image (Fig. 13.d) for comparison. The SSIM score for images Fig. 13.c and 13.d is 0.73.

(a) Content Image   (b) Style Image   (c) Resulting Image   (d) Ground truth Image

Fig. 13. Stylization example with the real image for comparison.

5   Conclusion and What's Next

We presented a study case for a visual prediction system based on photorealistic style transfer. That is, from a user-selected photo, the system must realistically stylize it to indicate current weather conditions. In this case study, we used a photorealistic method for style transfer known as Fast Photo Style, with some modifications. We observed that the technique has very satisfactory results when the content and style images have correct semantic segmentation and matching of their elements. We added a sharpness process to the end of the method, which brought a greater sense of realism to the final result. The experiments used the images provided by Laffont et al. in [6] as a content image, providing

ground truth images for comparing the similarity with the obtained results. The experiments reach an average SSIM index above.

For the style images to have proper semantic segmentation and to present all possible climatic situations, we have developed a set of 3D images with various outdoor scenarios and weather conditions. We produced the images in such a way as to maximize the matching of elements with the most common outdoor scenes. However, some situations were left out (e.g., cities with cars and people). We believe this is our first future work, i.e., increasing the range of scenarios to expand the possibilities of semantic matching.

Another interesting study is to verify the possibility of performing semantic matching with more than one style reference image. Thus, if a single developed scenario does not contain all the elements to perform semantic matching, the system would use a second reference image. Another possibility to study is to verify the generating a set of elements with different stylizations (related to the weather), such as sky (cloudy, clear, starry, etc.) or street (covered with snow, clear). Thus the system would no longer use a style reference image, but the style of the elements of that set.

References

1. Cambridge in Color - a learning community for photographers (2005–2020). https://www.cambridgeincolour.com. Accessed 12 Dec 2020
2. Structural similarity. Wikipedia, February 2021. https://en.wikipedia.org/wiki/Structural_similarity. Accessed 12 Feb 2021
3. Fix, E., HodgesHodges, J.L.: Discriminatory analysis. Nonparametric discrimination: consistency properties. Int. Statist. Rev./Revue Internationale de Statistique **57**(3), 238–247 (1989). http://www.jstor.org/stable/1403797
4. Fraser, B., Schewe, J.: Real World Image Sharpening with Adobe Photoshop, Camera Raw, and Lightroom, 2nd edn. Peachpit Press, Berkeley (2009)
5. Karacan, L., Akata, Z., Erdem, A., Erdem, E.: Manipulating attributes of natural scenes via hallucination. ACM Trans. Graph. **39**, 1–17 (2019)
6. Laffont, P.Y., Ren, Z., Tao, X., Qian, C., Hays, J.: Transient attributes for high-level understanding and editing of outdoor scenes. ACM Trans. Graph. (Proceedings of SIGGRAPH) **33**(4), 1–11 (2014)
7. Li, Y., Liu, M.Y., Li, X., Yang, M.H., Kautz, J.: A closed-form solution to photorealistic image stylization. In: ECCV (2018)
8. Min, K., Sim, D.: Confidence-based adaptive frame rate up-conversion. EURASIP J. Adv. Signal Process. **2013**, 13 (2013). https://doi.org/10.1186/1687-6180-2013-13
9. wiredfool, A.C., et al.: Pillow: 3.1.0, January 2016. https://doi.org/10.5281/zenodo.44297
10. Zhou, B., et al.: Semantic understanding of scenes through the ade20k dataset. Int. J. Comput. Vis. **127**, 302–321 (2018)
11. Wang, Z., Bovik, A.C., Sheikh, H.R., Simoncelli, E.P.: Image quality assessment: from error visibility to structural similarity. IEEE Trans. Image Process. **13**(4), 600–612 (2004). https://doi.org/10.1109/TIP.2003.819861

Automatic Generation of Machine Learning Synthetic Data Using ROS

Kyle M. Hart$^{(\boxtimes)}$ [iD], Ari B. Goodman, and Ryan P. O'Shea

Naval Air Warfare Center – Aircraft Division – Lakehurst, Lakehurst, NJ 08757, USA
{kyle.m.hart,ari.b.goodman,ryan.oshea3}@navy.mil

Abstract. Data labeling is a time intensive process. As such, many data scientists use various tools to aid in the data generation and labeling process. While these tools help automate labeling, many still require user interaction throughout the process. Additionally, most target only a few network frameworks. Any researchers exploring multiple frameworks must find additional tools or write conversion scripts. This paper presents an automated tool for generating synthetic data in arbitrary network formats. It uses Robot Operating System (ROS) and Gazebo, which are common tools in the robotics community. Through ROS paradigms, it allows extensive user customization of the simulation environment and data generation process. Additionally, a plugin-like framework allows the development of arbitrary data format writers without the need to change the main body of code. Using this tool, the authors were able to generate an arbitrarily large image dataset for three unique training formats using approximately 15 min of user setup time and a variable amount of hands-off run time, depending on the dataset size. The source code for this data generation tool is available at https://github.com/Navy-RISE-Lab/nn_data_collection

Keywords: Machine learning · Data generation · ROS

1 Introduction

Data labeling is such a time intensive and tedious task that many data scientists turn to existing datasets or outsource their data labeling process to others. However, there are occasions where scientists may require brand new datasets due to mission needs. Creating these custom datasets is a time-consuming process due to the time involved with data acquisition and data labeling. The time further increases if the researchers explore different networks or new scenarios after collecting the dataset. For example, if researchers wish to investigate scenes with different objects, they must create and label more data containing these new objects. Furthermore, researchers must make sure to account for any potential statistical changes in their dataset when they add or remove data. Often due to limited time and financial resources, researchers must be careful to

This is a U.S. government work and not under copyright protection in the U.S.; foreign copyright protection may apply 2021
H. Degen and S. Ntoa (Eds.): HCII 2021, LNAI 12797, pp. 310–325, 2021.
https://doi.org/10.1007/978-3-030-77772-2_21

select parameters to succinctly create their dataset. As the number of different parameters increase, so too does the dataset size and therefore the cost as well.

Because of this, a variety of tools are available to aid with the data labeling process. These tools assist in the labeling process with varying levels of automation. This work introduces a new fully automated tool targeted towards the creation of data sets using Robot Operating System (ROS) [1] and Gazebo [2], which are common tools within the robotics community. Using ROS and Gazebo, this tool generates perfectly labeled data from arbitrary user-specified scenes. By using common robotic toolsets, this tool provides a familiar means by which robotics researchers can build labeled image datasets for their work.

This tool is designed for minimal user involvement during the data generation process and maximum flexibility. The user can specify arbitrary motion plans for various objects in the scene, the camera position, simulated frame rate, and more. Additionally, the tool allows the addition of arbitrary new data formats through a plugin-like structure. This provides extensibility to new use cases not originally considered. Once configured, the tool automatically runs without requiring user involvement, dramatically reducing hands-on time.

The paper structure is as follows. First is an overview of other data labeling tools. Then, the authors describe their first version of this tool, which focused on a specific use case. Next, the paper describes the improved version of the tool. A discussion on usability follows, including steps required by the user, information on creating new data format writers, and a small vignette to illustrate the tool's effectiveness. Lastly, the authors identify several follow-on steps to further improve the tool.

Additionally, the source code for the tool described here is available at https://git hub.com/Navy-RISE-Lab/nn_data_collection.

2 Related Work

There is a wide array of techniques used to obtain labeled data for neural network training. These include the use of existing labeled datasets, leveraging data augmentation techniques on unlabeled data, and manually creating and labeling new datasets. Roh et. al. [3] provide a comprehensive overview of the various techniques. Many current techniques use existing datasets for training. However, there are cases where a project requires the creation of a new dataset through manual image labeling.

The traditional way to create labeled datasets involves manually labeling each image. There are several tools, such as Yolo_mark [4] and LabelImg [5], that assist a user with labeling. These tools typically provide user interfaces to aid in quickly drawing bounding boxes and assigning labels to an image. The tools then write the labels to file in a specified format. While these tools speed up the data labeling process, they still require the user to draw the boxes and assign the labels or require user confirmation that an automated guess is correct. Depending on the size of the dataset, labeling can be a laborious process. For example, the authors used Yolo_mark to label a video approximately a minute and a half in length in about three hours. The process can also introduce human error into the dataset if an image is accidentally mislabeled. Ideally, multiple people will label each image; however, having multiple people label each image significantly increases

the amount of time and money involved in creating the dataset. Additionally, these tools typically target only one or a few networks. Extending datasets to additional networks in the worst case can be impossible without relabeling missing data, and in the best case requires deep knowledge of the tool's code or a post-processing conversion script.

There is also a growing body of research on using neural networks to create and label data for training and testing other neural networks. Sixt et al. [6] and Pfeiffer et al. [7] both use generative adversarial networks (GANs) to take ideal 3D models of objects of interest and place them into realistic synthetic images. They use example images to mimic the lighting conditions, blur, and other characteristics. Similarly, Lee et al. [8] use a neural network to perform style transfer. These approaches allow for automatically generating large datasets. However, the examples the authors found were limited to single objects found within an image and focused on image labels instead of multiple objects within an image.

Besginow et al. [9] propose an alternative approach that generates labeled data for a single object. They use a hardware setup to capture an object from multiple angles. They also supply simple interfaces to allow some user customization for a semi-automated approach. The tool then generates the object detection labels. However, this approach is limited to objects that fit within the hardware setup. Additionally, they only target a single network output format.

Additionally, Dutta et al. [10] provide an overview of a number of automatic labeling methods that leverage machine learning techniques. This includes nearest neighbor approaches, neural network approaches, and SVM classifiers. The nearest neighbor approaches use already labeled images to match based on visual similarity. This requires a sufficiently diverse collection of labeled images, which may not exist for new datasets. All approaches also potentially result in some mislabeling, depending on the accuracy of the model used.

Some researchers have also begun to use other sensor modalities to automatically label training data. Kuhner et al. [11] propose the use of LiDAR in driverless cars to automatically create semantic labels for the image data generated by the car's cameras. Their approach quickly annotates images of roads and curbs for use in training neural networks used to detect and navigate around these objects. This approach offers fast generation of labeled real world data but does not generalize to other problem domains.

Lastly, the authors were unable to find any examples of data generation tools using ROS and Gazebo for scene simulation. Some synthetic generation methods use general purpose simulation environments. For example, Lee et al. [8] use the 3D game engine Unity to generate synthetic training data of wrenches in an industrial environment. Others, like the LGSVL simulator [12], are domain specific.

3   Version 1

The first version of this utility originated in a previous project that used YOLOv3 [13] and the Darknet [14] framework. Part of that project explored the impact of several setup parameters, such as image resolution, on network accuracy. This required multiple datasets composed of several simulated robots in arbitrary poses within a simulated environment. Performing this labeling by hand was too time consuming. To speed this

up, the team created a ROS package that automatically generated the necessary data in the correct format. By using perfect knowledge of the simulated world, the labels are guaranteed to be accurate. Additionally, the automatic nature of the tool means the authors could run the tool during off hours without the extensive time commitment needed to hand label the data.

The general algorithm is below. The project focused on object detection within single images, so the algorithm uses the entire available space within the simulated environment for robot placement. Additionally, because of the authors' familiarity, they chose to write this version of the tool in C++.

```
Version 1 Procedure:
Load user parameters
For each required datapoint:
  image <- capture new image
  Write to file(image)
  For each object:
    selected_pixel <- select pixel from map pixels (1)
    If selected_pixel is not free:
      GOTO (1)
    radius <- load user specified safety radius
    outer_pixels <- Bresenham's Circle Algorithm(selected_pixel,
        radius)
    For each outer_pixel in outer_pixels:
      line_pixels <- Bresenham's Line Algorithm(selected_pixel,
          outer_pixel)
      For each line_pixel in line_pixels:
        If line_pixel is not free:
        GOTO (1)
    position <- transform selected_pixel to coordinate
    orientation <- select value from (-pi, pi]
    Move object in simulation(position, orientation)
    bounding_rectangle <- Project Shape into Image
    Write to label file(bounding_rectangle, class id)
```

To execute the program, the user first specifies several settings via ROS's parameter server, which is a standard way to customize routines in ROS. Specifically, this includes information about where to save files, which robots to use, and each robot's footprint, height, and a safety radius that ensures no accidental collisions during object placement. Additionally, the user must publish a map of the environment indicating which areas are free or occupied. This map also follows a standard ROS convention and uses pixel values to indicate if a space's occupancy status. After providing this information and starting the simulation environment, the user can then run the generation utility and let it auto-generate until finished.

While running, the tool selects a random pose for the robot. To do this, it picks a free pixel on the map, then uses the robot's safety radius, Bresenham's Line drawing algorithm [15], and Bresenham's Circle drawing algorithm [16] to identify if the robot can fit at the selected spot. The drawing algorithms identify which pixels the robot could potentially occupy for a given pose. It then checks if each pixel is already occupied or

not. If any pixel is occupied, the algorithm selects a new position, and the process begins again. If the space is free, the algorithm selects a random orientation for the robot. The algorithm then requests Gazebo to move the robot to that pose. This process then repeats for each robot.

After each robot is in position, the tool captures the image data from a simulated camera within the environment, using ROS's publisher/subscriber model. When capturing the image, it also captures the camera information, such as its intrinsic parameters.

Using this information, the tool then generates the labels using the algorithm shown below. It uses the robot's pose and its user-specified shape to construct a rectangular cuboid that circumscribes the robot in the robot's frame of reference. It then transforms the vertices of this cuboid into the image using known positional data and the camera's parameters. Next, it circumscribes a bounding box around the projected vertices using OpenCV [17]. Because these are vertices of a bounding shape, this bounding box is guaranteed to encompass the robot on the image. As proof, consider the case where the object has some feature outside of the bounding rectangle on the image. If this is true, then the vertices projected on the image would also extend outside the rectangle, as the vertices circumscribe the object in 3D space. However, the bounding rectangle is constructed to circumscribe all the projected vertices, which contradicts the original premise of this case. Therefore, this bounding box on the image contains the entire robot within it.

```
Project Shape into Image(object):
vertices <- load user specified object shape
transform <- lookup transform to camera frame(object)
projected_vertices <- apply transform(vertices, trans
    form)
camera_matrix <- lookup camera parameters
projected_points <- apply camera projection(projected_vertices,
camera_matrix)
bounding_rectangle <- OpenCV.BoundingRect(projected_
    points)
Return bounding_rectangle
```

Lastly, the tool writes the values to file in the correct format. Darknet datasets consist of two elements. The first is the set of raw, unlabeled images. The second element is an associated text file for each image. Each line of the text file contains the information for one object within the image. This information includes the class id for the class of the object, and the size and location of the bounding box, expressed as fractions of the overall width and height of the image. For example, a bounding box highlighting an object with class id 1, centered on the image with width and height one quarter of the respective width and height for the image would be labeled as *"1 0.5 0.5 0.25 0.25"*.

This entire process can then repeat until the tool creates the desired amount of data. Additionally, the user can easily rerun the process with new settings to create validation or test datasets, or to explore new scenarios. Rerunning takes no added user input other than the time required to change settings.

Using this method, the authors were able to generate a fully labeled 1,000 image synthetic dataset with about 10 min of setup time and 30 min of data generation time.

The setup time is the only time that requires user operation and does not increase as the size of the generated dataset increases. The data generation time runs without the user and scales approximately linearly with the number of samples. Once generated, the data was immediately used for network training without any additional post-processing. This allowed the team to quickly explore several elements to the project without spending work time on data creation.

However, there are some drawbacks to this tool. As mentioned previously, the algorithm targets the Darknet format and design considerations, such as moving the robot to arbitrary positions on the map, were made because of it. Any new format would require extensive rewrites. Additionally, there are several required dependencies for the package to work. While some are standard ROS packages found in any ROS installation, some dependencies are less common. Some were even lab specific ROS packages, preventing widespread use of this tool. When the main project began work on a second phase, it became clear that the team required a tool with greater flexibility.

4 Version 2

After completion of the first phase of the project, the underlying research expanded to include improved neural network methods. This included new network types, more detailed detection, and the use of video data instead of isolated images. Adapting the existing synthetic generation tool would require significant work. Therefore, the team took the opportunity to redesign the tool to promote greater modularity and reduce the need for future rewrites. The four primary goals for this version were as follows.

- Allow the user to specify and label video frames and sequential images to capture motion between images.
- Use a modular structure to allow the easy creation of new data formats as needed.
- Generate data to train instance segmentation networks.
- Reduce the required dependencies to run the tool, ideally to only a few that are likely to already exist on a computer with ROS and neural network frameworks installed.

Additionally, the team decided to rewrite the entire package in Python. Python is a primary language amongst both data scientists and ROS users, so any potential users are more likely to be familiar with Python than they would be with C++.

4.1 Motion Generation

The first version of the generation tool created sets of unrelated images, so the tool moved objects to arbitrary locations at each successive image it generated. The later stage of the project relies on video data fed in as successive frames to exploit state information between images. Therefore, the tool needs to generate data the shows an object's motion across a series of images. To support modularity, this requires the ability to specify any possible motion path in an intuitive way, to allow the exploration of a variety of motion scenarios.

Achieving this flexibility uses a built-in capability of ROS. ROS uses packages called tf and tf2 [18] to track relative pose information between every frame of reference in the scene throughout the entire simulation. This structure is called the TF tree. Additionally, ROS supplies a way to record any information published during simulation using a concept known as bag files. Bag files are a common means by which researchers record their experiments for use later. The algorithm simply reads and stores a TF tree from a pre-recorded bag file to generate a record of each objects pose throughout the simulation.

To generate this bag file, the user can use whatever methods they want. They can use real or simulated robots, joystick control or autonomous control, or any other means. As TF trees are extremely common in ROS robotic simulations, most setups already produce the required information, so it is a simple matter to record the data. Additionally, trees are represented as text information, so the file size is small, even for large simulations.

When run, the algorithm reads in the TF tree from the bag file and uses it as a set of instructions to recreate the scene. It follows the algorithm shown below. Using a user specified frame rate, the algorithm steps through the scene and finds the pose of each object at that moment in time, based on the TF tree. It then moves the objects to those same poses in the Gazebo simulation, recreating the recorded scene. It is important to note that this recreation does not have to occur in real time. Since the bag file contains the positions for the entire simulation, the algorithm can spend as much time processing information as needed before moving on to the next frame, similar to the use of Claymation in movies.

```
Scene Generation Procedure:
Load bag file
start_time <- identify valid start time from bag file
end_time <- identify valid end time from bag file
frame_rate <- load user specified frame rate
current_time <- start_time
While current_time <- end_time:
  For each object to control:
     pose <- lookup object pose in bag file
     move object in simulation(pose)
  scene_data <- Capture all scene data
  For each format writer:
     format_writer.WriteScene(scene_data)
  current_time <- current_time + (1.0 / frame_rate)
```

An added user benefit of this approach is the ability to create new scenes. When recreating a frame using the TF tree, the algorithm does not care about the visual appearance of the object nor the position of the simulated camera. This means a user can create a single motion plan, then generate data for several different setups. These alternate setups can include different environments, lighting, and camera positions. Alternate setups can also include new types of objects following the same motion paths. The algorithm is agnostic to each and does not require an updated bag file. For instance, the authors developed some example data in an empty environment, then used the same motion plan in an environment with walls and a new camera position. The authors did not need to change the recorded bag file, so created new datasets without any significant user involvement.

This motion recording framework provides a robust means to create motion scenes. The users can move objects through a scene with whatever means they typically use. Then, by recording standard ROS published information, the algorithm has enough data to recreate these scenes in simulation. The user can also explore different scenes using the same recorded information without the need to modify the recorded data.

4.2 Modular Formats

Another goal for the tool is to allow arbitrary label formats. If a user desires a new labeling scheme to support some new network, they should be able to quickly write the specific code they need to parse the information, without concern for the rest of the system. This increases the usefulness of the tool across a range of use cases.

To accomplish this, the package uses a plugin style scheme. Specifically, the package defines a base class for all potential format writers using the Abstract Base Classes (ABC) library [19]. ABC enables the declaration of abstract methods that any inheriting class must implement. Each new format writer is implemented as an inheriting class, thus ensuring that each format writer interfaces correctly with the main algorithm. There are three specific methods that each inheriting class must implement. The main algorithm keeps a list of all the classes and calls the appropriate methods for each one at the correct time.

The first called method occurs at each successive step through the scene, when the objects are at their recorded poses for that instance in time. The complete scene information, including object poses, the raw image, and pixels masks for each object are all passed to the method. The method is then only responsible for extracting the specific information the format needs and writing it to file in the correct manner.

The next is a function called at the very end of the entire execution. This allows the format writers to perform any final steps required by the format. For instance, Darknet uses a main text file that contains a list of each image in the dataset. The final abstract method is one simply used to indicate if a particular format requires instance segmentation. As discussed later, this is only to help improve processing time.

The base class also offers two helper functions that are of use to most formats. It offers a method that transforms points between frames of reference and another that projects points into an image. A common helper function avoids the need to reimplement this transform functionality for each writer.

By defining the format writers in this manner, users can add new formats quickly and without reprogramming the core data collection process. For example, adding a Darknet format involves only a few steps. First, is the scene writing method. When the method is called, the Darknet format writer uses the provided information about the objects, including their location and vertices of bounding shapes, to transform the vertices into the image. From there, the format writer simply determines the pixel bounding box using these projected vertices and writes this bounding box to the right file. At the end of the execution, the Darknet format writer then writes the main list. Because the base class and main algorithm manages most functionality, this Darknet module is around 100 lines of code with no step more complicated than calling functions or finding an average. Implementing this functionality took considerably less time than implementing it in Version 1.

Following this approach, there are currently three formats defined within the package. They are Darknet [14], a network implementation known as PVNet [20] that uses projected keypoints, and a custom format that uses the vertices projected into an arbitrary frame of reference.

By using this modular format structure, users can easily add additional formats without unnecessary rewriting of the core functionality of data collection. This allows users to extend this tool to new formats without the need for extensive rework.

4.3   Instance Segmentation

An entire subclass of networks performs instance segmentation on image data. This requires labeled data that features unique identifiers for each instance of an object class and indicates which pixels contain an object in a given image. To support labeling data for these networks, this algorithm generates pixel masks to indicate which pixels belong to each object. The general algorithm is as shown below. Figure 1 Shows an example observation generated by the algorithm, along with the raw image of the scene.

```
Instance Segmentation Initialization:
subtractor <- OpenCV.CreateBackgroundSubtractorMOG2()
Remove all objects from scene
For a user specified number of times:
  image <- capture image
  subtractor.apply(image)
Return subtractor


Instance Segmentation:
If instance segmentation required:
  For each object under control:
    store pose
    move out of camera view
  For each object under control:
    move back to pose
    image <- capture image
    mask <- subtractor.apply(image)
    mask <- apply image post-processing(mask)
    store mask for use by data writers
    Move out of camera view
```

Prior to any steps in the above algorithm, the tool polls each format writer. If no formats require instance segmentation, the tool skips this entire routine. This helps increase processing speed by avoiding unnecessary image manipulation.

During initialization, the algorithm first creates a background subtractor. It uses OpenCV's Gaussian Mixture-based implementation [17]. A few of the parameters of the subtractor can be user specified to tune performance. It then moves every object out of view of the camera to capture background images. The algorithm collects multiple images to account for noise. As each image is captured, the tool applies them to the background subtractor to generate a reference background.

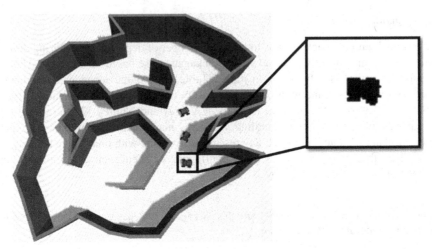

Fig. 1. A sample raw image collected during the data generation process and associated pixel mask for one of the objects in the scene. The colors of the mask are inverted for readability.

The initialized subtractor is then used during scene generation. At each step, the tool moves all objects out of view of the camera. Then, one at a time, it moves them back to their correct position within the scene. It then captures the image and runs it through the background subtractor. Because the object is the only thing in the scene besides the background, the results are a pixel mask indicating which pixels belong to the object. This process continues for each object. After capturing each object's pixel mas, the algorithm then moves all objects back into the scene in their correct spots to capture the raw image for that scene.

Right after image collection, the algorithm performs some post processing on the pixel masks. Using the known 3D bounding shape of the object and the correct transforms into the image, it constructs a bounding box around each object. It then uses this box as a filter for the pixel mask. No pixel outside of the box is set in the mask, since the box defines a conservative outer bound on possible pixels associated with the object. This helps reduce noise in the pixel mask.

When the tool calls each format writer to label a single scene, it provides these generated pixel masks to the writers. Currently, layering the masks to correctly match model occlusion in the scene remains an open question and is left to each data writer to manage. While this is an important functionality to include in the core algorithm or data writer base class, the authors chose to defer implementation due to project constraints as none of the formats currently in use by the team require a layered pixel map. However, one potential approach is to use the object's Euclidian distance to the camera to decide the ordering layer for the combined mask.

With this capability, the automatic generation of synthetic data can now extend to even more types of neural networks.

4.4 Dependencies

The last goal with the package was to limit the number of dependencies. During development, the team carefully selected which packages to use based on what the average ROS data scientist is likely to have installed. Table 1 shows the dependencies. Almost all utilized packages come with the standard ROS installation and the default Python installation. The only two that do not, OpenCV [17] and NumPy [21], are used for background subtraction, image handling, and array manipulation. While these are not default packages, the team felt that any researcher working with image data is likely to already have these packages installed. By limiting these dependencies, the tool is easier to integrate into anyone's workflow.

Table 1. A list of dependencies required for this package. Bold dependencies do not come with the default Python and ROS installation.

ROS packages	Python packages
cv_bridge	abc
gazebo_msgs	**cv2**
geometry_msgs	**numpy**
rosbag	os
rospy	
sensor_msgs	
tf2_ros	

5 Usability

The previously discussed goals for this tool ensure that it is straightforward to use for anyone familiar with ROS. In general, a user will need to configure the initial setup, run the application, wait for it to finish, then start using the results. Each step only involves a few actions, if any, to complete. Additional steps are needed if new formats need incorporated.

5.1 Setup

While setting up the package for a specific job is the most involved step, it is still straightforward. The package comes with a detailed README that walks the user through the steps. Additionally, example files document all settings and describe their default behavior. Customizing these settings and completing setup involves three main steps.

First, the user must generate the necessary bag file. They can do this at any time prior to running the package. As discussed above, there are very few constraints for this bag file. The user need only record the TF tree while they move objects in the desired motion

patterns. This is a standard procedure for any ROS user, so most users will already be familiar with this step.

After creating the bag file, the user must also create a simulated world in Gazebo. While in most cases, this world will mimic the one used to generate the bag file, that is not necessary. This environment can be as simple as an empty world or as complicated as a realistic office setup. The user must place a camera for image capture somewhere within the environment. They must also instantiate a number of objects representative of the objects controlled during bag file creation. The object identifiers must match the ones used when creating the bag file. However, visual appearances can differ.

Once the user records the bag file and has the Gazebo world is running, the last setup step is configuring the parameters. ROS uses YAML files to configure parameters from a single file. This is a standard way of customizing ROS packages. The package supplies a default YAML file to use as a template. It also has documentation on each parameter, many of which have default values if not provided. These parameters specify things such as which bag file to use, the list of objects to control, and where to place output files. Each data writing format may have its own parameters as well that the user can specify in the same YAML file.

After configuration of the parameters, the data generation is ready to run.

5.2 Execution and Performance

After completing setup, the user is ready to run the program. This is a simple, one line command to start the entire data generation process. After starting, the user need only wait while the program executes.

Starting the program uses a ROS concept called launch files, which is a standard way to start programs with user specified parameters. When started, the program checks the various user specified values. It ensures that each required parameter is set and warns the user if an optional parameter is unspecified. During execution, the package will periodically print the percent complete to the command line to update the user. No user interaction is required during execution.

During development, the authors measured performance of the generation tool. They explored runtimes across different numbers of objects, collections of data writers, and time. Figure 2 features a summary of results. There were two types of formats considered, one that requires instance segmentation and one that does not. The authors measured these results using Ubuntu's time command on a laptop with an i9 processor and 32 GB of memory.

As currently designed, this package is not fast. To promote modularity, some efficiency was sacrificed. For example, depending on the format specifications, multiple format writers might have to manipulate the data in the same way. The result is a running time that scales with the number of data writers. Additionally, the tool moves each object at each iteration. This increases the running time as the number of objects increase, as shown in Fig. 2.

The figure also shows the relative impact of performing instance segmentation. The format that requires it grows at a faster rate as the number of robots increase. This is due to the background subtractor. As written, it runs once for each robot in a scene.

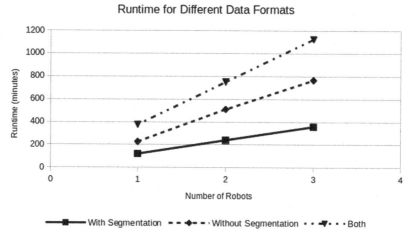

Fig. 2. A comparison of runtimes across different numbers of objects and different format writers.

However, the main algorithm also queries each format writer and skips this step if not needed, resulting in the faster execution rate seen in the figure.

While the data generation time is lengthy, it is important to note that it is fully automated. Once the algorithm begins to collect data, no user intervention is necessary. This is one of the primary benefits to this package.

5.3 After Generation

Once execution is complete, the user has one or several complete sets of labeled data. This data is already in the correct format for training and can be used straight away without further user effort required. The only limited post processing a user might need is splitting out a test or validation set.

5.4 New Formats

When a new data format is required, the user must develop the code to write the format correctly. Following the above-mentioned structure, the user must implement a new class with a few required methods. Then, the user includes this new format class in the list of formats to call during execution. The included README documents this entire process.

To start making a new format, the user defines a new Python class. This class should inherit the base class offered by the package. Because this base class has abstract methods, the user must implement them and will receive an error if they do not and try to run the code anyway.

Within this new class, the user should initialize any parameters that the format requires. The base class already looks up the user specified output location, but the inheriting class can expand this to include any configuration settings needed. Next, the user implements the method called at each scene. This is the primary method used to write data to file. Each time it is called, the method is provided a list of all objects, their

locations, object specific information such as keypoints, and image data. This method should use the provided information to translate positional information into appropriate labels. It should also write the necessary label files, as dictated by the format. After that, the user implements a final method called at the end of data generation. This provides an opportunity for any cleanup operations or writing metainformation, such as lists of label file locations. Lastly, the user should specify if the format requires instance segmentation pixel masks. As discussed previously, this allows the main algorithm to reduce computational time if no format requires segmentation.

After defining the class, the user then includes it in the library and list of classes called by the algorithm. These are both single line steps. Inclusion in the library involves importing the class into the overall data format module. Adding it to the tool involves instantiating an object within the main program. The documentation illustrates these steps using previous examples. After completion, the main code will automatically call the correct methods at the appropriate time.

By creating the data writer class, defining the methods, and including it in the main tool, the user can quickly add new data formats. Creating these new definitions does not require changing the main code. By including this functionality, the usefulness of the tool can continue to expand as the library of existing formats grows.

5.5 Vignette

As an example of the efficiency of the tool, consider the following anecdote. The authors were exploring a new network. They used this tool to generate data for a simple setup just to ensure the network was working. After creating the data, they found that the camera placement and resolution did not have sufficient coverage of specific parts of the scene. The authors adjusted the camera information in simulation and still used the same settings and recorded bag file to generate new data. The entire process of creating a new dataset took approximately 10 min of human effort compared to an estimated 1–2 h for manual labeling. The authors then repeated this process a few times with new setups and simulated frame rates with similar levels of effort required. Additionally, the authors explored a new type of network. The authors quickly wrote up an additional plugin to write data for this new format. They completed the plugin and integrated it into the existing list with the others within an hour. Data generation then proceeded as usual. The result was a smooth data generation process that allowed the authors to spend minimal effort to create the data and more time exploring network optimization.

6 Remaining Questions

This synthetic data generation tool is a flexible approach to generating data for a range of neural network formats. However, there are still future improvements that would further enhance useability. This includes increasing speed, expanding the format library, and GUI development.

As discussed above, the algorithm suffers from poor scalability. While modularity typically introduces some overhead, code changes can likely improve the runtime. For example, some of the underlying algorithms can be rewritten to store and manipulate

data more efficiently. As many routines are called multiple times, any performance gain is likely to have significant impact on overall runtime.

Additionally, expanding the data format library increases useability. Since the tool allows easy development of format writers, users can quickly incorporate new formats into the available list. This has the added benefit of testing underlying assumptions on the algorithm to ensure it is truly useable across a wide range of networks. The authors are already exploring Mask-RNN [22] and format writers to mimic common dataset formats, such as the Common Objects in Context (COCO) dataset [23].

Lastly, the inclusion of a GUI simplifies setup. While the current approach follows familiar ROS conventions, a well-designed GUI offers a means to guide the user through the setup process and address potential issues prior to running the tool. This would reduce errors and further decrease setup time.

The primary benefits of this tool are its simplicity and minimal user effort. Increasing the runtime, adding additional formats, and developing a GUI all contribute to these benefits and make the tool more useful.

7   Conclusion

This paper proposes an automated tool for generating annotated image training data for various object detection networks. The proposed package is the first to use ROS and Gazebo for the purpose of automatically generating synthetic annotated machine learning training data. Flexibility in package parameters like the simulated world, recorded motion plans, and camera parameters allows for rapid generation of diverse datasets with minimal effort from users. Additionally, the ability to easily add more data format plugins allows users to tailor the package to their specific data generation needs. Using this toolset, the authors created three fully annotated datasets in three separate formats with about 15 min of manual setup and several hours of hands-off running time, depending on the specific configuration. This package drastically reduces hands-on data collection and labeling time while ensuring the generation of accurate data in a modular way to allow various use cases.

References

1. Quigley, M., et al.: ROS: an open-source robot operating system. In: Proceedings of the IEEE International Conference on Robotics and Automation (ICRA) Workshop on Open Source Robotics, Kobe, Japan (2009)
2. Koenig, N., Howard, A.: Design and use paradigms for Gazebo, an open-source multi-robot simulator. In: 2004 IEEE/RSJ International Conference on Intelligent Robots and Systems (IROS), vol. 3, pp. 2149–2154 (IEEE Cat. No.04CH37566) (2004)
3. Roh, Y., Heo, G., Whang, S.E.: A survey on data collection for machine learning: a big data -- AI integration perspective. ArXiv181103402 Cs stat (2019)
4. Alexey: Yolo_mark
5. Tzutalin: LabelImg (2015)
6. Sixt, L., Wild, B., Landgraf, T.: RenderGAN: generating realistic labeled data. Front. Robot. AI. **5**, 66 (2018). https://doi.org/10.3389/frobt.2018.00066

7. Pfeiffer, M., et al.: Generating large labeled data sets for laparoscopic image processing tasks using unpaired image-to-image translation. ArXiv190702882 Cs stat (2019)
8. Lee, Y.-H., Chuang, C.-C., Lai, S.-H., Jhang, Z.-J.: automatic generation of photorealistic training data for detection of industrial components. In: 2019 IEEE International Conference on Image Processing (ICIP), Taipei, Taiwan, pp. 2751–2755. IEEE (2019)
9. Besginow, A., Büttner, S., Röcker, C.: Making object detection available to everyone—a hardware prototype for semi-automatic synthetic data generation. In: Streitz, N., Konomi, S. (eds.) Distributed, Ambient and Pervasive Interactions. LNCS, vol. 12203, pp. 178–192. Springer, Cham (2020). https://doi.org/10.1007/978-3-030-50344-4_14
10. Dutta, A., Verma, Y., Jawahar, C.V.: Automatic image annotation: the quirks and what works. Multimedia Tools Appl. **77**(24), 31991–32011 (2018). https://doi.org/10.1007/s11042-018-6247-3
11. Kuhner, T., Wirges, S., Lauer, M.: Automatic generation of training data for image classification of road scenes. In: 2019 IEEE Intelligent Transportation Systems Conference (ITSC), Auckland, New Zealand, pp. 1097–1103. IEEE (2019)
12. Rong, G., et al.: LGSVL simulator: a high fidelity simulator for autonomous driving. ArXiv200503778. Cs Eess (2020)
13. Redmon, J., Farhadi, A.: YOLOv3: an incremental improvement. ArXiv180402767 Cs (2018)
14. Redmon, J.: Darknet: open source neural networks in C (2013)
15. Bresenham, J.E.: Algorithm for computer control of a digital plotter. IBM Syst. J. **4**, 25–30 (1965)
16. Bresenham, J.: A linear, incremental algorithm for digitally plotting circles. Tech Rep (1964)
17. Bradski, G.: The openCV library. Dr Dobbs J. Softw. Tools. **25**, 120–125 (2000)
18. Foote, T.: tf: The transform library. In: IEEE International Conference on Technologies for Practical Robot Applications (TePRA), 2013, pp. 1–6 (2013)
19. Rossum, G. van, Talin.: Introducing Abstract Base Classes (2007)
20. Peng, S., Liu, Y., Huang, Q., Bao, H., Zhou, X.: PVNet: pixel-wise voting network for 6DoF pose estimation. ArXiv181211788 Cs (2018)
21. Harris, C.R., et al.: Array programming with NumPy. Nature **585**, 357–362 (2020). https://doi.org/10.1038/s41586-020-2649-2
22. He, K., Gkioxari, G., Dollár, P., Girshick, R.: Mask R-CNN. ArXiv170306870 Cs (2018)
23. Lin, T.-Y., et al.: Microsoft COCO: common objects in context. ArXiv14050312 Cs (2015)

A Questionnaire Data Clustering Method Based on Optimized K-Modes Algorithm

Wen-jun Hou[1,2], Jia-xin Liu[1,2(✉)], and Xiang-yuan Yan[1,2]

[1] School of Digital Media and Design Arts, Beijing University of Posts and
Telecommunications, Beijing 100876, China
liujiaxin_bj@126.com
[2] Beijing Key Laboratory of Network Systems and Network Culture, Beijing University of
Posts and Telecommunications, Beijing 100876, China

Abstract. When conducting user research, questionnaires are often used to collect user characteristics, attitudes, and other information. Data cluster analysis is often used to divide user groups. The traditional clustering algorithms are mostly only suitable for numerical attributes or disordered categorical attributes. However, questionnaire data is mainly composed of disordered and ordinal categorical attributes. To solve the questionnaire data clustering problem, based on the traditional K-Modes algorithm, a method that combines subjective weighting and objective clustering for questionnaire data analysis is proposed. This method first performs multiple-choice clustering questions to reduce dimensionality and then re-weighting ordinal categorical attributes to rationalize the distance measurement. An optimized mixed K-Modes algorithm for questionnaire data clustering is proposed. The dissimilarity measure between objects according to the two types of disorder and ordinal categorical attributes. In order to evaluate the clustering results, an effective cluster validity index is also defined in this paper. Using a bank user survey questionnaire as a case proved the effectiveness of this data clustering method.

Keywords: Questionnaire data clustering · K-modes algorithm · Mixed categorical data · Ordinal categorical data weighting

1 Introduction

In user research and product design, it is often necessary to understand user characteristics and corresponding needs and divide user groups. Researchers can make personalized designs for different groups of users, and at the same time, understand the primary users of the product and make more targeted designs. The premise of these analyses is to have a certain amount of user data. Whether it is scientific research or commercial surveys, a large part of this data is obtained through questionnaires. On a current online crowdsourcing platform in mainland China, tens of thousands of questionnaires and more than 1 million answer sheets are generated every day. The questionnaire method is a common method used for user demand research, personas establishment, and product design evaluation [1]. It can conduct pre-surveys when there is no mature product. Experts often use

© Springer Nature Switzerland AG 2021
H. Degen and S. Ntoa (Eds.): HCII 2021, LNAI 12797, pp. 326–342, 2021.
https://doi.org/10.1007/978-3-030-77772-2_22

questionnaires to collect user information, such as demographic information, attitudes, behaviors, and other related information. However, the original data obtained after the questionnaire is often issued complex and distributed. At present, most of the questionnaire's data analysis only stays on simple frequency analysis and does not explore the value behind the data. However, these surface data analyses cannot directly reflect user characteristics and cannot directly see information such as user group division through data. Therefore, the data processing of the questionnaire is essential [2].

Therefore, starting from the original data of the questionnaire, this paper proposes a clustering method suitable for the questionnaire's data processing. In the second part of the related work, this paper briefly describes the current common questionnaire processing methods and the optimization results and limitations of the clustering algorithm in domestic and foreign papers. In the third part, an optimized mixed K-Modes algorithm for questionnaire data clustering is proposed, an effective cluster validity index is also defined. In the fourth part of the paper, a bank user questionnaire is used as a case to prove the effectiveness of the questionnaire data analysis method. Finally, in the sixth part, the work of this paper is summarized and prospected.

2   Related Work

The analysis methods of questionnaire data mainly include basic descriptive statistics [3], correlation analysis, and significance analysis, such as independent-samples t-test, paired-samples t-test [4], ANOVA, correspondence analysis [5], principal component analysis [6], etc., as well as influencing factor analysis and modeling binary logistic regression analysis, multiple logistic regression analysis [7], etc. Most of these analysis methods are cleaning and verifying questionnaire data. In order to explore the characteristics of the questionnaire and the user group information reflected behind it, cluster analysis of the questionnaire data is required. The current clustering analysis methods using SPSS have certain limitations on the data format [1], and the calculation speed is slow, which is not suitable for processing large amounts of data. Therefore, the processing of questionnaire data requires the use of clustering algorithms based on machine learning.

So far, a large number of clustering algorithms have been proposed in the literature, such as the K-Means algorithm. This type of algorithm is based on Euclidean distance for clustering analysis, but it is only suitable for numerical attributes [8, 9]. The questionnaire data is mostly qualitative data, categorical attributes, such as 'gender', 'occupation', and other common survey information. In order to solve the clustering problem of categorical attributes, Huang proposed the K-Modes algorithm in 1998 [10], which uses a simple matching method for dissimilarity measurement. After this, for different application scenarios, many clustering algorithms suitable for mixed numerical and categorical data have been proposed [11], but the categorical data considered in the mixed clustering algorithm are all disordered categorical attributes. Although many scholars have given optimization the improvement method of the sample distance of the categorical attribute [12], for the questionnaire data, the distance of the categorical data is not only one except for zero. For attributes such as "age group" and "consumption amount," it belongs to ordinal categorical attributes. As an ordinal categorical attribute,

a dissimilarity measurement method suitable for ordinal attributes should be used for cluster analysis. Yuan proposed an algorithm distance formula for mixed categorical data of ordinal and disordered data in 2020 [13]. Although it is suitable for the UCI data set, it is not comprehensive enough for questionnaire data processing. There are subjective factors in the design of the questionnaire options, which will be designed according to the distribution of the target group. Therefore, when conducting data analysis, subjective factors still need to be added for data processing.

This paper's main work is to propose a method suitable for clustering analysis of questionnaire data to solve the problem that questionnaire data has both disordered categorical attributes and ordinal categorical attributes, and the corresponding clustering evaluation indicators are given. At the same time, the subjective weighting method for the distance of ordinal attribute options is also given. Through subjective and objective data cluster analysis, the questionnaire results are more precise and more effective.

3 Method

3.1 Process

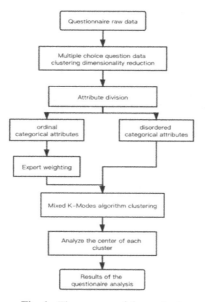

Fig. 1. The process of the method

The method of questionnaire data clustering is implemented with the following steps (Fig. 1).

- Step 1: Firstly, the data of each question is processed separately, and the traditional K-Modes clustering algorithm is used to reduce the dimensionality of multiple-choice questions.

- Step 2: Select the categorical attributes that need to be clustered, and divide them into disordered attributes and ordinal attributes.
- Step 3: Determine each ordinal categorical attribute's option weight through the expert questionnaire survey method and calculate the weight range of each ordinal categorical attribute.
- Step 4: Use the mixed K-Modes algorithm proposed in this paper for clustering, and get the desired clustering results.
- Step 5: Analyze the clusters and the value of each attribute of each cluster's center to obtain the required results of the questionnaire analysis.

3.2 Multiple Choice Questions Dimensionality Reduction Processing

The design of multiple-choice questions and the design of matrix multiple-choice questions can cause multiple variables to appear in the same dimension, so we need to preprocess the multiple-choice questions and reduce their dimensionality.

The source data format of multiple-choice questions obtained from the questionnaire results is shown in the following Table 1:

Table 1. The source data of multiple-choice questions obtained from the questionnaire results

	Option A	Option B	Option C	Option D	Option E
User 1	0	0	0	1	0
User 2	0	1	1	0	0
User 3	1	0	0	1	0
	...	...	...	...	...
User 199	0	0	0	0	1
User 200	0	0	0	0	0

In previous studies, the clustering of multiple-choice questions used rule-based clustering [15]. However, this clustering method needs to manually check each object, compare them one by one, check whether the opinions are consistent, and classify them. When the number of objects is large, this method is time-consuming and laborious.

For example, user one and user three have the following answer source data for a multiple-choice question:

User One: 0 0 0 1 0 0.

User Three: 1 0 0 1 0 0.

Experts focus on the same and different numbers of corresponding options in the two objects when clustering. The more the same number, the more consistent their opinions are. This kind of method is consistent with the idea of Hamming distance so that we can use the clustering method based on Hamming distance, that is, the traditional K-Modes algorithm to cluster multiple-choice questions to reduce the dimensionality.

K-Modes Algorithm

The traditional K-Modes algorithm is an extension of the K-Means algorithm. Because

the K-Means algorithm is only suitable for numerical data, that is continuous attributes. It is not appropriate to use Euclidean distance as the sample distance for discrete attributes. Therefore, Huang proposed the K-Modes algorithm [10] in 1998. The center's selection uses the mode instead of the mean, and the Hamming distance is used to calculate the sample distance, which is suitable for discrete data sets, that is, categorical data.

The Dissimilarity Measurement
Suppose a categorical data set $X = \{x_1, x_2, \ldots, x_n\}$ has n categorical objects, and each object described by m categorical attributes, which can be expressed as $\{A_1, A_2, \ldots, A_m\}$. According to the idea of simple matching, the dissimilarity measure between any two objects $x_i = \{x_{i1}, x_{i2}, \ldots, x_{im}\}$ and $x_j = \{x_{j1}, x_{j2}, \ldots, x_{jm}\}$ can be expressed as:

$$D(x_i, x_j) = \sum_{r=1}^{m} D(x_{ir}, x_{jr}) \qquad (1)$$

Where,

$$D(x_{ir}, x_{jr}) = \begin{cases} 1, x_{ir} \neq x_{jr} \\ 0, x_{ir} = x_{jr} \end{cases} \qquad (2)$$

Algorithm Steps
The algorithm steps of K-Modes are as follows:

- Step 1: Determine the number of clusters k, randomly select k objects as the initial center of each cluster.
- Step 2: Calculate the distance from the remaining objects to each cluster center according to formula (1), and reallocate the object to the nearest cluster.
- Step 3: After all objects have been allocated to existing clusters, update the center of each cluster (use the mode of the cluster as the center), and reset the dissimilarity of objects against the current modes according to Step 2. If the new distance is less than the current distance, the objects are reallocated to closer clusters.
- Step 4: Repeat Step 3 until the clusters to which all objects belong no longer change or the cluster centers no longer change.

3.3 Questionnaire Disordered Categorical Attribute and Ordinal Categorical Attribute Clustering

In the overall questionnaire data, the distance relationship between attributes is not just disorderly. For example, for the question "What is your age?", the options are set to {"Under 18", "18–25", "26–35", "35–50", "Over 50 "}. The corresponding value converted into data analysis is {1, 2, 3, 4, 5}. For this question, in actual analysis, the distance between option 1 "under 18" and option 2 "18–25" is different from the distance between option 1 and option 3 "26–35" the distance relationship is shown in the figure. However, the traditional K-Modes algorithm classifies it as the same distance. That is, as long as the categories are different, the distance is set to 1. This clustering distance calculation method is not comprehensive enough for questionnaire analysis (Fig 2).

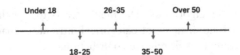

Fig. 2. The distance relationship of age.

Let the mixed categorical data set $X = \{x_1, x_2, \ldots, x_n\}$, in which there are a total of n objects. Each object is A-dimensional, namely $x_i = \{ord_i^1, ord_i^2, ord_i^3, \ldots, ord_i^{a_o}, nom_i^1, nom_i^2, \ldots, nom_i^{a_n}\}$, where $a_n + a_o = a$. $\{ord_i^1, ord_i^2, ord_i^3, \ldots, ord_i^{a_o}\}$ represents the ordinal attribute, $\{nom_i^1, nom_i^2, \ldots, nom_i^{a_n}\}$ represents the disordered attribute part.

The Dissimilarity Measure Between Disordered Categorical Attribute

For disordered attributes, this paper adopts the dissimilarity measurement of the traditional K-Modes algorithm. The simple matching idea is that the distance between two samples is equal to the number of attributes with unequal attribute values. That is, the disordered attribute distance between samples x_i and x_j is shown in the following formula (a_n is the number of disordered attributes):

$$D_{A^n}(x_i, x_j) = \sum_{r=1}^{a_n} D_{nom}(x_{ir}, x_{jr}) \tag{3}$$

Where,

$$D_{nom}(x_{ir}, x_{jr}) = \begin{cases} 1, x_{ir} \neq x_{jr} \\ 0, x_{ir} = x_{jr} \end{cases} \tag{4}$$

The Dissimilarity Measure Between Ordinal Categorical Attribute

For the distance of ordinal attributes, the order relationship and difference between attributes need to be considered.

Here, the attribute difference is used as the distance. In order to be consistent with the disordered attribute range, the value range of the distance is $[0, 1]$, so the range value is used as the denominator. The ordinal attribute distance between objects x_i and x_j is shown in the following formula (a_o is the number of ordinal attributes):

$$D_{A^o}(x_i, x_j) = \sum_{r=1}^{a_o} D_{ord}(x_{ir}, x_{jr}) \tag{5}$$

Where,

$$D_{ord}(x_{ir}, x_{jr}) = \begin{cases} \frac{|x_{ir} - x_{jr}|}{m_r - 1}, x_{ir} \neq x_{jr} \\ 0, x_{ir} = x_{jr} \end{cases} \tag{6}$$

m_r is the maximum value in the value range of the attribute.

Combination of Ordinal Attribute and Disordered Attribute Distance

In 1998, Huang proposed the K-Prototypes algorithm [10], which is a simple integration

of K-Means and K-Modes algorithms. The main idea is to calculate the distance of numerical attributes and the distance of categorical attributes separately and perform simple matching according to their relative contributions.

The dissimilarity between two objects x, y can be measured by

$$D(x, y) = D_{A^r}(x, y) + \gamma D_{A^c}(x, y) \tag{7}$$

Where γ is used to representing the relative contribution of numerical attributes and categorical attributes, but the selection of this γ value is difficult.

For this reason, Liang et al. [16] proposed a method to calculate the γ value. The modified formula is shown in the figure below:

$$D(x, y) = \frac{|A^r|}{A} D_{A^r}(x, y) + \frac{|A^c|}{A} D_{A^c}(x, y) \tag{8}$$

Since the distance formula of this mixed K-Modes algorithm is determined by both disordered attributes and ordinal attributes, the relative contribution of disordered attributes and ordinal attributes must also be considered. That is, the selection of γ value still needs to be considered. Here, choose Liang's γ value definition method.

The dissimilarity measurement of mixed K-Modes algorithm in this paper is defined as follows:

Let x_i and x_j be two independent objects in data set X. The dissimilarity between two objects can be measured by

$$D(x_i, x_j) = \frac{|a^n|}{|a|} D_{A^n}(x_i, x_j) + \frac{|a^o|}{|a|} D_{A^o}(x_i, x_j) \tag{9}$$

Where, $a = a^n + a^o$, is the total number of attributes.

Assume that $Z^k = \{z_1, z_2, \ldots, z_k\}$ is the set of centers of each cluster, and k is the number of clusters. Select the center $z \in Z^k$, and the distance between the objects x and the center z is as follows:

$$D(x, z) = \frac{|a^n|}{|a|} \frac{D_{A^n}(x, z)}{\sum_{i=1}^{k} D_{A^n}(x, z_i)} + \frac{|a^o|}{|a|} \frac{D_{A^o}(x, z)}{\sum_{i=1}^{k} D_{A^o}(x, z_i)} \tag{10}$$

Based on this dissimilarity measurement, a modified mixed K-Modes algorithm is proposed, which is as follows.

- Step 1: Determine the number of clusters k, randomly select k objects as the initial center of each cluster.
- Step 2: According to formula (3) and formula (5), calculate the disorder attribute distance and the ordinal attribute distance from the remaining objects to each cluster center and sum them according to formula (10). Allocate the object to the nearest cluster.
- Step 3: After all objects have been allocated to existing clusters, update the center of each cluster (use the mode of the cluster as the center), and reset the dissimilarity of objects against the current modes according to Step 2. If the new distance is less than the current distance, the objects are reallocated to closer clusters.
- Step 4: Repeat Step 3 until the clusters to which all objects belong no longer change or the cluster centers no longer change.

3.4 Ordinal Attribute Distance Expert Weighting

In the questionnaire design, considering the distribution of users and the number of options, the options are usually not set at equal intervals. For example, in the survey of consumption amount, there will be the following set of options {below 5,000, 5,000–10,000, 10,000–20,000, 20,000–100,000, more than 100,000}, and you can find the distance between the options are not equal. We need to assign weights to options for such ordinal attributes and use numerical values to quantify the distance between them. Here, a method to determine the weight of each attribute based on an expert questionnaire survey is proposed.

Assume that the initial value range of an ordinal categorical attribute is $Dom(ord^p) = \{a_0, a_1, a_2, \ldots, a_n\}$. Through the expert questionnaire survey method, each expert's score on the attribute weight is obtained. Synthesize the weight scores of the attribute given by the experts, and take the mode of the weight scores of all experts to get the weight set of the ordinal attribute, $w^p = \{w_0, w_1, w_2, \ldots, w_n\}$. Then, use the weight set to assign a weight to the ordinal attribute to obtain a new value range is

$$Dom(ord^p) = \{w_0 a_0, w_1 a_1, w_2 a_2, \ldots, w_n a_n\} \tag{11}$$

After that, this new value range is used as the value selection of ordinal attributes for classification.

3.5 Clustering Effect Evaluation

Clustering is a kind of unsupervised learning. Its purpose is to divide the target object into several clusters and ensure that the distance between the objects in each cluster is as close as possible. The distance between objects in different clusters is as far as possible. Therefore, like unsupervised learning, the evaluation of the clustering effect is very necessary. At present, the evaluation of the clustering effect will be measured by two types of indicators. One is the internal cluster validity index [17], which measures the compactness and separation of clusters, and the other is the external cluster validity index, which is a measure of agreement between two partitions: one given by a clustering algorithm and the other defined by external criteria.

Internal Cluster Validity Index

However, most of the current internal cluster validity indicators are suitable for numerical variables or disordered categorical variables [18, 19]. For the mixed clustering method proposed in this paper, the existing internal cluster validity indicators cannot be used for evaluation [20]. Liang et al. proposed the CUM evaluation index for the K-prototypes clustering algorithm [16]. The main idea is to calculate the cluster dispersion of numerical attributes and categorical attributes, respectively, and then linearly add them according to their relative contributions to obtain the overall cluster dispersion. Based on this idea, this paper proposes the CUC index for evaluating the internal clustering dispersion of the mixed K-Modes clustering algorithm in this paper.

The clustering validity index needs to calculate the within-cluster distance and between-cluster distance. Assuming that all objects are divided into k clusters, $C = \{c_1, c_2, c_3, \ldots, c_k\}$ is the cluster set, $Z^k = \{z_1, z_2, \ldots, z_k\}$ is the cluster center set.

Then the within-cluster distance of the ordinal attributes of the cluster i is as follows:

$$D_{inner}^{ord}(c_i) = \frac{1}{n}\sum_{j=1}^{n} D_{A^o}(x_j, z_i) \tag{12}$$

Where, n is the number of objects in the cluster, and z_i is the center of the cluster.

Let $c_i, c_j \in C$, the between-cluster distance is the distance between the two clusters' centers.

$$D_{outer}^{ord}(c_i, c_j) = D_{A^o}(z_i, z_j) \tag{13}$$

Then the degree of dispersion between c_i and other clusters can be calculated by,

$$DBI^{ord}(c_i) = \max_{1 \leq j \leq k, j \neq i} \left(\frac{D_{inner}^{ord}(c_i) + D_{inner}^{ord}(c_j)}{D_{outer}^{ord}(c_i, c_j)} \right) \tag{14}$$

Let CUO represents the clustering dispersion of the ordinal attributes of each cluster. CUO is as follows:

$$CUO = \frac{1}{k}\sum_{i=1}^{k} DBI^{ord}(c_i) \tag{15}$$

In the same way, the cluster dispersion degree CUN of disordered attributes can be obtained as shown in the following formula:

$$CUN = \frac{1}{k}\sum_{i=1}^{k} DBI^{nom}(c_i) \tag{16}$$

Where,

$$D_{inner}^{nom}(c_i) = \frac{1}{n}\sum_{j=1}^{n} D_{A^n}(x_j, z_i) \tag{17}$$

$$D_{outer}^{nom}(c_i, c_j) = D_{A^n}(z_i, z_j) \tag{18}$$

$$DBI^{nom}(c_i) = \max_{1 \leq j \leq k, j \neq i} \left(\frac{D_{inner}^{nom}(c_i) + D_{inner}^{nom}(c_j)}{D_{outer}^{nom}(c_i, c_j)} \right) \tag{19}$$

According to the relative contribution of ordinal attributes and disordered attributes, a validity index CUC can be obtained to evaluate the mixed K-Modes algorithm's clustering validity.

$$CUC = \frac{|a^n|}{|a|}CUN + \frac{|a^o|}{|a|}CUO \tag{20}$$

It is clear that the lower the value of CUC above, the lower the clustering dispersion and the better the clustering results.

External Cluster Validity Index

Since the data results of this questionnaire are pre-classified by experts, the Adjusted Rand Index (ARI) can be used to evaluate the clustering validity [20]. The definition of Adjusted Rand Index (ARI) is as follows:

Suppose n_{ij} represents the number of objects in the same cluster X_i and cluster Y_j, a_i is the number of objects in the cluster X_i, and b_i is the number of objects in the cluster Y_j, as shown in the following Table 2:

Table 2. The data of cluster X_i and cluster Y_j

Class/Cluster	Y_1	Y_2	...	Y_s	Sums
X_1	n_{11}	n_{12}	...	n_{1s}	a_1
X_2	n_{21}	n_{22}	...	n_{2s}	a_2
...	...		...	...	...
X_r	n_{r1}	n_{r2}	...	n_{rs}	a_r
Sums	b_1	b_2	...	b_s	

Then,

$$ARI = \frac{\sum_{ij}\binom{n_{ij}}{2} - \left[\sum_i \binom{a_i}{2} \sum_j \binom{b_j}{2}\right] / \binom{n}{2}}{\frac{1}{2}\left[\sum_i \binom{a_i}{2} + \sum_j \binom{b_j}{2}\right] - \left[\sum_i \binom{a_i}{2} \sum_j \binom{b_j}{2}\right] / \binom{n}{2}} \tag{21}$$

The value range of ARI is [−1, 1]. The higher the value, the more consistent the clustering result and the real result, and the better the clustering results.

4 Case Study

In order to establish the personas of a bank [21], we designed and distributed a questionnaire [14], collecting 200 real and effective user data. The questionnaire includes a single choice and multiple-choice questions. The above methods are used for analysis and treatment.

4.1 Preprocessing of Multiple-Choice Questions

We first use the traditional K-Modes algorithm to cluster multiple-choice questions to reduce the dimensionality.

Here we use the question "The disadvantage of this bank in the user's mind" in the questionnaire as an example. The choices settings of this question are shown in the following Table 3:

Table 3. The disadvantage of this bank in the user's mind's question and options

What do you think is the current disadvantage of this bank?
Long queue
Bad service attitude
Slow speed
Bad hardware facilities (such as machines, seats)
Bad business
Far from home
Others

The data collected for this question is shown in the Table 4 below:

Table 4. The data collected for the above question

	long_queue	bad_service	low_speed	bad_hardware	bad_business	home_far	bad_others
1	0	0	0	1	0	0	0
2	0	1	1	0	0	0	0
3	1	0	0	1	0	0	0
...	...	...	...	...	...	...	...
199	0	0	0	0	1	0	1
200	0	0	0	0	0	0	1

Using traditional K-Modes algorithm to cluster this question, the three sets of data obtained are described in the following Table 5:

Table 5. The three sets of data

Serial number	Cluster feature	Meaning
0	The number of "bad_service", "long_queue" and "home_far" is almost 0; The options are concentrated in "low_speed", "bad_hardware" and bad_business";	Users who pay more attention to business level
1	The options are concentrated in "long_queue" and "home_far"	Users who pay more attention to convenience
2	The options are concentrated in "long_queue" and "low_speed"	Users who pay more attention to time efficiency

In the same way, the questions in the questionnaire such as "the advantages of the bank in the user's mind", "the type of business that the user often handles", and "the way the user handles the business" all use the same method to perform clustering and dimensionality reduction.

So far, there are 14 attributes in the current questionnaire, which are divided into 5 ordinal categorical attributes and 9 disordered categorical attributes. The results are shown in the following Table 6:

Table 6. Classification of all questions and attributes of the questionnaire

Serial number	Question	Attribute	Data type
1	What is your age?	Age	Ordinal categorical attribute
2	What is your gender?	Gender	Disordered categorical attributes
3	What is your job?	Job	Disordered categorical attributes
4	Among the following seven consumption categories, what is your monthly consumption proportion?	Custom patterns	Disordered categorical attributes
5	What brand of sneakers do you often buy?	Consumption level	Disordered categorical attributes
6	How long have you been a customer of this bank?	User bank age	Ordinal categorical attribute
7	How do you usually handle business?	Handling method	Disordered categorical attributes
8	What kind of business do you often handle?	Business category	Disordered categorical attributes
9	How long does it take on average from your entry to leaving this bank?	Spend time	Ordinal categorical attribute
10	Why did you initially choose this bank?	User source	Disordered categorical attributes
11	How long do you want to shorten the queue time in this bank?	Shorten the queue time	Ordinal categorical attribute
12	What is the average amount of your business in this bank?	Average business amount	Ordinal categorical attribute
13	What do you think is the current advantage of this bank?	Advantage category	Disordered categorical attributes
14	What do you think is the current disadvantage of this bank?	Disadvantage category	Disordered categorical attributes

4.2 Ordinal Attribute Distance Weighting

We selected the ordinal categorical attributes, gave the experts the questions and options to assign weights, and synthesized the experts' opinions to get the following weighting Table 7, 8.

Table 7. The weight of the length of time a user joins the bank

How long have you been a customer of this bank?	Raw data	Expert weight	Final value
Within a year	1	1	1
1–3 years	2	1	2
4–6 years	3	1	3
6–10 years	4	1.25	5
More than 10 years	5	1.4	7

Table 8. The weight of the average amount of business

What is the average amount of your business in this bank?	Raw data	Expert weight	Final value
Below 5,000	1	1	1
5,000–10,000	2	1.5	3
10,000–20,000	3	2	6
20,000–100,000	4	2.5	10
More than 100,000	5	3	15

Take "how long have you been a user of this bank" and "what is the average amount of business you handle in the bank" as examples here. We surveyed a total of 5 user research experts and asked them to assign weights based on the content of the options. For example, for the length of time a user joins the bank, experts generally think that the value represented by this option in 6–10 years is larger than the option in 1–3 years and its weight needs to be increased. And the value corresponding to more than 10 years needs to have a higher weight, thereby expanding the value range of this attribute. As for the average amount of business handled by users, experts also think that the range and amount involved in 20,000–100,000 and more than 100,000 are far beyond the scope of the previous options, so the value range needs to be greatly increased, giving it a higher weight.

4.3 Mixed K-Modes Clustering and Cluster Results Evaluation

After re-weighting all ordinal attributes and determining the new final value, all attributes are respectively used in the traditional K-Modes algorithm (Huang's K-modes), the

mixed K-Modes algorithm without re-weighting (My K-modes), and the mixed K-modes algorithm with weighting considered (My K-modes weight) for clustering.

Use the internal cluster validity index CUC and the external cluster validity index ARI to evaluate these three clustering methods' effects.

The clustering validity index results obtained are shown in the following table, and the relationship between them is shown in the Fig. 3 (Table 9):

Table 9. Evaluation index value of three clustering methods

Method name	CUO	CUN	CUC	ARI
Huang's K-modes	0	1.629524545	1.629524545	0.42119889
My K-modes	4.836867453	1.589842221	2.749494089	0.662559097
My K-modes weight	3.66241322	1.590836941	2.304069535	0.793407829

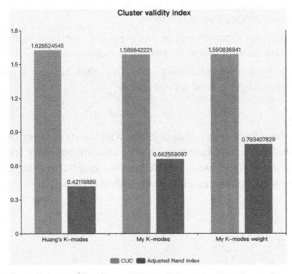

Fig. 3. Evaluation index value of three clustering methods' figure

The lower the value of CUC above, the better the clustering results. The higher the value of ARI above, the better the clustering results.

From the data, we can see that the Adjusted Rand Index of the mixed K-Modes algorithm is higher than that of the traditional K-Modes algorithm for the external clustering evaluation index - Adjusted Rand Index. Moreover, the clustering effect after weighting is better, which also proves the rationality of expert weighting. At this time, the Adjusted Rand Index can reach about 0.79.

As for the internal clustering evaluation index CUC, since the CUC index proposed in this paper is only for mixed data, for the traditional K-Modes algorithm, the cluster dispersion of ordinal attributes is 0, and the total dispersion is also relatively low. But in

terms of the dispersion of disordered attributes, the dispersion of disordered attributes of the mixed K-Modes algorithm is smaller than that of the traditional K-Modes algorithm. Through the table, we can also find that the cluster dispersion after weighting is also significantly smaller than before weighting, which means that the algorithm effect after weighting is better.

In summary, the weighted mixed K-Modes clustering algorithm proposed in this paper has a significant improvement effect than the traditional K-Modes algorithm in questionnaire processing.

Through the analysis of this questionnaire, the users are grouped into 3 clusters. Here, the centers of each cluster and the frequency of attribute data of each cluster are checked, and the three types of users are given corresponding labels (Table 10):

Table 10. The three types of users

Category	Labels
1	Young, student, fashion consumption, less business, low business frequency, multi-use mobile phone for business;
2	Young, technical staff, strong purchasing power, high business frequency, more bank card business, pay more attention to business speed;
3	Middle-aged, saver, more business, do more offline business, do more passbook business, care more about service attitude;

We compare these three types of user labels with the three types of users given by experts and find that the personas are basically the same. In summary, from the results of objective data (Adjusted Rand Index) and subjective data (label comparison), it can be concluded that the weighted mixed clustering method proposed in this paper is feasible and effective.

5 Conclusions

This paper proposes a clustering method for categorized questionnaire data, which includes the following parts:

1. Multiple-choice questions dimensionality reduction. A preprocessing dimensionality reduction method for multiple-choice questions is proposed. Through traditional K-Modes clustering, the dimensionality of the data is reduced, and the classification results of the corresponding attributes of the multiple-choice questions are obtained.
2. Attribute division. The questionnaire questions are classified as user attributes, and the user attributes are divided into ordinal attributes and disordered attributes.
3. Ordinal categorical attribute weighting. A method of using an expert survey method to weight ordinal attributes is proposed, which makes the distance of ordinal attributes more reasonable.

4. Mixed clustering method. The traditional K-Modes clustering algorithm is improved, and the attributes are calculated according to the two types of ordinal and disordered attributes so that the clustering effect is better.
5. Internal clustering validity evaluation indicator CUC. A cluster dispersion degree CUC, which is a validity evaluation index for mixed K-Modes clustering, is proposed, which is composed of ordinal categorical attribute cluster dispersion degree CUO and disordered categorical attribute cluster dispersion degree CUN.

In this paper, through the improvement of traditional K-Modes clustering algorithm, combined with expert weighting, a clustering method of questionnaire data is proposed for the questionnaire analysis in the user research field.

Through the analysis of the bank user questionnaire, it can be seen that the questionnaire data clustering method proposed in this article is fast and effective and can be applied to common user pre-survey questionnaire scenarios to cluster users and obtain personas.

However, in addition to the questionnaire to which this case belongs, there are other types of questionnaires, such as screening questionnaires and interview questions. The applicability of the research methods proposed in this article to other questionnaire types needs to be further verified. The assignment of ordinal attribute weights currently still depends on the expert's choice and professionalism. How to scientifically select the weights is worthy of discussion in the next step.

References

1. Brickey, J., Walczak, S., Burgess, T.: A comparative analysis of persona clustering methods. In: Sustainable It Collaboration Around the Globe Americas Conference on Information Systems DBLP (2010)
2. Tu, N., Dong, X., Rau, P., Zhang, T.: Using cluster analysis in persona development. In: International Conference on Supply Chain Management and Information Systems (2010)
3. Yanlong, D., Jian, W.: Cross border e-commerce in the context of the epidemic: non market factors, logistics bottlenecks and agglomeration dynamics: based on a questionnaire analysis of 300 enterprises in "Yiwu Business Circle." J. Mudanjiang Univ. **10**, 1–6 (2020)
4. Li, G.: Construction and research of precision learning model based on multi-model sample t-test. J. Shaoguan Univ. **41**(06), 13–17 (2020)
5. Laporte, L., Slegers, K., De Grooff, D.: Using correspondence analysis to monitor the persona segmentation process. In: Proceedings of the 7th Nordic Conference on Human-Computer Interaction: Making Sense Through Design (NordiCHI 2012). Association for Computing Machinery, New York, NY, USA, pp. 265–274 (2012)
6. Greaney, J., Riordan, M.: The use of statistically derived personas in modelling mobile user populations. In: Chittaro, L. (ed.) Human-Computer Interaction with Mobile Devices and Services. LNCS, vol. 2795, pp. 476–480. Springer, Heidelberg (2003). https://doi.org/10.1007/978-3-540-45233-1_50
7. Shiwei, Y., et al.: Correlation between bullying and anxiety symptoms in boarding middle school students in Anyang City. School Health in China, 1–4 (2020)
8. Mocanu, V., Dang, J.T., Switzer, N., Madsen, K., Birch, D.W., Karmali, S.: Sex and race predict adverse outcomes following bariatric surgery: an MBSAQIP analysis. Obes. Surg. J. Metab. Surg. Allied Care **30**(5), 1093–1101 (2020)

9. HuaXu, J., Liu, H.: Web user clustering analysis based on KMeans algorithm. In: 2010 International Conference on Information, Networking and Automation (ICINA). IEEE (2010)
10. Huang, Z.X.: Extensions to the k-means algorithm for clustering large data sets with categorical values. Data Min. Knowl. Disc. **2**(3), 283–304 (1998). https://doi.org/10.1023/A:100976 9707641
11. Cheung, Y.-M., Jia, H.: Categorical-and-numerical-attribute data clustering based on a unified similarity metric without knowing cluster number. Pattern Recogn. **46**(8), 2228–2238 (2013)
12. Ahmad, A., Dey, L.: A method to compute distance between two categorical values of same attribute in unsupervised learning for categorical data set. Pattern Recogn. Lett. **28**(1), 110–118 (2007)
13. Fang, Y., Youlong, Y.: Improved distance formula of K-modes clustering algorithm for mixed categorical attribute data. Comput. Eng. Appl. **56**(6), 186–193 (2020)
14. Hou, W.J., Yan, X.Y., Liu, J.X.: A method for quickly establishing personas. In: Artificial Intelligence in HCI (2020)
15. Liang, J., Zhao, X., Li, D., et al.: Determining the number of clusters using information entropy for mixed data. Pattern Recogn. **45**(6), 2251–2265 (2012)
16. Liu, Y., Li, Z., Xiong, H., Gao, X., Wu, J.: Understanding of internal clustering validation measures. In: 2010 IEEE International Conference on Data Mining (2010). https://doi.org/10.1109/icdm.2010.35
17. Lian-Jiang, Z., Bing-Xian, M.A., Xue-Quan, Z.: Clustering validity analysis based on silhouette coefficient. J. Comput. Appl. **30**(2), 139–141 (2010)
18. Lukasik, S., Kowalski, P.A., Charytanowicz, M., et al.: Clustering using flower pollination algorithm and Calinski-Harabasz index. In: 2016 IEEE Congress on Evolutionary Computation, Vancouver, 24–29 July 2016, pp. 2724–2728. IEEE (2016)
19. Halkidi, M., Vazirgiannis, M.: A density-based cluster validity approach using multi-representatives. Pattern Recogn. Lett. **29**(6), 773–786 (2008)
20. Steinley, D.: Properties of the Hubert-Arable adjusted rand index. Psychol. Methods **9**(3), 386–396 (2004)
21. Cooper, A.: The Inmates Are Running the Asylum: Why High- Tech Products Drive Us Crazy and How to Restore the Sanity. Sams - Pearson Education, Indianapolis (2004)

A Deep Learning Based Multi-modal Approach for Images and Texts Recommendation

Haowu Hu and Zhao Huang[(✉)]

School of Computer Science, ShaanXi Normal University, Xi'an 710119,
People's Republic of China
{haowu.hu,zhaohuang}@snnu.edu.cn

Abstract. Nowadays, the Internet is filled with massive and unstructured image and text resources, which makes users difficulties in obtaining target information. Recommendation system can support user information seeking, providing useful information to better meet their interests. However, different information modalities, such as image and text, have their own characteristics, which can be relevantly used together to provide complementary information. Therefore, we propose a deep learning based multi-modal approach for images and texts recommendation. To conduct the study, first, we develop two intra-modal representations. Second, Multimodal DBN (Deep Belief Network) is used to learn inter-modal representation of image and text. Third, we combine inter-modal representation with the two intra-representations respectively. Then the higher-order representations are obtained. Finally, the model is built in recommendation system to recommend image-text pairs that might be of interest to the user based on his or her history. Wikipedia data set is selected to verify the effectiveness of our proposed approach. This study can contribute to develop a multi-modal recommendation system with higher accuracy and more in line with users' preferences.

Keywords: Deep learning · Cross-modal retrieval · Recommendation system

1 Introduction

There is a large amount of data available on the Internet today, presenting in a variety of forms, such as ratings, reviews, graphs, images, etc. Meanwhile, people are facing difficulties in finding useful information or knowledge from those data [1]. To address these challenges, recommendation systems appear, which is helpful to display the most appropriate items, products or services to users based on their history records [2, 3]. Recommendation systems have been widely used to recommend music [4], movies [5], books [6], new articles [7], and so on. Literature in this area can be broadly grouped as collaborative filtering-based or content-based recommendation system [8]. The former does not need any previous knowledge about users or items, instead, it makes recommendations based on interactions information between users and items [9]. Although collaborative filtering-based method is efficient and simple, it suffers from a number of problems, such as cold start [10] and prediction accuracy [11]. Whereas, the latter

© Springer Nature Switzerland AG 2021
H. Degen and S. Ntoa (Eds.): HCII 2021, LNAI 12797, pp. 343–358, 2021.
https://doi.org/10.1007/978-3-030-77772-2_23

uses item information, represented as attributes, to calculate the similarities between items and recommend users items which are similar to their history [12]. Its prediction accuracy is relatively higher on the calculation of items similarity.

Currently, most of recommendation systems focus on providing single-modal content. For example, they recommend pictures based on pictures, or recommend texts based on texts, or recommend music based on music. In fact, different modalities, such as pictures and texts, have unbalanced and complementary relationships when describing the same semantics. More specific, image can usually contain more details that text cannot display, whereas text has advantages in expressing a high-level of meaning [13]. Users need more information resources that combine with multi-modes, which involves the field of cross-modal retrieval. Cross-modal retrieval is about that the user input a mode of information. It can return the information combined with multiple modes. For instance, when picture or text data is input, the system can return information resources combined with pictures and texts [14]. At present, cross-modal retrieval methods are mainly used in the field of images and texts retrieval. However, the distributions of information in different modalities vary. It is difficult to establish connections between different modalities and measure their similarity [15]. Within the current methods, a common way is to convert different forms of data into a common subspace, and then in this subspace, the similarity between different modes can be directly measured by a predefined similarity measurement method. Traditional methods [16, 17] use pairs of statistical co-occurrence information (e.g., correlation and covariance) to make linear prediction. However, these methods [18] are based on linear mapping models. If there is no linear subspace structure in the original data, these systems may not be able to process the real data [19]. Although some kernel-based methods have been proposed [20], they have difficulties in choosing the right kernel and dealing with large-scale problems, which hinders their use in many practical applications. Since deep learning has achieved great success in feature extraction [21], various deep learning-based cross-modal retrieval methods [13, 22–25] have emerged. They take advantages of the powerful ability of deep neural networks (DNN) to capture nonlinear relationships in data and have achieved encouraging performance. They usually consist of two phases, the first phase is obtaining information within various modes, and the second phase is establishing relationships between the information of various modalities. However, many current deep learning-based cross-modal retrieval methods have many disadvantages, such as the underutilization of image information and the insufficient number of network layers. Moreover, they usually only model the intra-modal information of each modes, but ignore the correlation between them in the first stage. The above reasons lead to these cross-modal retrieval methods that cannot effectively mine the correlation relationships between different modalities.

To solve above issues, this paper proposes a cross-modal retrieval method, called Multi-Modal Deep Neural Network (MMDNN), and applies it to recommendation system to provide image-text pairs to users based on their history records. Figure 1 shows the framework of our proposed multi-modal recommendation system. In MMDNN, we first use MobileNetV3 and DAE (Deep AutoEncoder) to get two intra-modal representations of image and text. Second, Multimodal DBN is used to obtain the inter-modal information of image and text. Third, two RBMs (Restricted Boltzmann Machine) are

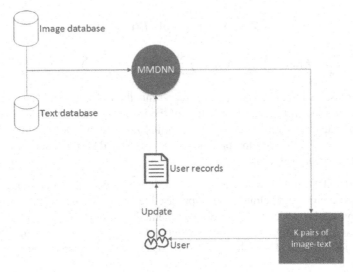

Fig. 1. The structure of our recommendation system using cross-modal retrieval method (MMDNN).

adopted to combine inter-modal representation and above two intra-modal representations respectively. Finally, extracting higher-order representations of image and text and establishing relationships between them are performed by SCAE and association constraint functions.

The remainder of this paper is organized as follows: we briefly introduce related works in Sect. 2. In Sect. 3, our proposed multi-modal retrieval approach and recommendation system are presented in detail. Then Sect. 4 reports the experimental results as well as analyses. Finally, Sect. 5 concludes this paper and describes some future directions of our work in the future.

2 Related Work

In this section, we will introduce some models involved in this paper and review the representative cross-modal retrieval methods.

2.1 Some Models Involved in This Paper

RBM: It is a two-layer neural network structure based energy and used to learn the probability distribution of samples [38]. It contains visible layer (which are also called input layer) $v = \{v_i | i \in [1, m]\}$ and hidden layer $h = \{h_j | j \in [1, n]\}$, the related formulas are shown as formulas (1, 2, 3). b_i represents visible layer bias, c_j represents hidden layer bias, and w_{ij} represents the weight between visible layer node i and hidden layer node j. There is no connection between nodes in the visible layer and also in the hidden layer, but the nodes between the two layers are fully connected without direction.

$$P(v, h) = \frac{1}{z} \exp\left(-\sum(v, h)\right) \tag{1}$$

$$Z = \sum_v \sum_h exp\{-E(v, h)\} \tag{2}$$

$$\sum(v, h) = -\sum_{i=1}^{m} b_i v_i - \sum_{j=1}^{n} c_i h_i - \sum_{i=1}^{m} \sum_{j=1}^{n} v_i w_{ij} h_j \tag{3}$$

where v represents the input layer, and h represents the hidden layer. The formula (1) represents the joint probability distribution of the input layer and the hidden layer, z is the partition function which is also called normalization constant. Formula (3) is the energy function. For solving the parameters b, c, w in RBM, this paper uses the CD (Contrast Divergence) algorithm.

AE (AutoEncoder): It is a kind of neural network, which contains three or more layers of network structure, and belongs to unsupervised learning. Its goal is to reconstruct the original input of the model. Gradient descent method is commonly used for training AE according to reconstruction errors. AE is usually divided into two parts: encoder f and decoder g. The process of encoding and decoding can be expressed as follows:

$$h = f(x) = l_f\left(w_f x + b_h\right) \tag{4}$$

$$r = g(h) = l_g\left(w_g x + b_r\right) \tag{5}$$

$$L(x, r) = ||r - x||^2 \tag{6}$$

where h represents the hidden layer, r represents the reconstruction layer. f and g are activation functions, w is the weight and b is bias. The $L(x, r)$ function is the reconstruction error of the auto encoder, that is the error between the input data x and the output r of the reconstruction layer. For this model, we use the stochastic gradient descent method to train it.

2.2 Some Representative Cross-modal Retrieval Methods

There are two categories methods of cross-modal retrieval: traditional methods and deep learning based methods.

The traditional methods are mainly to map the features of different modalities into a common space and generate a common representation [15]. The representative method is Canonical Correlation Analysis (CCA) [16] and there are many extended models based on CCA. For example, the using kernel functions [26], integrating semantic category labels [27] taking high-level semantics as the third view [28], and considering the semantic information in the form of multi-label annotations [29].

As for deep learning based methods, the powerful ability of nonlinear correlation modeling is used to build correlation relationships between different modalities. Ngiam et al. [30] proposed a Bimodal Autoencoders, which is an extension of RBM. It is used to model the cross-modal correlation at the shared layer, and then followed by some similar network structures, such as [31, 32]. Multimodal DBN [33] is proposed to model the distribution of different modal data through joint RBM and learn the cross-modal correlation. Feng et al. [34] proposed a Correspondence AutoEncoder (Corr-AE), which

learn representations of different modalities and establish correlation between them. Deep canonical correlation analysis used in (DCCA) [35, 36] and attempted to combine deep network with CCA.

The above method mainly consists of two sub-networks, one is for getting representations of different modalities and the other is for building relationships between them. However, traditional methods cannot obtain the nonlinear relationships of image. Some of deep learning based methods ignore the inter-modal information between different modalities and some only have a few of hidden layers. So they cannot effectively mine the correlation relationships between different modes. To solve abo problems, we propose our cross-modal retrieval method and apply it to the field of recommendation system, trying to recommend users information resources that they may be interested in.

3 MMDNN Model and Recommendation System

This section consists of two parts. The first part introduces our MMDNN model, and the second part introduces the process of recommending image-text pairs to users based on their history records.

3.1 MMDNN Model

This section first introduces some basic modules involved in this model, and then introduces our MMDNN architecture in detail.

Multimodal DBN:Multimodal DBN: DBN is a structure with multi-layer RBM. The Multimodal Deep Belief Network (Multimodal DBN) uses two separate DBNs to model the data of each mode. Currently, it has been widely used to learn shared representations of multimodal data. Multimodal DBN has two inputs in this paper, one are image features extracted by MobileNetV3, the other one are the text features modeled by the RSRBM (Replicated Softmax RBM) model. Replicated Softmax RBM is applied when the input of the network is sparse discrete value and its energy function is as follows:

$$\sum(v, h; \theta) = \sum_{i=1}^{I} \sum_{j=1}^{J} v_i w_{ij} h_j - \sum_{i=1}^{I} b_i v_i - M \sum_{j=1}^{J} a_j h_j \qquad (7)$$

where v_i is the value of i-th node in the input layer, h_j is the value of the j-th node in the hidden layer, w_{ij} is the weight between them, b_i is the bias of the i-th node in the input layer. And a_j is the bias of the j-th node in the hidden layer, M is the sum of the discrete values of the visible layer. For DBN training, we first use CD algorithm to train layer by layer, then fine-tune the network with the BP (Back Propagation) algorithm.

MobileNetV3: MobileNetv3 is a deep convolutional neural network, proposed by Google team [37] in 2019. It has beaten multiple convolutional networks such as VGGNet, RESNet etc. in target detection and semantic segmentation. It also has the characteristics of small model volume, less number of parameters, fast training speed and high extraction accuracy. Therefore, this paper selects the MobileNetv3 model to conduct experimental study on cross-modal retrieval. MobileNetV3 evolved on the basis

of mobilenetv1 and mobilenetv2. MobileNetv1 uses packet convolution to reduce the amount of network calculations, and equalizes the number of network packets to the network channels. Then, it uses a 1*1 convolution kernel to merge channels. MobileNetv2 introduces the bottleneck structure, turns it into a spindle shape, and removes the last ReLU function of the Residual Block. Based on these two models, MobileNetV3 has made the following improvements:

The SE (Squeeze and Excite) structure is introduced. Because the SE structure consumes a certain amount of time, the author changes the channel of the expansion layer to 1/4 of the original in the structure containing SE.

It modifies the structure of the tail of MobileNetV2. MobileNetV3 first uses avg-pooling to reduce the feature map size from 7 x 7 to 1 x 1. Then, it adopts the 1x1 convolution kernel to increase the dimension of the data. Finally, it directly remove the two 3 * 3 and 1 * 1 convolution kernels in the previous spindle structure.

It replaces swish with h-swish. The swish and h-swish formulas are shown as formula (8, 9). Since the calculation of the sigmoid function takes a long time, the author uses the ReLU6(x + 3)/6 function to approximate the sigmoid function.

$$swishx = x * \sigma(x) \tag{8}$$

$$h - swish[x] = x\frac{ReLU(x+3)}{6} \tag{9}$$

There are two versions of MobileNetV3, one is MobileNetV3-large and the other is MobileNetV3-small. Although the former consumes more time, the former is more accurate than the latter. For higher accuracy, this paper chooses MobileNetV3-large as one part of our model for extracting image features. Its structure is shown in Table 1. After the image is processed by MobileNetV3, we let it pass through an AE to make the dimension of the image representation equal to the dimension of the image representation in DBN.

Deep Auto Encoder (DAE): It is composed of multi-layer autoencoder. The autoencoder of the bottom layer uses the original modal features as input, the latter uses the output of the previous autoencoder as input. DAE is trained by BP algorithm. In our proposed model, DAE is used for representation learning within the text mode and its original input is the tf-idf values of texts.

Stacked Correspondence Auto Encoder (SCAE): The correspondence auto encoder [39] is a network structure based on two single-mode autoencoders, and each autoencoder is responsible for learning the feature representation of its corresponding mode. By introducing similarity constraints, SCAE can build relationships between different modalities while learning their representations. We will use SCAE in the second stage of our model and it is trained layer by layer and then fine-tuned overall.

The Structure of Our Model: As shown in Fig. 2, our proposed method is divided into two stages. In the first stage, we use Multimodal DBN to learn the complementary information between different modalities. Then, we use MobileNetV3 and DAE to model

Table 1. Specification for MobileNetV3-Large.

Input	Operator	Expsize	#out	SE	NL	S
224^2*3	conv2d	–	16	–	HS	2
112^2*16	bneck 3*3	16	16	–	RE	1
112^2*16	bneck 3*3	64	24	–	RE	2
56^2*24	bneck 3*3	72	24	–	RE	1
56^2*24	bneck 5*5	72	40	✓	RE	2
28^2*40	bneck 5*5	120	40	✓	RE	1
28^2*40	bneck 5*5	120	40	✓	RE	1
28^2*40	bneck 3*3	240	80	–	HS	2
14^2*80	bneck 3*3	200	80	–	HS	1
14^2*80	bncck 3*3	184	80	–	HS	1
14^2*80	bneck 3*3	184	80	–	HS	1
14^2*80	bneck 3*3	480	112	✓	HS	1
14^2*112	bneck 3*3	672	112	✓	HS	1
14^2*112	bneck 5*5	672	160	✓	HS	2
7^2*160	bneck 5*5	960	160	✓	HS	1
7^2*160	bneck 5*5	960	160	✓	HS	1
7^2*160	cond2d 1*1	–	960	–	HS	1
7^2*960	pool 7*7	–	–	–	–	1
1^2*960	conv2d 1*1	–	1280	–	HS	1
1^2*1280	NBN conv2d 1*1 NBN	–	k	–	–	1

intra-modal representations of image and text. Finally, two RBMs are used to fuse the complementary information and the intra-modal representations of each mode. In the second stage, we use SCAE to obtain the final representations of different modalities and mine relationships between them.

In the first stage, for the intra-modal features of image, we first trained MobileNetv3 by classifying the image. When the classification effect was optimal, we kept the internal parameters unchanged. Then we remove the last classification layer, and use the remaining structure to extract the image features. For features in text modal, we use DAE to extract it. Its input is the same as the text input of the Multimodal DBN. Its objective function is shown as formula (10):

$$L(x) = L_r(x, x_{2h}) + \alpha \sum_{p=1}^{h} \left(||w_e^p||_2^2 + ||w_d^p||_2^2 \right) \tag{10}$$

This paper trains DAE by minimizing the objective function. $L_r(x, x_{2h})$ is the reconstruction error of the text, w_e and w_d represent the weights of the encoder and decoder of the DAE. By minimizing the reconstruction error, we can get the text intra representation.

For the common representation in the first stage, this paper uses Multimodal DBN to complete. It first uses a DBN to perform feature extraction on the text data modeled

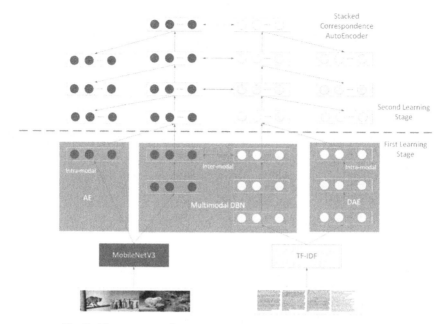

Fig. 2. The structure of our cross-modal retrieval method (MMDNN).

by RSRBM. Then, a shallow DBN is applied to perform feature extraction on the image data processed by MobileNetV3. Finally, we apply formula (11) and (12) to carry out alternating Gibbs sampling in the joint layer of DBN, in this way, we can mine the inter-modal information between the two modes. $h_t^{(3)}$ and h_i^3 are used to generate the distribution on each modal data, and the final output is the representation of the image and text with the correlation information between the modalities.

$$P\left(h_t^{(3)}|h_i^{(1)}\right) = \sigma\left(w_i^{(2)}h_i^{(1)} + a_t\right) \tag{11}$$

$$P\left(h_i^{(2)}|h_t^{(2)}\right) = \sigma\left(w_t^{(3)}h_t^{(2)} + a_i\right) \tag{12}$$

$$\sigma(x) = 1/\left(1 + e^{-x}\right) \tag{13}$$

In the last layer of first stage, we use two joint-RBMs to perform feature fusion between the inter-modal representation and intra-modal representations of the image and text. The joint distribution function is defined as follows:

$$P(v_1, v_2) = \sum_{h_1^{(1)}, h_2^{(1)}, h^{(2)}} P\left(h_1^{(1)}, h_2^{(1)}, h^{(2)}\right) * \sum_{h_1^{(1)}} P\left(v_i|h_1^{(1)}\right) * \sum_{h_2^{(1)}} P\left(v_t|h_2^{(1)}\right) \tag{14}$$

where v_1 and v_2 are the inter-modal representation and intra-modal representation of image or text.

In the second stage, we use image and text representations obtained in the first stage as the input of SCAE, which is used to establishing association relationships between different modalities while learning their representations. The association constraint function of the j-th level can be expressed as formula (15):

$$C\left(x_j^i, x_i^t; w_j^i, w_j^t\right) = \left\| f\left(x_j^i, w_j^i\right) - g\left(x_j^t, w_j^t\right) \right\|_2^2 \tag{15}$$

where x_j^i and x_i^t represent the input of image and text, w_j^i and w_j^t represent the weight and bias of image and text. $f\left(x_j^i, w_j^i\right)$ and $g\left(x_j^t, w_j^t\right)$ are the represents of hidden layer of image and text. Then the loss function of the j-th layer of SCAE can be defined as follows:

$$L\left(x_j^i, x_j^t; \theta\right) = L_j^i\left(x_j^i; \theta\right) + L_j^t\left(x_i^t; \theta\right) + L_j^c\left(x_j^i, x_i^t; w_j^i, w_j^t\right) \tag{16}$$

$$L_j^i\left(x_j^i, y_j^i; \theta\right) = \left\| y_j^i - x_j^i \right\|_2^2 \tag{17}$$

$$L_j^t\left(x_j^t, y_j^t; \theta\right) = \left\| y_j^t - x_j^t \right\|_2^2 \tag{18}$$

$$L_j^c\left(x_j^i, x_i^t; w_j^i, w_j^t\right) = C\left(x_j^i, x_i^t; w_j^i, w_j^t\right) \tag{19}$$

where L_j^i and L_j^t represent the reconstruction error of the autoencoder of the image and text respectively, L_j^c represents the associated constraint error between image and text. The optimization goal of training phase is formula (16), which is to minimize the sum of the reconstruction error and the correlation error of SCAE during the training process. We optimize the model with the strategy of layer-by-layer training and fine-tuning overall finally. And the objective function of the overall fine-tuning stage is shown in formula (20).

$$L(x_0, y_0) = L_r^i(x_0, x_{2h}) + L_r^t(y, y_{2h}) + L_d(x_h, y_h) + \delta(q) \tag{20}$$

where x_0 and y_0 are the input feature vectors of image and text, x_{2h} and y_{2h} are their corresponding reconstruction feature representations. $\delta(q)$ is the L_2 regularization of all parameters in the SCAE.

The Process of Model Training: The MMDNN model we proposed includes multiple neural networks, and its training process is shown in Table 2:

3.2 Our Recommendation System

We integrate the MMDNN model into our recommendation system, and recommend information resources combined with image and text to users based on their historical records. First, we select n users' history records (images or texts), and use MMDNN model to get their final representations P_i or P_t. Second, we take the average of these n

Table 2 The training process of MMDNN.

Input: The data of images and texts.
Output: The final representation of image and text data in MMDNN.
The training phase:
  First stage:
    for i in n do:
      Train MobileNetV3, update its parameters.
      Train AE, update its parameter.
      Train multimodal DBN(it's input data are pairs of representations of image
and text.),update its parameters.
      Train DAE, update its parameters.
    end for
  Second stage:
    Take formula 15 as the objective function to train SCAE (its input data are pairs of rep-
      resentations of image and text).
    for i in n do:
      Take formula 19 as the objective function to fine-tune SCAE.
    end for

---

representations as their unified representation S_i or S_t. Third, we pass all the remaining images or texts in the database through MMDNN model to obtain their final representation P_i or P_t. Fourth, we use the cosine similarity (the formula is shown as formula (21)) to find the k images or texts most similar to S_i or S_t. Finally, we use MMDNN to find the texts or images corresponding to above k images or texts, and recommend them to users in pairs. The specific process is shown in Table 3. For convenience, the table shows the process of recommending image-text pairs resources based on images, and the process of recommending based on text is similar.

$$S_i = \frac{1}{n} * \sum_{i=1}^{n} P_i \tag{21}$$

$$C(P_i, S_i) = (P_i - S_i)^2 \tag{22}$$

Table 3. The process of recommending.

Input: Image and text database, n user history records.
Output: k pairs of image-text resources.
The process of recommending:
  Step one: Randomly select n images that the user has viewed from user's history records.
  Step two: Use MMDNN to get the unified representation S_i of these n images.
  Step three: Use MMDNN and cosine similarity to find the first k pictures most similar to S_i
from the image database after removing the above n pictures.
  Step four: Use MMDNN to find k texts corresponding to the above k pictures from the text
database.
  Step five: Recommend these k pairs of image-text resources to user.
  Step six: Update user history based on user feedback.

---

4   Experiments

The experiments in this study consist of two parts. The first part is aimed at MMDNN, and the second part is for applying MMDNN to a recommendation system. The environment used in this article is i9-10900k CPU, 64G, NVIDIA Quadro RTX4000 8G, windows 10.

4.1   Experiment for MMDNN

Dataset: Wikipedia dataset as the most widely-used dataset for cross-modal retrieval, is selected from "featured articles" in Wikipedia with 10 most populated categories. It totally consists of 2,866 image-text pairs. We divide the dataset into 3 subsets: 2,173 pairs in training set, 231 pairs invalidation set and 462 pairs in testing set. For image processing: we first scale it to a unified format of 224 * 224. Then we uniformly convert the black and white images into a 3-channels image. Finally we send it to MobileNetV3 to obtain 1280-dimensional data. For text processing, we use the tf-idf value of each text to represent itself, and its dimension is compressed to 3000 by Auto Encoder.

Compared Methods: To evaluate the effectiveness of our MMDNN, some comparison experiments are conducted with 4 state-of-the-art approaches. The compared approaches are: 1) traditional cross-modal retrieval approaches including CCA [16], LGCFL [18], and 2) deep learning-based methods including DCCA [35], MCSM [13]. Their introductions are presented as follows:

CCA [16] learns project matrices to maximize the correlation between the projected features of different modalities in one common space.

LGCFL [18] jointly learns basis matrices of different modalities, by using a local group based priori in the formulation to fully take advantage of popular block based features.

DCCA [35] is a nonlinear extension of CCA. The correlation is maximized between the output layers of two separate sub-networks.

MCSM [13] builds a separate semantic space for each mode, and uses an end-to-end framework to directly generate modal-specific cross-modal similarities.

Evaluation Metric: Mean Average Precision (MAP) score is adopted as the evaluation metric on Wikipediat dataset. It is widely used by cross-modal retrieval methods, which is the mean value of Average Precision (AP) of each query. AP is defined as formula (23):

$$\mathrm{AP} = \frac{1}{R} \sum_{k=1}^{n} \frac{R_k}{k} * rel_k \tag{23}$$

where n is the number of total instances, R is the number of relevant instances. R_k is the number of relevant instances in the top k returned results. rel_k is set to be 1 when the k-th returned result is relevant, otherwise, rel_k is set to be 0.

Model Parameters: The hidden layer's neurons number of DBN used to learn text features is set to 2048, 1024, and this number for learning image features is 1024. There

are two layers in DAE, and the number of them are 2048 and 1024. The hidden layer dimensions of the AE are set to 1024. In the second stage, the number of hidden layers of SCAE is set to 3, which is the best choose that we obtained through experiments, and they are 512, 256, and 64 respectively.

Experimental Results and Related Analysis for MMDNN: Our comparative experiment is aimed at the two tasks of retrieving images by text and retrieving text by images. The specific results are shown in Table 4: It can be seen that our modal has the best effect on every task. Taking the traditional classic model CCA as the benchmark, our model increased 46.2%, 48.5% and 47.6% respectively. Taking the deep learning classic model DCCA as the benchmark, our modal increased by 15.5%, 17.8 and 16.7% respectively. This is due to 1): We use convolutional network to learn the feature representation of the image; 2): We use the TF-IDF value of the first 3000 important words of each document to represent each document. 3): We integrate the intra-modal and inter-modal representations between images and text; 4): We use experiments to find the optimal number of hidden layers in the SCAE.

Table 4. Comparison of other Baselines on the Wikipedia dataset.

Method	Image to text	Text to image	Average
CCA [16]	0.357	0.326	0.341
LGCFL [18]	0.466	0.431	0.449
DCCA [35]	0.452	0.411	0.431
MCSM [13]	0.516	0.458	0.487
MMDNN (ours)	**0.522**	**0.484**	**0.503**

4.2 Experiment for Recommendation System

In this experiment, we still use the Wikipedia database. Because the original data set of Wikipedia did not have the attribute of user, we artificially created 100 "people" based on the similarity of pictures. In order to simulate real application scenarios, we have created 10–20 historical records for each user. At the same time, we treat the text corresponding to these pictures as the user's historical record. Finally, we recommend 5 image-text pairs resources to users based on 5 user history records. Because we have not found that other researchers have applied cross-modal retrieval to the recommendation system, there is no other comparative experiment. This result lists the mean accuracy of pairs of image-text recommended based on images, texts and their average value. Its evaluation metric is Mean Accuracy score, which is the mean of n Accuracy values and it is shown as formula (24, 25).

$$\text{macc} = \left(\frac{1}{n}\sum_{i=1}^{n} acc_i\right) * 100\% \tag{24}$$

$$acc = \frac{t}{k}*100\% \qquad (25)$$

where acc_i is the accuracy of the result recommended by the system to the i-th user, n is the total number of users. t is the number of paired resources both in recommendation list and user's history records. k represents the total number of paired resources recommended to users.

Result: The experimental results are shown in Table 5:

Table 5. The result of experiment for recommendation system.

Recsys	Image to pairs of image-text	Text to pairs of image-text	Average
Ours	0.353	0.302	0.328

We can see that the average accuracy of the system is not very high. We think that the main reason is that there are two errors in the processing of recommending. One is in the processing of finding images based on images, the other is in the processing of finding corresponding text based on images. The accumulation of these two errors leads to the decrease in the accuracy of its recommendation. Although the recommendation accuracy of our system is not very high, this is also an attempt to integrate cross-modal retrieval into the recommendation system, which can provide some research ideas and directions for later comers.

5 Conclusion and Future Work

This paper proposes a cross-modal retrieval method. We first use a convolutional neural network MobileNetV3 and DAE to obtain the representations of the image and text in their respective modalities. Second, we use Multi-DBN to obtain the inter-modal representation. Third, two RBMs are used to fuse the inter-modal representation and the intra-modal representations of these two modes. Fourth, we apply association constraint functions and SCAE to obtain the final representations of image and text while establishing connection between them. Finally, the obtained cross-modal retrieval model is applied to the recommendation system. This is a great attempt to provide researchers with a way of thinking and direction. Experiments on the public data set Wikipedia show that our model performs best compared with the state-of-the-art cross-modal retrieval methods. When this model is used in the recommendation system, our recommendation system also shows good performance.

However, there are some limitations of this study. For example, we did not consider the alignment between the same semantics of different modal data. In addition, the TD-IDF model did not consider the relationship between text contexts and the attribute of users in the dataset is too single. Therefore, in the future: 1) We will explore ways to align the internal semantic information of images and text; 2) we will explore the use of

deep learning related algorithms to build better representations of text modal data; 3) In cross-modal recommendation systems, we will add user attributes, enrich the data set, make user's data more suitable for real application scenarios, and achieve more accurate recommendations.

Acknowledgement. This study was supported by research grants funded by the "National Natural Science Foundation of China" (Grant No.61771297).

References

1. Patel, K., Patel, H.B.: A state-of-the-art survey on recommendation system and prospective extensions. Comput. Electron. Agric. **178**, 105779 (2020)
2. Bobadilla, J., Ortega, F., Hernando, A., Gutiérrez, A.: Recommender systems survey. Knowl.-Based Syst. **46**, 109–132 (2013)
3. Adomavicius, G., Tuzhilin, A.: Toward the next generation of recommender systems: a survey of the state-of-the-art and possible extensions. IEEE Trans. Knowl. Data Eng. **17**(6), 734–749 (2005)
4. Ali, M., Johnson, C., Tang, A.: Parallel Collaborative Filtering for Streaming Data. University of Texas Austin, CiteSeerX (2011)
5. Bell, R.M., Koren, Y.: Lessons from the Netflix prize challenge. SIGKDD Explorations: Newsletter of the Special Interest Group (SIG) on Knowledge Discovery & Data Mining. **9**(2), 75–79 (2007)
6. Vaz, P.C., Martins de Matos, D., Martins, B., Calado, P.: Improving a hybrid literary book recommendation system through author ranking. In: Proceedings of the 12th ACM/IEEE-CS joint conference on Digital Libraries - JCDL 2012. ACM Press, New York (2012)
7. Li, L., Chu, W., Langford, J., Schapire, R.E.: A contextual-bandit approach to personalized news article recommendation. In: Proceedings of the 19th International Conference on World Wide Web, pp. 661–670. ACM (2010)
8. Achakulvisut, T., Acuna, D.E., Ruangrong, T., Kording, K.: Science concierge: a fast content-based recommendation system for scientific publications. PLoS ONE **11**, e0158423 (2016)
9. Nassar, N., Jafar, A., Rahhal, Y.: A novel deep multi-criteria collaborative filtering model for recommendation system. Knowl.-Based Syst. **187**, 104811 (2020)
10. Cleger-Tamayo, S., Fernández-Luna, J.M., Huete, J.F.: Top-N news recommendations in digital newspapers. Knowl.-Based Syst. **27**, 180–189 (2012)
11. Wasid, M., Ali, R.: An improved recommender system based on multi-criteria clustering approach. Procedia Comput. Sci. **131**, 93–101 (2018)
12. Son, J., Kim, S.B.: Content-based filtering for recommendation systems using multiattribute networks. Expert Syst. Appl. **89**, 404–412 (2017)
13. Peng, Y., Qi, J., Yuan, Y.: Modality-specific cross-modal similarity measurement with recurrent attention network. IEEE Trans. Image Process. **27**, 5585–5599 (2018)
14. Wang, X., Hu, P., Zhen, L., Peng, D.: DRSL: Deep relational similarity learning for cross-modal retrieval. Inf. Sci. (Ny) **546**, 298–311 (2021)
15. Peng, Y., Qi, J., Yuan, Y.: CM-GANs: cross-modal generative adversarial networks for common representation learning. IEEE Trans. Multimedia **20**, 405–420 (2017)
16. Hotelling, H.: Relations between two sets of variates. Biometrika **28**, 321–377 (1936)
17. Sharma, A., Jacobs, D.W.: Bypassing synthesis: PLS for face recognition with pose, low-resolution and sketch. In: 2011 IEEE Conference on Computer Vision and Pattern Recognition (CVPR), pp. 593–600 (2011)

18. Kang, C., Xiang, S., Liao, S., Xu, C., Pan, C.: Learning consistent feature representation for cross-modal multimedia retrieval. IEEE Trans. Multimedia **17**, 370–381 (2015)
19. Peng, X., Feng, J., Xiao, S., Yau, W.-Y., Zhou, J.T., Yang, S.: Structured AutoEncoders for subspace clustering. IEEE Trans. Image Process. **27**, 5076–5086 (2018)
20. Lai, P.L., Fyfe, C.: Kernel and nonlinear canonical correlation analysis. Int. J. Neural Syst. **10**(5), 365–377 (2000)
21. Jin, L., Li, S., La, H.M., Luo, X.: Manipulability optimization of redundant manipulators using dynamic neural networks. IEEE Trans. Industr. Electron. **64**, 4710–4720 (2017)
22. Wang, B., Yang, Y., Xu, X., Hanjalic, A., Shen, H.T.: Adversarial cross-modal retrieval. In: Proceedings of the 2017 ACM on Multimedia Conference, California, USA, pp. 154–162 (2017)
23. Zhen, L., Hu, P., Wang, X., Peng, D.: Deep supervised cross-modal retrieval. In: 2019 IEEE/CVF Conference on Computer Vision and Pattern Recognition (CVPR), Long Beach, pp. 10394–10403 (2019)
24. Hu, P., Wang, X., Zhen, L., Peng, D.: Separated variational hashing networks for cross-modal retrieval. In: Proceedings of the 27th ACM International Conference on Multimedia, pp. 1721–1729. ACM, New York (2019)
25. Wang, X., Hu, P., Liu, P., Peng, D.: Deep semisupervised class- and correlation-collapsed cross-view learning. IEEE Trans. Cybern. 1–14 (2020)
26. Hardoon, D.R., Szedmak, S., Shawe-Taylor, J.: Canonical correlation analysis: an overview with application to learning methods. Neural Comput. **16**, 2639–2664 (2004)
27. Rasiwasia, N., et al.: A new approach to cross-modal multimedia retrieval. In: Proceedings of the International Conference on Multimedia - MM 2010, pp. 251–260. ACM Press, New York (2010)
28. Gong, Y., Ke, Q., Isard, M., Lazebnik, S.: A multi-view embedding space for modeling internet images, tags, and their semantics. Int. J. Comput. Vis. **106**, 210–233 (2014)
29. Ranjan, V., Rasiwasia, N., Jawahar, C.V.: Multi-label Cross-Modal Retrieval. In: 2015 IEEE International Conference on Computer Vision (ICCV), Santiago, USA, 4049–4102 (2015)
30. Ngiam, J., Khosla, A., Kim, M., Nam, J., Lee, H.Y., Ng, A.: Multimodal deep learning. In: International Conference on Machine Learning (ICML), pp. 689–696 (2011)
31. Jungi, K., Jinseok, N., Iryna, G.: learning semantics with deep belief network for cross-language information retrieval. In: Proceeding of the 24th International Conference on Computer Linguistics (COLING 2012), Mumbai, pp. 579–588 (2012)
32. Wang, D., Cui, P., Ou, M., Zhu, W.: Deep multimodal hashing with orthogonal regularization. In: Proceedings of the Twenty-Fourth International Joint Conference on Artificial Intelligence, pp. 2291–2297 (2015)
33. Srivastava, N., Salakhutdinov, R.: Learning Representations for Multimodal Data with Deep Belief Nets. In: Presented at the ICML Representation Learning Workshop, Edinburgh, Scotland, UK (2012)
34. Feng, F., Wang, X., Li, R.: Cross-modal retrieval with correspondence autoencoder. In: Proceedings of the ACM International Conference on Multimedia - MM 2014, pp. 7–16. ACM Press, New York (2014)
35. Andrew, G., Arora, R., Bilmes, J., Livescu, K.: Deep canonical correlation analysis. In: Proceedings of 30th International Conference on Machine Learning, Atlanta, Georgia, USA, p. 28 (2013)
36. Yan, F., Mikolajczyk, K.: Deep correlation for matching images and text. In: 2015 IEEE Conference on Computer Vision and Pattern Recognition (CVPR), Boston, MA, USA, pp. 3441–3450 (2015)
37. Howard, A., et al.: Searching for MobileNetV3. http://arxiv.org/abs/1905.02244 (2019)

38. Rumelhart, D.E., McClelland, J.L.: Information processing in dynamical systems: foundations of harmony theory. In: Parallel Distributed Processing: Explorations in the Microstructure of Cognition, pp. 194–281 (1986)
39. Srivastava, N., Salakhutdinov, R.: Multimodal learning with deep boltzmann machines. In: Proceedings of the 26th Annual Conference on Neural Information Processing Systems, Lake Tahoe, NV, United states, pp. 2222–2230 (2012)

Pet Dogs' and Their Owners' Reactions Toward Four Differently Shaped Speaking Agents: A Report on Qualitative Results in a Pilot Test

Haruka Kasuga[✉] and Yuichiro Ikeda

Hokkaido University, Hokkaido 060-0814, Japan
felisfelis@eis.hokudai.ac.jp, y.ikeda@ist.hokudai.ac.jp
https://www.ist.hokudai.ac.jp/eng/

Abstract. Social agents, whose primary role is to communicate with people, are becoming increasingly prevalent. Given this, dogs, being one of the most popular companions, may become jealous owing to the resulting triad relationship or their fear of agents. To aid future investigations for the design of a communication agent that can build a good relationship with both dog owners and dogs at home, we herein conducted two pilot tests to explore the types of words dog owners use in front of their dogs to determine whether a positive owner attitude toward the agents would affect how the dogs behaved with these communication agents. Furthermore, we determined whether the owners' words and dogs' behavior were affected by the agents' physical forms (primitive, smaller/bigger humanoid, and dog-shaped). Through an analysis of video recordings of the behaviors of 29 dog owners and 34 dogs, the following suggestions are put forth: i) Tests involving dogs that have to confront unfamiliar speaking agents in an unfamiliar environment should occur in the presence of the dog owners, as conducted in previous studies, to evaluate the dogs' behavior in the presence of the agents. ii) Dog owners' descriptions of the agents differ on the basis of the agents' physical forms. For example, owners not only mentioned the agents' physical attributes, but also used adjectives such as "cool." Furthermore, the owners engaged in more intelligent conversations with more complex agents. iii) Dog-shaped agents can garner attention from dogs.

Keywords: Social agents · Pet · Dog · Embodiment · Speaking agents · Pilot test

1 Introduction

In the last ten years, an increasing number of communication agents, such as Google Home or Siri, have become prevalent in homes. Communicative humanoid robots, such as Peppers, have also been found in an increasing number of places, such as in airports or stores.

© Springer Nature Switzerland AG 2021
H. Degen and S. Ntoa (Eds.): HCII 2021, LNAI 12797, pp. 359–376, 2021.
https://doi.org/10.1007/978-3-030-77772-2_24

As these communication agents become a more common part of our daily lives, they cannot avoid interacting with non-human members of our society, a prime example of which are dogs, who have been beloved friends, partners, and family members to humans for over 10,000 years [25,32]. Dogs not only aid in maintaining human mental [27,36] and physical [8,9] health, but can also be employed to provide medical services and serve as professional service and therapy animals.

As a result of the relationship that humans and dogs have developed over an extended period, dogs have become adept at interpreting human social signals including not only verbal cues, such as commands, but also non-verbal ones, such as pointing gestures [13] or gazes. Studies indicate that pet dogs tend to prefer the company of humans over that of other dogs [35].

Thus, various experiments have been conducted to determine whether human-like or non-human-like social agents are more suitable for cases where dogs are present in a home [23,30]. However, when communication agents are introduced into a household and communicate with dog owners, negative canine behavior, including jealousy [14] towards communication agents, can evoke a triad relationship. However, if dog owners behave positively towards agents, dogs will accept them more easily based on social references [22]. Therefore, communication agents that are appropriate for both dogs and owners must be considered. Previously, quantitative studies on coding canine behavior and analyzing owner impressions via a questionnaire were conducted [15]. However, few qualitative studies have investigated canine behavior during agent–owner interactions and dog owner preferences with regard to communication agents.

The purpose of this study was therefore to explore the type of words dog owners used in front of their dogs to demonstrate positive attitudes toward the agents, and how their dogs behaved in the presence of these communication agents. Furthermore, this work examined whether owner language and the dogs' behavior were affected by the agent' physical forms (primitive, smaller/bigger humanoid, and dog-shaped agents, as shown in Fig. 1).

To investigate how positive owner attitude towards agents could mediate agent–dog interactions and the type of behaviors dog owners and dogs displayed, we conducted pilot tests that featured the owner speaking to the agent with a positive attitude, followed by the dog being spoken to by the agent while the owner remained still.

2   Related Work

For several decades, agents such as robots or other artificial objects have been used to investigate the social behaviors of animals. For example, Leaver et al. used a medium-sized dog figurine with a motor-driven wagging tail to investigate whether dogs recognize wagging tails as a social signal for intra-species interaction [18]. In the last ten years, an increasing number of studies have revealed that dogs that develop social cognition for interspecies interactions with humans also recognize interactivity with primitive-shaped agents after observing the agent movements that appear to have an aim [1,10–12].

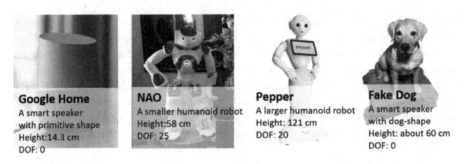

Google Home
A smart speaker
with primitive shape
Height:14.3 cm
DOF: 0

NAO
A smaller humanoid robot
Height:58 cm
DOF: 25

Pepper
A larger humanoid robot
Height: 121 cm
DOF: 20

Fake Dog
A smart speaker
with dog-shape
Height: about 60 cm
DOF: 0

Fig. 1. Speaking agents used for the pilot tests.

2.1 Reaction of Dogs to Canine-Figurine Agents

Kubiny et al. observed canine behavior during encounters with four different agents in non-feeding and feeding situations. As a result of the observations of adult dogs (N = 24) and 4–5 month old puppies (N = 16), AIBO, which is a robot in the form of a furry canine figure, was differentiated from a car in social behaviors, such as in terms of "sniffing the rear of the agent," which suggested that the shape of an agent can affect the dogs' reactions to the agent. On the other hand, in the feeding situation, neither of the AIBO variants (with and without fur) was recognized as a potential rival by the adult dogs; moreover, the adult dogs did not growl as significantly as living puppies the size of the AIBO did. This result implies that although dog-shaped agents can elicit social behavior to a certain extent when compared to agents with a primitive shape, the agent having a dog shape does not prove sufficient for its recognition as a social partner or rival by dogs [16].

2.2 Reaction of Dogs to Speaking Agents

Because dogs can recognize human verbal cues, several studies have investigated speaking social agents for dogs.

Lakatos et al. set the social and asocial conditions of an agent with hands on the basis of dogs observing social interactions between a human experimenter and the agent, where the dogs were called by their names when the agent pointed toward a bowl in pointing tests. As a result, although there was no significant difference in the score of choice in the pointing test, the dogs chose the pointed bowl to a level exceeding the chance level in the social condition, and to a level less than the chance level in the asocial condition. In addition, the dogs looked at the bowl pointed at for a significantly longer duration in the social condition. The results imply that "calling a dog's name" can be a social factor towards dogs, although the experiment included the other factor–that is, observing human-agent interactions [17].

Shaw and Riley reported that dogs can respond appropriately to a speaker with the following characteristics: i) Position at a height equal to that of the

human eye, ii) ability to output speech using pre-recorded voices of dog owners, and iii) a function as a food (reward) dispenser. With these factors, the dogs' response to the speaker was as reliable as that to the dog owner in person, although it should be noted that the subjects are probably not representative of ordinal dogs because many of them were well-trained dogs in sports and agility [33]. Qin et al. experimented with NAO, a humanoid robot speaking naturally, and a speaker to investigate if a human-shaped was better than a speaker in terms of eliciting a response from the dogs to the command "Sit down!" [30].

2.3   Behavior of Dogs in a Dog—Owner—Object Triad Relationship

When an agent enters a home, this agent enters the already established animal-owner relationship, which cannot be ignored when a relationship with a third object or person is developed. Dogs and cats can distinguish between owners and strangers, and they behave according to this distinction [3,7,31]. In the owner-animal–object triad relationship, when both dogs and cats are simultaneously confronted with a novel object with their owners, social references gathered by means of listening to the owner's reactions are observed in more than 80% of animals [20–22]. In addition, studies on triad relationships have been conducted to determine the effects in terms of the frequency of dogs not receiving food from humans, not helping their owners [5], and showing the tendency to display jealousy when their owners focused on objects or other dogs [2,14,28,29].

3   Explorational Experiment

In the future, dog owners could introduce humanoid robots into their households as communication agents. Therefore, we examined whether agents could create a triad relationship among owners, dogs, and agents, by allowing agents to interact with the dogs and their owners. We performed two within-subject experiments and changed the order of the conditions among the participants in both Pilot Tests I and II. A total of 24 possible permutations of the four conditions were assigned to each dog-owner pair. Three video cameras (two video cameras only for PD1) were set up to record the experiment.

Ethics. In advance, the dog owners agreed and provided written consent to record the experiment and were informed that they could opt out of the experiment at any time if they felt that their dog was too frightened or had other problems that made it impossible to continue. Only participants who completed and submitted an informed consent form were included in the experiment.

3.1   Participants

Dog Participants. Three pet dogs participated in Pilot Test I, and 31 pet dogs of different genders, ages, and breeds participated in Pilot Test II (N = 34, 15 males, 19 females; mean age ±SD 4.35 ±3.41; mean length of living with the owner ±SD 4.28 ±3.46), as shown in Table 1.

Table 1. Dog profiles.

ID	Age	Sex	Breed	Size (SML)	Length of living together with the owner	Whether or not they live with other dogs
PD1	1 years	Male	American Pit Bull Terrier	Medium	1 years	Yes
PD2	2 years	Male	American Pit Bull Terrier	Medium	2 years	Yes
PD3	1 years	Male	Whippet	Medium	1 years	Yes
D1	2 years	Male	Labrador Retriever	Large	2 years	Yes
D2	5 months	Male	Labrador Retriever	Large	4 months	Yes
D3	4 years	Male	Miniature Dachshund	Small	3 years	Yes
D4	1 years	Female	German Shepherd Dog	Large	1 years	Yes
D5	5 years	Female	German Shepherd Dog	Large	5 years	Yes
D6	3 years	Male	Long Coat Chihuahua	Super small	3 years	Yes
D7	10 years	Female	Long Coat Chihuahua	Super small	10 years	No
D8	1 years	Female	Border Collie	Large	1 years	—
D9	2 years	Female	Toy Poodle	Super small	1 years	No
D10	2 years	Female	Pug	Small	2 years	No
D11	4 years	Female	Labrador Retriever	Large	3 years	Yes
D12	4 years	Female	Border Collie	Medium	4 years	Yes
D13	4 years	Male	Border Collie	Medium	4 years	Yes
D14	4 years	Female	Toy Poodle	Super small	4 years	—
D15	5 years	Male	Toy Poodle	Super small	5 years	No
D16	1 years	Female	Mix (Shiba)	Small	1 years	No
D17	12 years	Male	Toy Poodle	Super small	12 years	—
D18	2 years	Female	Long Coat Chihuahua	Super small	2 years	Yes
D19	10 years	Male	French Bulldog	Medium	10 years	No
D20	0.2 years	Male	American Pit Bull Terrier	Medium	2 weeks	Yes
D21	10 years	Female	Long Coat Chihuahua	Super small	10 years	No
D22	9 years	Male	Border Collie	Medium	9 years	Yes
D23	1 years	Female	Bichon Frise	Small	1 years	No
D24	9 years	Male	Toy Poodle	Small	8 years	Yes
D25	4 years	Female	Toy Poodle	Super small	4 years	Yes
D26	2 years	Male	American Pit Bull Terrier	Medium	2 years	Yes
D27	9 years	Female	Mix (Corgi)	Medium	9 years	No
D28	3 years	Male	Mix (Corgi and Maltese)	Small	2 years	No
D29	7 years	Feale	Miniature Schnauzer	Small	7 years	No
D30	10 years	Female	Mix (Japanese)	Small	10 years	No
D31	1 years	Female	Mix (Maltese and Toy Poodle)	Small	1 years	No
Mean	4.28 years	—	—	—	4.13 years	—
SD	3.46 years	—	—	—	3.48 years	—

Human Participants (Dog-Owners). Three dog owners (PO1–PO3) participated in Pilot Test I, and 27 owners (O1–O27) participated in Pilot Test II. The owners (N = 29; nine males, 20 females; nine househusbands/wives, eight office workers, two part-time workers, nine other professions, and one with an unknown profession; mean age ±SD 45.59 ±9.75) participated in the experiment with their dogs. There were no owners who owned robots, but one had a smart speaker. However, eleven owners had talked to communication robots (e.g., Pepper, AIBO, cleaning robots like Roomba) on the street or in shops, and three owners had talked to smart speakers.

3.2 Apparatus

The experiment was performed in a 2.7 m × 3.6 m × 2.4 m room. To reduce external noise, the ceiling and windows were covered with layers of insulation and sound-absorbing materials for soundproofing. We fixed the positions of the agents and a chair in a room with plastic tape. Three video cameras were used in the experiment (Sony FDR-AX45 Handycam) for recording during the experiment. We remotely controlled each camera through a smartphone app called PlayMemories.

3.3 Speaking Agents

We set the following four conditions as communication agents: The appearances of the agents are shown in Fig. 1. We used VOICEROID+ Tamiyasu Tomoe EX as the agent voice. NAO and Pepper talked while nodding and making other small movements using functions set in an API for NAO and Pepper called NAOqi, developed by Softbank Robotics.

- **Google Home:** A smart speaker with a pillar-shaped, primitive body (diameter: 9.64 cm; height: 14.3 cm; weight: approximately 0.5 kg).
- **NAO:** A small humanoid robot designed and developed by Softbank Robotics (height: 58 cm, weight: approximately 5.4 kg, degrees of freedom: 25).
- **Pepper:** A large humanoid robot designed by Softbank Robotics (height: 121 cm, weight: approximately 29 kg, degrees of freedom: 20).
- **Fake Dog:** A communication agent with the appearance of a dog, as shown in Fig. 1. Harris et al. [14,29] performed experiments on the three-way relationship between owners, dogs, and a third party, and observed how dogs responded differently to different types of third parties. It was found that the canine response was greater when owners and other humans paid attention to the "fake dog" as compared to other simple-shaped puppets and books.

We used the Wizard of Oz (WoZ) method to explore the dog owners' and dogs' reactions toward agents with different physical appearances. The WoZ method is a research method in which subjects interact with systems for the subjects to believe the systems work autonomously; however, these systems are actually operated or partially operated by an unseen human being [6,19]. Even without complete development of the systems, the WoZ method can work if the purpose of the experiment is to evaluate the human factors of users.

3.4   Procedure

To explore what can occur in a triad relationship experiment in the owner– dog– agent space, we conducted the following two pilot tests with (Pilot test I) and without (Pilot test II) the phase involving dogs staying without their owners. The owners and their dogs waited in the waiting room; after the owners signed the consent form, they answered a questionnaire that included their profiles as well as that of their dogs.

Pilot Test I. This experiment was conducted according to the following procedure:

1. To familiarize the dog with the experiment room, the owner and dog simultaneously entered the room and explored it for 2 min. In most experiments with the dogs, the time taken for familiarization toward the laboratory was approximately 30 s to a few minutes; thus, we chose a duration of approximately 2 min for this experiment [2, 11].
2. The owner re-entered the laboratory and was seated in front of the agent.
3. The experimenter allowed the dog to enter the room immediately after the owner began interacting with the agent.
4. The owner responded to the agent's instructions and began patting the agent's head with an affirmative voice and phrases, such as "Good boy!". This step ended in 1 min without any reaction to the dog (**agent-owner phase**). For 1 min, the agent provided auditory signals indicating a happy emotion, such as "Yippee!", every 10 s.
5. In the agent's direction, the owner remained quiet and did not respond to the dog for 1 min. Every 10 s, the agent uttered the dog's name (that the dog was referred to in positive situations at home) three times and the agent's name three times in a random order (**agent–dog phase**).
6. The agent then asked the dog owner to leave the room and leave the dog inside the room. After the owner left the room, the agent called the dog's name once and the agent's name in a random order after a 10 s pause. During this time, the dog owner watched the dog from outside the room through a video camera and iPad application while being beside the experimenter.
7. As directed by the agent, the owner re-entered the room and brought the dog outside.
8. Steps 2–7 were repeated for four agents.
9. After the end of each pilot test, the experimenter cleaned the floor of the room using a deodorant and cleaning cloth.

Pilot Test II. This experiment was conducted involving all but Step 6 from Pilot Test I to prevent the dogs from staying in the room while their owners were not present.

3.5 Evaluation

During the pilot tests, every word spoken by dog owners, and their dogs' behaviors, were recorded by the three cameras as indicated in the Apparatus section. The video recordings from the three different angles were automatically synchronized to merge the multi-camera clips into one screen based on the waveform of sounds using a function for multi-camera clips in Adobe Premiere Pro. The authors watched the synchronized recordings to determine what the dog owners said, and which parts of agents the dog owners and the dogs touched. We further used these recordings to determine the categories of canine behavior that could be recognized relatively definitively and would be useful to investigate canine behaviors in the cases where speaking agents are used in future studies.

3.6 Personality of Dog

Before conducting Pilot Tests I and II, we asked the dog owners about the personalities (dominant/subordinate and jealousy level) of their dogs because we thought that personality traits could be related to boldness and aggressiveness toward unfamiliar objects that attract owner attention. This step was conducted to verify that the subject groups did not have extreme tendencies.

Evaluating the Dog Subject Dominance Rank. We assessed the dog subjects' rank (dominant/subordinate) at home using a questionnaire [2] used in the study of dogs' jealousy behavior in a triad relationship among the owner, dog, and third party (using different experimental objects such as a toy-car and other dogs as rivals) conducted by Abdai et al. The questionnaire items were as follows:

- Q1. When a stranger comes to the house, which dog starts to bark first (or if they start to bark together, which dog barks more or longer)?
- Q2. Which dog licks the other dog's mouth more often?
- Q3. If the dogs get food at the same time and at the same spot, which dog starts to eat first or eats the other dog's food?
- Q4. If the dogs start to fight, which dog usually wins?

The dog-subject dominance rank was evaluated mainly based on the answer in Q4. If the name of the dog was not mentioned in Q4 and the dog appeared in at least two of the items in Q1–Q3, the dog was categorized as *dominant*. If the dog's name was not mentioned in Q4 and the name of another dog was mentioned in at least two of the items in Q1–Q3, the dog was rated as *subordinate*. The results are shown in Table 2; there were 12 dominant individuals, 14 subordinate individuals, and eight unidentified individuals. There was no extreme tendency as per the dog subject dominance ranking in the experiment. Because the original evaluation by Pongrácz et al. [26] was conducted for multiple breeds, the evaluation for single breed dogs in this study is enclosed in parentheses in Table 2 for reference.

Table 2. Owners' answers to the questionnaire about jealousy in subjects.

ID	Level of jealousy behaviour (10-point)	Object of jealousy behaviour	Jealousy at Home (5-point)	Jealousy at stranger environment (5-point)	Whether or not they live with other dogs	Dominant/Subordinate
PD1	4	Other dog	3	3	Yes	Dominant
PD2	7	Other dog	4	4	Yes	Subordinate
PD3	7	Other dog and human	3	3	Yes	Subordinate
D1	8	Other dog	4	2	Yes	Dominant
D2	6	Other dog	4	3	Yes	Subordinate
D3	8	Other dog	3	4	Yes	Dominant
D4	10	Owner	5	5	Yes	Dominant
D5	7	Owner	4	3	Yes	Dominant
D6	6	Other dog	4	4	Yes	Subordinate
D7	4	Other dog	3	3	Yes	Dominant
D8	8	Other dog	4	4	—	(Subordinate)
D9	5	Other dog	2	2	No	Unknown
D10	5	—	1	1	No	Unknown
D11	3	Other dog	2	2	Yes	Dominant
D12	5	Other dog	3	3	Yes	Unknown
D13	4	Other dog	4	3	Yes	Subordinate
D14	4	Children	3	3	—	(Dominant)
D15	1	—	1	1	No	Unknown
D16	5	Relationship with the owner	4	—	No	Unknown
D17	8	Other dog and cat	4	3	—	(Dominant)
D18	2	Other dog	5	5	—	(Subordinate)
D19	5	Children	3	3	No	Unknown
D20	1	—	1	1	Yes	Dominant
D21	1	—	1	1	No	(Subordinate)
D22	1	—	1	1	Yes	Subordinate
D23	7	Other dog	4	4	No	(Subordinate)
D24	2	—	1	1	Yes	Dominant
D25	10	Other dog and human	5	5	Yes	Subordinate
D26	7	Owner	4	2	Yes	Dominant
D27	7	Other dog and cat	4	2	No	Unknown
D28	8	Other pet	4	3	No	(Subordinate)
D29	8	Other cat	4	2	No	(Subordinate)
D30	1	—	1	1	No	Unknown
D31	2	Other dog	2	2	No	(Subordinate)
Mean	5.21	—	3.09	2.70	—	—
SD	2.68	—	1.29	1.22	—	—

Evaluating the Dog Subjects' Jealousy Levels. Because jealousy levels could affect canine behavior during an agent interacting with their owner, it was meaningful to determine whether the subject group had a strong bias of jealousy. Therefore, we determined the jealousy level of the dog subjects using a questionnaire that Abdai et al. [2] used for 631 dogs. The questionnaire items were as follows:

- Q1. How jealous do you think your dog is compared to the average dog (scale from 1 to 10) (mean ±SD 5.68 ±2.67 in the previous survey [2])?
- Q2. Who does the dog usually gets jealous of?
- Q3. Where does your dog get jealous (at home, at unfamiliar places; on a scale from 1 to 5) (mean ±SD 3.19 ±1.43, 2.36 ±1.33 in the previous survey [2])?

Owners that filled the questionnaire awarded a mean ±SD 5.21 ±2.68 jealousy score to their dogs, a mean ±SD 3.09 ±1.29 score at home, and a mean ±SD 2.70 ±1.22 score in unfamiliar places, although one owner did not provide a score. There was no extreme tendency indicating a bias in the dog subject jealousy level in the experiment (details are shown in Table 2).

4 Result

4.1 Causes of Failures in the Experiments

Although it may not be necessary to describe the failures caused by the experimental equipment or the experimenter, we present those that prohibited us from quantitatively evaluating the results. Furthermore, we describe the failures resulting from the dog owners or dogs because these cases can be helpful for HCI researchers when conducting future experiments with dogs, especially when studying the relationship between owners and dogs.

Failures Caused by Dog-Owners. A few dog owners did not follow our directions for repeating the same behaviors and speaking with the same tone of voice among the four agents. For example, O8 patted the head of an agent while uttering positive words to the agent for only approximately 20 s in the first condition. Then, O8 continued the behavior for 60 s per the agent's instructions from the owner–agent phase to the agent–dog phase. There were also examples in which O22 stood up to touch the head of the agent only in the Pepper condition, as shown in Fig. 2. O22 remained seated in a chair in the other three conditions, which could have caused a difference in the heights of the agents.

Fig. 2. Failures caused by O22 and D3.

Failures Caused by Dogs. We found that the dogs could break the agents. For example, D4 flipped the Google Home unit by entangling her foot in the cable while running into a room, as shown in Fig. 2. D4 also jumped toward Pepper's head after Pepper called her name, and the dog owner (O3) moved her hand to support D4 to avoid breaking Pepper, as shown in Fig. 2. This occurred despite our instruction for owners to remain still in the **agent–dog phase**. In addition, smaller dogs could also damage the experimental setup. D29 moved the Google Home unit after the experimenter set its position according to the vinyl tape marks.

4.2 Dog-Owners' Behaviors (Pilot Test II)

The following parts of each agent were touched by the dog owners:

– **Google Home**: the side of the body, cardboard box
– **NAO**: hand, head, shoulder, chest, foot, arm, back

– **Pepper**: hand, convex on foot, waist, abdomen, touchscreen
– **Fake Dog**: rear, paw, chest, head, shoulder/neck, abdomen, hind leg, back

Considering the words used for each agent, terms such as "Nice!", "Good boy/girl!", "Amazing!", "You are smart!", "Cute!" and "You can keep quiet, you're a good boy/girl," were used for all four agents. Owners characterized Google Home and NAO as "small", and Pepper as "big." For NAO, Pepper, and the Fake Dog, dog owners complimented the agent and mentioned parts of their bodies. For example, O26 said, "Your eyes, face, and paws are all cute! All of you is cute!" to the fake Dog. Regarding the humanoid robots (NAO and Pepper), a few dog owners, especially O26, not only used "Cute!", but also "Cool!", to complement the agent. A few owners started conversations by asking the humanoid robots questions. For example, O25 said to Pepper, "It's cold today. Isn't it cold? Your body is cold."

In addition, owners mentioned agents' names. For example, O22 said, "Your name is Yuta (as introduced by the agent), isn't it?" O22 and O25 said, "I have forgotten his name...", in front of their dogs.

4.3 Dogs' Behaviors

Separation Anxiety (Pilot Test I). In Pilot Test I, we set the phase in which the dogs stayed with each agent without their owner for 30 s. PD1, in particular, sniffed the chair that the owner sat in after whining in front of the door, as shown in Fig. 3. All dogs whined in front of the door, which appeared to be a result of separation anxiety from their owners and discomfort of being with the agents that they perceived to be strangers in the unfamiliar or strange environment.

Fig. 3. Dogs' behaviors during absence of their owners (Pilot Test I).

The Parts of Agents Touched by Dogs. The parts of each agent touched by the dogs that could be relatively easily recognized were the following:

– **Google Home**: the side of the body, cardboard box
– **NAO**: hand, head, shoulder, chest, foot, arm, back
– **Pepper**: hand, convex on foot, waist, abdomen
– **Fake Dog**: rear, paw, chest, head, shoulder/neck, abdomen, hind leg, back

Vocal Reactions. Whining and growling are recognized relatively easily. However, barking, which was used to evaluate the dogs' jealousy [14], was difficult to differentiate from sneezing (PD1) or sighing (D26).

Staying Position and Wandering Proxy. During Pilot Test II, most dogs demonstrated at least one of the following behaviors that were evaluated as either a staying position or a wandering proxy:

- Standing in front of the door to the room (left in Fig. 4)
- Standing, sitting, or lying by the owner (middle of Fig. 4)
- Exploring the room by sniffing the floor or wall
- Repeatedly wandering around their owner and stopping to gaze at the agent (right in Fig. 4)
- Gazing at the agent near it
- Sniffing the agent

Case-Based Remarkable Behavior. Though Harris et al. counted the number of dogs "Getting between Owner/Objects" to evaluate jealousy [14], it was difficult for a coder to differentiate between "Getting between Owner/Objects" and just "Passing through between Owner/Objects." Instead of "Getting between Owner/Objects," the dogs in the experiment more often attempted

Fig. 4. Examples of dogs' staying position and wandering proxies (D1, PD1, and D26).

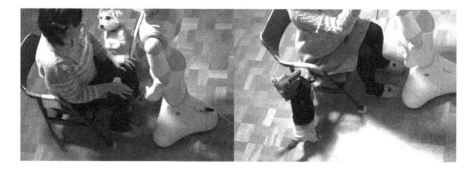

Fig. 5. Examples of dogs putting thier paws on the chair (D20 and D3).

to place their paws on a chair and stretch their bodies toward the owner to gain the owner's attention. These behaviors are recognized more easily, as shown in Fig. 5.

Some dogs also placed themselves between their owners/objects, especially D13, as shown in Fig. 6. D13 stepped between his owner and the agent, except in the case of Pepper. D13 simply passed by Pepper while sniffing Pepper's hand in the **owner–agent phase** and staring at Pepper while maintaining a distance in the **agent–dog phase**, as shown in the left and middle of Fig. 7. However, there were also dogs that approached Pepper without hesitation, such as D4 in Fig. 2. D4 licked Pepper's hand, as indicated on the right side of Fig. 7.

Fig. 6. D13 getting between the owner and agent in the Google Home, NAO and Fake Dog conditions.

Fig. 7. Example of the different attitudes toward Pepper between D13 and D4.

The fake Dog evoked sniffing behavior or gained attention from the dogs more than the other agents. D24 for the NAO and Pepper conditions, and D30 in the NAO condition stepped back dramatically as the agent moved in the **agent–dog phase**. D27 continued to orient towards the fake Dog, while D27 oriented towards an agent in the other conditions. D31 did not wander and sniff the floor and stared at the fake Dog, while D31 wandered and sniffed the floor in the other conditions.

The positive attitudes of dog owners towards agents can lead to dogs feeling less scared of the agents. Especially for D28 in both the NAO and Pepper conditions, and D30 in the NAO condition, they stepped back when the agent called their name while moving during the **agent–dog phase**.

5   Discussion

First, separation anxiety can be a critical impediment in test execution because both the environment and objects are unfamiliar to the dogs. Researchers have observed dogs being managed by a familiar assistant to avoid the presence of dog owners during the experiment to separate the performance of the dog from the performance of the team (dog and owner) [4]. To effectively conduct the tests, when unfamiliar agents are featured without dog owners present, it is necessary for dogs to become familiar with the experimental environment.

Second, the vocabulary used by the owners to praise the agents in a way that the dogs could understand could also have changed the dogs' perceptions of the speaking agents. Pepper, NAO, and the fake Dog were described as "cute", while their faces, nice eyes, and other physical attributes were mentioned in front of the dog subjects. The terms "cute," "quiet and nice," "good boy/girl," "amazing," and "you can keep quiet; so, you are a good boy/girl", were used by all agents. However, "cool" was only used to describe Pepper and NAO, but not the fake Dog and Google Home. For Sentences that resembled conversations and not just compliments were communicated with Pepper. For example, certain dog owners said, "it's cold today" and "your body is cold." This may have been due to the fact that human-like agents are more likely to be perceived as intelligent rather than animal-like agents [34]. Google Home's lack of a complex physical structure may have made it easy for it to have been praised as being "quiet and nice," which implies that it was difficult for the dog owners to introduce the agent in a manner that dogs could understand.

In addition to the word choices, dog owners seemed to think that the names of agents could play a role in making their dogs display a positive attitude toward the agents. Owners who remembered the agent's name praised the agent while referring to the agent's name. For example, one owner said, "Your name is Yuta (as introduced by the agent), isn't it?"; and in the case the owner forgot the agent's name, the owner said, "I have forgotten your name..." In particular, D4, which jumped on Pepper's head, jumped as Pepper called her name from the speakers on both sides of Pepper's head. This supports the results of a previous study, which claimed that calling dogs' names can play a role in dogs' perceptions of an object/subject's sociality [17]. Therefore, dog owners consider names to play a role in mediating positive feelings between their dogs and agents.

Moreover, although it is unclear to what extent canines are aware of the agents' physical forms, significant contact was made with the hands, face, rear, and protruding parts of the feet. Dogs are very sensitive to reading the pointing gestures of humans, and a "human-shaped hand" could also play a role in eliciting social behaviors. The hands of humanoid agents may be the most sensitive to all bulges. Regarding the moving agents, it has been reported that dogs often stare at the appropriate agent when obtaining food from an unreachable place. Thus, they recognize the differences in the roles of crane-type and sedan-type remote control cars. In the future, these protruding parts of the speaking agent may be useful in making the canine aware of the differences in their roles [24].

Considering the evaluation of canine behavior for future studies, it can be difficult to distinguish whether an intervening move [14] between the owner and the agent is just a passing move or an interruption, which is significant in a triad relationship. When the owner is sitting in a chair, assessing the behavior of the dog stretching towards the owner with a paw on a chair, or behaviors to gain the attention of the person in the chair, may be used for future evaluations, considering these behaviors are easy to quantitatively measure. In addition, analyzing the approach and orientation to evaluate canine behaviors toward other objects was useful; however, in an experimental environment where distance and orientation cannot be measured accurately, the following behaviors seemed to be used: "waiting by the door," "waiting by the owner," "sniffing the room," "gazing at the agent near the agent," and "not repeatedly wandering around the owner and stopping to gaze at the agent." For example, if a dog wanted to quickly finish the experiment, the dog would stay in front of the door of the experiment room. If a dog does not care about the agent's presence and does not eagerly want the experiment to finish, the dog would explore the room by sniffing the room with curiosity in the new environment without paying attention to the agent. If a dog wanders repeatedly around its owner and stops to gaze at the agent or remains gazing at the agent for a long time, the dog would be both interested and scared by the agent; if the dog is interested in the agent and is not afraid of it, the dog would approach the agent to investigate the agent by sniffing or touching the agent.

While evaluating vocal cues, in certain cases, whining and grunting were relatively easy to distinguish, while barking was difficult to distinguish from sighing and sneezing. To improve the accuracy of the evaluations, a bone conduction microphone or similar device attached to the collar of the test dog was used to ensure that environmental noises were excluded and to obtain the voices of dogs with better quality.

6   Conclusion

In the future, social agents whose main role is to communicate with people will become increasingly prevalent at home, where risks of dogs displaying jealousy resulting from a triad relationship or feeling scared of the agent are possible. In this study, to aid in the future research of investigating the design of a communication agent that can build a good relationship between dog owners and dogs at home, we conducted two pilot tests to explore the types of words dog owners use in front of their dogs to show the owners' positive attitude toward the agents and how their dogs behave before these communication agents. Furthermore, we determined whether the owners' words and the dogs' behaviors could be affected by the agents' physical form (primitive, smaller/bigger humanoid, and dog-shaped). By analyzing video recordings of the behaviors of 29 dog owners and 34 dogs, the following three points are suggested: i) Tests with dogs being confronted with unfamiliar speaking agents in an unfamiliar environment should occur with the dog owner present, as in previous studies, to evaluate

dogs' behavior in front of the agents. ii) Dog owners' words used to describe the agents differed based on the physical appearance of agents; they mentioned not only physical attributes but also adjectives such as "cool." Furthermore, they engaged in more intelligent conversations. iii) Dog-shaped agents can gain attention from dogs. This study will be useful for HCI researchers who are planning to work with dogs.

References

1. Abdai, J., Gergely, A., Petró, E., Topál, J., Miklósi, Á.: An investigation on social representations: inanimate agent can mislead dogs (Canis familiaris) in a food choice task. PloS one **10**(8) (2015). https://doi.org/10.1371/journal.pone.0139531
2. Abdai, J., Terencio, C.B., Fraga, P.P., Miklósi, Á.: Investigating jealous behaviour in dogs. Sci. Rep. **8**(1), 8911 (2018). https://doi.org/10.1038/s41598-018-27251-1
3. Adachi, I., Kuwahata, H., Fujita, K.: Dogs recall their owner's face upon hearing the owner's voice. Anim. Cogn. **10**(1), 17–21 (2007). https://doi.org/10.1007/s10071-006-0025-8
4. Call, J., Bräuer, J., Kaminski, J., Tomasello, M.: Domestic dogs (Canis familiaris) are sensitive to the attentional state of humans. J. Comp. Psychol. **117**(3), 257 (2003). https://doi.org/10.1037/0735-7036.117.3.257
5. Chijiiwa, H., Kuroshima, H., Hori, Y., Anderson, J.R., Fujita, K.: Dogs avoid people who behave negatively to their owner: third-party affective evaluation. Anim. Behav. **106**, 123–127 (2015)
6. Dahlbäck, N., Jönsson, A., Ahrenberg, L.: Wizard of Oz studies - why and how. Knowl.-Based Syst. **6**(4), 258–266 (1993). https://doi.org/10.1016/0950-7051(93)90017-N
7. Ellis, S.L.H., Thompson, H., Guijarro, C., Zulch, H.E.: The influence of body region, handler familiarity and order of region handled on the domestic cat's response to being stroked. Appl. Anim. Behav. Sci. **173**, 60–67 (2015)
8. Friedmann, E., Katcher, A.H., Thomas, S.A., Lynch, J.J., Messent, P.R.: Social interaction and blood pressure: influence of animal companions. J. Nerv. Ment. Dis. **171**(8), 461–465 (1983). https://doi.org/10.1097/00005053-198308000-00002
9. Friedmann, E., Thomas, S.A.: Pet ownership, social support, and one-year survival after acute myocardial infarction in the cardiac arrhythmia suppression trial (cast). Am. J. Cardiol. **76**(17), 1213–1217 (1995). https://doi.org/10.1016/s0002-9149(99)80343-9
10. Gergely, A., Abdai, J., Petró, E., Kosztolányi, A., Topál, J., Miklósi, Á.: Dogs rapidly develop socially competent behaviour while interacting with a contingently responding self-propelled object. Anim. Behav. **108**, 137–144 (2015). https://doi.org/10.1016/j.anbehav.2015.07.024
11. Gergely, A., Compton, A.B., Newberry, R.C., Miklósi, Á.: Social interaction with an "unidentified moving object" elicits a-not-b error in domestic dogs. PloS one **11**(4) (2016). https://doi.org/10.1371/journal.pone.0151600
12. Gergely, A., Petró, E., Topál, J., Miklósi, Á.: What are you or who are you? The emergence of social interaction between dog and an unidentified moving object (UMO). PloS one **8**(8), e72727 (2013). https://doi.org/10.1371/journal.pone.0072727
13. Hare, B., Brown, M., Williamson, C., Tomasello, M.: The domestication of social cognition in dogs. Science **298**(5598), 1634–1636 (2002). https://doi.org/10.1126/science.1072702

14. Harris, C.R., Prouvost, C.: Jealousy in dogs. PLoS ONE **9**, 7 (2014). https://doi. org/10.1371/journal.pone.0094597
15. Kasuga, H., Ikeda, Y.: Gap between owner's perceptions and dog's behaviors toward the same physical agents: using a dog-like speaker and a humanoid robot. In: Proceedings of the 8th International Conference on Human-Agent Interaction, HAI 2020, pp. 96–104. Association for Computing Machinery, New York (2020). https://doi.org/10.1145/3406499.3415068
16. Kubinyi, E., Miklósi, Á., Kaplan, F., Gácsi, M., Topál, J., Csányi, V.: Social behaviour of dogs encountering AIBO, an animal-like robot in a neutral and in a feeding situation. Behav. Process. **65**(3), 231–239 (2004). https://doi.org/10.1016/j.beproc.2003.10.003
17. Lakatos, G., et al.: Sensing sociality in dogs: what may make an interactive robot social? Anim. Cogn. **17**(2), 387–397 (2013). https://doi.org/10.1007/s10071-013-0670-7
18. Leaver, S.D.A., Reimchen, T.E.: Behavioural responses of Canis familiaris to different tail lengths of a remotely-controlled life-size dog replica. Behaviour **145**(3), 377–390 (2008). https://doi.org/10.1163/156853908783402894
19. Maulsby, D., Greenberg, S., Mander, R.: Prototyping an intelligent agent through wizard of Oz. In: Proceedings of the INTERACT 1993 and CHI 1993 Conference on Human Factors in Computing Systems, CHI 1993, pp. 277–284. Association for Computing Machinery, New York (1993). https://doi.org/10.1145/169059.169215
20. Merola, I., Lazzaroni, M., Marshall-Pescini, S., Prato-Previde, E.: Social referencing and cat-human communication. Anim. Cogn. **18**(3), 639–648 (2015). https://doi.org/10.1007/s10071-014-0832-2
21. Merola, I., Prato-Previde, E., Marshall-Pescini, S.: Dogs' social referencing towards owners and strangers. PloS One **7**(10), e47653 (2012)
22. Merola, I., Prato-Previde, E., Marshall-Pescini, S.: Social referencing in dog-owner dyads? Anim. Cogn. **15**(2), 175–185 (2012). https://doi.org/10.1007/s10071-011-0443-0
23. Morovitz, M., Mueller, M., Scheutz, M.: Animal-robot interaction: the role of human likeness on the success of dog-robot interactions. In: Proceedings on 1st International Workshop on Vocal Interactivity in-and-between Humans, Animals and Robots (VIHAR), pp. 22–26 (2017)
24. Petró, E., Abdai, J., Gergely, A., Topál, J., Miklósi, Á.: Dogs (*Canis familiaris*) adjust their social behaviour to the differential role of inanimate interactive agents. Anim. Cogn. **19**(2), 367–374 (2015). https://doi.org/10.1007/s10071-015-0939-0
25. Pollinger, J.P., et al.: Genome-wide SNP and haplotype analyses reveal a rich history underlying dog domestication. Nature **464**(7290), 898–902 (2010). https://doi.org/10.1038/nature08837
26. Pongrácz, P., Vida, V., Bánhegyi, P., Miklósi, Á.: How does dominance rank status affect individual and social learning performance in the dog (Canis familiaris)? Anim. Cogn. **11**(1), 75–82 (2008). https://doi.org/10.1007/s10071-007-0090-7
27. Poresky, R.H.: Companion animals and other factors affecting young children's development. Anthrozoös **9**(4), 159–168 (1996). https://doi.org/10.2752/089279396787001437
28. Prato-Previde, E., Nicotra, V., Fusar Poli, S., Pelosi, A., Valsecchi, P.: Do dogs exhibit jealous behaviors when their owner attends to their companion dog? Anim. Cogn. **21**(5), 703–713 (2018). https://doi.org/10.1007/s10071-018-1204-0
29. Prato-Previde, E., Nicotra, V., Pelosi, A., Valsecchi, P.: Pet dogs' behavior when the owner and an unfamiliar person attend to a faux rival. PloS One **13**(4), e0194577 (2018). https://doi.org/10.1371/journal.pone.0194577

30. Qin, M., Huang, Y., Stumph, E., Santos, L., Scassellati, B.: Dog sit! domestic dogs (Canis familiaris) follow a robot's sit commands. In: Companion of the 2020 ACM/IEEE International Conference on Human-Robot Interaction, pp. 16–24. https://doi.org/10.1145/3371382.3380734

31. Saito, A., Shinozuka, K.: Vocal recognition of owners by domestic cats (felis catus). Anim. Cogn. **16**(4), 685–690 (2013). https://doi.org/10.1007/s10071-013-0620-4

32. Savolainen, P., Zhang, Y.P., Luo, J., Lundeberg, J., Leitner, T.: Genetic evidence for an East Asian origin of domestic dogs. Science **298**(5598), 1610–1613 (2002). https://doi.org/10.1126/science.1073906

33. Shaw, N., Riley, L.M.: Domestic dogs respond correctly to verbal cues issued by an artificial agent. Appl. Anim. Behav. Sci. **224**, 104940 (2020). https://doi.org/10.1016/j.applanim.2020.104940

34. Terada, K., Jing, L., Yamada, S.: Effects of agent appearance on customer buying motivations on online shopping sites. In: Proceedings of the 33rd Annual ACM Conference Extended Abstracts on Human Factors in Computing Systems, CHI EA 2015, pp. 929–934. Association for Computing Machinery, New York (2015). https://doi.org/10.1145/2702613.2732798

35. Tuber, D.S., Hennessy, M.B., Sanders, S., Miller, J.A.: Behavioral and glucocorticoid responses of adult domestic dogs (Canis familiaris) to companionship and social separation. J. Comp. sychol. **110**(1), 103 (1996). https://doi.org/10.1037/0735-7036.110.1.103

36. Virués-Ortega, J., Buela-Casal, G.: Psychophysiological effects of human-animal interaction: theoretical issues and long-term interaction effects. J. Nerv. Ment. Dis. **194**(1), 52–57 (2006). https://doi.org/10.1097/01.nmd.0000195354.03653.63

Investigating Viewer's Reliance
on Captions Based on Gaze Information

Wen-Hung Liao[✉], Chiao-Ju Chen, and Yi-Chieh Wu

Department of Computer Science, National Chengchi University, Taipei, Taiwan
whliao@nccu.edu.tw

Abstract. Subtitles are present in almost all TV programs and films in Taiwan. Are Taiwanese more dependent on subtitles to appreciate the content of the film compared to people of other nationality? What happens if subtitles are removed or replaced by unfamiliar languages? In this research, we use Tobii EyeX to collect eye movement data from 45 native-speakers while they watch different films, and propose appropriate indicators to analyze their viewing behavior. To facilitate subsequent data analysis, certain areas of interest (AOI), such as the caption region and human face, are automatically detected using techniques including Canny edge detector and Faster R-CNN.

Experimental results indicate that auditory language is the most critical factor. Subjects in Group #1 (English, Chinese, English and Chinese) have a higher tendency to focus on the face area. Subjects in Group #2 (Chinese, English, Chinese and English) appear to read the subtitles more often. The initial behavior seems to determine the viewing pattern subsequently. For subjects in Group #2, preference for caption is clearly observed than those in Group #1. This habitual preference continues in follow-up movies, resulting in an immersion phenomenon. We also observe that when unfamiliar texts appear, the subjects exhibit 'escaping' behavior by avoiding the text region. It is worth noting that the video at the beginning of Group #2 is the native language of the testee, and the result demonstrates that the subject develops preferences toward viewing subtitles. Therefore, we can partially confirm that Taiwanese people have a certain degree of dependence on subtitles.

Keywords: Eye movement · Eye tracker · Subtitles

1 Introduction

As the cost of making video is getting lower and the platforms for sharing the content have become prevalent, it is essential to understand the preference of the audience to enhance overall user experience. Undoubtedly, the audience's viewing habit on broadcasting or cable TV can provide useful information. Take the media and entertainment industry in Taiwan as an example, subtitles are present in almost all TV programs and films. In addition to the regulation of the Radio and Television Law [1], it's beneficial to clarify the specific Chinese characters that were spoken with subtitles, as observed in the following:

© Springer Nature Switzerland AG 2021
H. Degen and S. Ntoa (Eds.): HCII 2021, LNAI 12797, pp. 377–393, 2021.
https://doi.org/10.1007/978-3-030-77772-2_25

1. Since Mandarin (Chinese) is one of the "tone" language [2], there exist distinct pitches of same pronounced syllable representing different lexical or grammatical meanings.
2. Local accent and pronunciation affect the accuracy of information received by the audience.

Under such long-term immersion, the spectators in Taiwan have regarded captions as an integrated part of a video. Are Taiwanese more dependent on subtitles to appreciate the content of the film compared to people of other nationality? What happens if subtitles are removed or replaced by unfamiliar languages? Does subtitles assist or interfere with auditory information? Finally, as the channel for video viewing has shifted to on-demand services such as YouTube and Netflix, has the audience behavior changed?

To address the above issues, this research investigates how the presence of subtitles, faces and auditory language in a video will affect viewer's attention based on eye-tracking data obtained during the viewing process. The contributions of this research are summarized below:

- We design an experimental protocol for collecting the viewing behavior of participants. A total of 45 records containing eye-tracking information along with the video content have been gathered, which can be used for the study of subtitle reliance, as well as research regarding other areas of interest within the whole video frame.
- We employ computer vision techniques to automatically detect the regions that contain subtitles and human faces. These areas of interest are overlaid with the eye-tracking data to examine viewer's dependency on subtitles under different settings.
- We identify the auditory language used in the film to be a key factor in regulating viewer's reliance on subtitles. For videos dubbed in Mandarin (the mother tongue of most Taiwanese), viewers tend to pay more attention to the subtitles. For videos dubbed in other languages, users pay more attention to the face area. Such viewing preference carries over to the whole length of the film, even when the dubbing language has been changed in between, suggesting the formation of an immersion phenomenon.

The remainder of this paper is organized as follows. In Sect. 2 we review related work, covering topics such as eye tracking and computer vision techniques for text and face detection. Section 3 elucidates our experimental design, including system workflow, data collection, pilot study, and evaluation metrics. In Sect. 4, we compare and discuss experimental results. Section 5 concludes this paper with a brief summary.

2   Related Work

Eye tracking devices record and measure the movement of pupils to obtain important information such as fixation and saccade that reflect viewer's attention or

preference. Eye tracker has been widely employed in human-computer interaction research, and serves as an indispensable tool in gathering user data for quantitative analysis. The introduction of low-cost equipment including Tobii EyeX and Eye Tribe has made this technology more accessible.

In [3], the authors collected and analyzed both gaze and mouse movement data to perform usability study of various visualization tools. Liao *et al.* [4] classified five type of reading patterns from fixation data using machine learning techniques in an attempt to understand and evaluate the reading and learning process. Both projects confirm the feasibility of acquiring reliable experimental data using low-cost eye trackers. Built upon the past experience, we use Tobii EyeX [5] to collect eye movement data while the subjects watch different films, and analyze their viewing patterns using proper indicators.

O'Bryan [6] experimented with different media formats and presentation styles. Based on eye tracking data, three modes of attention have been observed, and the most appropriate settings for young viewers have been identified. Subtitles are not the only materials getting user's attention. Instead, different viewing patterns were perceived depending on the content being consumed.

Most traditional eye movement research employs fixed AOI to simplify the experimental design. In this work, we focus on two dynamic AOIs extracted using computer vision method, namely, the caption region and the human face. These AOIs are automatically detected using techniques including Canny edge detector [7] and Faster R-CNN [8]. After extracting all AOIs in the video, the collected eye movement data can then be analyzed along with these dynamic areas. Through the integration of macro and micro analysis, we can better understand the subject's preference and dependence on subtitles and other areas of interest (such as faces).

3   Methodology

We employ eye tracking and object detection techniques to investigate viewing behavior under different settings. Figure 1 depicts the detailed process of this work. The experimental environment and settings are illustrated in Fig. 2.

Experimental materials need to be selected with great care to eliminate unnecessary interference during the viewing process. We chose TED videos [9] in this work as TED presentation usually features a single speaker without complex background or animations. Captions in different language are also provided to help spread the ideas outlined in the talk. Even so, there exist different genres of TED video, and a pilot study has been conducted to explore whether the presentation style will have an impact on viewing behavior. Details of the formal experimental design can also be crafted at this stage.

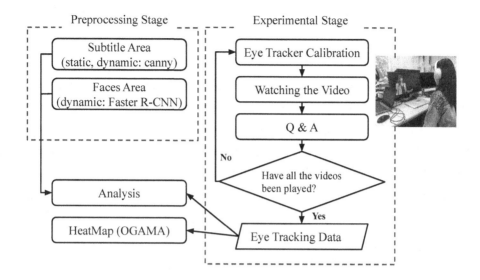

Fig. 1. Overview of the analysis architecture

Fig. 2. The experimental setup. (a) The monitor for playing videos (b) Tobii EyeX (c) The host for data collection

3.1 Pilot Study

TED videos can be roughly categorized into two types:

1. Speech: features a single background with multiple perspectives, focusing mainly on the speaker. An example is shown in Fig. 3-(a) [10].
2. Animation: contains frequent switch of diverse content. Figure 3-(b) [11] belongs to this category.

(a)   (b)

Fig. 3. Two types of video on TED. (a) Speech [10] (b) Animation [11]

To quantify the difference between these two types of videos, we partition the whole video frame into 3×3 blocks, as illustrated in Fig. 4. We then compute the optical flow [12] in each sub-region. The amount of movement (or change in video content) in one minute is accumulated and compared. Three other speech type videos from [13–15] have also been analyzed. The results are demonstrated in Fig. 5.

(a)   (b)

Fig. 4. Content update (a) Speech [10] (b) Animation [11]

The analysis indicates that animation type films exhibit much larger variations in the content at most sub-regions, except for areas 2, 5 and 8 which form

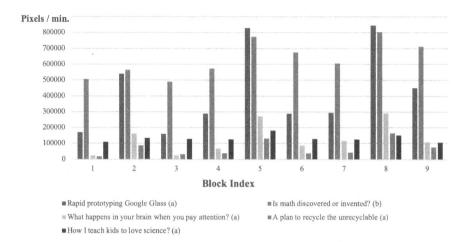

Fig. 5. Average content change in one minute: (a) Speech type [10, 13–15] (b) Animation type [11]

the central column of the picture. Frequent transitions in animations contribute to more pronounced optical flow, resulting in possible switch of attention. To restrict factors that can affect viewing behavior, we chose speech type TED videos as our experimental materials.

3.2 Materials

In selecting the final TED presentations for our experiments, both the length and the subject of the video are taken into consideration. The video must be long enough to cover an entire topic, yet not too long in order to keep the viewer focused. Two videos of length 6 min [16, 17] and two videos of length 9 min [18, 19] are employed in this investigation. Figure 6 depicts the representative frames in our selected films.

Fig. 6. TED videos used in the experiments. (a) How books can open your mind? [16] (b) Forget shopping. Soon you'll download your new clothes [17] (c) My 12 pairs of legs [18] (d) Protecting twitter users (sometimes from themselves) [19]

3.3   Area of Interest (AOI) Extraction in Videos

As the objective of this research is to investigate viewing behavior with the presence of subtitles and human faces, we need to extract these areas of interest in an efficient manner, as elucidated below.

Subtitle Area. We perform Canny edge detection [7] in each video frame and estimate the bounding box covering the subtitles. We dilate the rectangle by 1 pixel in each direction to obtain the final subtitle area. As a reference, we employ a conventional setting that defines the bottom 20% of the screen as the subtitle region. The results are shown in Fig. 7.

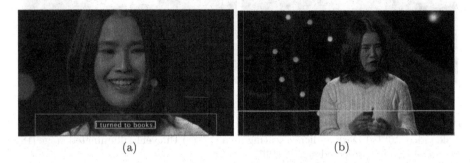

(a)   (b)

Fig. 7. Extracting subtitle regions. (a) Retrieved dynamically from Canny edge detection results, as marked by a blue frame (b) Fixed area as marked by a cyan frame

Face Areas. Faster R-CNN [8], a robust two-stage object detection method, is used to extract human faces in the video. Figure 8 demonstrates face detection results in videos of various lighting conditions.

3.4   Experimental Design

Viewing behavior is our major interest in this research. However, we are also concerned with non-visual cues that might affect user preference. In order to investigate the auditory factor, the participants were divided into two groups. In each video stage, the participant will not see any subtitles initially. After that, English, Russian, and Chinese subtitles appear in the order listed in Table 1.

Eye Tracking Data Validity. As depicted in Fig. 1, we re-calibrate the eye tracker before the start of each video in order to obtain high quality data. Nonetheless, the tracker sometimes fails to capture the eye movement accurately, generating invalid entry at certain time steps. If the overall invalid ratio exceeds 0.2, the data from that particular subject will be excluded. Out of the 61 participants, a total of 45 valid records have been attained. Table 2 reports the data validity in each experimental group.

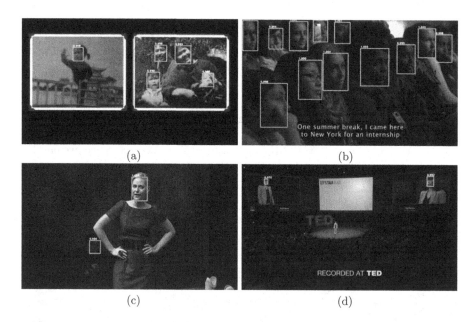

Fig. 8. Face detection using Faster R-CNN, as marked by the cyan frames. (a) Gray-scale video [16] (b) Dark audience seating area [17] (c) Unevenly illuminated stage [18] (d) Speakers in a projected screen [19]

Table 1. Experimental design for the participant

		Video 1				Video 2				Video 3				Video 4			
Auditory	Group #1	English				Chinese				English				Chinese			
Language	Group #2	Chinese				English				Chinese				English			
Video length		6-min.				6-min.				9-min.				9-min.			
Subtitle language		N*	E*	R*	C*	N	E	R	C	N	E	R	C	N	E	R	C

** The subtitles of different language can be retrieved on TED.
N: None
E: English
R: Russian
C: Chinese

Table 2. Overview of eye tracking participants

	Valid		Invalid	
	Male	Female	Male	Female
Group #1	22		13	
	15	7	8	5
Group #2	23		3	
	16	7	3	0
Total	45		16	
	31	14	11	5

Behavior Indicators. Fixations on specific regions in a video frame indicate user's attention or interest. As the length and content of the videos vary, proper indicators derived from the fixation data should be developed. Here we focus on two metrics, namely, fixation count and fixation duration within different regions of interest. Formal definitions are given as follows:

- Subtitle collision ratio (SCR): defined as the fixation count within the subtitle area divided by the total gaze count.
- Subtitle duration ratio (SDR): defined as the gaze duration in the subtitle area divided by the total gaze duration.
- Face collision ratio (FCR): defined as the fixation count within the face area divided by the total gaze count.
- Face duration ratio (FDR): defined as the gaze duration in the face area divided by the total gaze duration.
- Traditional text collision ratio (TTCR): defined as the fixation count within the traditional text area divided by the total gaze count.
- Traditional text duration ratio (TTDR): defined as the gaze duration in the traditional text area divided by the total gaze duration.

Using the above indicators, we can analyze the eye movement patterns and viewing behavior of each participant. We also compute the heat maps using OGAMA software [20], which provide a high level view of all users' attention area.

4   Experimental Results

This research attempts to examine viewer's reliance on caption when watching videos. On a micro-level analysis, we discuss changes in viewer's attention using correlation measure, within group and between group comparison On a macro-level analysis, heat maps are presented and compared to reveal overall user preference.

4.1   Results with Correlation

We compute the six behavior indicators defined previously and obtain their correlations with respect to the English skill level, gender, test score (test is given right after viewing the film) and the group number Table 1 for each video. The results are summarized in Table 3.

The following observations are made:

1. In video #1, both subtitle and traditional text indicators exhibit weak to strong positive correlation with grouping, suggesting that the viewers in Group #2 (the dubbing language being Chinese) pay more attention to the subtitles. On the other hand, indicators concerning face regions demonstrate weak negative correlation, implying that users watching video dubbed in English (i.e., Group #1 users) spent more time in the face area than those who watched video dubbed in Chinese (i.e., Group #2 users).

Table 3. Overall correlation

		English level (1–3)	Gender (M = 1; F = 2)	Score (0–4)	Group (1 or 2)
Video 1	SCR	0.051	0.041	0.082	0.377
	SDR	−0.044	0.004	0.091	0.41
	FCR	0.107	0.112	0.178	−0.281
	FDR	0.129	0.068	0.207	−0.171
	TTCR	0.066	0.017	0.089	0.371
	TTDR	−0.008	0.01	0.108	0.423
Video 2	SCR	0.059	0.045	0.286	0.489
	SDR	0.024	0.059	0.22	0.424
	FCR	0.179	0.176	0.238	−0.25
	FDR	0.15	0.187	0.262	−0.158
	TTCR	0.1	0.088	0.268	0.48
	TTDR	0.048	0.106	0.193	0.397
Video 3	SCR	0.047	0.182	−0.041	0.158
	SDR	0.051	0.233	−0.062	0.295
	FCR	−0.017	0.147	0.292	0.549
	FDR	−0.036	0.098	0.279	0.468
	TTCR	0.074	0.186	−0.033	0.208
	TTDR	0.082	0.232	−0.033	0.324
Video 4	SCR	0.117	0.065	0.175	0.418
	SDR	0.105	0.032	0.149	0.338
	FCR	0.147	0.387	−0.049	−0.27
	FDR	0.188	0.388	−0.026	−0.191
	TTCR	0.134	0.068	0.2	0.425
	TTDR	0.116	0.034	0.172	0.345

corr. ≤ −0.5 : ⬛   −0.5 < corr. ≤ −0.4 : ⬜
−0.4 < corr. ≤ −0.3 : ⬜   −0.3 < corr. ≤ −0.2: ⬜
0.2 ≤ corr. < 0.3: ⬜   0.3 ≤ corr. < 0.4: ⬜
0.4 ≤ corr. < 0.5: ⬜   0.5 ≤ corr.: ⬛

2. In video #2, both subtitle and traditional text indicators still exhibit weak to strong positive correlation with grouping. Notice that the dubbing language has now changed from Chinese to English, suggesting that Group #2 viewers spent more time on subtitle than Group #1 viewers. Face count indicator shows weak negative correlation, implying that Group #1 users tend to watch the faces more.

3. In video #3, both subtitle and traditional text indicators still exhibits positive correlation with grouping, although not as strong as those seen in videos #1 and #2. The dubbing language in this video has switched again, suggesting that Group #2 viewers focused more on the subtitles than Group #1 viewers. Face area indicators display strong positive correlation, implying increased duration in the face regions for Group #2 viewers. As for the subtitle area, the count has decreased, yet reliance on subtitles is still observed.
4. In video #4, similar results on the two subtitle indicators are again observed. Face region indicators roll back to weak negative correlation, implying more times spent on faces for Group #1 viewers. In addition, face indicators show strong correlation with gender, implying that female participants focus more on the face areas in this video.

The above discussions point to the presence of immersion phenomenon. That is, initial viewing pattern will sustain even video content and dubbing language have changed. Additionally, user attention is constantly fluctuating as a result of changes in video content, dubbing or subtitle language. The consideration of auditory is thus essential.

4.2   Within-Group Results

The discussion here is centered on the influence of subtitles that come in different languages (Chinese, English, Russian or None) at different time intervals. To exclude auditory factor, we report the results for each user group separately. Figure 9 and Fig. 10 depict the change of attention in subtitle and face regions for Group#1 and Group #2 subjects.

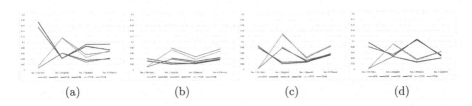

     (a)   (b)   (c)   (d)

Fig. 9. Group #1 (Dubbing order: English-Chinese-English-Chinese) Within-group trend: red for face indicators, blue for subtitle indicators and green for traditional text indicators. (a) Video 1 (b) Video 2 (c) Video 3 (d) Video 4 (Color figure online)

Group #1 viewers focus more on the subtitle region when the dubbing language is English (video 1 and 3) than when the dubbing language is Chinese (video 2 and 4). Attention is persistent when the subtitle changes from Russian to Chinese for the users in this group. Furthermore, in video 1 and 4, the viewers focus more on the face area when the subtitle is Russian, an unfamiliar language for all the participants in this research. Finally, the results indicate users' lack of attention when watching video 2.

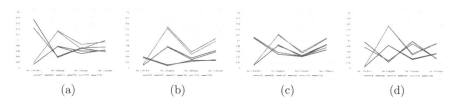

(a)   (b)   (c)   (d)

Fig. 10. Group #2 (Dubbing order: Chinese-English-Chinese-English) Within-group trend: red for face indicators, blue for subtitle indicators and green for traditional text indicators. (a) Video 1 (b) Video 2 (c) Video 3 (d) Video 4 (Color figure online)

Subjects in Group #2 tend to pay more attention to the subtitles regardless of the dubbing language except when the subtitle is presented in Russian. Face indicators are similar to those observed for Group #1 users, except in video 3 where the indicators show a higher value. This explains the positive correlation w.r.t. grouping as Group #1 users direct some attention to the subtitles while Group #2 users focus more on face areas.

To summarize, indicators related to face regions are quite similar for both groups. When no subtitle is present, users will generally focus on faces. When English subtitle appears, the attention turns to the text area. When the subtitle text changes to Russian (an unfamiliar language), attention is again shifted to faces. Finally, Chinese subtitle attracts users attention back to the text region. The levels of attention, however, are a bit different in the two groups. Is the difference significant? Is the difference caused by the dubbing language? We will explore these issues in the next subsection.

4.3   Between-Group Results with Auditory Language Factors

Statistical tests are administered to study the relationship between various behavior indicators and auditory language. The results are summarized in Table 4, from which we can look up the fixation count and duration in each subtitle segment.

In video 1, significant difference in subtitle indicator can be confirmed between two groups for Russian and Chinese subtitle segments. As Group #1 users are listening to English, Russian subtitle is regarded as irrelevant and the attention switches to faces. Such significant difference carries on to the Chinese subtitle segment, and become less apparent when English subtitle re-emerges.

In video 2, Group #2 subjects spend more time on subtitles in both English and Chinese subtitle segments compared to Group #1 users. During English subtitle segment, Group #1 subjects focus more frequently on face areas, which is consistent with the correlation analysis reported in Sect. 4.1. However, the overall viewing behavior in video 2 differs from that in video 2. If auditory language indeed plays a crucial role, why do we see opposite results in video 1 and 2? The observed phenomenon may be attributed to the order of video and subtitle presentation. After watching the first video, the subject might already

Table 4. Auditory language groups test (2 tails)

			Group #1 (μ/σ)	Group #2 (μ/σ)	Test F	T
Video 1	Sec. 1 (None)	FCR	0.155/0.031	0.144/0.031	0.8806	0.2909
		FDR	0.173/0.034	0.174/0.041	0.4739	0.9283
		TTCR	0.014/0.008	0.019/0.011	0.1099	0.1263
		TTDR	0.008/0.004	0.013/0.008	8.2E − 04	0.016
	Sec. 2 (Eng.)	SCR	0.117/0.062	0.135/0.037	0.0297	0.1697
		SDR	0.062/0.035	0.078/0.024	0.0894	0.037
		FCR	0.043/0.016	0.04/0.018	0.4086	0.501
		FDR	0.043/0.023	0.038/0.02	0.5182	0.2865
		TTCR	0.116/0.062	0.132/0.037	0.0255	0.189
		TTDR	0.062/0.034	0.077/0.023	0.0775	0.0454
	Sec. 3 (Rus.)	SCR	0.056/0.03	0.085/0.033	0.6774	3.1E − 03
		SDR	0.035/0.027	0.061/0.033	0.4287	4.3E − 03
		FCR	0.092/0.032	0.073/ 0.034	0.7069	0.0388
		FDR	0.086/0.032	0.071/0.039	0.3822	0.1376
		TTCR	0.046/0.029	0.069/0.032	0.7074	0.0108
		TTDR	0.029/0.027	0.051/0.03	0.6325	9.6E − 03
	Sec. 4 (Ch.)	SCR	0.066/0.037	0.095/0.037	0.833	7.3E − 03
		SDR	0.04/0.03	0.06/0.027	0.7428	0.0127
		FCR	0.093/0.033	0.076/0.033	0.8836	0.0592
		FDR	0.073/0.031	0.059/0.03	0.941	0.1006
		TTCR	0.07/0.038	0.1/0.037	0.9379	5.5E − 03
		TTDR	0.043/0.031	0.066/0.028	0.6166	8.2E − 03
Video 2	Sec. 1 (None)	FCR	0.041/0.012	0.04/0.017	0.1306	0.6947
		FDR	0.032/0.012	0.037/0.019	0.0413	0.3916
		TTCR	0.013/9.8E − 03	0.013/8.2E − 03	0.3569	0.9046
		TTDR	0.011/0.011	9.6E − 03/7.6E − 03	0.1055	0.6486
	Sec. 2 (Eng.)	SCR	0.079/0.051	0.149/0.055	0.8354	9.5E − 05
		SDR	0.043/0.027	0.078/0.031	0.5574	2.1E − 04
		FCR	0.026/0.01	0.012/9E − 03	0.6533	6E − 06
		FDR	0.021/9.5E − 03	8.4E − 03/8.4E − 03	0.5025	4E − 05
		TTCR	0.072/0.054	0.144/0.056	0.9828	1.2E − 04
		TTDR	0.039/0.029	0.077/0.032	0.7595	2.3E − 04
	Sec. 3 (Rus.)	SCR	0.047/0.036	0.058/0.032	0.4582	0.2666
		SDR	0.031/0.027	0.036/0.022	0.2594	0.4167
		FCR	0.022/6.8E − 03	0.024/0.013	8.6E − 03	0.543
		FDR	0.024/9.9E − 03	0.028/0.019	5.8E − 03	0.4473
		TTCR	0.038/0.037	0.05/0.031	0.3835	0.2711
		TTDR	0.026/0.029	0.031/0.021	0.1441	0.5578
	Sec. 4 (Ch.)	SCR	0.075/0.042	0.108/0.034	0.3359	4.3E − 03
		SDR	0.044/0.03	0.063/0.024	0.4117	0.0116
		FCR	0.037/0.014	0.03/0.015	0.8656	0.1044
		FDR	0.037/0.018	0.027/0.015	0.4531	0.037
		TTCR	0.067/0.04	0.098/0.033	0.3351	5.1E − 03
		TTDR	0.04/0.028	0.058/0.023	0.4335	0.0115

(continued)

Table 4. (*continued*)

			Group #1 (μ/σ)	Group #2 (μ/σ)	Test F	T
Video 3	Sec. 1 (None)	FCR	0.079/0.019	0.113/0.035	0.0147	3.7E − 04
		FDR	0.085/0.025	0.11/0.045	0.0117	0.0326
		TTCR	4.5E − 03/5.3E − 03	0.013/8.9E − 03	0.0302	6.2E − 04
		TTDR	4.2E − 03/5.6E − 03	0.011/7.7E − 03	0.1744	3.7E − 03
	Sec. 2 (Eng.)	SCR	0.13/0.064	0.124/0.042	0.0453	0.7247
		SDR	0.081/0.043	0.086/0.03	0.0908	0.6561
		FCR	0.027/0.017	0.057/0.025	0.0663	5.8E − −05
		FDR	0.022/0.015	0.051/0.023	0.0503	2.4E − 05
		TTCR	0.126/0.064	0.121/0.041	0.0342	0.7399
		TTDR	0.078/0.043	0.082/0.03	0.0836	0.6707
	Sec. 3 (Rus.)	SCR	0.042/0.027	0.055/0.021	0.1902	0.0676
		SDR	0.031/0.021	0.042/0.021	0.9962	0.0619
		FCR	0.03/0.013	0.044/0.02	0.0397	0.0129
		FDR	0.028/0.012	0.043/0.024	2.3E − 03	0.0149
		TTCR	0.046/0.028	0.06/0.022	0.2103	0.0582
		TTDR	0.034/0.023	0.0464/0.022	0.93	0.069
	Sec. 4 (Ch.)	SCR	0.083/0.045	0.107/0.035	0.2663	0.0328
		SDR	0.053/0.038	0.085/0.034	0.5851	3.2E − 03
		FCR	0.057/0.022	0.075/0.031	0.1002	0.0403
		FDR	0.052/0.02	0.067/0.032	0.0283	0.1062
		TTCR	0.085/0.045	0.11/0.035	0.3208	0.031
		TTDR	0.054/0.038	0.087/0.034	0.6436	3.1E − 03
Video 4	Sec. 1 (None)	FCR	0.081/0.023	0.076/0.029	0.2841	0.4247
		FDR	0.096/0.031	0.096/0.036	0.4799	0.9245
		TTCR	4.1E − 03/6E − 03	8.2E − 03/0.01	0.0293	0.128
		TTDR	3.1E − 03/5.6E − 03	5.8E − 03/7.7E − 03	0.1757	0.219
	Sec. 2 (Eng.)	SCR	0.092/0.065	0.154/0.064	0.8579	3.2E − 03
		SDR	0.046/0.035	0.084/0.04	0.6406	2.4E − 03
		FCR	0.054/0.029	0.028/0.025	0.5443	1.1E − 03
		FDR	0.047/0.032	0.023/0.023	0.1738	2.6E − 03
		TTCR	0.091/0.064	0.154/0.063	0.8722	2.9E − 03
		TTDR	0.046/0.035	0.084/0.04	0.6424	2.1E − 03
	Sec. 3 (Rus.)	SCR	0.036/0.038	0.053/0.031	0.25	0.1185
		SDR	0.027/0.033	0.035/0.025	0.1553	0.3945
		FCR	0.105/0.041	0.09/0.046	0.4971	0.1694
		FDR	0.108/0.042	0.099/0.059	0.0984	0.4535
		TTCR	0.034/0.038	0.049/0.029	0.1726	0.1405
		TTDR	0.025/0.032	0.032/0.023	0.1114	0.4167
	Sec. 4 (Ch.)	SCR	0.064/0.034	0.092/0.038	0.6778	0.0141
		SDR	0.041/0.027	0.054/0.028	0.9528	0.1016
		FCR	0.051/0.022	0.037/0.023	0.61	0.0269
		FDR	0.048/0.022	0.036/0.025	0.4197	0.0542
		TTCR	0.063/0.034	0.09/0.038	0.772	0.0165
		TTDR	0.04/0.028	0.054/0.027	0.8697	0.1007

Significant ($\alpha < 2.5\%$): ☐

have some familiarity of the subtitle sequence, and respond accordingly when expected content shows up.

In video 3, we discovered a substantial increase in face area attention for Group #2 subjects, resulting in significance for face indicators during English and Russian subtitle segment. Additionally, subjects in Group #1 grow higher attention in English subtitle segment. Overall statistics indicate that Group #2 users spend more time in the subtitle region that Group #1 viewers.

In English subtitle segment of video 4, Group #2 subjects focus more on subtitles while Group #1 subjects pay more attention to faces. In Chinese subtitle segment, Group #2 viewers switch attention to subtitles. This is similar to the viewing behavior for video 1.

4.4   Heat Maps

We also compute the heat-maps using OGAMA software [20] and compare the eye movement patterns between Group #1 and #2, as shown in Fig. 11.

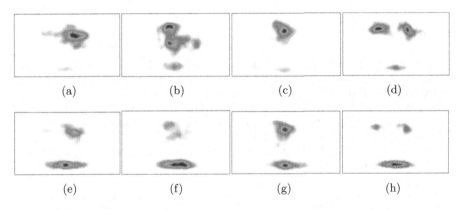

(a)   (b)   (c)   (d)

(e)   (f)   (g)   (h)

Fig. 11. Heat maps analyzed by OGAMA [20]. Group #1: (a)(b)(c)(d); Group #2: (e)(f)(g)(h); Video #1: (a)(e); Video #2: (b)(f); Video #3: (c)(g); Video #4: (d)(h)

The macro-level inspection shown above matches the analysis given in Sect. 4.1. To reiterate, Group #1 users focus more on faces, while Group #2 users pay more attention to the subtitles (except for video 3). It should be noted that the conclusions are drawn in accordance with the type of video chosen for the experiment. Area of attention may change if other types of video are employed.

5   Conclusion

We record eye movement data and study the viewing behavior of 45 participants when videos containing different combinations of subtitle and dubbing languages are presented.

Experimental results indicate that auditory language is the most critical factor. Subjects in Group #1 (English, Chinese, English and Chinese) have a higher tendency to focus on the face area. Subjects in Group #2 (Chinese, English, Chinese and English) appear to read the subtitles more often. The initial behavior seems to determine the viewing pattern subsequently.

For subjects in Group #2, preference for caption is clearly observed than those in Group #1. This habitual preference continues in follow-up movies, resulting in an immersion phenomenon. We also observe that when unfamiliar texts appear (Russian), the subjects exhibit 'escaping' behavior by avoiding the text region. It is worth noting that the video at the beginning of Group #2 is the native language of the testee, and the result shows that the subject develops preferences to read subtitles. Therefore, we can speculate that Taiwanese audiences exhibit a certain degree of dependence on subtitles.

In future work, more types of AOI can be extracted using deep learning approaches such as YOLO v4. The type, location, and area of these AOIs can be analyzed in accordance with the eye movement data to dynamically monitor what attracts the spectators' eyeballs.

References

1. National Communications Commission. Radio and Television Act (2020). Accessed 1 Dec 2020. https://law.moj.gov.tw/Eng/LawClass/LawAll.aspx?PCode=P0050001
2. Maddieson, I.: Tone. In: Dryer, M.S., Haspelmath, M. (eds.) The World Atlas of Language Structures Online. Max Planck Institute for Evolutionary Anthropology, Leipzig (2013)
3. Peng, C.-F., Liao, W.-H.: Evaluation of interactive data visualization tools based on gaze and mouse tracking. In: 2016 IEEE International Symposium on Multimedia (ISM), pp. 431–434. IEEE (2016)
4. Liao, W.-H., Chang, C.-W., Wu, Y.-C.: Classification of reading patterns based on gaze information. In: 2017 IEEE International Symposium on Multimedia (ISM), pp. 595–600. IEEE (2017)
5. Tobii Technology AB. An introduction to Tobii EyeX (2020). Accessed 1 Dec 2020. https://developer.tobii.com/an-introduction-to-the-tobii-eyex-sdk/
6. O'Bryan, K.G.: Eye movements as an index of television viewing strategies (1975)
7. Canny, J.: A computational approach to edge detection. IEEE Trans. Pattern Anal. Mach. Intell. **6**, 679–698 (1986)
8. He, K., Zhang, X., Ren, S., Sun, J.: Deep residual learning for image recognition. In: Proceedings of the IEEE Conference on Computer Vision and Pattern Recognition, pp. 770–778 (2016)
9. LLC. TED Conferences. TED: Ideas worth spreading (2020). Accessed 1 Dec 2020. https://www.ted.com/
10. Chi, T.: Rapid prototyping Google Glass (2013). Accessed 1 Dec 2020. https://ed.ted.com/lessons/rapid-prototyping-google-glass-tom-chi
11. Dekofsky, J.: Is math discovered or invented? (2014). Accessed 1 Dec 2020. https://www.ted.com/talks/jeff_dekofsky_is_math_discovered_or_invented
12. Horn, B.K.P., Schunck, B.G.: Determining optical flow. Artif. Intell. **17**(1–3), 185–203 (1981)

13. Ordikhani-Seyedlar, M.: What happens in your brain when you pay attention? (2017). Accessed 1 Dec 2020. https://www.ted.com/talks/mehdi_ordikhani_seyedlar_what_happens_in_your_brain_when_you_pay_attention/
14. Cofer, A.: A plan to recycle the unrecyclable (2016). Accessed 1 Dec 2020. https://www.ted.com/talks/ashton_cofer_a_plan_to_recycle_the_unrecyclable
15. Harada, C.: How I teach kids to love science? (2015). Accessed 1 Dec 2020. https://www.ted.com/talks/cesar_harada_how_i_teach_kids_to_love_science
16. Bu, L.: How books can open your mind? (2013). Accessed 1 Dec 2020. https://www.ted.com/talks/lisa_bu_how_books_can_open_your_mind
17. Peleg, D.: Forget shopping. Soon you'll download your new clothes (2015). Accessed 1 Dec 2020. https://www.ted.com/talks/danit_peleg_forget_shopping_soon_you_ll_download_your_new_clothes
18. Mullins, A.: My 12 pairs of legs (2009). Accessed 1 Dec 2020. https://www.ted.com/talks/aimee_mullins_my_12_pairs_of_legs
19. Harvey, D.: Protecting twitter users (sometimes from themselves) (2014). Accessed 1 Dec 2020. https://www.ted.com/talks/del_harvey_protecting_twitter_users_sometimes_from_themselves
20. OGAMA. Open gaze and mouse analyzer (2016). Accessed 4 Feb 2016. http://www.ogama.net/sites/default/files/pdf/OGAMA-DescriptionV25.pdf

KLSI Methods for Human Simultaneous Interpretation and Towards Building a Simultaneous Machine Translation System Reflecting the KLSI Methods

Kevin Lin[1] and Ming Qian[2]($\boxtimes$)

[1] KL Communications Ltd., Link House, 140 Broadway, Surbiton KT6 7HT, UK
kevin@klcom.uk
[2] Pathfinders Translation and Interpretation Research, Cary, NC, USA
qianmi@pathfinders-transinterp.com

Abstract. Simultaneous machine translation aims to maintain translation quality while minimizing the delay between reading input and incrementally producing the output. KL Simultaneous Interpreting (KLSI) is a set of methods developed to deliver non-revisable translation from English into Chinese with one second as the benchmark latency. To achieve this, it trains the human brain to stop thinking and execute commands instead. It has developed a range of formulaic techniques to be applied mechanically. This paper presents some of the key features and techniques of KLSI and explores its implications for machine simultaneous interpreting. The techniques include convergence, the concept of interpreting within three words heard at any given moment in time, co-texting, defaulting, and sequential translation techniques such as repeat, replace, reverse logic, and SAI (Skip, Add, Insert). Multiple English-to-Chinese examples and video recordings are listed in the paper to illustrate the KLSI features and techniques. In the second part of this paper, we describe computational methods related to KLMI techniques: wait-k policy with and without anticipation; word-based and phrase-based alignment and mapping; constrained context for neural machine translation. Commercial machine translation requires at least several gigabytes (or millions of words) of language pair documents for the training. It is not realistic to obtain enough parallel texts in English and Chinese reflecting KLSI techniques for training a neural machine translation system. However, a lot of parallel texts exist that do not reflect KLSI techniques. We propose to use a rule-based approach to modify available parallel texts using KLSI rules to generate a large enough KLSI-based corpus. The main contribution of this paper is to propose a novel rule-based approach—with the rules reflecting human interpretation traits—to revise training corpus to enable short latency and non-revisable machine translation.

Keywords: Simultaneous interpreting · Simultaneous machine translation · Wait-k model · Latency metric · Neural machine translation · Short latency · Non-revisable · Rule-based training corpus revision

© Springer Nature Switzerland AG 2021
H. Degen and S. Ntoa (Eds.): HCII 2021, LNAI 12797, pp. 394–412, 2021.
https://doi.org/10.1007/978-3-030-77772-2_26

1   Introduction

To rival the best in human simultaneous interpreting, machine translation needs to deliver non-revisable translation with one second latency. Non-revisability is minimum requirement as one cannot listen to what is being said at the same time of listening to a revision of what has just been said.

The one second latency is the highest level of human simultaneous interpreting. It is considered by many even in the interpreting profession to be impossible to sustain. Many interpreters working from English into Chinese routinely lag behind words heard three seconds and more.

KL Simultaneous Interpreting (henceforth KLSI) however is one of a kind. It has been developed by KL Communications Limited based in London – www.klcom.uk. Its Founder Managing Director is Dr Kevin Lin OBE, the Official Interpreter of the UK Foreign Office for twenty years and a co-writer of this paper.

KLSI is the culmination of two decades of professional practice with several thousand days of simultaneous interpreting successfully delivered[1]. Subject matters interpreted range from business, finance, education, health care, to law, technology, bioscience and agriculture among many others.

Over the past two decades, it has trained several hundred Chinese graduates into frontline professional interpreters through collaborative master degree programmes at universities[2] and vocational training courses[3].

KLSI sets one second as the benchmark latency. To achieve that, it trains the human brain to stop thinking and execute commands instead. It has developed a range of formulaic techniques to be applied mechanically.

As the techniques are all word-based, KLSI has reduced simultaneous interpreting to word resolution in the order they are heard within one second. As such, KLSI is more akin to computing than translation.

This paper will present some of the key features and techniques of KLSI and explore their implications for machine simultaneous interpreting. Four video recordings from the World Economic Forum at Davos (henceforth WEF) will be referred to as evidence as well as illustration of the points made in this paper.

The first recording is an interview of Jack Ma the Founder of Alibaba. It is in English and available at: https://www.weforum.org/videos/davos-2015-an-insight-an-idea-with-jack-ma. The second recording is the same interview but in Chinese Mandarin (henceforth Chinese) delivered by KLSI. It is available on Youtube at: https://www.you tube.com/watch?v=WFxdLf2am1o.

The third recording is an interview of Richard Liu the Founder of JD.com available from the official website of WEF at: https://www.weforum.org/events/world-economic-forum-annual-meeting-2018/sessions/an-insight-an-idea-with-richard-liu.

The final recording is a speech by Hollywood movie star Leonardo DiCaprio at an WEF award ceremony. It is available on Youtube at: https://www.youtube.com/watch?v=5lSjR64A1tU.

Although only four video recordings, they are sufficiently representative of the kind of interpreter mediated verbal communication this paper focuses on. Because of that, they are robust evidence of the points made in this paper.

2   KLMI Methodologies for English-to-Chinese Simultaneous Interpreting

2.1   The One Second Latency

The benchmark latency is set at one second because that is the most common and natural time lag in speech as well as conversation. As long as interpreting maintains the one second latency, it will sound as if there were no gaps between when a word is heard in the source language and when interpreted version is heard in the target language.

In the English version of the interview of Jack Ma, for example, when the interviewer finished his question at 2'52", Jack Ma replied at 2'53". Although there was a gap of one second, it sounded as if his reply followed the interviewer question without a pause.

In the Richard Liu interview, he listened through KLSI via his headset and answered questions approximately two seconds after each question ended. The extra second was caused by the interpreter having to switch the output channel between English and Chinese on the equipment used. Without the delay caused by the equipment, the latency of interpreting itself would have been just one second.

In the Chinese version of the Jack Ma interview, the one second latency by KLSI can be heard quite distinctly. Even if one does not understand Chinese, one gets an idea of how closely the Chinese version follows the English version audible underneath it.

In neither interview referred to above, the interpreter's performance was perfect. Simultaneous interpreting is like playing a live concert without having rehearsed. There are factors beyond the control of the interpreter but impacting the interpreter's performance.

2.2   Convergence

Central to the one second latency is the concept of convergence. It describes the reality that many English words require fewer translations than their number of definitions. The issue of multiple meaning words is therefore overstated.

Take the word 'for' for example. The Oxford online dictionary provides a total of thirteen definitions. They are however translated into only three terms in Chinese with a fourth as null translation – when the word should not be translated.

In the following, bold and underline words each represents one of the three versions of translation. The null translation is indicated by the word 'Omitted'.

1   In support of – They fought for Napoleon. 他们**为**拿破仑而战.
2   In respect of – She is responsible for the team. 她**为**这个团队负责.
3   On behalf of – I got a present for you. 我<u>为</u>你获得一个礼品.
4   Employed by – She is a tutor for Oxford University. 她是<u>为</u>牛津大学工作的辅导老师.
5   As a purpose of – Networks for the exchange of information. **为**交换信息的网络.
6   As reason of – Aileen is proud of her family for their support. 爱玲<u>为</u>家人的支持感到自豪.
7   As a destination – They are leaving for London. 他们正在去伦敦. (Omitted)

8   In place of – They have swapped these two bottles for that one. 他们<u>用</u>这两个瓶子换那个.

9   Charged as – Copies are available for £1.20. <u>用</u>1.2英镑可以得到拷贝.

10 In relation to the expected norm of – She was tall for her age. 按年龄, 她很高.

11 Indicating the length of – He was jailed for twelve years. 他被监禁12年. (Omitted)

12 Indicating the extend of – He crawled for 300 yards. 他爬了300米. (Omitted)

13 Indicating an occasion in the series of - The camera failed for the third time. 照相机第三次失败了. (Omitted)

KLSI exploits convergence. For one example, there are two equally viable translations in Chinese for the word 'exciting' – 激动人心 and 令人振奋. There is practically no difference in the meaning of the two options. KLSI has designated the latter 令人振奋 as the choice of convergence as soon as 'exciting' is heard.

For another example, KLSI has assigned a default translation to a large number of English words. The default for 'visitor' is 到访者 instead of the potential range of 来访者, 参观者, 访客 and 游客.

How KLSI achieves accuracy will transpire in the following sections.

2.3 Within Three Words

Central to one second latency is also the concept of interpreting within three words heard at any given moment in time—a hallmark of KLSI. It requires the translation of word A when hearing no more than words A, B and C. There are specific commands for fulfilling the requirement:

Command 1, if A can be translated without any help from B or C, it is translated on its own without taking into account of B or C. Once A has been translated, the wordcount re-starts.

In the following, 'we' is translated as soon as heard.

*We **have been** considering the importance of the project.*

我们

After translation, the word count re-starts so that 'have' becomes A. 'been' and 'considering' become B and C:

*We **have been considering** the importance of the project.*

我们

Command 2, if A cannot be translated on its own, B will be taken into account. In the above sentence, 'have' and 'been' are translatable together without taking into account of 'considering':

*We **<u>have been</u>** considering the importance of the project.*

我们一直在

Command 3, if B is still felt insufficient for translation, C will be taken into account:

*We **<u>have been considering</u>** the importance of the project.*

我们一直在考虑

Command 4, if either A or B or C is a functional word without its own meaning such as the article 'the' or preposition 'of', it is not counted as one of the three.

*We have been considering **the importance of the project** .*

我们一直在考虑   ***项目的重要性***

A word is translatable on its own when its translation is not affected by any words that follow. "We" is always translated as "we" is not affected by whichever word follows it. A word is also translatable on its own when it has become clear what it should be translated into up to that point.

In the example here, the moment the word "policy" is heard, it can be translated into 保单 - an insurance contract. There is no need to hear the next word.

*Our insurance company phoned and told me that my **policy** was due for renewal.*

The four Commands cover all eventualities of the English language. There are no exceptions.

Within three words is underpinned by specially designed linguistic techniques to ensure accuracy. The following sections will present some of these techniques used by KLSI.

2.4 Co-texting

A co-text in KLSI refers to a word or words that determine the translation of another or others. Given any English word, there is only a relatively very small number of very stable co-text relationships. The most typical co-text are words that appear closely to each other. In the following example:

Did you wipe the table?

The word 'wipe' is the co-text of 'table'. Given that, table is translated into a piece of furniture instead of a set of facts or figures systematically displayed. Sometimes co-text is before the word to be translated. Sometimes after and within two words. It can also be both before and after as in the following:

Did you wipe the table thoroughly?

Co-texting reduces an otherwise multi-option question into a binary decision. The interpreter either translates into the co-text relationship identified or in the absence of co-text, into the default version of the word in question. The technique of Defaulting will be presented in the next section.

The example of 'for' is a case in point. The word is to be omitted in translation if it is followed by the co-text of a noun denoting a location or measurement or duration or the words that refer to time:

As a destination – They are leaving for London.
他们正在去伦敦。（Omitted）
Indicating the length of – He was jailed for twelve years.
他被监禁12年。（Omitted）
Indicating the extend of – He crawled for 300 yards.
他爬了300米。（Omitted）
Indicating an occasion in the series of - The camera failed for the third time.
照相机第三次失败了。（Omitted）

The word should be translated into 用 in Chinese if it is followed by the co-text of an amount of money identifiable by the currency sign and/or a numeric number.

In place of – They have swapped these two bottles for that one.
他们用这两个瓶子换那个瓶子。
Charged as – Copies are available for £1.20.
用1.2英镑可以得到拷贝。

The word should be translated into 按 in Chinese if it is proceeded by the co-text of 'be + adjective' and followed by a reference to age or a numeric number.

In relation to the expected norm of – She was tall for her age.
按年龄，她很高。

If none of the above co-text is present, the word 'for' needs to be defaulted into the Chinese word 为.

Given any English words, co-text is finite, exhaustive and formulaic as a result of convergence. No thinking is required in its determination. One of the key properties of co-text is its insensitivity to sentences. Co-text is identified from words heard earlier. There is no reference to sentence demarcation. Here is an illustration–

Our premiums are competitive. This makes our policies attractive.

It is the co-text of 'premium' that decides for the interpreter that the word 'policies' refers to insurance contracts. Whether 'premium' is in the current or previous sentence is irrelevant. The speaker may well have said instead –

Our premiums are competitive, and they make our policies attractive.

The fact that co-text is found across sentences in the first example but within the same sentence in the second makes no difference to KLSI.

2.5 Defaulting

Defaulting is key to the resolution by KLSI of multiple definition words. It is the catch-all translation when there is no co-text. The version used in such translation is the default. It is often the result of convergence discussed in the previous section.

With a large glossary of defaults built over years of interpreting practice, a KLSI trained interpreter does not think about how to translate a word. Instead, a binary decision is made. If there is co-text, co-text it. If there is not, default it. It is a closed loop.

If a speaker says "The letter is about increasing the premium of our policy", there is a co-text of 'premium' for 'our policy'. 'Policy' needs to be translated as an insurance

contract. If however, the speaker says "we do need to consider our policy" and the interpreter cannot recall an insurance co-text from words said earlier, 'policy' will be defaulted into "a course of action or principle".

Defaulting also helps resolve synonyms in English – a challenge in simultaneous interpreting. There is simply no time for consideration when synonyms are used together as in the following illustration:

*It is **important** that the **significant** increase is taken into account as an **essential** part of our planning.*

The words in bold are difficult to distinguish in simultaneous interpreting from English into Chinese unless defaulted. Many apparently difficult to distinguish synonyms can be resolved effectively and efficiently through defaulting–

- main, major, important, significant, critical, vital, essential, fundamental
- amazing, fantastic, wonderful, great
- hundreds of such sets in the English language

To sum up the discussion so far, the One Second Latency and Within Three Words form the framework of KLSI. The framework is underpinned by Convergence, Co-texting and Defaulting.

The next two sections will present how KLSI is performed in real interpreting.

2.6 Sequential Translation

KLSI interprets English words heard into Chinese in the order they are heard and within three words at any given moment in time maintaining one second latency throughout. This process is enabled by a set of linguistic techniques. Four of the most frequently used ones are –

- Repeat
- Replace
- Reverse logic
- SAI (Skip, Add, Insert).

All four techniques are formulaic to be applied mechanically. Their execution is triggered by pre-set commands. There is no thinking required.

Repeat. Repeat is an effective and mechanical way of resolving post-modifiers as well as a range of English expressions or sentence patterns. Post-modifiers are English words or phrases or clauses that modify a word or a phrase and are placed after the word or phrase they modify. Hence the name.

The following provides one example of each of the three main types of post-modifiers underlined for clarity. The bold words are the part being modified and arrows indicate the modifying relationship.

They **finished** the work quickly.

The **children** with flowers in their hands are from London.

The **children** that we saw earlier are now here.

In Chinese, modifiers are usually placed before the word or phrase they modify. In that sense, they are pre-modifiers. In conventional translation, the post modifiers in English need to be brought forward and translated into pre-modifiers in Chinese.

Instead, KLSI repeats the word or phrase being modified. The examples earlier will be translated into the following - the second word in bold is the repeat.

They **finished** the work, quickly **finished**.
他们**结束**工作很快地**结束**。

The **children**, in their hands with flowers **children** are from London.
孩子们手里拿着花的**孩子们**来自伦敦。

The **children**, that we saw earlier **children** are now here.
孩子们我们稍早看到的**孩子们**现在在这里。

In addition to post-modifiers, the technique of Repeat can be used to resolve a range of linguistic issues. The Chinese equivalent of 'there is/are' for example comes after the location. Conventional translation would require reversing the order and placing 'there is/are' after the location. With Repeat, KLSI translates 'there is/are' as it is heard and repeat it after the location is heard.

> **There is** a tree in the garden.
> 有棵树，花园里**有**。

If 'the garden' is followed by its own post-modifier, the same technique of Repeat applies –

> There is a tree in the **garden** of my house.
> 有棵树，**花园**有，我家的**花园**。

The string of 'how + adjective' as in 'how important…' will also require restructuring in Chinese. That too can be avoided by Repeat –

> This shows **how important** a healthy meal is.
> 这显示出**多么重要**健康的餐食**多么重要**。

There is significantly more use of the Passive Voice in English than in Chinese. Conventional translation would require restructuring. Repeat avoids that.

> The house was **designed** by a leading architect.
> 房子**被设计**，被一位领军的设计师**设计**。

The technique of Repeat is triggered by key words. They are finite, exhaustive, and no more than about twenty. When hearing one, KLSI repeats the word or phrase in question. The execution is formulaic and mechanical. No thinking is required.

Replace.   Replace means to replace a preposition with a verb. This enables the interpreter to repeat the verb at the beginning of a preposition clause instead of at its end as is the case in Repeat. It improves the quality of output and saves time.

There are several applications of Replace. The first is the default - replace the preposition with the same verb as used earlier. In the illustrations provided here, there are two versions in Chinese for comparison. The first version uses Repeat. The second version uses Replace.

> *We'll take a holiday in Greece.*
> 我们将去度假，在希腊**度假**。
> 我们将去度假，**去**希腊。

Here is another –

> *We submitted our application to the city council.*
> 我们提交我们的申请，向市政理事会**提交**。
> 我们提交我们的申请，**提交**市政理事会。

A second application of Replace is preferred when the verb has to do with quantity or numeric numbers – speed, growth or measurement. In this application, the preposition is replaced by the Chinese word denoting 'reaching' – 达到.

> *The high speed train is travelling at 350 kilometres per hour.*
> 高铁正在行驶，以350公里每小时**行驶**。
> 高铁正在行驶，**达到**350公里每小时。
> *Our business is growing at 20% per year.*
> 我们的业务正在增长，以每年20%**增长**。
> 我们的业务正在增长，**达到**每年20%。

Whilst the technique of Repeat works in practically all situations, Replace works in only some. It nevertheless significantly reduces repetition given any length of speech.

Reverse Logic.   This refers to a formula that translates an English word or phrase into a version in Chinese that is its opposite in logic. This eliminates the need for restructuring required in conventional translation.

When 'before' or 'after' leads a clause in the second half of a sentence, conventional translation would require the clause to be brought forward to the first half of the sentence. Having already translated the first half of the sentence and when hearing 'before' or 'after' leading a clause, KLSI will translate 'before' into 'after which' – 之前 and 'after' into 'before which' – 之后. Here is one example of each.

> *They had bought a chair **before** they found the desk.*
> 他们买了椅子**之后**发现了桌子。
> *They bought a chair **after** they had found the desk.*
> 他们买了椅子**此前**发现了桌子。

The technique of Reversing the Logic is used with less than a handful of specific words. Another one is 'never' and in a small number of cases. When it is followed immediately by either 'stop' or 'seize' or 'pause', KLSI will translate the two words together into 'always'.

Thereafter, if it is followed by 'to be' plus an adjective, the adjective is translated after reversing the logic:

> *They never stop/cease/pause to be amazed at the speed.*
> 他们总是惊讶速度。

If 'never stop' is followed by 'to do something', KLSI will insert the Chinese word 'not' 不 and then translate 'do something':

> *They never stop to think about the future.*
> 他们总是不想未来。

The technique of Reverse Logic is formulaic. It does not require thinking. As soon as the trigger words such as 'never stop' are heard, the technique is applied.

SAI. SAI stands for Skip, Add and Insert. Being the most sophisticated technique in KLSI, it resolves many otherwise sentence level issues by dealing with words only.

SAI can be used to resolve the issue of a modifying clause instead of Repeat to achieve better quality output. In an earlier example -

*The **children** that we saw earlier are now here.*

When hearing 'children', the interpreter will have heard '...that we...'. It is clear by now that this is a post-modifying clause. The interpreter skips 'children' to interpret the modifying clause. At the end of the clause, the interpreter adds the Chinese word 的 to indicate this is a modifier before inserting the word 'children'.

> *The **children** that we saw earlier are now here.*
> **我们稍早看到的孩子们**现在在这里

If the post-modifier is an Advert modifying a verb and is within three words of the verb, SAI can also be used. There is no need to add any words. KLSI will skip the verb initially but insert it after the post-modifier -

> *They finished it **quickly**.*
> 他们**很快地**结束了它。

Comparatives. When a sentence contains a comparison, conventional translation will require re-structuring. SAI turns the sentence into a narrative statement instead -

> *We have **more** market share **than** the rest of the brands combined.*
> 我们有**很多**市场份额**超过**其余品牌总和。

The formula is such that as soon as 'more ...' is heard, it is skipped and the word 'very' is added. Subsequent words are translated as they are. When reaching 'than', the skipped word 'more' is inserted. The formula is triggered by the word 'more' and executed mechanically. No thinking is required.

The same methodology is applied to the translation of '…less…than…'. If 'less' is followed by an adjective or adverb –

*We are **less** lucky **than** our competitors.*
*我们**不那么**幸运**不如**我们的竞争对手。*

If '…less…than…' is followed by a noun or 'the' or 'a' that proceed a noun -

*We have **less** market share **than** our competitors.*
*我们有**比较少的**市场份额**不如**我们的竞争对手。*

The linguistic techniques presented so far is not a complete list from KLSI but a selected few. They are meant to demonstrate how some of the most impossible hurdles in simultaneous interpreting are resolved through mechanical application of formulaic techniques. The next section will demonstrate how KLSI sequentially translates a chunk of speech with one second latency.

2.7   KLSI Demonstration

A 56" video recording of from the World Economic Forum at Davos in 2016 is selected for the demonstration. The video is available on Youtube[1]. The transcript is here (17" to 1'13" of the video) –

There is no doubt to the world scientific community this is a direct result of human activity and the effects of climate change will become astronomically worse in the future. Last week President Obama told those who continue to deny the irrefutable science behind climate change they will find themselves increasingly lonely. But studies also show us that those who deny the reality of climate change will also find themselves less economically successful. We simply cannot afford to allow the corporate greed of the coal, oil and gas industries to determine the future of humanity. Those entities with a financial interest in preserving this destructive system had denied and even covered up the evidence of our changing climate.

It is essential to bear in mind that listening to live simultaneous interpreting is very different from reading its transcription in text. What sounds coherent and smooth in listening may appear to be awkward when read in a transcript. Therefore, to get an idea of what it would have sounded like, one needs to read the transcript out loud.

Given the Chinese language in text looks far more compact than its English equivalent with the same font size and line spacing, hyphens are used to connect Chinese words that are said continuously in the following presentation. This is to avoid the misimpression that there are gaps between those words. Real pauses in interpreting are shown by blank spaces.

For illustration, Repeat is highlighted in bold SimHei. SAI with underlining. 'Less…than…' is in italics. Replace in bold italics. Words that are not highlighted are interpreted without any particular techniques applied. The breaking of a sentence is due to the limited width of a page. The speech is continuous.

---

[1] https://www.youtube.com/watch?v=5lSjR64A1tU.

There is no doubt to the world scientific community…
没有疑问----------对于国际---科学----社区**没有疑问**

this is a direct result of human activity…
--------这是直接的结果--------人类活动**的结果**

and the effects of climate change…
----------而且---------------------气候变化的影响

will become astronomically worse in the future.
--------将-----变得-------------------极端恶化，在未来**恶化**

Last week President Obama told those…
-------------上周---------------奥巴马总统告诉那些

who continue to deny the irrefutable science behind climate change…
----------------继续---否认-----------不可否认的科学在气候-------变化背后**的科学**
的人

they will find themselves increasingly lonely.
--------他们将发现--------自己----------越来越孤独。

But studies also show us…
-----但是研究也---------向我们显示

that those who deny the reality of climate change…
--------------那些-----否认---------------------------气候变化现实的人

will also find themselves less economically successful.
-----------还将发现--------他们自己----------经济上*不那么成功*。

We simply cannot afford to allow the corporate greed…
-----我们根本------无法担负得起允许----------------公司贪婪

of the coal, oil and gas industries to determine the future of humanity.
--------------煤、石油、天然气行业**的贪婪**----决定--------------------人类的未来。

Those entities with a financial interest…
-----------------那些实体-----*有*财务---利益

in preserving this destructive system…
-----------------*要*保存---------这个破坏性体系**的利益**

had denied and even covered up the evidence of our changing climate.
--------------他们否认并且甚至-----掩盖------证据---我们-----变化中的气候**的证据**

3 Computational Linguistics Methods Corresponding to KLSI Methods and KLSI-Based Training Corpus Preparation to Enable Short Latency and Non-revisable Simultaneous Machine Translation

Computational Linguistics answers linguistic questions using computational tools. Section 3.1 describes computational methods related to KLMI techniques.

Simultaneous machine translation aims to maintain translation quality while minimizing the delay between reading input and incrementally producing the output. Section 3.2 discusses a rule-based approach to establishing a customized training corpus reflecting the KLSI sequential translation techniques (e.g. repeat, replace, reverse logic, and SAI) for a simultaneous machine translation system.

3.1 Computational Linguistics Methods

In this section, we identify computational methods related to KLMI techniques. Some of these methods have existed for a long time and can be used to achieve similar performance. Some methods can be improved by applying KLMI principles.

Wait-k Policy for Simultaneous Machine Translation. The mainstream machine translation platforms (e.g. Google Translator or Bing Microsoft Translator) aim towards written text translation, meaning they are only suitable for full-sentence translation and do not work well in simultaneous translation scenarios where the input text is made incrementally available. Another major challenge is the diverging word order between the source and target languages (e.g. post modifiers in English versus pre-modifiers in Chinese).

Reference [2] summarized several computational approaches to address the waiting and latency for simultaneous translation:

(1) Baseline: waiting until the whole sentence finishes which is not practical for simultaneous translation.
(2) Wait as long as it is necessary: for example, wait until the post-modifier is spoken.
(3) Wait-k policy with anticipation: translates concurrently with the source sentence, but always k words behind. It predicts the word that has not been said based on patterns learned from a training set. If the prediction is wrong, this approach produces nonsensical translation.
(4) Wait-k policy without anticipation: translates concurrently with the source sentence, but always k words behind. Without prediction, this approach produces nonsensical translation when there is a diverging word order problem.

KLMI emphasizes one-second latency and at most being three words behind. That is a Wait-3 policy. KLMI, however, avoids anticipation/prediction, instead relying on other tactics such as repeat, replace, reverse logic, and SAI. Guessing in the sense of anticipating is likely to be wrong if words are said out of context—in reality, words being said out of context is a typical feature of live speech – conversations. Also, in live speech, speakers often say a word and then explain what it means—Context is provided afterwards instead of beforehand. Guessing is likely to be wrong if a speaker uses language differently from most other people. Guessing is statistical. 1% in statistics can be 100% error when given a speech.

Table 1 compares the time lag between using and not using KLSI—the approach of waiting as long as necessary without guessing. Each row represents what non-KLSI tends to require before translation can begin.

Timing is based on a speaking speed of three words per second. Therefore, a latency of 0.33" means translation is just one word behind. A latency of 1" means translation is three words behind. For example, Row 2 shows non-KLSI needs to wait for 10 words before translation.

Table 2 is an as-being-translated view of Table 1. Each row is a decision to 'translate' instead of 'waiting' for more. The KLSI techniques used are provided in brackets when applied.

The target paragraph is a speech by a senior executive of the Visa credit card company about China: *"Throughout the financial crisis China has provided leadership in the region and its continued economic stability will be critical to the region and indeed the world. Let me begin today by reflecting on China's recent economic reforms before moving on to discuss the role of consumer spending in helping China meet its economic goals. No matter how often I revisit Shanghai, I never cease to be amazed with the level of a construction, changing the renewal that is occurring here. Visa is the clear industry leader. Visa cards have more market share than all other car brands combined."*

This example illustrates the benefits of KLSI over non-KLSI in terms of latency minimization: on average, the KLSI reduces the latency by 75% by applying various techniques and avoiding guessing.

Table 1. Time lag comparison between using and not using KLSI.

Row	Speech	KLSI	Non-KLSI	Faster by
1	Throughout the financial crisis	0.33"	1.33"	1"
2	China has provided leadership in the region and	1"	2.33"	1.33"
3	its continued economic stability	0.33"	1.33"	1"
4	will be critical to the region and indeed the world.	1"	3.33"	2.33"
5	Let me begin today by reflecting on China's recent economic reforms	1"	4"	3"
6	before moving onto discuss the role of consumer spending in helping China meet its economic goals.	1"	4"	3"
7	No matter how often I revisit Shanghai	0.33"	2"	1.66"
8	I never cease to be amazed with the level of construction, change and renewal that is occurring here.	1.33"	5"	3.66"
9	Visa is the clear industry leader	0.33"	1"	0.66"
10	visa cards have more market share than all other card brands combined.	0.33"	4"	3.66"

Convergence. The concept of convergence in KLSI states that many English words can be mapped to several possible Chinese words. A similar concept was well-known in rule-based and statistical machine translation [3, 4]: a word in a source language can be mapped to several possible words/phrases in a target language, and probabilities can be associated with every possible mapping. This applies to both word-based and phrase-based systems.

Co-texting. Co-text is conventionalized, idiosyncratic combinations of words that appear closely to each other. In the field of phraseology such word combinations are sometimes called phrasemes, while the computational linguistics community uses the term multiword expressions for them[5]—both can be found using statistical methods from corpora—for phrase-based machine translation [4].

Table 2. Time lag comparison between using and not using KLSI (as-being-translated view).

KLSI	Non-KLSI
Throughout	Throughout the financial crisis
the financial crisis	
China	China has provided leadership in the region
has provided	
leadership	
in the region (Replace)	
and	and its continued economic stability
it's	
continued	
economic	
stability	
will be critical	will be critical to the region and indeed the world.
to the region (Replace)	
and indeed	
the world.	
Let me begin	Let me begin today by reflecting on China's recent economic reform
today	
by reflecting on	
China's	
recent	
economic	
reform	
before (Reverse Logic)	before moving on to discuss
moving on	
to discuss	
the role of (Skip)	the role of consumer spending in helping China meet its economic goals.
consumer spending	
in helping	
China	
meet	
it's	
Economic	
goals. (Add and Insert)	

(continued)

Table 2. (*continued*)

KLSI	Non-KLSI
No matter	No matter how often I revisit Shanghai,
how often	
I	
revisit	
Shanghai,	
I	I never cease to be amazed with the construction, change and renewal that is occurring here.
never cease (Reverse Logic)	
to be amazed	
with the construction (Replace)	
change	
and	
renewal	
that is occurring here. (Repeat)	
Visa	Visa is the industry leader.
is	
the industry leader.	
Visa	Visa cards have more market share than all other car brands combined.
cards	
have	
more market (Comparative)	
share	
than	
all other	
Card	
brands	
Combined.	

However, there are three key differences between phraseme and co-text. The first one is that whilst co-text is an issue between two languages, phraseme is always within just one given language. The second difference is that phraseme is about the sum meaning of its component words whereas co-text is about how one word is to be understood given the presence of another. The meaning of a phraseme is either different from the meaning of its component words (they turned up) or more detailed or stronger than the meaning of its components (infinite laugh). Co-text does not have a meaning of its own. The final difference is that words forming a phraseme must occur together whereas words forming a co-text is frequently across several words and often across sentences.

For neural machine translation [6], the representation of the source sentence and the representation of the partially generated target sentence (translation) at each position are referred to as source context and target context [7]. Due to the unique characteristics of simultaneous translation, co-text can be regarded as the constrained context containing only the few words to the immediate left of the word for translation and no more than three words to its right (Wait-3). For simultaneous machine translation, the limitation of only a few words to the immediate left can be relaxed because, unlike humans, a machine has no cognitive limitations. On the other hand, the Wait-3 rule to the immediate right still needs to be maintained. The generation of a target word is determined jointly by the constrained source context and target context.

3.2 Customized Training Corpus for Simultaneous Machine Translation Reflecting KLSI Sequential Translation Techniques

For previous work in simultaneous translation, whether they were full-sentence based [8, 9] or following a prefix-to-prefix framework [2], they all relied on word prediction to overcome word-order difference (Wait-k policy with anticipation). But as we pointed out before, the prediction/guess method is error prone. A better approach is to adopt KLSI's Wait-k policy supported by sequential translation techniques (repeat, replace, reverse logic, and SAI). By adopting these techniques, we can avoid error-prone predictions.

For data-driven machine translation (statistics or neural MT), there has been some works on domain customization. Such customization includes terminology, formatting and some syntactic idiosyncrasies [10]. To incorporate KLSI's sequential translation techniques, we need to have a customized training corpus to reflect these techniques.

Commercial machine translation requires at least several gigabytes (or millions) of language pair documents for the training. It is not realistic to obtain enough parallel texts in English and Chinese reflecting KLSI techniques in order to train a neural machine translation system. On the other hand, a lot of parallel texts exist without reflecting KLSI techniques. We propose to use a rule-based approach to modify available parallel texts using KLSI rules in order to generate a large enough KLSI-based corpus (Fig. 1). Table 3 illustrates some of the conditions to apply the rules.

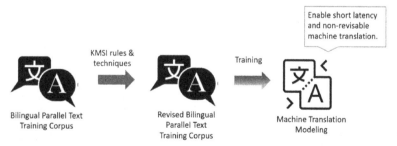

Fig. 1. KLSI Rule-based approach to modify available parallel training corpus in order to generate a revised KLSI-based corpus and enable short-latency and non-revisable machine translation.

Table 3. Conditions to apply sequential translation techniques.

KLSI sequential translation techniques	Conditions to apply the rules
Repeat	Key word detection: prepositions
Replace	Keyword detection: 1. Verbs and prepositions 2. Verbs associated with quantity or numeric numbers – speed, growth or measurement
Reverse logic	Keyword detection: 'before' and 'after' leading a clause, or 'never' when it is followed immediately by either 'stop' or 'seize' or 'pause'
SAI	Keyword detection 1. That/who/whom followed by an Article or Adjective or Noun 2. '…more…' or '…less…than…'

4 Conclusion

Simultaneous machine translation aims to maintain translation quality while minimizing the delay between reading input and incrementally producing the output.

The KL Simultaneous Interpreting (KLSI) is a set of methods developed by human experts to deliver non-revisable translation from English into Chinese with one second as the benchmark latency. To achieve this, it trains the human brain to stop thinking and execute commands instead. It has developed a range of formulaic techniques to be applied mechanically. This paper presents some of the key features and techniques of KLSI and explores their implications for machine simultaneous interpreting. The KLSI techniques include convergence, the concept of interpreting within three words heard at any given moment in time, co-texting, defaulting, and sequential translation techniques such as repeat, replace, reverse logic, and SAI (Skip, Add, Insert). Multiple English-to-Chinese examples and video recordings are listed in the paper to illustrate the KLSI features and techniques. In the second part this paper, we describe computational methods related to KLSI techniques: Wait-k policy with and without anticipation; word-based and phrase-based alignment and mapping; constrained context for neural machine translation. Commercial machine translation requires at least several gigabytes (or millions of words) of language pair documents for training. It is not realistic to obtain enough parallel texts in English and Chinese reflecting KLSI techniques in order to train a neural machine translation system. However, a lot of parallel texts exist that do not reflect KLSI techniques. We propose to use a rule-based approach to modify available parallel texts using KLSI rules in order to generate a large enough KLSI-based corpus. The main contribution of this paper is to propose a novel rule-based approach—with the rules reflecting human interpretation traits—to revise training corpus to enable short latency and non-revisable machine translation.

References

1. Lin, K.: Field Simultaneous Interpretation. China Translation & Publishing House, Beijing (2012)
2. Ma, M., et al.: STACL: Simultaneous translation with implicit anticipation and controllable latency using prefix-to-prefix framework. arXiv preprint arXiv:1810.08398, Accessed 19 Oct 2018
3. Wikipedia page on dictionary-based machine translation. https://en.wikipedia.org/wiki/Dictionary-baed_machine_translation
4. Wikipedia page on Statistical machine translation. https://en.wikipedia.org/wiki/Statistical_machine_translation
5. Parmentier, Y., Waszczuk, J.: Representation and Parsing of Multiword Expressions: Current Trends. Language Science Press, Berlin (2019)
6. Wikipedia page on Neural machine translation. https://en.wikipedia.org/wiki/Neural_machine_translation
7. Tu, Z., Liu, Y., Lu, Z., Liu, X., Li, H.: Context gates for neural machine translation. Trans. Assoc. Comput. Linguist. **5**, 87–99 (2017)
8. Bangalore, S., Sridhar, V.K.R., Kolan, P., Golipour, L., Jimenez, A.: Real-time incremental speech-to-speech translation of dialogs. In: Proceedings of NAACLHLT (2012)
9. Gu, J., Neubig, G., Cho, K., Li, V.O.K.: Learning to translate in real-time with neural machine translation. In: Proceedings of the 15th Conference of the European Chapter of the Association for Computational Linguistics, EACL 2017, Valencia, Spain, 3–7 April 2017, vol. 1: Long Papers, pp. 1053–1062 (2017). https://aclanthology.info/papers/E17-1099/e17-1099
10. Richardson, S.D., Dolan, W.B., Menezes, A., Corston-Oliver, M.: Overcoming the customization bottleneck using example-based MT. In: Proceedings of the ACL 2001 Workshop on Data-Driven Methods in Machine Translation (2001)

Development of Real Environment Datasets Creation Method for Deep Learning to Improve Quality of Depth Image

Masahiro Murayama[1]([✉]), Yuki Harazono[2], Hirotake Ishii[2], Hiroshi Shimoda[2], Yasuyoshi Taruta[3], and Yuya Koda[3]

[1] Faculty of Engineering, Kyoto University, Kyoto, Japan
murayama@ei.energy.kyoto-u.ac.jp
[2] Graduate School of Energy Science, Kyoto University, Kyoto, Japan
[3] Fugen Decommissioning Engineering Center, Japan Atomic Energy Agency, Ibaraki, Japan

Abstract. High quality depth images are required for stable and accurate camera tracking and 3D modeling. The method to improve the depth image quality using deep learning requires large and accurate datasets in advance. Datasets created by Middlebury Datasets which are typical depth image datasets do not always represent the feature of noise caused by depth cameras, and the number of datasets is not enough. In this study, the method for creating datasets for deep learning was developed and evaluated. The proposed method can improve the accuracy of the distance for each pixel by aligning the positions of pixels that capture the same part of the real world with multiple frames. In addition to super-resolution and denoising, the images are preprocessed such as patch division and data augmentation to eliminate the holes in correct depth images. By using this method, large number of real environment datasets can be automatically created. Two neural networks using Middlebury datasets and the datasets generated by the proposed method were trained respectively, and produced high quality depth images with them. In order to compare the Middlebury Datasets and the proposed method, we visually evaluated the hole filling and the smoothness of edges and surfaces of objects from the results. The result showed the network using the datasets created by the proposed method can remove noise rather than that using Middlebury Datasets since they include noise features caused by the performance limits of depth cameras.

Keywords: Depth image · Noise removal · Super-resolution · Deep learning

1 Introduction

Depth images can be acquired by depth cameras and applied to various computer vision technologies such as camera tracking [1] and 3D modeling [2]. Depth

H. Degen and S. Ntoa (Eds.): HCII 2021, LNAI 12797, pp. 413–426, 2021.
https://doi.org/10.1007/978-3-030-77772-2_27

images captured by depth cameras are often noisy and have low resolution. There are several methods to improve the quality of depth images: using only the depth image to reduce noise by bilateral filter [3], using the RGB image by trilateral filter [4], or dividing the space into three-dimensional grids and projecting the depth image onto the grids by tracking to remove noise from each grid, as in KinectFusion [5]. Recently, deep learning has been widely used to improve the quality of depth images. This method can improve the accuracy of depth images on a voxel-by-voxel basis from only one set of RGB images and depth images. In order to utilize this method, it is necessary to prepare a dataset of low quality depth images (low resolution and degraded by noise) and correct images (high quality depth images) as input images for training the network in advance.

Datasets of a deep learning for depth images processing are often created based on Middlebury Datasets [6–9] which are typical depth image datasets. In the datasets, low quality depth images are created by adding noise that simulates the degradation caused by depth cameras to the correct images of the datasets [10]. However, this noise does not have the features that would be found in real depth cameras. If we use it as datasets for deep learning that processes depth images obtained by shooting the real environment, it will learn wrong noise features and will not produce depth images that correctly represent the environment. In addition, there is also a problem that the number of published datasets is small [11].

On the other hand, if the dataset is created directly from depth images captured by depth cameras, it is necessary to create a high quality depth image as the correct image. However, when using the existing noise reduction filter such as [3,4], the details of the depth image are not enough accurate. In the case when removing noise for each 3D grid such as KinectFusion, the number of grids increases as the resolution increases, and the processing at high resolutions requires a lot of memory and computation time.

The purpose of this study is therefore to propose a new dataset creation method to replace Middlebury Datasets as a deep learning dataset for improving the quality of depth images of real-world environments, and to evaluate the usefulness of the proposed method. Using this method, it will be possible to produce a large number of accurate and high-resolution correct images with little effort, which will improve the performance of improving the quality of depth images using deep learning.

2   Proposed Method of Creating Datasets

2.1   Overview

Figure 1 shows the process flow of the proposed dataset creation method. In the proposed method, high quality depth images are created from the depth images captured by depth cameras. In this process, tracking is applied to infer the position and orientation of the camera when each image was taken. Then, after preprocessing to bring edges closer to the real ones, image resolution increases and noises of depth images are removed. Then, in order to eliminate the holes

in depth images unsuccessfully created by the super-resolution and denoising, the images are preprocessed such as patch division to separate images and data augmentation to increase the number of data.

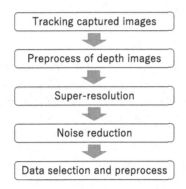

Fig. 1. Process flow of the dataset creation method.

2.2 Tracking Depth Images

In the creation of high-quality depth images, multiple frames are processed by aligning the pixels that capture the same part of the real world to reduce noise and interpolate between pixels. Therefore, it is necessary to estimate the camera pose in each frame by tracking.

This method applies our original natural feature-based tracking [12], which does not require any special equipment and can be used in a wide range. By using this tracking method, the relative positions of feature points and the camera are determined by feature point matching (ORB feature base [13]), which finds the combination of feature points with similar features among images. Then, the camera pose of all the images is estimated using Perspective-n-Point problem solver in the order of capture. Finally, Bundle Adjustment is applied to reduce the error of camera pose. Since this method cannot track images with very large blur or noise, such images are excluded from the dataset beforehand.

2.3 Preprocess of Depth Images

In areas which include large distance change, such as the boundary between two objects, it is difficult to obtain accurate infrared reflections from the depth camera, and measurement errors of the depth camera are likely to be large. In particular, depth images captured by the structured light depth camera (Xtion PRO LIVE [14]) used in this study do not measure the boundaries of objects correctly, and the objects in the foreground tend to appear larger than they actually are. Since the size of the captured area per pixel varies depending on the distance between the object and the camera, the edge pixels are deleted

according to the depth value. Specifically, the depth value which is over 5 m is not deleted, but the number of pixels to be deleted increase by one as the object approaches the camera by 1 m. This preprocessing brings the edges closer to the actual ones of the object, and the denoising process after this process attempts to obtain accurate distances.

2.4   Noise Reduction

In the method of our noise reduction, TSDF (Truncated Signed Distance Function) [15] of the grid is calculated for each pixel of the target depth image by using multiple sets of RGB images and depth images before and after the target image. First, the space on the pixels of the target depth image into equally spaced grids is divided, as shown in Fig. 2. Then, the 3D point of the grid to be evaluated are projected onto all images before and after target image using the tracking results. From the depth value of the projected pixel, the 3D coordinates of the point indicated by the pixel are calculated. TSDFs of these 3D coordinates and the 3D coordinates of the grid are calculated, and the sum of TSDFs is calculated for only those projections whose absolute value is less than a threshold value (30 mm in this study), and the sum is used as the evaluation value of the grid. We suppose that the boundary of an object is at the point where the positive and negative values of the grids are switched, as shown in Fig. 3. Then, the point with the evaluation value of zero is obtained by linear interpolation by using the evaluation values and the 3D coordinates of the two grids whose signs are different, and the distance between the point and the focal point of the camera is used as the new depth value of the pixel.

As shown in Fig. 4, when there are multiple points with zero evaluation values in the space of pixels, intensity images converted from RGB image are used in order to select the correct boundary. In the same way as above, the 3D coordinates of the grid are projected onto the intensity images taken before and after the target image, and the average of the intensities of the projected pixels is calculated to be the intensity of the grid. Then, the intensity of the grid located at the boundary of each object is compared with the intensity of the target pixel in the target intensity image corresponding to the target depth image, and the grid with the closest intensity value is selected. The above processes are performed repeatedly for all captured depth images to gradually improve the quality of depth images.

2.5   Super-Resolution

In this method, super-resolution can be achieved by the same process as the method of the noise reduction. In the process of the noise reduction, the target pixels that already existed in the depth image are processed. However, in the super-resolution, the resolution can increase by creating new pixels between the existing pixels and processing them in the same way as in the noise reduction. For each pixel including the newly created pixel, the linear space on the pixel is equally divided into a grid, and the boundary of the object is searched using

other images before the resolution increases. Then, the obtained depth value is used as the depth value of the newly created pixel.

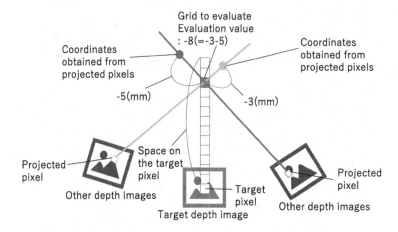

Fig. 2. Process of dividing into a grid and obtaining TSDF.

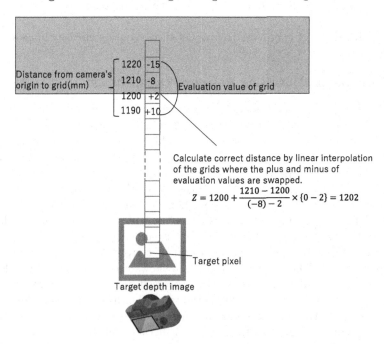

Fig. 3. Process of calculating distance from the evaluation value of the grid.

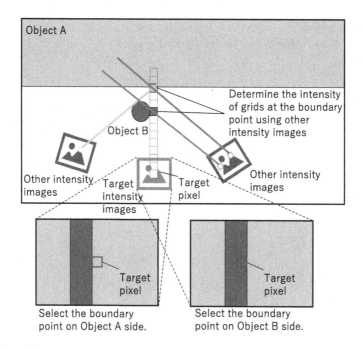

Fig. 4. Process of improving accuracy of edges by using intensity images.

2.6   Example of Processing Result of Depth Image

Figure 5 shows an example of the result of processing a depth image taken by Xtion PRO LIVE in the Fugen Decommissioning Engineering Center of the Japan Atomic Energy Agency (JAEA). In this example, the magnification of the super-resolution is 2×, the total number of pairs of before and after images used in the denoising process is 20, the grid space is 10 mm and the number of denoising iterations is 10. In the processed depth image, edges are smooth, and the holes near the boundary of objects have been filled. It can be seen that by improving the accuracy of the distance, the surface of the tank, which was uneven in the image before the process, has become smoother.

Figure 6 shows an example of the result of converting the processed image into a point cloud. The point cloud in Fig. 6 shows the tank seen from below in the RGB image on the left side. In the processed point cloud, we can see that the number of points has increased, so the resolution has increased. The surface of the tank in the point cloud before process is stepped, while the surface in the processed point cloud is smooth. As shown in the small bumps on the surface of the tank (in the red circle in Fig. 6), they are clearly defined in the processed point cloud.

Since each pixel is processed independently, this method is suitable for multiprocessing using GPU, and we have parallelized the processing using GPU. The average time to process 500 depth images of resolution 640 × 480 using the

PC (CPU: Corei7-9700K, memory: 64 GB, GPU: NVIDIA GeForce RTX 2070 SUPER, OS: Windows 10 Pro 64 bit) for five different image sets was 6681.9 s. It takes about 13 s per frame to create a high-quality depth image. Since dataset creation can take a lot of time before training, this method cannot be used for purposes that require real-time processing such as tracking for augmented reality, but it is practical enough to be used for dataset creation.

Fig. 5. Example of processing result of a depth image, from left to right: RGB image, depth image before processing and processed depth image.

2.7   Data Selection and Preprocess

Depth images are selected at regular frame intervals from the processed depth images in order to avoid including data with close frames in the dataset. The edges of the image are removed as they may have holes in the depth images to be processed. Then, as shown in Fig. 7, the image is cut into smaller images (patches) and the patches with large holes deleted. In the patch division, the image is split so that the neighboring patches are half covered. In the patches, the existing holes are filled by inpainting of OpenCV. For data augmentation, the patches are rotated with 90°, 180°, 270° and flipped them upside-down to obtain 8 more images. These preprocessing steps are performed on all of RGB images, the depth images before the quality enhancement process and the processed depth images.

3   Evaluation of the Proposed Method

3.1   Overview

In the evaluation, we compare the processing results using the Middlebury Datasets, which are widely used as deep learning datasets for depth image processing, and the datasets created by the proposed method. The images used for the datasets of the proposed method and the evaluation were taken by Xtion PRO LIVE in the Fugen Decommissioning Engineering Center of the Japan

Fig. 6. Example of processing result of point cloud, top: point cloud before processing and bottom: processed point cloud. (Color figure online)

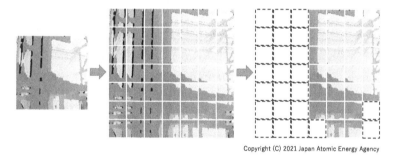

Fig. 7. Example of patch division and deleting patches with large holes.

Atomic Energy Agency. RGB images were converted to intensity images as the inputs of the network. The intensity of intensity images is represented by 8 bits and the depth by 16 bits, and these scales need to be unified for training. Since the maximum depth of the captured depth images was 10,000, the depth was divided by 10,000 and the intensity by 255 so that the values of both depth and intensity were unified to be 0 to 1. Because the depth of the Middlebury Datasets is expressed in 8 bits, the depth was divided by 255 to unify the value scale. After we trained each network using the Middlebury Datasets and the datasets created by the proposed method, we processed the depth images using the trained network and evaluated the results. In the evaluation, it is difficult to obtain true depth images that accurately represent the real environment, so it is not possible to evaluate quantitatively. Therefore, depth images processed using

deep learning were evaluated subjectively. We use a PC (CPU: Corei7-9700K, memory: 64 GB, GPU: NVIDIA GeForce RTX 2070 SUPER ×2 NVLink SLI, OS: Windows 10 Pro 64bit) using Python 3.7.7 and TensorFlow 2.1.0 for training the network and predicting the depth images with it.

3.2 Learning Methods

We use the network proposed by Zhu et al. [16] for improving the accuracy of depth images. Since this network does not increase a resolution, in this study, the resolution of depth images was doubled before inputting them into the network. In Zhu et al.'s study [16], the output of the network is the residual image. In this study however the output is the depth image because the edges in the depth images processed by the original network were obscured.

Although Zhu et al.'s study [16] employs the Euclidean distance-based loss function, it was not suitable because the accuracy is evaluated subjectively in this study. Therefore, we use the loss function [17] proposed by Voynov et al. In this loss function, the rendered images of the surfaces are compared and the mean square error of the normal map is calculated. The loss function combines of the mean square error of the normal map and mean absolute deviation of Laplacian pyramid Lap_1 [18] as a regularizer.

We used Adam [19] with bata 0.9 to train the network. A mini-batch size was 32 and we set the learning rate from 10^{-4} to 10^{-7} which was multiplied by 0.5 if the loss of validation data does not improve over 4 epochs. Train was terminated when number of epochs reached 50 or when overtraining occurred. A patch size of all image pairs was 100×100 to reduce memory consumption.

3.3 Evaluation Method

The networks were trained using the Middlebury Datasets and the datasets created by the proposed method respectively, and then the depth images of the real environment were processed by the networks and compared. The Middlebury Datasets which have noisy depth images [11] consist of 30 pairs of images, 27 of which were used as training data and 3 as validation data. Patch division and data expansion as in the proposed method were applied and the size was standardized to 100×100. We randomly selected 27, 54, 108, 216, 432, and 848 pairs of images from the 848 pairs of images prepared by the proposed method, and created training data. Then, we compared the processing results of the proposed method with the number of training data. When the number of pairs is 27 pairs, we directly compared the depth images processed between by Middlebury Datasets and by those with the proposed method. On the other hand, the comparison between 54 pairs and 848 pairs is used as an evaluation of the usefulness of the proposed method, which can create a large number of datasets.

3.4   Results and Discussion

The average processing time per image was 0.22 [s/Frame], therefore, the network processing is sufficient for real-time processing such as tracking.

Figure 8, 9 show the results of the edges of the processed depth image. The white line in Fig. 9 is the color boundary detected by Canny method from the RGB image. It can be seen that the edges of the depth image processed using the dataset with 27 pairs created by the proposed method does not change much from the depth image before processing. On the other hand, the edges of the proposed method becomes smoother as the number of datasets increases, similar to the boundary of RGB images. Furthermore, the edges of the Middlebury Datasets are misaligned, while the position of the edges of the proposed method become correct as the number of data used increases. The reason for these results is that the datasets of the proposed method include features of depth images taken by a depth camera, such as the unevenness and the shift of the position of the edges and can learn according to the features of the camera used.

According to Fig. 10, both the Middlebury Datasets and the datasets of the proposed method are able to fill in small holes, though the distance is not always correct. As the number of data increases, the proposed method can improve the accuracy of the distance by filling in the holes. However, it can be seen that when processing depth images with large holes such as Fig. 11, neither the Middlebury Datasets nor the dataset of the proposed method can fill in the holes correctly. One of the reasons is that the dataset contains few data with large holes. In the proposed method, we reduced the number of patches which have large holes in the correct image when preprocessing. In this process, it is assumed that the patches in the noisy depth images with large holes are reduced. Another reason is that the learning model was not suitable for filling in the holes. Since the loss function used in this study calculates the loss based on the normal map of the object's surface, the distance information of pixels is not included in the loss. We need to come up with a loss function that can calculate the loss for holes, such as directly comparing the difference in depth values.

Figure 12 shows the result of converting the processed image into a point cloud using each dataset. When the number of data used is small, the surface of the proposed method is not smoother. However, it becomes smoother when the number of data is increased. The dataset of the proposed method using the large number of data can express the details (the area surrounded by the red circle, which is the nameplate of the tank and is a flat surface of about 0.1 m square). These results are due to the fact that the surface shape is used in the calculation of the loss. It was found that the proposed method requires a large number of data for improving the quality of depth images. The proposed method can create a large number of datasets if we have enough captured images, so the proposed method is considered to be useful.

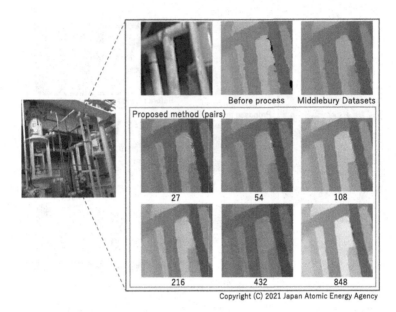

Fig. 8. Comparison of the appearance of the edges of the depth image.

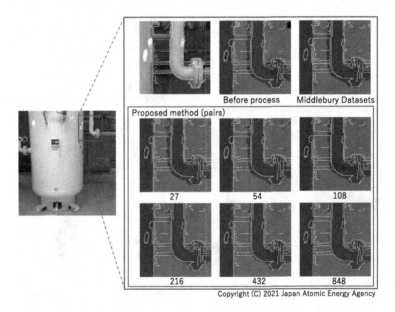

Fig. 9. Comparison of the position of the edges of the depth image.

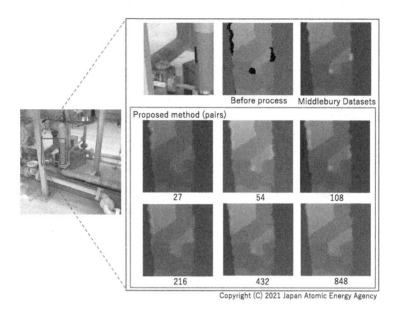

Fig. 10. Comparison of filling small holes.

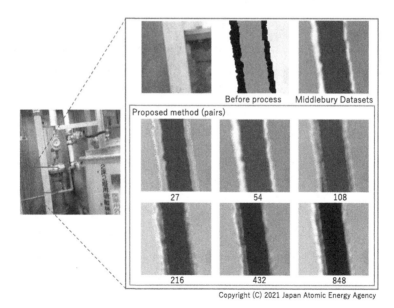

Fig. 11. Comparison of filling large holes.

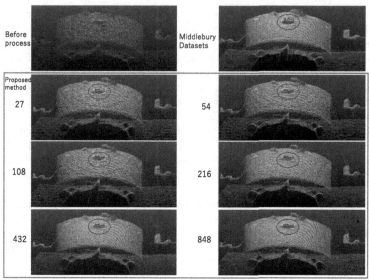

Fig. 12. Comparison of the point cloud.

3.5   Conclusion

In this study, we proposed and evaluated a method for creating training datasets for neural networks to improve the quality of depth images. In the proposed method, high quality depth images are generated from RGB images and depth images taken by a depth camera. As the result of evaluation, when the number of data was limited, Middlebury Datasets was superior to the proposed method, but when the number was increased to take advantage of the proposed method, the proposed method was superior since the dataset created by the proposed method includes the noise features of the depth camera. In the future, it is necessary to improve the dataset creation method and the network and loss function to properly handle holes.

References

1. Marchand, E., Uchiyama, H., Spindler, F.: Pose estimation for augmented reality: a hands-on survey. IEEE Trans. Visual Comput. Graphics **22**(12), 2633–2651 (2015)
2. Kähler, O., Prisacariu, V.A., Ren, C.Y., Sun, X., Torr, P., Murray, D.: Very high frame rate volumetric integration of depth images on mobile devices. IEEE Trans. Visual Comput. Graphics **21**(11), 1241–1250 (2015)
3. Tomasi, C., Manduchi, R.: Bilateral filtering for gray and color images. In: Sixth International Conference on Computer Vision (IEEE Cat. No. 98CH36271), pp. 839–846 (1998)
4. Choudhury, P., Tumblin, J.: The trilateral filter for high contrast images and meshes. In: Proceedings of the 14th Eurographics Workshop on Rendering, pp. 186–196 (2003)

5. Newcombe, R.A., et al.: Kinectfusion: real-time dense surface mapping and tracking. In: 2011 10th IEEE International Symposium on Mixed and Augmented Reality, pp. 127–136 (2011)

6. Scharstein, D., Szeliski, R.: A taxonomy and evaluation of dense two-frame stereo correspondence algorithms. Int. J. Comput. Vision **47**(1–3), 7–42 (2002)

7. Scharstein, D., Szeliski, R.: High-accuracy stereo depth maps using structured light. In: 2003 IEEE Computer Society Conference on Computer Vision and Pattern Recognition, vol. 1, pp. 195–202 (2003)

8. Scharstein, D., Pal, C.: Learning conditional random fields for stereo. In: 2007 IEEE Conference on Computer Vision and Pattern Recognition, pp. 1–8 (2007)

9. Hirschmuller, H., Scharstein, D.: Evaluation of cost functions for stereo matching. In: 2007 IEEE Conference on Computer Vision and Pattern Recognition, pp. 1–8 (2007)

10. Lu, S., Ren, X., Liu, F.: Depth enhancement via low-rank matrix completion. In: Proceedings of the IEEE Conference on Computer Vision and Pattern Recognition (CVPR), pp. 3390–3397 (2014)

11. Lu, S., Ren, X., Liu, F.: Depth enhancement via low-rank matrix completion. http://web.cecs.pdx.edu/~fliu/project/depth-enhance/Middlebury.htm. Accessed 22 Dec 2020

12. Yuki, H., Naoya, M., Toyohiro, H., Hirotake, I., Hiroshi, S., Yuya, K.: Performance evaluation of scanning support system for constructing 3D reconstruction models. In: IEEE 5th International Conference on Computer and Communications (ICCC) (2019)

13. Ethan, R., Vincent, R., Kurt, K., Gary, R.B.: ORB: an efficient alternative to sift or surf. In: International Conference on Computer Vision (ICCV), pp. 2564–2571 (2011)

14. ASUS: Xtion pro live. https://www.asus.com/jp/3D-Sensor/Xtion_PRO_LIVE/. Accessed 26 Dec 12

15. Curless, B., Levoy, M.: A volumetric method for building complex models from range images. In: Proceedings of The 23rd Annual Conference on Computer Graphics and Interactive Techniques, pp. 303–312 (1996)

16. Zhu, J., Zhang, J., Cao, Y., Wang, Z.: Image guided depth enhancement via deep fusion and local linear regularizaron. In: 2017 IEEE International Conference on Image Processing (ICIP), pp. 4068–4072 (2017). https://doi.org/10.1109/ICIP.2017.8297047

17. Voynov, O., et al.: Perceptual deep depth super-resolution. In: Proceedings of the IEEE International Conference on Computer Vision, pp. 5653–5663 (2019)

18. Bojanowski, P., Joulin, A., Lopez-Pas, D., Szlam, A.: Optimizing the latent space of generative networks. In: Proceedings of the 35th International Conference on Machine Learning, vol. 80, pp. 600–609, 10–15 July 2018

19. Kingma, D.P., Ba, J.: Adam: a method for stochastic optimization. arXiv preprint arXiv:1412.6980 (2014)

Assisting Text Localization and Transcreation Tasks Using AI-Based Masked Language Modeling

Ming Qian[1]([⊠]) and Jessie Liu[2]

[1] Soar Technology, Ann Arbor, MI, USA
ming.qian@soartech.com
[2] Translation and Interpretation Program, Middlebury Institute of International Studies at Monterey, Monterey, CA, USA
jessiel@middlebury.edu

Abstract. Localization refers to the adaptation of a document's content to meet the linguistic, cultural, and other requirements of a specific target market—a locale. Transcreation describes the process of adapting a message from one language to another, while maintaining its intent, style, tone, and context. In recent years, pre-trained language models have pushed the limits of natural language understanding and generation and dominated the NLP progress. We foresee that the AI-based pre-trained language models (e.g. masked language modeling) and other existing and upcoming language modeling techniques will be integrated as effective tools to support localization/transcreation efforts in the coming years. To support local-ization/transcreation tasks, we use AI-based Masked Language Modeling (MLM) to provide a powerful human-machine teaming tool to query language models for the most proper words/phrases to reflect the proper linguistical and cultural characteristics of the target language. For linguistic applications, we list exam-ples on logical connectives, pronouns and antecedents, and unnecessary redundant nouns and verbs. For intercultural conceptualization applications, we list exam-ples of cultural event schema, role schema, emotional schema, and propositional schema. There are two possible approaches to determine where to put masks: a human-based approach or an algorithm-based approach. For the algorithm-based approach, constituency parsing can be used to break a text into sub-phrases, or constituents, after which typical linguistic patterns can be detected and then finally masking tasks can be attempted on the related texts.

Keyword: Human-machine teaming · Language modeling · Masked language modeling · Translation assistance tool · Writing assistance tool · Cross-lingual writing assistance · Post editing · Localization · Transcreation · Cross-lingual · Inter-culture

1 Introduction

Machine translation post-editing (MTPE) refers to the process where a human translator edits or revises text that has been translated by machine translation software. MTPE is

© Springer Nature Switzerland AG 2021
H. Degen and S. Ntoa (Eds.): HCII 2021, LNAI 12797, pp. 427–438, 2021.
https://doi.org/10.1007/978-3-030-77772-2_28

composed of two major components: (1) fixing grammatical and other linguistic flaws with a focus on delivering the meaning of the source text correctly; (2) localization and transcreation with a focus on human touch.

Localization, as defined by W3.org, refers to the adaptation of a document's content to meet the linguistic, cultural, and other requirements of a specific target market—a locale. In addition, transcreation, as defined in Wikipedia, describes the process of adapting a message from one language to another, while maintaining its intent, style, tone, and context. Most of the time, translators perform localization/transcreation tasks by applying their knowledge and judgment to capture the essence and meaning of the source text, and then adjusting the translation results to accommodate not only the difference in languages at a basic translation level, but also the cultural, political, and legal differences between the source and target markets. After localization/transcreation, the contents become both culturally appropriate and have a much higher chance of connecting with the target audience.

Most translators, however, are more familiar with one linguistic system and culture than the other. In this paper, we investigate the cases in which translators are very familiar with the linguistical and cultural characteristics of the source language—they fully understand the essence and meaning of the source text—but need some help to deliver localization/transcreation adjustments due to their inadequate knowledge on the linguistic and cultural characteristics of the target language.

In this paper, we used AI-based Masked Language Modeling (MLM) to provide translators with a powerful tool to input a detailed context description, and then use the masked task to query the language model for the most proper words/phrases to reflect the proper linguistical and cultural characteristics of the target language. The approach is a human-machine teaming approach because a human translator builds a masked task by giving a detailed description on the context and sets up the query as a mask or multiple masks. Then the MLM solves the puzzle by applying its linguistic and cultural knowledge learned from a huge training corpus (e.g. BERT model trained on Wikipedia and Book Corpus, a dataset containing + 10,000 books of different genres).

2 Methodologies

2.1 Approach

We adopt an experimental approach to test our hypothesis that machine translation engine, AI-based Masked Language Modeling (MLM), and human inputs can work together to perform localization/transcreation tasks.

Machine translation can generate results with high-speed and low-cost. Human experts still need to post-edit the output to finish the so-called TEP (translate—edit—proof) process. One of the post-editing tasks can be localization/transcreation because the machine translation can be direct and literal, and thus fail to reflect the underlying linguistic and cultural characteristics of the target language. At the same time, the human expert might lack the related linguistic knowledge or cultural awareness to perform the localization/transcreation task. To solve this issue, we propose the new approach for a human expert design a masked word task and let the AI-based masked language modeling tool solve the task. The rational is based on the fact that the masked language model

has been trained on a large training corpus. Consequently, it reflects rich linguistic and cultural knowledge. Therefore, the provided solution can be either accurate or at the very least help human experts to gain some insight (Fig. 1).

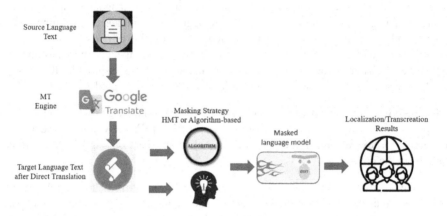

Fig. 1. The direct translation results generated by the machine translation engine might contain contents that need to be localized/transcreated. Human experts or AI-algorithm can apply masking strategies to generate highly relevant masked word tasks for the AI-based MLM. The MLM can either provide localization/transcreation results or useful insight by leveraging language models trained on massive corpus in target language.

We drew most linguistic examples from the book "*The translator's Guide to Chinglish*" [4], and cultural examples from the paper "*Unpacking cultural conceptualizations in Chinese English*" [5].

2.2 Masked Language Modeling

The objective of masked language modeling (MLM) is to mask one or more words in a sentence and have the Natural Language Processing (NLP) model identify those masked words given the other words (representing context) in a sentence [1, 3, 7]. By training the model with this objective, it can learn certain statistical properties of word sequences embedded in a training corpus. BERT is the first large transformer architecture to be trained using this masked language modeling task [2] (Fig. 2).

Fig. 2. An example of MLM task.

3   Results and Discussion

3.1   Linguistic Element Examples

Translators need to recognize linguistic elements in a target language impacted by a source language (e.g. Chinglish—English language that is influenced by Chinese language) and revise them to eliminate those elements. This kind of revision is part of the localization, defined as adapting the collateral to the target language, and transcreation, defined as the creation of content in the target language that is inspired by the source, but highly adapted for the language and culture where it will be used.

Chinglish Example: Logical Connectives. The Chinese language often omits logical connectives that in English are considered essential [4]. While Chinese readers can understand a text without these connectives, native speakers of English expect more help from the connective words.

Table 1 shows an example of a missing logical connective. The logical relation between two clauses is cause-effect. Figure 3 shows the MLM task and the top choice is the connective word "because". The result shows that the language model is able to capture the cause-effect relation.

Table 1. Chinglish example on missing logical connectives.

Original Chinese	湖南的大多数煤矿都位于偏远的丘陵地带, 该省鼓励那里的下岗工人改行农业谋生
English translation generated by Google Translator	Most coal mines in Hunan are located in remote hilly areas, and the province encourages laid-off workers there to switch to agriculture to earn a living

Sentence:

[MASK] most of Hunan's coal mines are lcoated in remote, hilly land, the province has encouraged laid-off workers there to switch to agriculture to earn a living.

Mask 1 Predictions:
36.8% **Because**
19.9% **Although**
14.0% **While**
13.0% **Since**
9.3% **As**

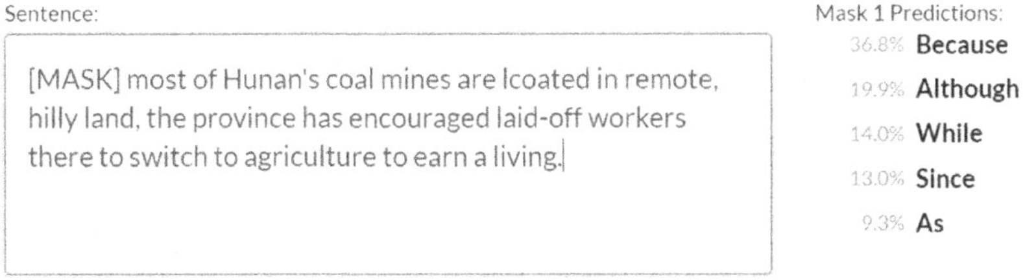

Fig. 3. MLM-based interpretation.

Chinglish Example: Pronouns and Antecedents. Chinglish also tends to make repeated references to the same thing [4]. While Chinese readers can understand a text without these connectives, native speakers of English expect more help from the connective words.

Table 2 shows an example of a repeated reference. The phrase "两种类型的工作"—"both types of work" —was repeated. Figure 4 shows the MLM task and the top choice is the pronoun "them". The result shows that the language model is able to capture the use of the pronoun.

Table 2. Chinglish example on repeated reference.

Original Chinese	我们未能同时重视这两种类型的工作, 并且两种类型的工作之间没有适当的协调
English translation generated by Google Translator	We fail to attach importance to both types of work at the same time, and there is no proper coordination between the two types of work

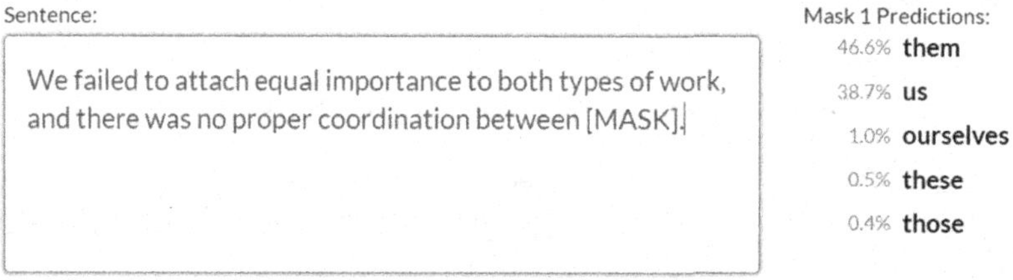

Sentence:

> We failed to attach equal importance to both types of work, and there was no proper coordination between [MASK].

Mask 1 Predictions:
46.6% **them**
38.7% **us**
1.0% **ourselves**
0.5% **these**
0.4% **those**

Fig. 4. MLM-based interpretation.

Chinglish Example: Unnecessary Redundant Nouns. Unnecessary nouns add nothing to the meaning of the sentence. when they are deleted, the sense is not diminished, only clarified. The example below is listed in the foreword section in [4].

Table 3 shows an example of unnecessary redundant nouns where the word "people's" is an unnecessary redundant word because the phrase "living standards" itself is sufficient. Figure 5 shows the MLM task and the top choice is the determiner ("the"), that shows there is no need for another noun here.

Table 3. Chinglish example on unnecessary redundant nouns.

Original Chinese	随着社会的发展和技术的进步, 人民生活水平显著提高
English translation generated by Google Translator	With the development of society and technological progress, people's living standards have improved significantly

Sentence:

With social development and technological progress, [MASK] living standard have improved significantly.

Mask 1 Predictions:
60.7% **the**
8.1% **their**
5.4% **our**
3.6% **human**
0.8% **its**

Fig. 5. MLM-based interpretation.

Chinglish Example: Unnecessary Redundant Verbs. Unnecessary verbs add nothing to the meaning of the sentence. When they are deleted, the sense is not diminished, only clarified. The example is from [8].

Table 4 shows an example of unnecessary redundant verbs where the word "promote" and "construct" are redundant to each other. Figure 6 shows the MLM task and most top choices are associated with a single verb.

Table 4. Chinglish example on unnecessary redundant verbs.

Original Chinese	新时代的中国能源发展, 积极适应国内国际形势的新发展新要求, 更好推动建设清洁美丽世界
English translation generated by Google Translator	China's energy development in the new era actively adapts to the new development and new requirements of the domestic and international situations, and better promotes the construction of a clean and beautiful world

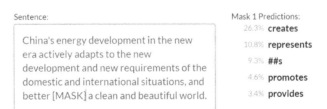

Sentence:

China's energy development in the new era actively adapts to the new development and new requirements of the domestic and international situations, and better [MASK] a clean and beautiful world.

Mask 1 Predictions:
26.3% **creates**
10.8% **represents**
9.3% **##s**
4.6% **promotes**
3.4% **provides**

Fig. 6. MLM-based interpretation.

3.2 Cultural Conceptualization Examples

Cultural conceptualizations are defined as conceptual structures such as schemas, categories, and conceptual metaphors, which not only exist at the individual level of cognition, but also develop at a higher level of cultural cognition, where they are constantly negotiated and renegotiated through generations of speakers within a cultural

group, across time and space [5]. We use multiple cultural schema examples listed in [6] for experimenting with our approach, including event schema, role schema, emotional schema, and proposition schema. Unlike the linguistic element examples, for cultural conceptualization examples, the masking tasks are designed in a more free-form format that could be totally separated from the original text. Instead, it serves as a tool for a user to build a context and query the language model for localization/transcreation clues.

Cultural Schema: Event Schema. Event schemas are conceptual structures abstracted from human experience of certain events [6].

Table 5 shows an example of event schema: for a wedding in China, the term "red envelope" is a cultural reflection of 'face' (the more money it contains, the more 'face' the sending family gains and offers to the newlyweds), and 'politeness' (a small amount in the red envelope that is lower than the commonly accepted standard is considered an act of impoliteness or insult). Face and politeness are Chinese cultural schemas for a wedding event [6]. Figure 7 shows that the MLM task helps to find the closest interpretation in the English culture: wealth and love. For localization/transcreation purpose, instead of using "My family has a lot of face", a translator could use "My family shows a lot of love and shows off our wealth."

Table 5. Cultural schema: event schema example.

Original Chinese	今天晚上的婚礼, 爸爸给他们很大的红包. 我家很有面子, 但是我觉得他给的太多了
English translation generated by Google Translator	For the wedding tonight, my father gave them a big red envelope. My family has a lot of face, but I think he gave too much

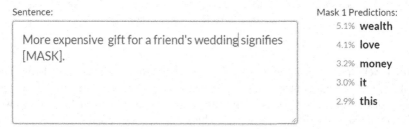

Fig. 7. MLM-based approach helps to identify western counterparts of the implications of a cultural event.

Cultural Schema: Role Schema. A Chinese family is usually perceived vertically in terms of the relationships among its members, e.g., a married woman with a child would see herself as a mother first, instead of a wife, while in certain other cultures, a married woman may perceive herself first as the wife of her husband rather than the mother of her child [6], as illustrated by the example in Table 6.

Figure 8 shows the MLM task helps to find the correct interpretation for the Western culture. Based on the two top statistics (61.8% for mother, and 23.9% for wife), the mother's role seems to be more dominant than the wife's role in the Western culture.

Table 6. Cultural Schema: role schema example.

Original Chinese	R: 您还履行家庭角色, 例如…… I-2: 是的, 比如母亲, 妻子……
English translation generated by Google Translator	R: You also fulfill the family role, like… I-2: Yeah, like a mother, a wife

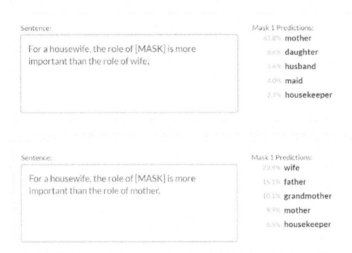

Fig. 8. MLM-based interpretation: the confidence levels seem to show that in the Western culture, the role of mother is more important than the role of wife as well.

Cultural Schema: Emotional Schema. A Chinese emotional (sadness) schema, listed in Table 7, can be elaborated as: X experiences sadness when he/she senses a gap or lacking ability, because of a lack of exposure, awareness and sufficient knowledge, when he/she compares himself/herself to others. This Chinese cultural sadness schema indexes inferiority, self-pity, and helplessness [6].

Figure 9 shows the MLM task helps to find the interpretation in the Western culture. The top emotions found are 'awkward', 'inferior', and 'embarrassed'.

3.2.1   Cultural Schema: Proposition Schema

Every culture has value propositions such as family values or how to be a "good man." A Chinese proposition schema listed in Table 8 illustrates a Confucius saying: The order of the world is rooted in every family [6].

Table 7. Cultural Schema: emotional schema example. The emotion of 'sadness' is unique to Chinese culture when a person senses a gap or lacking ability. Therefore, a proper emotion for Western culture needs to be identified.

Original Chinese	我的英语词汇量太小了, 每次跟英语母语的美国人比我都觉得伤心
English translation generated by Google Translator	My English vocabulary is too small, and I feel sad every time I compare myself to a native English speaker

Sentence:

My English vocabulary is too small, and I feel [MASK] every time I compare myself to a native English speaker.

Mask 1 Predictions:
10.3% **awkward**
5.6% **sick**
5.3% **inferior**
4.4% **embarrassed**
3.7% **stupid**

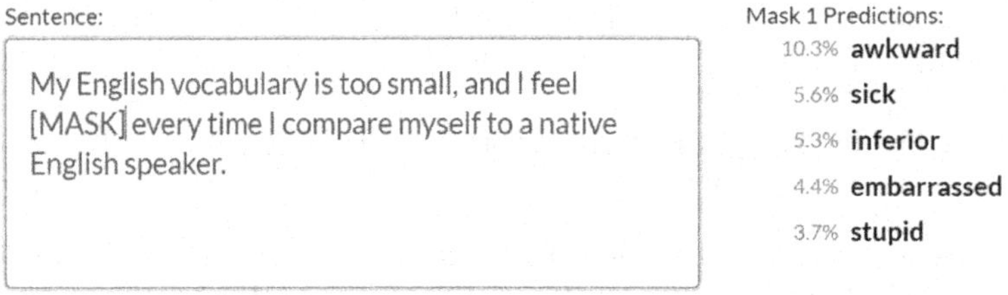

Fig. 9. MLM-based interpretation of a unique emotion.

Figure 10 shows that we can design the MLM tasks to identify the closest interpretation/metaphor in Western culture. In this case, the results are not very good. That could mean that the ideal counterpart in Western culture is hard to find.

Table 8. Cultural Schema: propositional schema example.

Original Chinese	家和万事兴
English translation generated by Google Translator	Home and everything flourish (It is a faulty direct translation. A better direct translation maybe "Maintain harmony at home, and everything flourishes".)

3.3 Discussion on Masking

Masking Strategy. Masked Language Modeling is a fill-in-the-blank task. As we have shown in multiple examples in previous sections, MLM can be used to perform localization/transcreation tasks to bridge cross-lingual and inter-cultural differences.

The next question is how to determine where to put masks. There are two possible approaches: one is human decision and the other is to move a mask across and let an algorithm determine which masking task is valuable.

Fig. 10. MLM-based interpretation. Western culture has different

The naïve approach is to let human users determine what to mask. For example, a human user could notice that there is a logic connectives problem or a redundant nouns problem, and he/she can determine what to mask. However, with this approach, the assumption is that human users already know where the problem is. Arguably, if they already know where the problem is, human users can fix the problem themsevles instead of relying on the MLM output. It is possible that the MLM can provide better solutions, but we rely on human judgment to detect and be aware of the problem.

A more sophisticated approach is to run an algorithm and try masking tasks at different locations in the text. Then the results can be provided to human users for further decisions. The benefit of this approach is that humans may not be aware of a problem (e.g. missing logic connectives, or redundant nouns), but the algorithm can detect a potential problem and provide masking task results to enlighten a human user.

Constituency parsing is the task of breaking a text into sub-phrases, or constituents. There are existing AI tools to perform constituency parsing tasks (e.g. [9]). Figure 11 shows the result of running constituency parsing on the example of unnecessary redundant nouns we used in Sect. 3.1. The sub-phrase "people's living standards" is extracted as a sub-phrase and we can run a masking task on it as we show in Sect. 3.1. Figure 12 shows the result of running constituency parsing on the example of unnecessary redundant verbs we used in Sect. 3.1. The sub-phrase "promotes the construction" is extracted as a sub-phrase and we can run a masking task on it as we show in Sect. 3.1.

Unique patterns can be found on the direct translation results for various categories. For example, redundant nouns are usually associated with two nouns related to possessive form ("people's living standard") and redundant verbs are usually associated with verb nominalization. Algorithms can be designed to detect these patterns and perform masking tasks on the texts related to those patterns.

Two Different Goals for Masking Tasks. For cross-lingual applications, as illustrated in Sect. 3.1, the masking tasks are performed on the direct translation of the original text. The goal is to find the correct fill-in words to replace or revise the direct translation results.

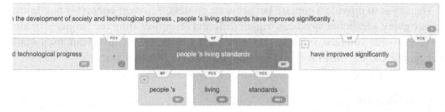

Fig. 11. Constituency parsing result on the example of unnecessary redundant nouns.

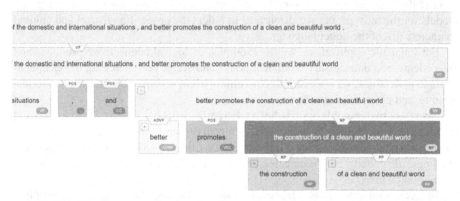

Fig. 12. Constituency parsing result on the example of unnecessary redundant verbs.

However, for intercultural applications, as illustrated in Sect. 3.2, the masking tasks are designed in a more free-form format that could be totally separated from the original text. Instead, it serves as a tool for a user to build a context and query the language model for localization/transcreation clues. For example, for the cultural event schema example listed in Sect. 3.2.1, the original text is about the amount of money in a "red envelope". However, the context for the masking task is "More expensive gift for a friend's wedding signifies [MASK].", which is totally separate from the original text. Nevertheless, it serves the purpose of localization/transcreation: identifying the equivalent conceptualization in the target culture.

Limitation of the Masking Approach. The benefits of using masked language models for localization/transcreation tasks come from the fact that the pre-trained language models have "read" a large training corpus of text materials, and as a result have tremendous linguistic and cultural knowledge. By feeding the direct translation results as the context or designing a separate context, the language model can fill the mask effectively and correctly in terms of linguistic and cultural accuracy.

However, there is no guarantee that the results are always accurate. The language model is highly dependent on the training corpus. There is a chance that a language model has no prior knowledge on a context, or a context is similar to many other contexts. Under these conditions, the solutions can be ineffective or useless.

The key is to provide a concise but accurate context. A longer context is not desired because every additional word can trigger different responses from the language model. The more words you have in the context, the higher chance some false responses would be triggered and the fill-in solution for the mask would be unsatisfactory.

4 Conclusion

To support localization/transcreation tasks, we use AI-based Masked Language Modeling (MLM) to provide a powerful human-machine teaming tool to query language models for the most proper words/phrases to reflect the proper linguistical and cultural characteristics of the target language.

For linguistic applications, we list examples on logical connectives, pronouns and antecedents, and unnecessary redundant nouns and verbs. For intercultural conceptualization applications, we list examples of cultural event schema, role schema, emotional schema, and propositional schema.

There are two possible approaches to determine where to put masks: human-based approach or algorithm-based approach. For the algorithm-based approach, constituency parsing can be used to break a text into sub-phrases, or constituents, and after which typical linguistic patterns can be detected, and then finally masking tasks can be attempted on the related texts.

In recent years, pre-trained language models have pushed the limits of natural language understanding and generation and dominated the NLP progress. We foresee that the AI-based pre-trained language models (e.g. masked language modeling) and other language modeling techniques will become integrated as effective tools to support localization/transcreation efforts in the coming years.

References

1. Wang, C., Li, M., Smola, A.: Language Models with Transformers, https://arxiv.org/pdf/1904.09408.pdf, Accessed Oct 2019
2. Devlin, J., Chang, M-W., Lee, K., Toutanova, K.: BERT: pre-training of deep bidirectional transformers for language understanding. https://arxiv.org/abs/1810.04805, Accessed Oct 2018
3. Masked language modeling demo, AllenNLP, Allen Institute for AI. https://demo.allennlp.org/
4. Joan Pinkham with the collaboration of Guihua Jiang, The translator's Guide to Chinglish, Foreign Language. Teaching and Researching Press, China (2000)
5. Sharifian, F.: Cultural linguistics. In: Chapelle, C.A. (ed.) The encyclopedia of applied linguistics (pp. 1590–1596). New Jersey, Boston, Oxford: Blackwell Publishing Ltd., 2013.
6. Xu, Z., Sharifian, F.: Unpacking cultural conceptualizations in Chinese English. J. Asian Pac. Commun. **27**(1), 65–84 (2017)
7. Donahue, C., Lee, M., Liang, P.: Models to Fill Enabling Language in the Blanks. arXiv preprint arXiv:2005.05339, Acessed 11 May 2020
8. Official website of the State Council Information Office, China. White Paper "China's Energy Development in the New Era" (2020)
9. Constituency Parsing demo, AllenNLP, Allen Institute for AI. https://demo.allennlp.org/constituency-parsing

A Preliminary Study for Identification of Additive Manufactured Objects with Transmitted Images

Kenta Yamamoto[1]([✉]), Ryota Kawamura[1], Kazuki Takazawa[2],
Hiroyuki Osone[1], and Yoichi Ochiai[1,2]

[1] University of Tsukuba, Tsukuba, Japan
kenta.yam@digitalnature.slis.tsukuba.ac.jp
[2] Pixie Dust Technologies, Inc., Tokyo, Japan

Abstract. Additive manufacturing has the potential to become a standard method for manufacturing products, and product information is indispensable for the item distribution system. While most products are given barcodes to the exterior surfaces, research on embedding barcodes inside products is underway. This is because additive manufacturing makes it possible to carry out manufacturing and information adding at the same time, and embedding information inside does not impair the exterior appearance of the product. However, products that have not been embedded information can not be identified, and embedded information can not be rewritten later. In this study, we have developed a product identification system that does not require embedding barcodes inside. This system uses a transmission image of the product which contains information of each product such as different inner support structures and manufacturing errors. We have shown through experiments that if datasets of transmission images are available, objects can be identified with an accuracy of over 90%. This result suggests that our approach can be useful for identifying objects without embedded information.

Keywords: Additive manufacturing · Transmitted image · Object identification

1 Introduction

The additive manufacturing (AM) process has greatly improved the ease of model customization. When managing products with a digital system, the addition of information is essential, such as a barcode. While management information has only to be attached to the exterior, additive manufacturing process makes it possible to embed information inside the product. Detection is then possible by various methods, for example, embedded information can be read using terahertz (Thz) sensing in InfraStructs [34], and using visible light projection in AirCode [14].

© Springer Nature Switzerland AG 2021
H. Degen and S. Ntoa (Eds.): HCII 2021, LNAI 12797, pp. 439–458, 2021.
https://doi.org/10.1007/978-3-030-77772-2_29

1. Fabricating Products   **2. Capturing Images**   **3. Identification of Products**

Fig. 1. System Overview. The system is a three step process: fabrication, capturing, and identification. Products are fabricated without embedding information internally. Image capture requires transmitted light in the near infrared region. Identification is performed using feature matching or deep learning with datasets.

Two problems exist with the embedded information approach. The first is that an information adding step must be introduced into the manufacturing process. This also constitutes an advantage as the integration of manufacturing and adding information enables overall reduction of processing time and cost, nevertheless it is not possible to identify objects that were manufactured without embedded information. The second problem arises when the embedded information is duplicated such that product batches cannot be uniquely identified. This can occur in the current item distribution system. In previous research, defects in the barcode surface produced by the manufacturing process have been used for unique identification of products [33].

In this study, a system of product identification was developed that does not use embedded information. All products have inherent structure combined with characteristics created by the manufacturing process, particularly manufacturing errors. Transmission images of product items are utilized to extract inherent product characteristics. This idea was inspired by research that uses the finger vein pattern to authenticate individuals. Finger vein pattern authentication is biometric, hence information is not added in advance of the authentication process; and, in order to visualize the internal blood vessels, an image of the finger transmitted by infrared light is used. Also in our system, information is not added, and we use images transmitted by infrared light.

To verify the identification system, two experiments were conducted: feature matching and deep learning; using two types of object group with identical exterior appearance. In one group, the internal structure of the object was varied. In the other group, all structures, including the internal structure, were unchanged; i.e., the objects differed only as a result of manufacturing errors.

The results suggest that feature matching is possible when there is a change in shape, and that deep learning can be used to identify objects that vary due to naturally occurring manufacturing errors. The purpose of this investigation is to examine the possibility of utilizing an identification system based on transmission images, to discriminate products in the absence of embedded information.

Table 1. Comparison with existing methods.

	InfraStructs [34]	AirCode [14]	LayerCode [17]	Our system
Information place	Inside	Inside	Exterior	everywhere
Pre embedding	Yes	Yes	Yes	No
Detection angle	One way	One way	All angle	All angle
Learning dataset	No	No	No	Yes
Wavelength	THz	VIS	VIS or NIR	NIR
Reflection or transmission	Reflection	Reflection	Reflection	Transmission

2 Related Works

2.1 Properties of Additive Manufacturing

Additive Manufacturing encompasses a variety of methods [35]. Fused Deposition Modeling (FDM), in particular, has made a great contribution to the uptake of the desktop 3D printer. The widespread adoption of FDM has been accompanied by examination of the characteristics of FDM manufactured products. Many previous studies have considered how the characteristics of objects are affected by the manufacturing method itself; e.g., product characteristics [3,9,12,25], mechanical properties [13,32], and material properties [5,20,26,27]. Several investigations have also been conducted to determine how objects change with different parameters; i.e., the relationship between each parameter and mechanical properties [6,22,24,31], and the correspondence between the internal structure and mechanical properties [1,2,8,11,23]. By referring to this body of literature concerning the FDM method, the fabrication characteristics of FDM could be established without the necessity to carry out exploratory research on the process.

2.2 Biometrics Using Transmitted Images

It has become very common to perform personal authentication using biometric characteristics such as fingerprint authentication and face recognition. Most biometric authentication methods use an image produced by reflection of light from a target. However, some biometrics acquire an image after light has passed through the target. Finger vein pattern recognition is a typical example. In 2004, Miura et al. [18] published a paper on personal authentication using a transmission image of a finger vein pattern obtained by illumination with infrared LEDs. The main challenge for this method lies in ensuring that the system is robust. It must be able to authenticate properly even if finger placement changes slightly, and it must not authenticate as another person. These problems led to research on how to automatically determine the region of interest [19,28,36] and improve imaging quality [30,37]. Therefore, research literature concerning transparent

biometric authentication was referred to in order to assist with the construction of an identification system using transmission images of AM products.

2.3   Embedding Information with Additive Manufacturing

Additive Manufacturing is a promising rapid prototyping method since it enables easy customization of shapes, thus prompting research on embedding barcodes during manufacturing. InfraStructs [34] is a famous early study that embodies this concept. During the manufacturing process, a structure was built inside the object that was detected and decoded as information, using Thz sensing. This information could not be captured with a visible light camera. Several subsequent studies successfully embedded and read information in the object interior using alternative methods: i.e., internal information read in the far infrared region [21], information embedded using highly reflective material [29], and information embedded in the form of air pockets and read using visible light projection [14]. Recently, Maia et al. [17] embedded information in the entire surface of an object in order to simplify reading of the information.

Thus, existing research is based on building a system that embeds information in an invisible form during manufacturing and reads it out. In contrast, a method based on transmitted images does not require design of embedding information and the reading angle is free. The method is similar to the recent study that utilizes slicing parameters to identify objects [7]. Table 1 summarizes these differences. The utility of the transmission imaging system for identifying a product without preprocessing has been verified.

3   Materials and Methods

3.1   System Overview

This section details the system overview. The system consists of three processes: fabrication, capturing images, and identification (Fig. 1). During fabrication, information is not intentionally embedded in products. Instead, inherent product characteristics are used for identification; i.e., visibly indistinguishable products may differ in their internal structure, and manufactured items hold intrinsic information such as dimensional values, which differ due to manufacturing errors. In the capturing process, transmission images are acquired. By using near infrared light, internal shape is captured in the transmitted image to enable identification of products on the basis of changes to internal structure. Inclusion of internal information also improves the accuracy of identifying changes resulting from manufacturing errors. Two methods were considered for the identification process: feature matching and deep learning. Feature matching does not require pre-learning for identification, allowing a simple system to be utilized. This contrasts with deep learning, which does require pre-learning, however discrimination accuracy is higher.

Table 2. Fabrication characteristics.

Main topic	Sub topic	Sub-sub topic
Shape Parameters	Printing width	
	Layer thickness	
	Inner support structure	Infill pattern
		Infill density
		Infill position
Condition Parameters	Material properties	Material
		Color
	Printing speed	
	Temperature of heat-bed	
	Temperature of printing head	
	Retraction speed	
	Environment	

3.2 Fabrication Characteristics

Table 2 lists the characteristics that can appear in the transmission image of the product. All characteristics are referred to in the existing white paper [9] and the software that runs the 3D printer. The characteristics were divided into two types, those related to shape and those related to condition. Of the shape-related properties, we paid particular attention to the inner support structure. Many previous studies intentionally modified the interior space of a product without affecting the exterior appearance. In contrast, this study did not include intentional redesign of the interior space, but instead focused on random internal changes that occur naturally in manufacturing. Condition-related parameters were fixed in this study as it was difficult to distinguish between the effects of changing these parameters and differences arising from manufacturing errors. The assumption was made that if a manufacturing error could be identified, a change in condition parameters could also be distinguished. Therefore, only manufacturing errors were considered.

Inner Support Structure: This parameter consists of three sub-parameters: infill pattern, infill density, and infill position (Fig. 2), which are all visible on the transmission image. Infill pattern is an important parameter that is related to the shape of the internal structure and affects the mechanical properties of the product. Modifying the infill pattern changes the durability of the product under stress; i.e., the resistance to stress and the strength of the product are modified. Infill density refers to the density of a given infill pattern. Optimization of infill density is always desirable in terms of manufacturing cost and cycle time, as the amount of material increases with density. Infill position refers to the placement of the inner support structure for the given infill pattern. As placement is often done automatically by the slicer software, in general the system user is less aware of infill position.

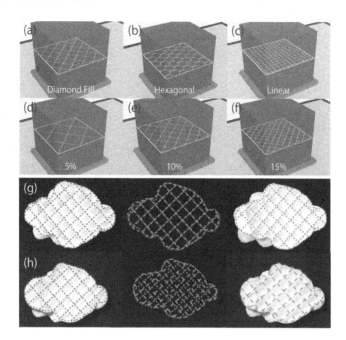

Fig. 2. Shape parameter variations used in the identification tests. (a)–(c) Different infill patterns. (d)–(f) Different infill densities. (g) and (h) show the two in infill positions. (g) Initial infill position. (h) Alternative infill position.

3.3   Transparent Imaging System

The transmissive imaging system operates by capturing the characteristics and internal structure of products, both primitive and complex, in images using light wavelengths that are transmitted through the target object.

Optical Setup. An optical system using near infrared (NIR) light was constructed to obtain an optimal transmission image of the object. The target object, NIR camera, NIR light source, and an optical element were arranged as shown in Fig. 3(a). In order to eliminate the influence of speckle, the halogen lamp was used as the near infrared light source, which was equipped with an infrared transmission filter. A dihedral corner reflector array (DCRA) [16] was used to irradiate the back of the object. As shown in Fig. 3(b), DCRA has the function of imaging the light source at a plane-symmetrical position. Similar functions can be realized with mirror and lens as shown in Fig. 3(c). In this experiment, because it was difficult to incorporate large mirror and lens, the amount of illuminating light was maximized by using DCRA.

Transmitted Images. In Sect. 3.2, it was explained that inner support structures differ in type, and that for a given support structure, the infill patterns,

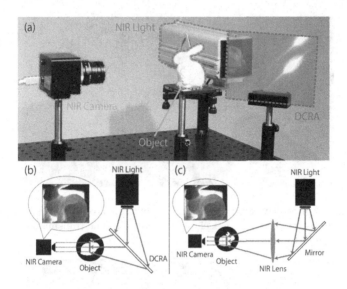

Fig. 3. Optical System. (a) A picture of the imaging system used in this study. (b) A plan view of (a), showing the path of light. (c) Alternative imaging system.

densities, and positions of structural elements are variable. In this section, the appearance of the transmission image is examined when the inner support structure is either changed by design, or left unchanged such that variation arises from manufacturing errors.

Infill Pattern: Cube and bunny shapes were manufactured using three types of infill pattern: Diamond Fill, Hexagonal, and Linear. Transmission image results are shown in Fig. 4. A different image was obtained for each infill pattern, for both cube and bunny.

Infill Density: Cube and bunny shapes were produced with Diamond Fill pattern densities of 5, 10, and 15% for the cube, and 10, 20, and 30% for the bunny. As the density increased, the amount of transmitted light decreased, hence the contrast in the transmissive images also decreased. These results are shown in Fig. 5.

Infill Position: The same inner structure was created at multiple positions, with a Diamond Fill inner pattern density of 10% for the cube, and 20% for the bunny. Transmission image results are shown Fig. 6. The cube images change as the position of the inner support structure changes. However, the transmission images of the bunny were almost unchanged due to the complexity of the model.

Same Inner Parameters (only Manufacturing Errors): We compared the transmission images when all the parameters are equal. In order to identify objects manufactured with the same parameters, it is necessary to utilize slight changes such as manufacturing errors. We observed whether such a small difference could be acquired within a transmission image. As shown in Fig. 7, we

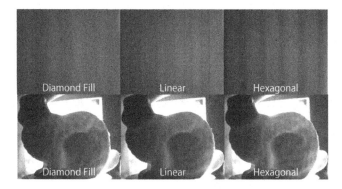

Fig. 4. Variation of additive manufacturing infill pattern. (top) The transmission image when the infill pattern of the cube is changed. (bottom) The transmission image when the infill pattern of the bunny is changed.

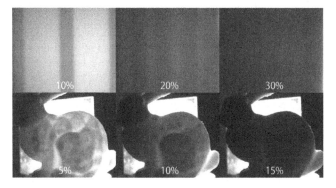

Fig. 5. Variation of additive manufacturing infill density. (top) The transmission image when the infill density of the cube is changed. (bottom) The transmission image when the infill density of the bunny is changed.

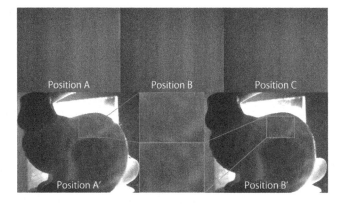

Fig. 6. Variation of additive manufacturing infill position (top) The transmission image when the infill position of the cube is changed. (bottom) The transmission image when the infill position of the bunny is changed.

Fig. 7. Random variation in items made with the same manufacturing parameters. (top) The comparison of transmission images of cubes with the same infill conditions. (bottom) The comparison of transmission images of additive manufactured bunnies with the same infill conditions.

prepared the cubes and bunnies with all same parameters listed in Table 2. Since very similar images are acquired, the region of interest is enlarged. In the enlarged image, it can be confirmed that trivial differences occur in both the cube and the bunny.

4   Results

In Sect. 3.3, the relationships between the characteristics of objects and their transmission images were verified. When parameters related to shape or structure differ between objects, clear differences are likely to appear between their transmission images. However, only minor differences are visibly discernible between transmission images of objects with the same parameters. To evaluate the level of difference that can be resolved between various images, two algorithm-based experiments were conducted: feature matching, and classification by deep learning.

4.1   Feature Matching

To verify whether or not differences between the NIR transmission images were detectable by image processing, feature extraction and feature matching were initially conducted. In this experiment, the SURF algorithm [4] was used to perform feature extraction.

The results of matching feature points are shown in Fig. 8. Three key features were extracted for use in feature point matching. The matching features in a reference image and a target image were obtained by the k-nearest neighbor

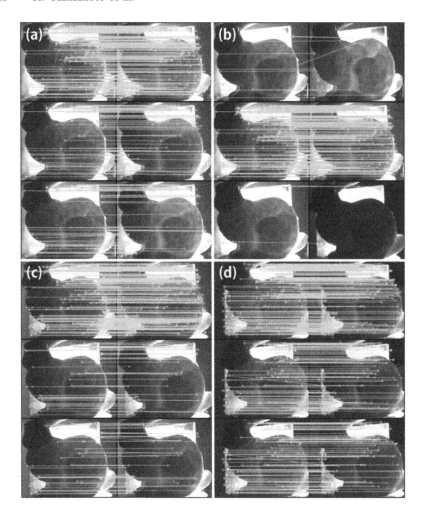

Fig. 8. The results of feature matching when: (a) infill pattern is changed; (b) infill density is changed; (c) infill position is changed; and (d) comparing three bunnies with the same infill condition.

method. The D. Lowe ratio test [15] was used to select the top two matched features, the remaining feature is drawn in Fig. 8. Feature matching was then performed between pairs of images that varied by infill pattern, or infill density, or infill position, or only by random differences due to manufacturing errors. Of the paired images showing matching results, the image on the left is the reference image, and the image on the right is the target image. The right hand images were changed and the results of feature matching were recorded as the number of matching points after the ratio test. These results are listed in Table 3. They quantify the change in matching with the change in a parameter; i.e., the degree to which the algorithm recognises the difference between images. A single

reference image is used for all variations of a parameter hence the locations and maximum number of feature matching points are retained in the parameter series, thus allowing quantification and direct comparison of the results.

The results in Fig. 8 and Table 3 demonstrate that it is possible to discriminate between objects with obvious shape or structural differences by feature extraction and feature point matching using the SURF algorithm. This is not the case when objects are manufactured with the same structural parameters and differ only by manufacturing errors. Quantitative differences in matching results from one target image to the next were not significant, therefore feature matching is unlikely to be able to distinguish between objects at a higher degree of similarity.

Table 3. Match rate of feature points. Since the same image is used to compare the same parameters, the match rate is 100%.

Infill pattern	Diamond	Linear	Hexagonal
Diamond	255 / 255	82 / 255	62 / 266
Linear	79 / 261	261 / 261	86 / 261
Hexagonal	67 / 305	81 / 305	305 / 305
Infill density	10%	20%	30%
10%	310 / 310	17 / 310	3 / 310
20%	11 / 255	255 / 255	3 / 255
30%	2 / 174	5 / 174	174 / 174
Infill position	Position A	Position B	Position C
Position A	258 / 258	58 / 258	54 / 258
Position B	42 / 215	215 / 215	35 / 215
Position C	49 / 265	39 / 265	265 / 265
Same position	Object 1	Object 2	Object 3
Object 1	235 / 235	128 / 235	107 / 235
Object 2	122 / 227	227 / 227	78 / 227
Object 3	104 / 204	88 / 204	204 / 204

Table 4. Dataset type.

	3D printer	Position	Pattern	Density	Amount
a	Makerbot	Random	Diamond fill	20%	10
b	Makerbot	Fixed	Diamond fill	20%	10
c	Makerbot	Fixed	Diamond fill	10%	10
d	Bellulo	Fixed	Honeycomb	10%	10

4.2 Deep Learning

Deep learning has contributed greatly to the improvement of classification accuracy in recent years. Following the partial success of feature matching as discussed in the previous section, the deep learning technique was adopted to distinguish between items that are made using the same parameters, but differ due to manufacturing errors. Four groups of products were manufactured and their transmission images acquired. Datasets were produced from the transmission images (Table 4). Dataset (a) consists of products with different infill positions. Datasets (b), (c), and (d) consist of products that differ only as a result of manufacturing errors.

In order to apply the deep learning method, training datasets must first be prepared. The imaging system described in Sect. 3.3 (Fig. 3) was used to obtain images for the datasets. In a commercial or industrial setting, the target may rotate slightly, or the set position can shift. To ensure that the identification system would be sufficiently robust to cope with this type of image variation, images with the target rotated and translated were included in the training dataset. As shown in Fig. 9, 25 points were set at 0.5 mm intervals. By moving to each point and rotating 360° at that location, images of objects transmitted at all angles were acquired. Transmission images were also obtained continuously, using the camera in video mode while the target was moving and rotating. Datasets were created by extracting an image from each frame of the video data.

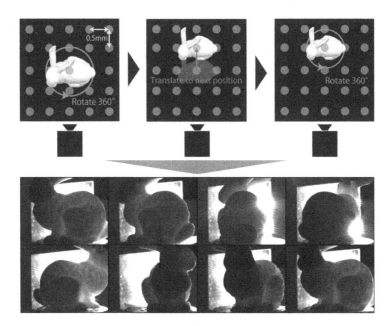

Fig. 9. Flow of dataset creation for deep learning. (top) How to move an object when creating a dataset. (bottom) An example of a transmission image that was learned.

Deep learning identification tests were performed using the datasets described above. ResNet [10] was employed as the learning model and was trained in advance to improve learning efficiency.

Datasets for the first learning test consisted only of the data from standard transmission images; i.e., no added rotation or noise. The progress of the accuracy rate is shown in Fig. 10(left). The accuracy of discrimination between similar images was high from the outset, due to the use of a trained neural network. As the classification rate reached 100% in some cases, a more difficult classification task was set for the next test.

For the second learning test, the dataset included pre-processed images (augmented data), e.g., Gaussian noise and rotation images (5, 10, and 15 degrees). The progress of the accuracy rate is shown in Fig. 10(right). Although the level of accuracy was lower than achieved in the first learning test, it was sufficient to distinguish between similar images. Hence, identification by deep learning is effective, even with noise added to the dataset.

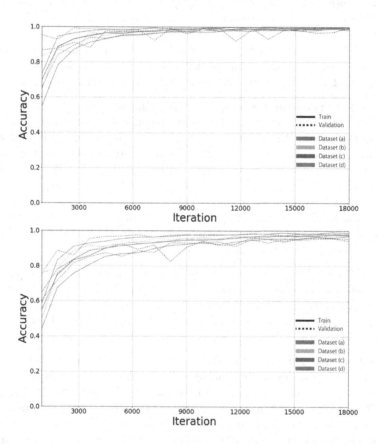

Fig. 10. Deep learning second test. (top) Graph of learning progress when there is no noise in the dataset. (bottom) Graph of learning progress when noise is added to the dataset.

5   Discussions

The proposed system was able to identify products with high accuracy in the identification trials, however, there are still many points to be discussed. The limitations of the study are described below.

5.1   Limitations and Improvements

Transmitted Light and Contrast: The experimental investigation described in this paper was conducted on the premise that the transmitted light and contrast resolution, as available from the optical setup, were sufficient to acquire the characteristics of the objects being imaged. In future, it will be necessary to determine the minimum required transmitted light intensity, and the threshold of the contrast required to resolve all details in transmitted images.

Product Shape: As a preliminary investigation, this study was limited, on the basis of time and cost, to using a cube as a simple model and a bunny as a complex model. Further verification of the deep learning method will require resources for a large number of products to be manufactured with specified conditions, and their transmission images to be acquired as datasets. Products that do not have a flat bottom will also require a fixture to enable NIR photography.

Unverified Parameters: Changes to the "Inner Support Structure" sub-parameters, and manufacturing errors were the only product modifications considered in this paper. While differences arising from both were verified as recognizable by the identification system, the effects of changing the remaining unverified parameters must also be evaluated. "Printing Width" and "Layer Thickness" are additional shape-related parameters (Table 2). Neither was tested in this investigation as both can affect the overall appearance of the product. However, in some product designs it is possible to change these parameters without affecting the appearance of the product. The effects of varying the parameters related to condition are also still to be tested.

Product Aging: In a real world distribution system, the product surface may be scratched or otherwise damaged during transportation, and product color can change over time. Although internal structure is less exposed to external impacts, the inner shape and the inner materials can undergo aging effects. The level of response of the detection system to such changes must be evaluated.

Optical System: The optical system is shown in Fig. 3. Although DCRA was used to increase the amount of backlight in the small scale optical setup, the optical system should not require such special optical elements. In this study, the DCRA often appeared in transmission images (Fig. 8). This problem needs to be addressed since transmission images are used as training data for the neural network, i.e. transmitted data must consist only of the object image, without any form of extraneous information. The optical system requires optimization in accordance with these issues.

Object Quantity: As shown in Table 4, 10 objects were prepared for each dataset. This is because it took a considerable amount of time from the production of objects to shooting videos for datasets, hence it was difficult to prepare a large number of objects. However, the actual item distribution system often handles more than 1000 items. In the future, it is necessary to study how to deal with such a large amount of data.

Deep Learning Downside: Deep learning was adopted to realize high-precision identification. This method always creates the necessity for preparation of datasets and pre-learning. Dataset creation is also inefficient within the current system. The process requires simplification.

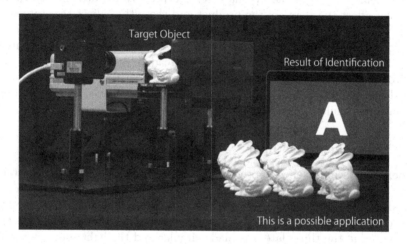

Fig. 11. This is a future application that, like a point of sales (POS) system, identifies an object in real time and returns the result when the object is placed.

6 Conclusion

A product identification system using transmission images was developed. Product parameters relevant to the system were defined, and identification experiments were conducted based on a subset of several selected parameters. An optical system for acquiring transmission images was constructed, and the acquired transmission images were observed. Two methods, feature matching and deep learning, were verified as identification methods, and their functionality and limitations were analyzed.

Future experiments will be conducted under more complex conditions, using parameters not yet verified and objects of various shapes. Improvements will be made to the system, enabling it to operate effectively under many conditions. It will then be a useful system for identifying additively manufactured products to which the information embedding method cannot be applied (Fig. 11).

A   A Layer Feature

The transmission image of the product is a photograph of the light intensity emitted from a certain point after passing through various paths. The smallest unit of a product that constitutes such a phenomenon can be defined as a "layer". Regardless of the shape, infill pattern, infill density, and infill position, the transmitted light has only passed through many layers. In other words, the stacked feature of each layer becomes a transmission image. We have already defined the characteristics of each layer in Sect. 3.2. Therefore, here we verified how the transmission image changes according to the change in the characteristics by changing the parameters of each layer.

The elements that make up the layer are broadly divided into two types: those that affect the shape of the product and those that affect the state of the product. The elements that affect the shape are "Layer Thickness" and "Printing Width". The elements that affect the state are "Printing Speed", "Temperature of Heat-Bed", "Temperature of Printing Head", "Retraction Distance", and "Environment". The reason why "Inner Support Structure" is not included in these elements is that "Inner Support Strucure" is established by the positional relationship of multiple layers.

A.1   Experimental Setup

In order to confirm the effect on the transmission image of one layer, we installed a near infrared light source, a thin plate equivalent to one layer, and a near infrared camera as shown in Fig. 12(a). To make it easier to see how the light source changed before and after passing through the layer, the shape of the light source entering the thin plate was made circular and the light source was collimated. However, since coherent light is used as a light source, the acquired image is accompanied by speckles. Therefore, in order to give priority to the measurement of the rough change of the light source shape, the influence of speckle was ignored and the saturation was allowed. Also, in order to observe how the light diffuses after passing through one layer, it is necessary to take a transmission image after passing through multiple layers. In addition, we prepared a foundation with grooves formed at intervals of 5 mm, and obtained a transmission image by fitting multiple sheets into the foundation while changing the distance between the sheets.

A.2   Effect of Parameters

Of the two types of elements that make up the layer, changes in shape are thought to have a large effect on the transmitted image. Therefore, we first conducted an experiment to observe changes in the transmission image when the parameters about the shape were changed. While "Printing Width" has the same value when using the same manufacturing equipment, "Layer Thickness" can be easily changed from the settings provided in the user interface. The transmission images were compared. Figure 12(b) shows what changes appear in

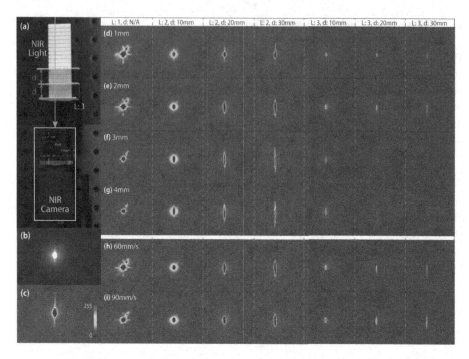

Fig. 12. Changes in transmission image due to layer characteristics. (a) a photograph showing the experimental setup for imaging. (b) the actually acquired transmission image. (c) an image of (b) displayed in pseudo color. (d)–(g) the change of the transmission image when "Layer Thickness" is changed. (h), (i) the change of the transmission image when "Printing Speed" is changed.

the transmission image when the layer thickness is changed. In this experiment, four layer thicknesses of 0.1 mm, 0.2 mm, 0.3 mm, and 0.4 mm were prepared. We confirmed that the transmission image changed according to the thickness and positional relationship of the layer.

Next, an experiment was conducted to observe the change in the transmitted image accompanying the change in the parameters affecting the state. "Printing Speed" was selected as an element related to the state. The reason for verifying only one of the many state parameters is that the change in the state is considered to have a lower influence on the transmission image than the change in the shape. Figure 12(h) and (i) show the imaging results. Compared to when the shape parameters were changed (Fig. 12(d)–(g)), the effect on the transmitted image was small.

As described above, we have compared the transmission images of elements related to shape and elements related to state. Although there was a considerable change in both, the shape change is considered to be easy to use as a clue for identification.

B   Robust Transmissive Images

When using a transmission image for object identification, it is desirable that the transmission image is not affected even if the orientation of the backlight or the position and orientation of the object changes. In order to achieve such robustness of the transmitted image, it is recommended that the transmitted light sufficiently passes through multipath. In general, reflected light, internal diffusion, and transmitted light are generated when light enters an object. If sufficient internal diffusion occurs, the transmission image obtained should be robust to the orientation and position of the target.

In order to verify the robustness of this system, we combined an XY scanning stage and a rotating stage to change the orientation and position of the target (Fig. 13(left)). The XY scanning stage is Thorlabs MLS203-1, and the rotary stage is Thorlabs PRMTZ8. In addition, for the object to be imaged, an object with a primitive shape was prepared: a cube with a size of $50 \times 50 \times 50$ [mm], an infill density of 10%, and an infill pattern of diamond fill. Figure 13(right) show the imaging results when the XY scanning stage is moved, the target is rotated, or the camera position is changed. Looking at the region of interest in each imaging result, it was found that the orientation and position of the target had little effect on the transmitted image.

Fig. 13. (left) a picture of an imaging system to verify robustness. (right top) the imaging results when the XY stage is moved in X and Y directions. (right middle) the imaging results when rotating the rotation stage. (right bottom) the result when the camera is moved in X direction and imaged toward the center of the object.

References

1. Al, C.M., Yaman, U.: Improving the strength of additively manufactured objects via modified interior structure (2017)
2. Baich, L.J., Manogharan, G., Marie, H.: Study of infill print design on production cost-time of 3D printed ABS parts (2016)

3. Bakar, N.S.A., Alkahari, M.R., Boejang, H.: Analysis on fused deposition modelling performance. J. Zhejiang Univ. Sci. A **11**(12), 972–977 (2010). https://doi.org/10.1631/jzus.A1001365

4. Bay, H., Tuytelaars, T., Van Gool, L.: SURF: speeded up robust features. In: Leonardis, A., Bischof, H., Pinz, A. (eds.) ECCV 2006. LNCS, vol. 3951, pp. 404–417. Springer, Heidelberg (2006). https://doi.org/10.1007/11744023_32

5. Cantrell, J., et al.: Experimental characterization of the mechanical properties of 3D-printed abs and polycarbonate parts. Rapid Prototyp. J. **23** (2017). https://doi.org/10.1108/RPJ-03-2016-0042

6. Chacón, J., Caminero, M., García-Plaza, E., López, P.: Additive manufacturing of PLA structures using fused deposition modelling: effect of process parameters on mechanical properties and their optimal selection. Mater. Des. **124**, 143–157 (2017). https://doi.org/10.1016/j.matdes.2017.03.065

7. Dogan, M.D., et al.: G-ID: identifying 3D prints using slicing parameters. In: Proceedings of the 2020 CHI Conference on Human Factors in Computing Systems, pp. 1–13. CHI 2020. Association for Computing Machinery, New York (2020). https://doi.org/10.1145/3313831.3376202

8. Fernandez-Vicente, M., Calle, W., Ferrandiz, S., Conejero, A.: Effect of infill parameters on tensile mechanical behavior in desktop 3D printing (2016)

9. Grimm, T.: Fused deposition modeling: a technology evaluation. Time-Compression Technol. **11**(2), 1–6 (2003)

10. He, K., Zhang, X., Ren, S., Sun, J.: Deep residual learning for image recognition. In: The IEEE Conference on Computer Vision and Pattern Recognition (CVPR) (June 2016)

11. KennyLAlvarez, C., RodrigoFLagos, C., Aizpun, M.: Investigating the influence of infill percentage on the mechanical properties of fused deposition modelled ABS parts (2016)

12. Kumar, P.R., Ahuja, I.S., Singh, T.: Application of fusion deposition modelling for rapid investment casting – a review (2012)

13. Lanzotti, A., Grasso, M., Staiano, G., Martorelli, M.: The impact of process parameters on mechanical properties of parts fabricated in PLA with an open-source 3-D printer (2015)

14. Li, D., Nair, A.S., Nayar, S.K., Zheng, C.: Aircode: Unobtrusive physical tags for digital fabrication. arXiv abs/1707.05754 (2017)

15. Lowe, D.G.: Distinctive image features from scale-invariant keypoints. Int. J. Comput. Vis. **60**(2), 91–110 (2004). https://doi.org/10.1023/B:VISI.0000029664.99615.94

16. Maekawa, S., Nitta, K., Matoba, O.: Transmissive mirror device with micro dihedral corner reflector array-the mirror forming a real image. ITE J. **30**, 49–52 (2006)

17. Maia, H.T., Li, D., Yang, Y., Zheng, C.: LayerCode: optical barcodes for 3D printed shapes. ACM Trans. Graph. **38**, 112:1–112:14 (2019)

18. Miura, N., Nagasaka, A., Miyatake, T.: Feature extraction of finger-vein patterns based on repeated line tracking and its application to personal identification. Mach. Vis. Appl. **15**(4), 194–203 (2004). https://doi.org/10.1007/s00138-004-0149-2

19. Miura, N., Nagasaka, A., Miyatake, T.: Extraction of finger-vein patterns using maximum curvature points in image profiles. IEICE Trans. Inf. Syst. **90**(8), 1185–1194 (2007)

20. Novakova-Marcincinova, L., Kuric, I.: Basic and advanced materials for fused deposition modeling rapid prototyping technology (2012)

21. Okada, A., Silapasuphakornwong, P., Suzuki, M., Torii, H., Takashima, Y., Uehira, K.: Non-destructively reading out information embedded inside real objects by using far-infrared light. In: SPIE Optical Engineering + Applications (2015)
22. Popescu, D., Zapciu, A., Amza, C.G., Baciu, F., Marinescu, R.: FDM process parameters influence over the mechanical properties of polymer specimens: a review. Polym. Test. **69**, 157–166 (2018). https://doi.org/10.1016/j.polymertesting.2018.05.020
23. Rangisetty, S., Peel, L.: The effect of infill patterns and annealing on mechanical properties of additively manufactured thermoplastic composites, p. V001T08A017 (Sept 2017). https://doi.org/10.1115/SMASIS2017-4011
24. Rankouhi, B., Javadpour, S., Delfanian, F., Letcher, T.: Failure analysis and mechanical characterization of 3D printed abs with respect to layer thickness and orientation. J. Fail. Anal. Prev. **16**, 467–481 (2016)
25. Roberson, D.A., Espalin, D., Wicker, R.B.: 3D printer selection: a decision-making evaluation and ranking model (2013)
26. Schofield, J., Dawson, M.: Evaluation of dimensional accuracy and material properties of the makerbot 3D desktop printer. Rapid Prototyp. J. **21**, 618–627 (2015). https://doi.org/10.1108/RPJ-09-2013-0093
27. Singh, S., Ramakrishna, S., Singh, R.: Material issues in additive manufacturing: a review. J. Manuf. Process. **25**, 185–200 (2017). https://doi.org/10.1016/j.jmapro.2016.11.006
28. Song, W., Kim, T., Kim, H.C., Choi, J.H., Kong, H.J., Lee, S.R.: A finger-vein verification system using mean curvature. Pattern Recogn. Lett. **32**, 1541–1547 (2011)
29. Suzuki, M., Matumoto, T., Takashima, Y., Torii, H., Uehira, K.: Information hiding inside 3-D printed objects by forming high reflectance projections. In: ICVIP (2017)
30. Tanaka, K., Mukaigawa, Y., Matsushita, Y., Yagi, Y.: Descattering of transmissive observation using parallel high-frequency illumination. In: IEEE International Conference on Computational Photography (ICCP), pp. 1–8 (2013)
31. Torrado, A.R., Roberson, D.A.: Failure analysis and anisotropy evaluation of 3D-printed tensile test specimens of different geometries and print raster patterns. J. Fail. Anal. Prev. **16**, 154–164 (2016)
32. Torres, J., Cole, M., Owji, A., DeMastry, Z., Gordon, A.P.: An approach for mechanical property optimization of fused deposition modeling with polylactic acid via design of experiments (2016)
33. Ueno, R., Mitsugi, J.: Barcode fingerprinting: Unique identification of commercial products with their jan/ean/ucc barcode, pp. 416–420 (Feb 2018). https://doi.org/10.1109/WF-IoT.2018.8355122
34. Willis, K.D.D., Wilson, A.D.: Infrastructs: fabricating information inside physical objects for imaging in the terahertz region. ACM Trans. Graph. **32**, 138:1–138:10 (2013)
35. Wong, K.V., Hernandez, A.: A review of additive manufacturing. ISRN Mech. Eng. **2012**, 10 (2012)
36. Yang, J., Shi, Y.: Finger-vein ROI localization and vein ridge enhancement. Pattern Recogn. Lett. **33**, 1569–1579 (2012)
37. Yang, W., Huang, X., Zhou, F., Liao, Q.: Comparative competitive coding for personal identification by using finger vein and finger dorsal texture fusion. Inf. Sci. **268**, 20–32 (2014)

AI Applications in Smart Environments

A Tactile User Device to Interact with *Smart Environments*

Pietro Battistoni[(⊠)] and Monica Sebillo

University of Salerno, Fisciano, SA 84084, Italy
{pbattistoni,msebillo}@unisa.it

Abstract. *Smart Environments* refers to ambients integrated with smart technologies and should provide an easy-to-use and best experience to users. With the introduction of Intelligence Ambient, where Artificial Intelligence contributes to the interaction between people and the ambient's Smart Devices, the Human-Computer Interaction can be more advanced. This paper proposes an advanced tactile small device that allows people to interact with smart devices in their proximity, without the need for any audio and visual interface.

Keywords: Human-Computer Interaction · Ambient Intelligence · Tactile interfaces · Tangible interfaces · Haptic interface

1 Introduction

In 1998 Eli Zelkha, with Simon Birrell, coined the Ambient Intelligence (AmI) term. It was the vision of the pervasive computers in daily life, influenced by the human-centered design paradigm. The *intelligence* term denotes the technology capacity to learn, and in this specific case, learn how to interact with humans and their environments. After more than two decades, with the growth of the Internet of Things (IoT), where the number of Internet Protocol (IP) connected devices will reach three times the world population in 2023 according to [8], the concept of pervasive computing becomes finally tangible.

Nowadays, AmI indisputably evokes Artificial Intelligence (AI) with Machine Learning (ML) to serve humans according to the ambient context through Smart Devices, which are electronic devices connected to other devices or networks. In most AmI scenarios, AI understands the context and the needs of humans and, as a consequence, enables the actions of devices to satisfy such needs. For example, after recognizing an apartment's inhabitant, the AI turns off the alarm system, opens the door, and lets her/him in. Furthermore, inside the apartment, the AI can identify the existing context, such as a light condition, ambient temperature, and person location, to decide whether turn on the light or open the curtains, arranging the heating if necessary.

This research was funded by MIUR, PRIN 2017 grant number 2017JMHK4F_004.

H. Degen and S. Ntoa (Eds.): HCII 2021, LNAI 12797, pp. 461–471, 2021.
https://doi.org/10.1007/978-3-030-77772-2_30

Such a situation could be desirable for a person and indispensable for some impaired ones, but although humans always desired machines to serve them as much as possible, such a self-governing AI capability raises some concerns [3,4]. Although some of the Sci-Fi plots, where machines with full autonomy outperform humans and become too powerful to control, have become more plausible, other reasons justify the emergence of the hybrid human-artificial intelligence research area. In [7] the authors well argue that an active human-AI interaction not only could permit better human control of resulting devices action but, as it happens when humans team up to perform a task that none of them could do alone, this collaboration could solve some of the AI weakness, and achieve solutions that not the AI nor the humans could achieve alone. Indeed, although AI well performs implicit knowledge or hidden patterns from large-scale data, it still lacks reasoning, inference, and instinct judgments on dynamic and multiple factors, which are instead well performed by humans.

To summarize, to have beneficial performances, the AI must acquire, as well as possible, the knowledge of context, the location of people in such a context, and some human inputs to contextualize the actions. Sensors supply the most information to the AI about the environmental context and people location, while a practical and straightforward Human-Computer Interface (HCI) can supply the required interaction.

There are still significant technical challenges about how the AI acquires the context and even more challenges on how the HCI allows the AI to understand the human requirements in a specific context.

The latter is not merely how to give commands to perform on a specific device, as it can be done by remote control, but it is about how to let AI understand the person's requirements to autonomously decide how to control the Smart Devices to satisfy them.

There are devices, such as Amazon Echo with Amazon Alexa inspired by the Star Trek computer [20], which allow people to communicate more naturally with computers and Smart Environment, thanks to the Conversational AI [21]. Amazon Echo uses large volumes of data, machine learning, and natural language processing to imitate human interactions by recognizing speech and text inputs and translating their meanings to the AI.

Aside from the obvious observation that voice commands are not inclusive to deaf people, there are other situations where audio or visual interfaces are not appropriate. Within noisy environments at home, and other AmI scenarios in working environments, such as offices, health care places, and factories, the voice commands could disturb other people or could not be well received by the device, so modalities different from them could be required. Furthermore, although visual interfaces are functional when integrated into the devices like smartphones or personal computers, there are some circumstances where the sight must not be on the interface while using them. The following examples should clarify the idea. When using a remote controller to switch the TV channel, the user prefers to look at the TV to get feedback on what is happening instead of looking at the remote controller. Likewise, driving a car or a crane, it is

preferable to look at the road and not at the steering wheel. Thus, envisioning places like homes, working environments, crowded public places, factories, and hospitals, this paper aims to propose an entirely tactile simple interface, which allows interacting with the AI of the AmI, by using only one hand without the need to talk or move the sight from what the user is doing.

Moreover, the tactile interaction proposed in this paper is bidirectional. It means that through the tactile sense it will also be possible to receive feedback from the intelligent ambient. Actually, the literature frequently describes AmI as invisible technologies, recalling the 1991 statement by Mark Weiser [24] 'The most profound technologies are those that disappear.' It implies that when entering in contact with AmI, the human needs a way to perceive the *Smart Environment* that cannot be seen otherwise.

This research was conceived by envisioning a guiding scenario as follows. A person carrying in (her)his pocket a small device could feel a tactile sensation when entering a *Smart Environment* governed by AmI solution. In this context (s)he can hold the device with one hand and start interacting with the smart devices in the proximity, just by using the fingers, without looking at the interactive device or earring any audio signal.

The paper proceeds by describing some related works in the next Section. Section 3 describes the HCI device, and Sect. 4 depicts one of its possible uses. Conclusions follow.

2   Related Work

This section reports on previous research on tactile interfaces, which somehow influenced or encouraged the research work proposed in this paper, and recalls some research on the automated environment discouraging automation complete independence without human interaction.

The tactile sense can be stimulated by feedback from the device, usually referred to as haptic feedback in literature, and it can be used as input to the device as a touch screen and a computer touchpad do. Common examples of haptic feedback are the vibration of a smartphone and a joystick force feedback.

In [19] the authors present a haptic display for small devices. The work highlighted the difficulty and the value of implementing a tactile output on small devices. Focusing mainly on haptic feedback since the device already has a touch screen as input, the work explains how useful it can be to receive action feedback without looking at the interface, which is one of the present paper motivations. However, it addresses the usage of a small screen that the present research work aims to avoid.

Ozioko et al. [18] present a wearable tactile communication interface with vibrotactile feedback for assistive communication. The interface demonstrates the effectiveness of the tactile communication method used not only by deafblind people. In this case, the work does not offer a tactile input device solution, as proposed in this paper.

Kashyap et al. in [12] emphasize the need for appropriate user interfaces and problems of full-automation, lack of control, and the complexity of the everyday smart devices environment, while [3] indicates that users do not accept a fully automated system. However, although several attempts have been made to provide solutions with complementary explicit interaction, these topics have remained little explored.

By [2], Becker et al. compare three scenarios, controlling Appliances Through Wearable Augmented Reality. The paper proposes the use of multiple devices to wear and three different modalities. From results it came out that a tangible interface has some preferred use, compared to the virtual gesture interfaces.

3   The Cube-Shaped User Device

The research activities done at the Laboratory of Geographic Information Systems (LabGIS) of the Department of Computer Science (University of Salerno) were to identify solutions that let humans interact with an AmI by a fully tactile interface device. The rationale behind this is that a tactile interface, if small enough, could be used by one hand only and could avoid audio or visual actions when these are not feasible.

3.1   Premises

Although it is possible to stimulate the real tactile perception by mechanical means, many research studies focus on the generation of the tactile illusion, i.e., the misleading sensation of tactile perception. It is a more flexible way to reproduce a piloted tactile sensation with electronic devices [25]. A haptic output artificially generated can produce a tactile illusion. For example, [1] studies an electrotactile feedback that can reproduce the texture sensation on a touch screen. It is an illusory sensation as the touch screen does not change its physical texture, but a piloted current passes on its surface, generating such an illusion on human fingers. Another example is given by Brewster et al. in [6], who studied vibrotactile messages, which can be used for non-visual information. The authors described various solutions to message with vibrotactile *Roughness* and *Rhythm* illusion, easily generated by electronic devices.

Tactile feedback such as force, pressure, and roughness are the primary sensory inputs presented to a user using a haptic display [14], but the human tactile sense has thermal receptors too [14], and some works studied how effective it could be to incorporate thermal feedback into haptic devices [11].

Following these research fields, the design of the device described in this paper addressed the generation of tactile illusions by electronic solutions to send user feedback.

Since our device requirements establish that it should be small enough to be manipulated by one hand only, the device should have reduced space for mechanical parts and the battery. Therefore, a tiny battery fixed severe restrictions on power consumption, and a small embodiment reduced the possibility

of using sophisticated mechanical actuators. Hence, the device uses a vibrotactile messaging solution with an additional thermal sensation to represent the AI feedback, and the users could message to the AI using manipulation of the device and tapping on its faces.

3.2 The Design

The work started with designing a new low power, wireless device electronic circuit and a wireless architecture that could be easy to deploy in a *Smart Environment* governed by AI.

The hardware components selected for the device were chosen considering its features as well as the size. They allowed the realization of a cube-shaped device with an edge of 38 mm.

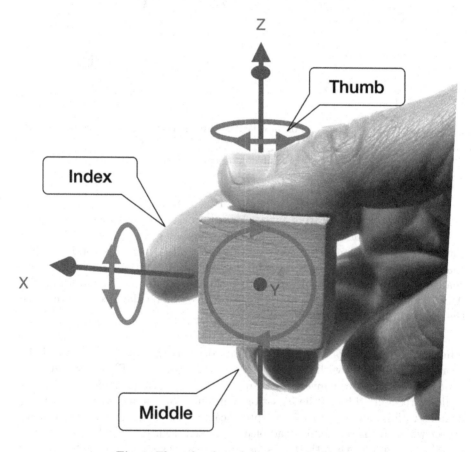

Fig. 1. The cube-shaped device manipulations

Figure 1 shows how it is possible to hold a cube with three fingers and rotate it.

Thumb and *Middle fingers*, hold the cube, while the *Index finger* is free to tap on the cube face below it. For each of the six cube faces that could stay under the *Index Finger*, it is possible to have four different faces under the *Thumb* just rotating the cube on the X-axis. It is so possible to have 6 ∗ 4 = 24 distinct positions of the cube. Furthermore, it is possible to give a distinct meaning to the single and the double-tap of the *Index Finger* for each of the twenty-four positions, reaching forty-eight discrete inputs.

Moving from one position to another is possible through cube rotation, and the rotation sequence can also be associate with various additional inputs.

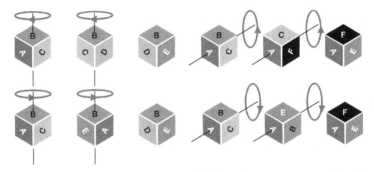

(a) Rotation sequences on face B   (b) Rotation sequences on face A

Fig. 2. Possible cube manipulation

Indeed, Fig. 2a shows that bringing the Face D to the position of Face A can be done in two ways, rotating twice clockwise or twice anticlockwise the face B. Each direction of rotation can assume a different meaning, even if reaching the same end position. The same consideration fits other axis rotations. For example, let us suppose that a person, located in an AmI environment, would change the heating conditions, then to communicate to the AI governing the AmI this request (s)he could rotate forward the cube on the face A to increase the temperature, or backward to decrease it (Fig. 2b).

The Embodiment for the prototype, once designed with a 3D CAD, was built by rapid prototyping with a Fused filament fabrication (FFF) 3D printer. The cube faces textures were investigated so that the cube positions under the fingers could be recognized by tactile sensation. Gibson in [9], conducted an experiment demonstrating that the tactile receptors can better recognize the form of an object when this is rotated instead of just pressed. Hence, the cube face texture is easier recognized when the face slides under the fingers, i.e., while the cube is rotated. The design focused mainly on a pattern that could enhance the tactile sensation while a face passes from one finger to another during a cube rotation. Other experiments by Lederman in [15] confirmed that the grooves and lands are recognized by the tactile receptors, better if the groves are large. Following such

indications, the six cube faces were designed to make consecutive face patterns unique for each possible rotation (Fig. 3a).

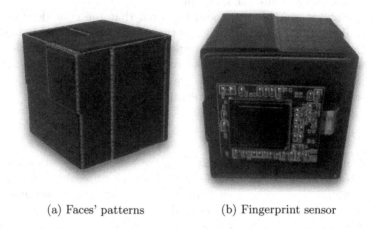

(a) Faces' patterns   (b) Fingerprint sensor

Fig. 3. Cube device prototype

One of the cube faces has a fingerprint sensor (Fig. 3b). The sensor allows user recognition and the choice of several personalized profiles. In this way, users start by holding the cube in a predetermined position, defining position zero. Starting always from a prefixed position gives an absolute origin to the cube possible manipulation and an easier way to associate the consecutive manipulations to a set of purposes.

4   Using the Cube

This section reports on a use case scenario to describe a possible cube application and facilitate understanding the device possible utilization. The scenario refers to the domotic environment to offer a familiar ambient for most of us.

A user entering an AmI environment feels the vibrating message by the cube kept in the pocket. This message informs the user of the presence of an AmI, which governs the smart devices present in the ambient. When the cube is in close proximity to a smart device, it warns the user of the possibility to interact, still vibrating but with a different pattern. If the user wants to interact with the device, s/he takes the cube and put (her)his finger on the cube fingerprint sensor, activating the interaction. Successively, the messages to AmI are given by tapping and rotating the cube faces.

For example, let us suppose that the user desires to change the room lighting by opening more the curtains. (S)he can inform the AmI by taking the cube and rotating forward the cube face in the curtains proximity. The AmI sends the correct command to the curtains controller to open them. The curtains are

now fully open, but the user turns forward the cube to inform the AmI that still (s)he desires more light. Then, the AmI has to take appropriate actions to satisfy the request. Probably it decides to dim up the lamp, and the user with the same action can regulate the lamp dimmer.

From the above example, it should be noted that the cube does not give the command to the devices but messages the requests to the AmI that should interpret them according to the context and circumstances. For example, if the previous scenario happens during the night, the AmI should understand that the cube forward motion cannot mean more light, but something else, to infer by the user's known habits or other profile information AmI know.

4.1   User's Actions

As already mentioned, after fingerprint scan, the cube has one established face under the user's *Index Finger*. The actions the user can perform are, *Tap* on the face under the *Index Finger* and rotate one or more steps the cube before tapping again. The tap can be single or double. Since there are available 24 distinct positions and for each of these, it is possible to tap a single time or double-times, a total of 48 distinct actions can be recognized univocally. By keeping the cube on the hand, it emerges that, referring to Fig. 1 the most instinctive manipulations that the user can perform to rotate the cube are those where the cube rotates forward, i.e., anticlockwise on the $Y - axis$, backward, i.e., clockwise on the $Y - axis$, leftward, i.e., anticlockwise on the $Z - axis$, and rightward, i.e., clockwise on the $Z - axis$. Although the rotations on $X - axis$ are possible, they require a little bit more dexterity.

Within this paper the term *manipulation* is used and not *gesture* because the latter is widely used for hand motion on free air without constraints, but the cube already set some constraints by its nature, and others are set to make the interaction as simple as possible. Indeed, the cube position is referred to the fingers touching the cube faces and not to the cube position in the space, thus offering more comfort and freedom of use.

4.2   Networking

The cube device communicates with the AmI through the Wi-Fi and recognizes the nearby smart devices by Bluetooth Low Energy (BLE) protocol. The BLE has the advantage that it uses very low power for short-range communication, allowing the battery-powered devices to last for years, even with small-sized batteries. Instead, Wi-Fi, used for long-range communication, needs accurate power management, hardware, and software, to achieve reduced battery consumption.

The cube device communicates to AmI by messages, using one of the most used IoT communication protocols, the Message Queuing Telemetry Transport (MQTT). It is a lightweight messaging protocol based on the publish/subscribe pattern. The publish/subscribe pattern [10] offers asynchronous interaction among network nodes in almost real-time.

Based on a client/server model, (the clients) can publish and subscribe to a specific *Topic* connecting to a known *Broker* (the server). The MQTT Broker will forward the messages published to a specific Topic to any subscribers of such a *Topic* [16].

The connection between publishers and subscribers is governed by the *Broker*. More that one *Subscriber* can *Subscribe* on the same *Topic* and more that one *Publisher* can *Publish* to the same *Topic*. The publishers and the subscribers do not need to know each other, they only need to know the address of the *Broker* to connect.

The client code of MQTT has a small footprint and, for this reason, can be deployed on high constrained devices, such as the cube hardware. Whenever the user taps the cube device or changes its position by rotating it with the fingers, the device publishes a message communicating the action.

The device running the Inference engine for the AI of the AmI had subscribed to the necessary *Topics* and then can receive those messages in almost real-time. Once received, the message is given as input to the Inference Engine, which translates it into the appropriate action to actuate the smart devices.

Since the cube device can be an MQTT subscriber, the AI device can send feedback messages to the cube device, publishing to the appropriate *Topic*. So, the cube device can receive the feedback messages and actuate its peripherals for the output.

The cube device has a vibrating motor activated to reproduce vibrotactile messages as feedback from the AI, and nichrome wire embedded in the cube faces (Fig. 4) getting heated when crossed by the electrical current, increasing the cube device haptic feedback ability.

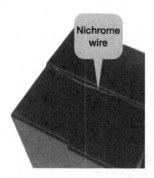

Fig. 4. Nichrome wire

4.3 The Cube Firmware

The cube device has a dual-core 32 bits microcontroller. One core is dedicated to the wireless communication protocols, BLE, and Wi-Fi stack, and the other is for the device programming, which includes the MQTT client and the drives of

Inputs and Outputs peripherals. Furthermore, the firmware, rendering the user cube manipulation, has to compose the message to publish, which essentially will communicate to the AI the user's intentions.

5   Conclusions

A completely tactile and tangible cube-shaped interface device was proposed to interact with AI in the AmI. The research aimed to offer an additional way of interaction within the AmI environments to worried users about the loss of control in fully automated environments, and improve the AI perception of the users' needs where audio or visual interfaces are not fitting.

The proposed cube device is small enough to be used with one hand only and able to get tactile inputs from users as well as to provide haptic feedback. It offers users interaction opportunities with smart devices in multiple *Smart Environments* governed by AmI solutions, avoiding visual and natural language speaking but keeping the conversational AI paradigm for AmI.

The device is not intended as a personal device but, recognizing the user by the biometric sensor, becomes full user-tailored.

The device produces two kinds of haptic feedback: vibrotactile messaging solution and bland thermal changes. While the vibrotactile messaging has much literature evidence [5,6,13,14,18,23], the temperature perception and Thermal-Interface [11,17] need further investigation, mainly because, for this work, it was inspired by the designer [22] and not by a user-centric approach.

References

1. Altinsoy, M.E., Merchel, S.: Electrotactile feedback for handheld devices with touch screen — texture reproduction. In: 2011 IEEE International Conference on Consumer Electronics (ICCE), pp. 59–60 (2011)
2. Becker, V., Rauchenstein, F., Sörös, G.: Connecting and controlling appliances through wearable augmented reality. Augmented Hum. Res. 5(1), 2 (2019)
3. Ben Allouch, S., van Dijk, J.A.G.M., Peters, O.: The acceptance of domestic ambient intelligence appliances by prospective users. In: Tokuda, H., Beigl, M., Friday, A., Brush, A.J.B., Tobe, Y. (eds.) Pervasive 2009. LNCS, vol. 5538, pp. 77–94. Springer, Heidelberg (2009). https://doi.org/10.1007/978-3-642-01516-8_7
4. Braun, A., Wichert, R., Maña, A. (eds.): AmI 2017. LNCS, vol. 10217. Springer, Cham (2017). https://doi.org/10.1007/978-3-319-56997-0
5. Brewster, S., Brown, L.M.: Tactons: structured tactile messages for non-visual information display. In: Proceedings of the Fifth Conference on Australasian User Interface - Volume 28. AUIC 2004, pp. 15–23. Australian Computer Society Inc., AUS (2004)
6. Brown, L., Brewster, S., Purchase, H.: Multidimensional tactons for non-visual information presentation in mobile devices. In: MobileHCI 2006, vol. 159, pp. 231–238, January 2006
7. Chen, L., Ning, H., Nugent, C.D., Yu, Z.: Hybrid human-artificial intelligence. Computer 53(08), 14–17 (2020). https://doi.org/10.1109/MC.2020.2997573

8. Cisco: Cisco annual internet report (2018–2023) white paper. Technical report, CISCO (2018)
9. Gibson, J.J.: Observation on active touch. Psychol. Rev. **69**(6), 477–491 (1962)
10. Jacobsen, H.A.: Publish/Subscribe, pp. 2208–2211. Springer, Boston (2009)
11. Jones, L.A., Berris, M.: The psychophysics of temperature perception and thermal-interface design. In: Proceedings 10th Symposium on Haptic Interfaces for Virtual Environment and Teleoperator Systems, HAPTICS 2002. pp. 137–142 (2002)
12. Kashyap, H., Singh, V., Chauhan, V., Siddhi, P.: A methodology to overcome challenges and risks associated with ambient intelligent systems. In: 2015 International Conference on Advances in Computer Engineering and Applications, pp. 245–248 (2015)
13. Lederman, S.J., Jones, L.A.: Tactile and haptic illusions. IEEE Trans. Haptics **4**(4), 273–294 (2011)
14. Lederman, S.J., Klatzky, R.L.: Haptic perception: a tutorial. Attention Percept. Psychophys. **71**(7), 1439–1459 (2009)
15. Lederman, S.J.: Tactile roughness of grooved surfaces: the touching process and effects of macro- and microsurface structure. Percept. Psychophys. **16**(2), 385–395 (1974)
16. Light, R.A.: Mosquitto: server and client implementation of the MQTT protocol. J. Open Source Softw. **2**(13), 265 (2017)
17. Manasrah, A., Crane, N., Guldiken, R., Reed, K.B.: Perceived cooling using asymmetrically-applied hot and cold stimuli. IEEE Trans. Haptics **10**(1), 75–83 (2017)
18. Ozioko, O., Karipoth, P., Hersh, M., Dahiya, R.: Wearable assistive tactile communication interface based on integrated touch sensors and actuators. IEEE Trans. Neural Syst. Rehabil. Eng. **28**(6), 1344–1352 (2020)
19. Poupyrev, I., Maruyama, S.: Tactile interfaces for small touch screens. In: Proceedings of the 16th Annual ACM Symposium on User Interface Software and Technology, UIST 2003, pp. 217–220. Association for Computing Machinery, New York (2003)
20. Prasad, R.: Alexa at five: looking back, looking forward. https://www.amazon.science/blog/alexa-at-five-looking-back-looking-forward
21. Ram, A., et al.: Conversational AI: The science behind the alexa prize (2018)
22. Saffer, D.: Designing for interaction: creating innovative applications and devices. New Riders (2010)
23. Spiers, A.J., Dollar, A.M.: Design and evaluation of shape-changing haptic interfaces for pedestrian navigation assistance. IEEE Trans. Haptics **10**(1), 17–28 (2017)
24. Weiser, M.: The computer for the 21st century. https://www.scientificamerican.com/article/the-computer-for-the-21st-century/
25. Yem, V., Kajimoto, H.: Comparative evaluation of tactile sensation by electrical and mechanical stimulation. IEEE Trans. Haptics **10**(1), 130–134 (2017)

Towards an Ambient Smart City: Using Augmented Reality to Geospatially Filter the Right Airbnb via Artificial Intelligence

Shreya Chopra[1,2(✉)] and Omar Addam[1]

[1] University of Calgary, Calgary, AB T2N 1N4, Canada
{shreya.chopra,okaddam}@ucalgary.ca
[2] Petrotranz - A WoodMackenzie Business, Calgary, AB T2P 0T9, Canada

Abstract. We investigate an augmented reality (AR) based method to help the user filter their preferred Airbnb home via artificial intelligence (AI) on a 3-dimensional street view style map in the context of a smart home and smart city ambient scenario. We are motivated by seeking to provide users a 3-dimensional AR interface that: provides equal representation of all data points and allows visual filtration of all filter dimensions directly on the map. We introduce a novel way of visually filtering information geospatially on an AR map by adding information instead of removing it called filtration by inclusion. We introduce the interface in its environmental context in terms of strategic locations for embedding it around the city. We demonstrate how a neural network model can be applied to our filtration technique to act as a human and make filtration decisions. We enforce how technology should enable, not limit the user and discuss how our system does not filter out a home if it does not match the filter criteria, but rather provides the visual information for the user to be able to make that decision on their own as human choices are fluid. We expand upon some limitations of our system and discuss future work on this technology. Our work has implications for user experience design in AI at its intersection with AR big data and situational interaction used most specifically for the context of smart communities and smart cities.

Keywords: Airbnb · Ambient environment · Artificial intelligence · Augmented reality · Big data · Geospatial · Human computer interaction · Internet of Things · Machine learning · Neural networks · Smart city · Smart community · Situational · User experience design

1 Introduction

Artificial Intelligence (AI), the broad science of mimicking human abilities, and *Machine Learning* (ML), the subset of AI in which machines are trained how to learn, have increasingly become a part of how we interact with our environment in recent times [2, 19]. *Ambient Environment* (AE) refers to of or relating to the immediate surroundings;

S. Chopra—Work in Petrotranz at the time of publication.

H. Degen and S. Ntoa (Eds.): HCII 2021, LNAI 12797, pp. 472–489, 2021.
https://doi.org/10.1007/978-3-030-77772-2_31

oftentimes specifying an all-encompassing aspect of the surroundings such as sound, light, temperature, etc. From an industrial point of view, one such example of AI in an ambient Environment is the autonomous Google Car [2]. In cases like this, AI has been implanted to process big data and formulate a resulting action from this. The recent Google Car processes a gigabyte of data per second and includes: a laser on top that determines where it is as well as a radar in the front that detects the speed and motion of all the other cars around it. Artificial Intelligence comes in when it takes all this data and predicts what potential drivers around the car are going to do and uses this information to formulate a successful course of action so that it can safely continue to its destination. Since AI is the science of logically thinking like a human mind. In specific terms, to create an AI algorithm, we first realize that even though human minds are able to think very non-linearly, usually, a flow chart/algorithm can be determined on how humans come to conclusions. In this way, we can teach a computer to learn binary decision-making and non-binary decision-making (with the addition of hidden neural network layers) to simulate a human brain. Machine learning can be very beneficial in that agents do not require programming to learn but can rather learn based on examples. These examples can either be picked up via specific descriptive labels (non-deep learning) or without humans having to specify labels (deep learning).

Augmented Reality (AR), technology that superimposes a computer-generated image on a user's view of the real world to provide a composite view, and the *Internet of Things* (IoT), the connection of everyday objects to the Internet, enabling them to send and receive data to each other [16], are increasingly intersecting at the point of ubiquitous computing [5, 6, 8, 11]. Examples of this being smart home and smart city research where sensor-driven scenarios are interacted with via headset AR such as a Microsoft HoloLens.

1.1 Motivation

Sifting through big data requires a big user interface: at least that is the supposed implication of some current examples of data interaction. What happens when that data is geographically linked to a location? Or can benefit from being linked to sensors in physical space? This is the case of real estate or Airbnb data: it is geographically linked and can benefit the smart community/smart city by being available at certain locations. We are inspired by the tourism aspect of big data and how that can benefit the smart city. Thus, we look at the current Airbnb website user interface [1] and how users are able to utilize the map to find a home based on what they are looking for.

The Airbnb website interface is an example of how maps are used in conjunction with listings to provide the user with information of available homes. We are motivated by aiming to improve this even further by providing filter information visually on the map as an alternative to users combining two pieces of information (map and listings) to decipher what they are looking for. For this purpose, we make use of headset augmented reality to be able to provide more than 2 dimensions of information on the map. Moreover, we seek to make use of the virtually unlimited real estate that augmented reality has to offer. We choose AR as a medium also because it allows the user the affordances of their physical environment in conjunction with a virtual augmentation that can be tied together via a sensor.

Fig. 1. [Left] Example map search on a generic webpage filter such as the one on the Airbnb webpage. Filters including price, rating, number of people, number of bedrooms, and type of room. [Top Right] Home listings that can dynamically appear/disappear based on the user's movement of the map to filter based on location. [Bottom Right] Number of pages that the user will have to go through to view all available listings under current filters: implying that all homes cannot be compared side by side.

1.2   Research Questions

Our research aims to answer the following questions:

1. How can we leverage headset augmented reality to further improve on generic map filtration that is currently interacted with via webpages?
2. How can we embed such a system into an ambient smart city or smart community environment?
3. How can we use artificial intelligence on our filtration technique to facilitate suggestions to the user with the benefit to the user as well as the smart city?

2   Related Work

We look at other **smart city applications** to determine what already exists especially in terms of Augmented reality and can perhaps be integrated into an ambient environment. Emergency response management is an example of this. Crowdsourcing via citizens in a smart city is another example of embedding information into an environment [21]. Utilizing AI for ER management is also examined [13]. Today, cities **looking to become smart cities** are pushing forward via open data dashboards available for anyone to use [3, 10, 22]. The aim is to increase citizen engagement and innovative collaboration such that the city can grow in the direction of a smart city. Cities also host map webpages on their governmental sites. Thus, maps and the data dashboard can be combined for the purpose of innovation. **Filtration techniques** are a majorly scrutinized aspect of big data, and visual filtration techniques are always being further researched on as a subset

of information visualization [17, 20]. Specifically, also in terms of smart community or government big data [20]. Agile methodologies of developing user experience design also exist [12]. Currently, **generic filters** exist in many consumerist applications such as shopping catalogues, Airbnb, real estate, etc. These filters are very similar in that they remove items that do not match the user criteria upon Boolean decisions of whether a criterion is matched or not. **Artificial Intelligence in filters** may come mostly in the form of Artificial Narrow Intelligence [ANI] such as what a user should watch on Netflix. Airbnb already makes use of multiple artificial intelligence algorithms to assist the user in finding an ideal home: especially in terms of what it thinks the user would prefer [14]. For example, listings do not appear in alphabetical order, but are ranked in terms of similarities with what the user has clicked on, how long the user has looked at each listing, and the places where it thinks the user is most likely to stay. We seek to augment this style of algorithm with a meta interface that lets the user filter geospatially in a visual manner. We start off in ANI and go towards more advanced AI to mimic user ability.

3 Augmented Reality Interface

3.1 Filtration by Inclusion

Human choices are **fluid**. Technology should **enable, not limit** humans while making choices. We introduce a geospatial filtration methodology that enables the user with flexibility while providing all visual filtration cues directly on the map [7]. We utilize the Airbnb dataset scraped on *Tom Slee's* website [23] to develop the required filters: maximum price per night (0–500), minimum average rating (0–5), minimum number of people that the home accommodates (0–5+), minimum number of bedrooms in the home (0–5+), and the room type (shared, private, or entire home/apartment). We create a holographic filtration panel to appear above a holographic map so that the user can directly reach forward and interact with the filters. Each filter is denoted by a colour that will be used to denote home qualities on the map. The interaction mode for the filters are sliders along with toggles for the home type (Fig. 2).

We make use of a holographic map library to display the map to the user in Google Earth fashion. Each available Airbnb accommodation is denoted by a holographic grey cube sitting on top of the map. As the user applies each filter, holographic cubes of the corresponding color are dynamically added on top of each home that matches that filter criteria. Each home becomes a stack of cubes: the tallest ones being the closest match to the user's filter criteria. In this way, users can always see all accommodations (even if they are filtered out) and make decisions based on which colours appear on each home as well as which home stacks are the tallest. In addition, the map library allows the user to zoom in or out. When the user does either, the interface reacts by displaying more or less homes at a time. When the user zooms in, each home stack appears bigger, and when they zoom out, each stack appears smaller.

Since our filtration technique *adds information* to the matching homes *instead of removing information* of the filtered-out homes, we call it **filtration by inclusion**. The benefit being that all homes and their information are always available to see: allowing the user to more easily change their criteria or compromise on some filtration criteria if they choose to do so. This is especially useful since this is geographical data and a

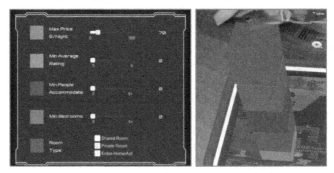

Fig. 2. [Left] The filter panel that holographically resides above a map in space and allows the user to select their filter criteria as they would on a generic webpage filter (we are inspired by the Airbnb one). Each home that matches each filter criteria would have the corresponding colour cube stacked on top of it to indicate a match. [Right] A home stack represents the filter criteria of a single home. The grey cube on the bottom denotes a home, the orange cube represents that the home has at least the minimum number of bedrooms that the user is looking for, the blue cube denotes that the home accommodates at least the minimum amount of people that the user has specified, the yellow cube specifies that the minimum average rating requested for by the user is met, and the purple cube denotes that the home fits within the price range that the user is looking for. There is no green cube on this home stack, and this means that the room type criteria that the user is looking for does not match this home. If it did match, a green cube would sit upon the grey one. Note that the left and right photos are not meant to match. (Color figure online)

Table 1. Example of the number of homes that are available for the user to see as each filter is progressively applied on generic filters versus the HoloLens app. The major takeaway being that filtered out homes are also always visible on the app.

Home quality filter	Web based filter	HoloLens App
Filter 1: At least 2 bedrooms	**200**	**200**
Filter 2: Rated at least 4/5	50	**200**
Filter 3: Shared home	2	**200**

major decision maker for users is location of the home. Users are oftentimes looking for homes around certain landmarks or points-of-interest and may switch their decision, if for example, a home is filtered-out due to one criterion but is close to a landmark. We design our interface such that the technology does not make decisions for the user but enables the user to make decisions (Fig. 3).

3.2 An Internet of Things and Ambient Environment Driven Design

Current research and trends project that a main feature that determines a successful future smart city is its ability to embed the Internet of Things. Sensor driven interfaces that allow users to interact with big data on the go are key players of smart community

Fig. 3. Our filtration interface. [Top Left] The user specifies a minimum of 2 bedrooms (orange cube). [Top Right] Orange cubes appearing on homes that have at least 2 bedrooms. [Center Left] The user steps forward to look down closely at the streets. They can also now clearly see the homes that appeared behind other ones. [Center Right] Details about a certain home can be viewed on the info panel if the user air taps on a home. [Bottom Left] More, but smaller home stacks appear if the map library we used is set to zoom out. [Bottom Right] Lesser, but bigger home stacks appear if the map library we used is set to zoom in. (Color figure online)

facilitation. We identify key aspects of how our application design fits into a ubicomp smart city.

Smart City Kiosks. A major advantage of headset AR is the ability of the user to be able to carry it virtually anywhere. However, cities currently provide visitor information around various specific locations: information centers, information kiosks, at points of interest, town hall, information boards at specific locations such as parks, etc. A smart city can benefit from providing AR enabled kiosks at specific locations: especially for user interaction with geospatial big data. Locations that benefit from our application include the following.

The *information center*, usually located at an entry point of a city with pamphlets of available hotels, is an ideal location when road travelers who have not pre-booked a stay want to find an Airbnb as they enter the city. A sensor or satellite driven segway to the interface would provide the user with a "you are here" entry into the application and they would first see nearby homes with ability to move to any location in the app including depending on points of interest.

Similarly, *information kiosks* at an airport or train station would serve visitors in the same way. The sensors at these locations would also be sensitive to which terminal or wing of the airport or train station that the user is currently located at and would provide the fastest way to get transport to the selected Airbnb.

Landmarks and points of interest also serve as ideal locations being mindful of the journey of a visitor in a smart city and being able to provide services at any point of it so that someone who made spontaneous plans is able to book an Airbnb. The sensors at this location would first provide nearby Airbnb homes on the map before the user expands their search further. The sensors here would be extra mindful of the fact that the visitor may be interested in other landmarks and would thus provide "hotspots" of Airbnb homes (those located near other landmarks in the city).

Neighbourhood entry point/community center is another location to provide smart kiosks. Our app at this location would be dedicated to highlighting the current neighbourhood Airbnb's and secondarily also suggest those in surrounding neighbourhoods. Of course, the user is always able to expand their search further. The benefit of neighbourhood kiosks being that visitors could drive by those homes and ask for viewings as required before booking.

In addition, *outside listed homes* are also good locations to make the app available to users. This would be best facilitated via Global Positioning System (GPS) satellite to track the user as they move (as is with Google Maps) and let users with personal hardware view the information for the current home. This would be supplemented with images of the inside of the home along with the ability to see similar homes in the surrounding area or anywhere else in the city. This could also be facilitated via sensors or QR codes outside Airbnb homes such that each one becomes a "smart kiosk". However, implementing and maintaining sensors outside each Airbnb could become cumbersome for the city and so, the GPS approach is ideal for this scenario.

Environmental Affordances. In Human Computer Interaction (HCI), designs are often crafted with the intention of taking advantage of action possibilities that the environment has to offer [18, 24]. These also make up what we characterize as the ambient environment. We determine the core affordances that present themselves in regard to our smart city app.

Headset Augmented Reality as a medium offers some generic affordances. The ability to expand the map virtually as big as the room that the user is in provides more **physical space** to move in while interacting. In our example, at locations where the user is outside, this is virtually unlimited space or real estate- as long as the user is safe from traffic and other pedestrians. At locations where the user is inside (such as at an airport or visitor information center), the space is limited to as much part of the room as the room allows without interfering with others in the environment. On the other hand, an expanded view of the interface could also allow for collaborators/ multiple viewers of the interface

(as many headsets are available). This could be beneficial in the case of if a family collectively wants to look through potential Airbnb's before making a decision. More space allows people to view the interface together as well as walk around and look down or squat to inspect various elements. The ability to walk around as well as the 3-dimensional depth perception allows for each home stack in a cluster of stacks to be seen distinctly (versus what appear via a 2D screen as seen in Fig. 4). The 3D interface also allows for equal representation of each home such that depth perception allows comparison of homes from each angle. 3D depth perception also alludes to the user being able to gauge height where a taller stack means more filters matched to the home.

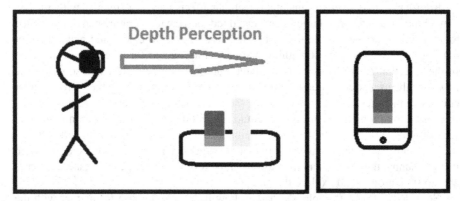

Fig. 4. Same 2 stacks as seen in 3D (app) versus on a 2D phone screen. What may appear as one unit on 2D are actually 2 separate items. A similar 2D vision applies on generic webpage maps such as the one referenced in Fig. 1. 3D depth perception allows for more clear distinction of items and equal representation of each item.

Headset AR also offers a **hands-free interaction** versus having to hold your phone or a mouse on a laptop. A hands-free interaction is somewhat in alignment with physical maps or info boards that are sometimes offered at points of interest or information centers in cities. A digital one, like an interactable directory in a mall, provides dynamic interaction as well as lets the user interact without having to hold any hardware. In addition to providing a dynamic and hands-free interaction, headset AR is additionally portable and unlimited in terms of virtual real estate. A hands-free interaction allows the visitor to be holding on to their suitcase, children, food, bicycle, or anything else and still be able to interact via voice control or with one hand. Additionally, headset AR does not discriminate between left-handed versus right-handed individuals or force left-handed individuals to alter the settings of the hardware such as is the case with a laptop mouse. Moreover, if the user is free to use both hands to interact in headset AR, it allows the user the ability to use both hands to view home information, etc. The **portability** of the headset allows any user who owns personal hardware to be able to view the app at any location or at the comfort of their own home (such as is with other personal devices like phones and laptops). This supports travel and movement.

Smart City Affordances are characterized by aspects of the environment that pre-exist in cities and are useful for how the user interacts with our interface. One being **benches**

or other seating that are oftentimes located around points of interest or transportation hubs such as airports. If our technology is made available via smart kiosks or sensors and these are located near seating, this would enable the user to make use of the technology either standing up or sitting down. Sitting down, this could become akin to flipping through a directory/book where the user can spend more time instead of feeling a rush to finish their task. In the case of collaboration, two users could be sitting side by side and interacting with the system together or a user could be sitting down, and another could be standing up where seating is limited. When being used portably via a personal headset, our app could serve as an augmented GPS/Google Map that can be used to **follow directions** to the Airbnb as a passenger in a car or a careful pedestrian.

Temperature is an environmental factor that affects how users interact with our system. In the case that the technology is located indoors, the temperature can be controlled to alter the experience of the user. Making it too hot or too cold could make the user not linger on for too long: something that can be accounted for when urban planning. In the case that the technology is available outdoors, temperature can also determine how long the user sticks around. Temperature can also be detected by the system to alter colours on the interface and set a touristic theme for a hot place versus a cold place. If it forecasted to rain or snow, the system could suggest the user a nearby accommodation. **Lighting** also affects user interaction similarly to temperature. As it gets dark outside, the hardware and software can work together to suggest a nearby accommodation. When it is too sunny, the system can increase brightness of the interface such that the user is able to stick around and explore the map.

Sound is a factor that plays a part in how users may choose to interact with the system. A lot of noise would indicate traffic or a slew of people in the smart city. In such a case the user may choose to interact in a small area or corner of the room/environment and perhaps be seated. In the case that there is not a lot of noise, the user may be free to use the interface as they please without the notion of disturbing other people in which case they may take up more space and think more freely about interacting via voice or gestures. In the case that the sound is music, this could affect user interaction depending on the type of music. The hands-free aspect of headset AR coupled with the availability of notepads/pens provided by the city at information desks/kiosks, allows the user to **make notes** of contact information/favourites as they use the interface. The user can also make notes digitally on their phone while interacting hands-free on the headset. Tourist spots such as **coffee shops and restaurants** could benefit from a smart city kiosk/sensor that includes the app being available in their environment where visitors would come in to be able to find an Airbnb at leisure. These cafés are already often coupled with tourist information centers and highway rest stops. When in this environment, the user would explore the interface for a bit longer while having a meal or coffee. This is synonymous to how people currently sit at café's and use the Wi-Fi for work.

Spontaneous discoverability is a key player in the adaptability of our interface. The point being that users need to feel convinced that they want/need to use such an interface- especially at the beginning stages of the technological transitioning into meta-user interfaces in general. A successful collaboration between smart city components via an ambient environment relies on the user being able to discover the interface on their own. As we transition into smart cities, physical objects will increasingly become

embedded with computers that are all producing multitudes of data and able to "talk" to each other [9]. In such a case, the discoverability design of each of those expandable interfaces needs to be prominent, yet subtle so as to not overstimulate the user with multiple smart city interfaces at the same time. One way of doing so is via cross-device discoverability. For example, we take the current Bluetooth, beacons, as well as Apple Airdrop models as an inspiration when we suggest this method of discoverability. In our setting, a user walking by our discoverability sensor may receive a notification on their personal device such as a phone to indicate the availability of our system. This could even provide information regarding if the system is currently being used by someone else or if it is free to use. Moreover, the phone may allow the user to digitally "line up" in a queue to use the system in a location hard-pressed with tourists. This also replaces the need for a human to monitor a physical line-up. Discoverability could also be driven by QR codes. They serve as subtle, yet visible ways to indicate additional technological interfaces to users. The QR codes could be coloured specifically to indicate various different interfaces. However, QR codes are not a component of dynamic technology that may appear at the user when they walk by as perhaps an indicator that is only visible to a user wearing a headset would. There are two major classes of spontaneous discoverability present to us: 1) that is visible/ usable in the physical world 2) that is visible in the holographic world. We should consider that methods of discoverability in the physical world would be visible to users who are both either: wearing or not wearing an AR headset. On the other hand, holographic ones would only be visible to those wearing a headset. Thus, at a time when most users possess their own headset, a holographic one would be considered, but while we still wait to transition into mainstream headset AR, a physical indicator on a cross-device or in the physical environment would be best to indicate to the user that there is a headset they can use to find an Airbnb.

3.3 A Design for Artificial Intelligence

HCI interfaces for smart cities really benefit from an artificially intelligent backend due to the abundance of big data that can be utilized in such cases. We identify specific aspects of our design that are created for an artificially intelligent backend.

Filtration by Inclusion, the filtration technique we contribute, implies a benefit from being plugged into a neural network on the backend. To reiterate, we have 5 filters on the filter panel: maximum price per night (0–500), minimum average rating (0–5), minimum number of people that the home accommodates (0–5+) , minimum number of bedrooms in the home (0–5+), and the room type (shared, private, or entire home/apartment). In addition, we have another factor: location on the map, which is sometimes a preference for users depending on landmarks or personal preferences. While the user is self-sufficient in performing the following:

1. Firstly, being able to select slider thresholds and toggle values on the filters and noting which home stacks are tallest (thus closest to their filters)
2. Next, deciphering if a specific home is in their preferred vicinity on the map
3. Lastly, being able to compromise between:

    a. home stacks that may be the tallest (closest to the filter criteria) but not in the preferred location

    b. home stacks that may be in their preferred location but not the tallest (not as close to their filter criteria)

This process being backed by a neural network can make the filter approach automated, help the user make decisions, and create customizable outcomes [15, 19]. The neural network, a computerized algorithm that mimics a human neural pattern (or way of thinking and making decisions), can be applied as specified below to our filtration technique. We demonstrate below our scenario in a simplistic neural network model based on linear regression.

We assume major factors in helping make the decision of finding the ideal Airbnb and represent them in binary notation:

1. Does the home reach the threshold of the filter:

    a. (X_1) Is the price per night within the specified maximum (Yes: 1, No: 0)

    b. (X_2) Is the average rating at least the specified minimum (Yes: 1, No: 0)

    c. (X_3) Does the home accommodate at least the specified minimum number of people (Yes: 1, No: 0)

    d. (X_4) Does the home have at least the specified minimum number of bedrooms (Yes: 1, No: 0)

    e. (X_5) Does the room type fall under one of the toggled-on categories where at least one is required to be toggled on (Shared, Private, Entire Home/Apartment) (Yes: 1, No: 0)

2. Is the home in the user's preferred location:

    a. (X_6) User preferred area/neighbourhood of the city (Yes: 1, No: 0)

    b. (X_7) Near (a) user preferred landmark(s) (Yes: 1, No: 0)

All values up to this point have been binary, but when it comes time to make a fluid choice (maybe peppered with some compromises) such as a human would, we assign weights of importance to each of the above points. We can do that generically like so:

1. Each of the 5 filter thresholds (**1a-1e** above) are set uniformly weighted at $(W_1 = 5)$ since we assume the user would highly value each of those since Airbnb chose to utilize those in their webpage interface.

2. A user preferred area or neighbourhood (**2a** above) is weighted at $(W_2 = 2)$ since we assume this makes a difference to the user, but the user is ok without it since tourism is their priority. Being in the vicinity of (a) preferred landmark(s) is weighted at $(W_3 = 3)$ since we assume these would be some of the most preferred locations to stay at in the city.

Additionally, we factor in a bias $= -$threshold value that is introduced via approximating a real-life problem which may be complicated. We set this value to (**bias $= -5$**). To

demonstrate an example, we arbitrarily assume a home with the following characteristics: our first three filters $(X_1 - X_3)$ and user preferred landmark vicinity (X_7) are satisfied and each equal to 1 while the last two filters $(X_4 - X_5)$ and user preferred neighbourhood/area locality (X_6) are not satisfied and each equal to 0. We assume that both "user preferred landmark vicinity" and "user preferred neighbourhood/area locality" refers to the system predetermining preferences based on where the user usually stays and general big data patterns. Preferences could be determined via "how far away from landmarks the user usually stays", "proportionately how populated the localities are where the user stays", etc. If the user is new, the system would make preference inferences based on general user behaviour and could make general assumptions based on preliminary demographic factors. We utilize a linear-regression-based formula and insert the values below.

$$\sum_{i=1}^{m} w_i x_i + bias = w_1 x_1 + w_2 x_2 + w_3 x_3 + w_4 x_4 + w_5 x_5 + w_6 x_6 + w_7 x_7 + bias$$

$$\sum_{i=1}^{m} w_i x_i + bias = (5)(1) + (5)(1) + (5)(1) + (5)(0) + (5)(0) + (2)(0) + (3)(1) + (-5) = 13$$

Since the answer is 13, greater than 0, this house will be one that can be selected. With this model, we will have multiple homes that would have a value greater than 0, and thus the system would be considered Artificial Narrow Intelligence (ANI) and either: have many narrowed down suggestions for the user **or** rely on further hidden neural network layers to determine a single home as the answer. The above is a single layer example, and we make some arbitrary assumptions about user preferences. However, if we factor in hidden neural network layers such as the following, we can get more specific answers and suggestions for the user:

1. What determines a preferred neighbourhood/area generically according to users?
2. What determines a preferred neighbourhood/area according to specific users?
3. What determines a preferred neighbourhood/area according to the city?
4. What determines a landmark?
5. What determines the correct number of landmarks to be surrounded by?
6. What determines a preferred landmark generically according to users?
7. What determines a preferred landmark according to specific users?
8. What determines a preferred landmark according to the city?

The hidden layers of the neural network that constitute of the above factors define the artificial intelligence of the system and its ability to make compromises and decisions like a human. All other factors (the 5 filters on the filtration panel) are Boolean values and thus easy to understand for a generic machine and do not require artificial intelligence. For example, homes are filtered out on a generic webpage filter if they don't fit one criterion (as seen in Table 1). When the above layers are factored in, the system starts moving away from Artificial Narrow Intelligence and becoming progressive towards Artificial General Intelligence (AGI) and even Artificial Super Intelligence (ANI). One factor that would move the system towards more intelligence is the bias. We initially select the bias arbitrarily and that can change based on the application of an optimizer and as the system learns about how humans make decisions. When the bias value becomes

more specific and the system also learns more about weights and each variable, we start moving towards a more intellectual and accurate system.

This neural network model, being a subset of deep learning, which is a subset of machine learning, which is a subset of artificial intelligence, presents itself in two ways in our scenario. If we train it via non-deep learning, it would be trained via humans: by labelling either natural user interaction data points or research scenario data points. Contrarily, if we train it via deep learning, it will train itself based on non-labelled big data. In either case, we are able to train the system such that it thinks, reacts, and suggests more closely to humans and can become a concierge in a touristic smart city scenario.

Other Artificially Intelligent Factors include a range of aspects that piece together for a system that understands itself more intuitively. **Real time data** for Airbnb can be used commensurate to how generic city-relevant datasets (such as transportation) are available on smart city data dashboards. In this case, AI can be utilized to pre-empt whether a home is a good choice for the user to select based on how long it usually takes before it gets booked, what time it usually gets booked, what time the price of that home goes up/down, etc. **Voice/Accent/Language Recognition** can be learned via AI as is already done for generic voice control apps such as SIRI. Voice control/ multimodal input would be vital control methods for a scenario such as ours where the user is a busy traveler who might have their hands full. Thus, implementation of custom voice for our app (in addition to generic ones available with hardware such as the Microsoft HoloLens) are a point of value for the end user. This custom voice command system would be optimized via AI to learn voice/accent/language patterns. A heavily sought-after AI feature is the implied ability of such a system to be able to make **user-customized recommendations** of datapoints (which would be homes in our case). Meaning that the system recommends homes to the user based on their *personal* previous behaviours: independent of the current scenario's city-specific factors as well as of other users' behaviours and preferences. This is a non-deep learning style model where the labels of a specific user on datapoints would be studied by the system. Similarly, **generalized recommendations based on users** would provide suggestions based on the behavior and preferences of all users collectively. This model would help predict and make suggestions on factors such as: what homes/areas in the city do users collectively prefer, what time do users book most, what combination of number of bedrooms and average home ratings do users prefer, etc. Lastly, **generalized recommendations specific to cities** would truly convert the app into one that is as beneficial to the smart city as it is to the end user. In addition to making better recommendations to the end user, this AI data would help facilitate city specific routines, events, and possibly even touristic monetization. Factors include: where and what dates city-specific events that bring in tourism usually occur, how price per night changes for homes on those days and in those areas, what is the economic make-up of the city, how transportation data ties into Airbnb data, etc. Thus, multiple factors of the app data benefit from artificial intelligence.

Feedforward and Feedback are both applicable in our scenario. Even though they are almost contradictory in nature, they can work complementarily in a system such as ours. Feedforward can be used via an AI model in our scenario. For example, big data allows us to track patterns and provides predictability into human behaviour, and we can apply this via feedforward interaction [4]. If due to data patterns, the system learns that users

who pick *"shared room type"* are likely to end up staying in the core downtown of a city, the system can make an inference that they are likely to do that and suggest that location before the user even moves the map there. In hindsight, this may be the case because shared room types may usually consist of hostels which are usually set in the downtown of cities, for example. However, the reason is not known to the system. All it knows is that users who select shared homes usually maneuver to the downtown locations to find an Airbnb, and thus, the system predicts and suggests this as soon they enter the room type but before they maneuver towards downtown themselves. In this way, many filtration patterns could be detected in such a big database like this and applied like so. On the other hand, feedback for a similar scenario may be applied such that the user is at an outskirt's location on the holographic map, and they enter *"Shared room type"* as their preferred filter. The current map scope does not show any/ many options for a shared room type. Thus, the system displays a holographic arrow indicating that the user should move the map in a direction towards where more shared room types are located. This consequently ends up being downtown. These are two variations of handling the same situation in which feedforward is more AI driven and pre-emptive to the user.

4   Limitations

There are some factors that may be limitations to our scenario.

4.1 Usability Issues

Hardware limitations in conjunction with some aspects of the software interface present some interaction tribulations. For example, sometimes it is hard for the system to detect a user entering slider input on the filter panel via either an air tap or drag. This is due to hardware limitation but is amplified by the size of the slider on the filter panel not being bigger and AR not being the ideal medium to interact with sliders. To minimize such a usability issue, a different design such as a checkbox could be implemented, but there would be pros and cons to all designs. Thus, both hardware and software limitations are something to consider when designing such an interface.

4.2 Accessibility

Our design relies heavily on colour to differentiate filters. An interface such as this should be mindful of choosing a colour-blindness-friendly palette. This is especially the case with the addition of more filters on an interface such as this. This also applies when designing spontaneous discoverability of software embedded around the city in that findability of the software should be mindful of colour-blindness. Other aspects of accessibility also need to be considered in terms of where and how to provide the software just like all aspects of urban planning are oftentimes examined for accessibility.

4.3 Outdoor Factors

The fact that we seek to embed such a system in a smart community/smart city comes with the implications that the user is expected to use the software outdoors. Thus, this exposes hardware such as sensors and headsets to the weather as well as vandalism. When creating such a system, city facilitators must be mindful of weather that may be too hot or cold and expose the hardware to damage. In addition, the vandalism aspect is one that pre-exists in cities across the world. Thus, safety measures must be taken to protect city assets.

4.4 User Willingness

Augmented reality in itself is being increasingly used on phones, but still not considered mainstream. Headsets are even further away from being heavily used commercially. This is something that users will have to get used to using if it becomes available in a scenario such as ours. In addition, users sometimes consider interacting with technology an intimate thing and are hesitant to perform inputs sometimes. Overstimulation is a risk we run when we increasingly embed technology into the user's physical environment. Thus, when we expand upon further features of our design as well as the feedback and spontaneous discoverability features, we must walk a fine line of providing the user with everything they need in a clean manner and make sure it is simple and subtle enough to minimize overstimulation that may turn the user away from using the technology.

4.5 Design and Software Limitations

There are always ways to further improve on designs and software methodology. The software could be modified to further optimize the system and output. Although our design is unique and unconventional in regard to keeping homes that would generically be "filtered out" on the interface so as to provide user decision flexibility, there are some counterarguments that could arise due to this approach. For one aspect, in some cases, it may leave the user with "too many options". For example, the user may find it hard to decide between 20 homes that are each matching 4 filters (even if they are different ones). Not visually seeing the removal of options that do not exactly match could increase user indecisiveness and frustration in some cases. However, such a frustration could also occur when only 2 homes remain after filters are applied on a generic filtration system and the user feels a lack of options. Thus, in order to provide more flexibility and information to a human to make their decision, a system like ours but with the additional hidden neural network layers to help alter filtration criteria weightages could be utilized as tie breakers in cases where too many homes are available. Moreover, since we are applying artificial intelligence to our system, we could also use generative design to improve some aspects of design via automation. However, this would come with pros and cons just like any other design. Since we are suggesting an AI model, we also have to consider that sometimes the agent can learn and conclude incorrectly; however, AI comes with this risk in general, but also provides a rich reward. Another limitation that falls under both software and design is that currently, if two or more homes geographically sit on the same longitude and latitude (for example in one building), there needs to be a way

to display a home stack for each of those. This could easily be solved for most cases mathematically and home stacks could either be shown next to each other if there is enough room on the interface or the stacks could be depicted as home clusters that the user is then able to visually expand and examine further.

4.6 Potential Study Limitations

If we were to perform a study on our system in its environmental context, it would be a field study and may become limited by current technology and situations. For example, this would probably be wizard-of-oz in terms of sensors and some aspects of research would be dependent on people's ability to discover our unfamiliar technology. In this way, many factors of such a field study would be aspects to consider and pre-plan. In such a case of unfamiliar technology, many unexpecting users may have to be primed due to which results would become influenced via the priming introduction.

5   Discussion and Future Work

Our application and techniques make use of big data to help facilitate a touristic infrastructure in the ambient smart city of tomorrow. We seek to make use of this big data that is already available to provide an even better user experience. Since we are embedding this system around and in context of an entire city, it is vital to keep track of the visitor's journey through the study. In addition to keeping track, it is important to take into consideration the nuances of situations that the traveler encounters in the city and how they act. For example, how does a traveler who has to handle a stroller interact differently than someone who has their hands full with luggage versus someone with their hands free. Furthermore, we have conducted a pedogeological pilot study of our system in a lab setting but recommend that a field study where participants are unexpectant could be a step further into deciphering the usefulness of the system as embedded around the city. An aspect of our system is that it could almost identically be used for real estate around the city. If this becomes the case, it would harbour the same technology and filtration, however, the narrative would change from a touristic one to providing for the citizens of the city: those that are familiar with it. In this case, the vitality of various aspects would become altered. For example, spontaneous discoverability may not be so spontaneous for users who are familiar with the city versus tourists. Another factor that can be probed upon is the cultural aspect of a smart city. Various parts of the world for this interaction would provoke a varying narrative in terms of: language, gathering places and practices, how and what they use technology for, as well as locations around the city. For example, it is a very North American concept to have information centers at entry points in the city and travel stops on the highways, and the same may not apply elsewhere. One more aspect to consider is that our interface is specialized for the guest user who is looking for a home rather than also for a host user who is looking to rent out. Due to this, our interface is designed keeping in mind solely the guest and does not seek to have umbrella aspects in the interface to accommodate both guest and host. Doing this has pros for the guest user as well as the city because embedding this software at specific locations around the city is implemented for the visitor to the city.

Moreover, a comparison feature could be added to exclusively select and compare two or more homes side by side. In such a case, the user would select two or more home stacks and they would appear on a holographic plate on which the user could visually examine the differences as well as compare information that appears on the information panel side-by-side. The visual as well as the informational differences and similarities would be visually highlighted by the system. Moreover, the system could also preserve some visual geographical cues on the comparison plate to keep the user informed about the geographical comparison of the locations of the homes.

We have conducted a pedogeological pilot study of two participants to compare our interface being interacted with on a laptop versus via headset AR. One future step would be to compare our headset AR interface with a generic webpage filtration one (such as the Airbnb website). This would provide us with a clear comparison of how users may transition from what is currently considered the most widely used technology to geospatially filter the same datapoints towards what may be used in the future as rapid technological growth encourages widespread adaptation of IoT embeddable systems in an ambient smart city.

6   Conclusion

We became motivated by wanting to augment generic geospatial filtration with the 3rd dimension in augmented reality to provide more optimal and flexible filtration. We aimed to use this design in a situational smart city and smart community context to facilitate touristic infrastructure. We contributed our filtration method: filtration by inclusion. We placed our filter scenario into a neural network model design to examine how it could make fluid choices like a human being. We probed on ambient environmental affordances factors such as spontaneous discoverability. We also discussed other aspects of AI such a feedforward and feedback of the interaction that can be leveraged here. We then identified limitations as well as opportunities for future work. We discussed various aspects of what can be probed further on our scenario. In general terms, through the demonstration and discussion of our system, we highlighted how our work has implications for UX design in AI at its intersection with AR big data and situational interaction used most specifically for smart communities and smart cities.

Acknowledgements. We would like to thank Dr. Frank Maurer, head of the Agile Software Engineering (ASE) group at the University of Calgary, who supervised us during our graduate studies and facilitated the lab resources for this research.

References

1. Airbnb Webpage. https://www.airbnb.ca, Accessed 25 Feb 2021
2. Artificial Intelligence- What it is and why it matters. https://www.sas.com/en_ca/insights/analytics/what-is-artificial-intelligence.html, Accessed 25 Feb 2021
3. Canadian Open Data Summit (2017). https://opendatasummit.ca, Accessed 25 Feb 2021
4. Chaboki, B., Van Oorschot, R., Torguet, R., Wu, Y., Yao, J.: Interaction Design Feedback and Feed Forward Framework: Making the Interaction Frogger Tangible (2012)

5. Chen, Q.: Immersive Analytics Interaction: User Preferences and Agreements by Task Type. Master's thesis. University of Calgary, Calgary, Canada (2018)
6. Chopra, S.: Evaluating User Preferences for Augmented Reality Interactions for the Internet of Things. Master's thesis. University of Calgary, Calgary, Canada (2019)
7. Chopra, S., Addam, O.: Big data geospatial filtration via information visualization: viewing 3D data information in augmented reality. In: Best Poster Presentation (2nd place) at Women in Data Science Conference Event (WIDS 2018). Calgary, Canada (2018)
8. Chopra, S., Maurer, F.: Evaluating user preferences for augmented reality interactions with the internet of things. In: Proceedings of AVI '20: International Conference on Advanced Visual Interfaces (2020)
9. Chung, N., Lee, H., Kim, J., Koo, C.: The role of augmented reality for experience-influenced environments: the case of cultural heritage tourism in Korea. J. Travel Res. **57**, 627–643 (2017)
10. City of Calgary Dashboard. https://data.calgary.ca/browse?category=Base+Maps, Accessed 25 Feb 2021
11. Coutaz, J.: Meta user interfaces for ambient spaces. In: Task Models and Diagrams for Users Interface Design (2006)
12. Da Silva, T.S., Silveira, M.S., Maurer, F., Silveira, F.F.: The evolution of agile UXD. Inf. Softw. Technol. **102**, 1–5 (2018)
13. Fard, F., Davies, C., Maurer, F.: Agile emergency responses using collaborative planning HTN. In: Proceedings of the International Conference on Information Systems for Crisis Response and Management (2017)
14. Heathman, A.: How AI is powering Airbnb's mission to change how we travel forever. https://www.standard.co.uk/tech/airbnb-artificial-intelligence-21st-century-travel-a38 16336.html, Accessed 25 Feb 2021
15. IBM Analytics- Linear Regression page. https://www.ibm.com/analytics/learn/linear-regres sion, Accessed 25 Feb 2021
16. Internet of Things|Definition of Internet of Things. https://www.lexico.com/en/definition/int ernet_of_things, Accessed 25 Feb 2021
17. Kapetanios G., et al.: Filtering Techniques for big data and big data based uncertainty indexes, Eurostat (2017). https://ec.europa.eu/eurostat/documents/3888793/8440791/KS-TC-17-007-EN-N.pdf
18. Kaptelinin, V.: Affordances in The Encyclopedia of Human-Computer Interaction, 2nd ed, vol. 44 (2014)
19. Kavlakoglu, E.: AI vs. Machine Learning vs. Deep Learning vs. Neutral Networks: What's the Difference? https://www.ibm.com/cloud/blog/ai-vs-machine-learning-vs-deep-learning-vs-neural-networks, Accessed 25 Feb 2021
20. Knudsen, S., et al.: Democratizing open energy data for public discourse using visualization. In: Extended Abstracts of CHI 2018: Conference on Human Factors in Computing Systems (2018)
21. Romano, M., Onorati, T., Aedo, I., Diaz, P.: Designing mobile applications for emergency response: citizens acting as human sensors. Sensors J. **16**(3), 406 (2016)
22. Singapore Government Open Data Portal. https://data.gov.sg, Accessed 25 Feb 2021
23. Tom Slee Airbnb Data Collection Page. https://tomslee.net/category/airbnb-data, Accessed 25 Feb 2021
24. Zeng, D., Tim, Y., Yu, J., Liu, W.: Actualizing big data analytics for smart cities: a cascading affordance study. Int. J. Inf. Manag. **54**, 102156 (2020)

Why Developing Simulation Capabilities Promotes Sustainable Adaptation to Climate Change

Gabriele De Luca[1]([⊠]) [iD], Thomas J. Lampoltshammer[1] [iD], and Shahanaz Parven[2] [iD]

[1] Department for E-Governance and Administration, Danube University Krems, Krems, Austria
gabriele.deluca@donau-uni.ac.at
[2] Department of Political Analysis and Administration, RUDN University, Moscow, Russia

Abstract. Simulations of social processes are a special category of interactions between humans and computers, characterized by the modeling through the latter of the behavior of the former. This paper proposes a framework for studying the development of simulation capabilities by society, as part of an ongoing process of adaptation by complex social systems against a more challenging environment. We begin by discussing the characteristics that the capability to simulate social systems confers in terms of increased effectiveness of the decision-making process of a society. Then, by framing this increased effectiveness as a problem of increased computational capacity by an intelligent agent, we describe the impact that this has on the fitness of an agent that is adapting to a changing environment. We thus provide a mathematical formalization for studying the development of computational sociology in terms of adaptive fitness. This formalization lets us draw the interesting conclusion that, if the adapting system records a decrease in its fitness over its recent past, the formulation of adapting decisions becomes an increasingly more self-referential problem. That is to say, it depends increasingly more on the computation of the outcome of the past actions of the agent, and increasingly less on the behavior of the environment around it. This promotes the theoretical generalization that the development of simulation capabilities for adaptation increases the role that the agent has in determining the status of the agent's world, and decreases the role that the environment has in the same process.

Keywords: Computational social sciences · Sustainable computing · Adaptation to climate change · Artificial intelligence · World modeling

1 The Relationship Between Adaptation by Society and the Development of Social Simulations

We observe that many social systems today are developing the capability to conduct simulations of themselves, and of the process of their adaptation against the environment [1]. This observation is valid regardless of the geographical location of those systems, and concerns transversally Europe [2], Africa [3], and Asia [4]. The underlying idea in developing simulations of climate adaptation, is that a social system that uses them

© Springer Nature Switzerland AG 2021
H. Degen and S. Ntoa (Eds.): HCII 2021, LNAI 12797, pp. 490–500, 2021.
https://doi.org/10.1007/978-3-030-77772-2_32

as part of its decision-making process is more capable to undertake actions that lead to positive adaptation to the environment, and to avoid those that do not [5]. The literature however presents a gap, concerning the formalization of the exact mechanism through which the development of simulation capabilities is conducive to adaptive behavior. One idea to this regard, is that simulations play the role of rhetoric tools that can be used by scientists, in order to legitimate a certain policy against policy-makers, and to avoid some other policies [6]. Another idea is that simulations allow the reduction of uncertainty about the environment [7], by mapping the space of possible worlds, and not just of the ones which are observed in practice. This, in turn, allows the navigation of the rugged landscape within which the social system is embedded. Another possible, more simple explanation, would suggest that simulation capabilities are being developed right now because of the availability of computational power in modern hardware, that allows complex simulations to be executed in practice, and not just being defined mathematically [8]. Notable, for that explanation, is the observation that High-Performance Computing is required to run any model that is not trivial in terms of the number of parameters, and in particular the models that are built upon real-world sociological data [9]. This type of explanation does not imply that societies which develop simulations are necessarily more adaptive, however; but rather, that the intellectual pursuit of scientific interest in mathematics or computer science is the explanation for the observed increase in the number and complexity of models that simulate climate adaptation.

In this paper, we want to study whether there is some deeper, underlying reason for the development of simulation capacities today, such that the process of adaptation by a biological human population to the environment could explain this development in this particular era. In a sense, we want to know whether the development of simulations is an evolutionary tipping point in the adaptation of our species to its environment [10]; and if it is, therefore, a contingent instantiation of the general problem of adaptation of a population through the reduction of uncertainty about the environment [11].

This problem concerns not only humans. There is, in fact, a general increase in the intelligence of the biosphere that is not specific to humans [12], and this suggests that intelligence may be a trait that provides selective advantages to any species that develops it [13]. If this is true, then intelligence as a trait undertakes a strong selective process in most species, and not in humans alone [14]. Human fitness, however, is highly conditioned by the capacity to reason and plan future actions because of the general poor adaptability of the human organism. As a consequence, in humans, more than in other animals, intelligence may play a crucial role in determining the fitness of the species [15]; and this may not be a shared problem with other animals.

Among the specific intellectual faculties of humans, lie the capacity to develop computational models of the individual and of the aggregate behavior. In this sense, computational social sciences provide, among other things [16], an increased capacity to reason and plan the actions of both the individual [17] and the collectivity [18]. The question that we want to address in this paper is, therefore, how to explain the development of simulation capabilities by a social system as part of a process of adaptation of the latter. To do so, we will draw from the sectors of computational social sciences, complex adaptive systems theory, environmental sciences, and the research on artificial intelligence, in order to explain the development of computational social sciences as an

adaptive evolution of the human social system against an increased complexity of the environment.

2   The Complexity of the Environment

A first theoretical approach to study climate change and the adaptation to it comes from the application of complexity science and complex adaptive systems theory to ecology and sociology [19]. In its foundational conceptual framework, a complex system is a system whose elements interact in a highly dependent manner, and whose systemic behavior is not reducible to the aggregation of the individual behavior of its composing elements [20]. The complex system can change the relationship or the identity of its components, as a reflection of the variation in complexity of an external environment. If it does, then the complex system is called a complex adaptive system [21]. In the process of studying them, complex adaptive systems are often modeled by means of computer simulations, and in particular by agent-based models. The application of complex adaptive systems to the study of social systems leads to the development of agent-based models for social sciences [22], whose construction and analysis is one of the primary subjects of study of computational sociology. The underlying hypothesis in the application of computational models to study sociology is that some kind of similarity exists between the world that is being simulated and the simulation itself [23]. If this hypothesis is valid, we can then treat the map, in this case the simulation, as the territory, in this case the world being simulated, and extend to the latter the conclusions that we draw by analyzing the former.

The application of complex adaptive systems theory to the study of the adaptation to climate change by a society is, indeed, useful. This is because it allows researchers to infer the general characteristics of the changes that a society undertakes, as part of the process of its adaptation to a changing environment. The literature on environmental sciences distinguishes climate variability from climate change [24], and treats variability and change as two related but independent aspects of any environmental impact assessment [25, 26]. Climate change can be seen as a generalized decrease in fitness [27]. Climate variability, on the other hand, corresponds to the empirical observation that the status of the climate becomes more varied over time, if we compare any two distinct time intervals [28]. If this is true, then it follows that climate change and climate variability correspond to two separated albeit related concepts. For the purposes of our paper, we can apply these concepts to the study of social systems in the following manner. If climate change corresponds to a generalized decrease in fitness of the human social system, climate variability corresponds instead to the empirical observation that the status of the environment and its variation becomes less predictable in recent years, with comparison to what was common in the past. We treat these two considerations on climate change and climate variability as assumptions for this paper, and we will see in the next section how to formalize them in terms of an adaptive agent.

3   Modeling the Adaptation of Complex Social Systems to Climate Change

We can now move towards a mathematical definition of the process of adaptation by a social system, which we treat as an agent, against an environment that becomes more

complex. We can model the process of adaptation by an agent to the environment as a process in which the agent selects an action a among many possible, in order to maximize its utility or, in this case, some kind of objective fitness function. First, however, we start by defining the agent and its relationship with the environment.

There is a world W in which the social system is embedded. In this world, the social system acts as an agent, and undertakes actions in response to the perception it has of the surrounding environment. The surrounding environment holds a status x, which at the discrete time t corresponds to x_t. The agent has a veridical perception of the environment [29], that we call P. Please note that, while this hypothesis is generally appropriate in the sector of robotics and artificial intelligence, this may not necessarily a good approximation for social systems. It has been proven through game-theoretical methods [30] that a biological system that evolves under natural selection develops a perceptual system that perceives its fitness function, and is otherwise a poor map of the underlying complexity of the surrounding environment. Since social systems are composed by biological organisms, and because populations that undertake natural selection are proven to not possess veridical perception, then the consequence is that social systems, too, do not possess a veridical perception of their environment. However, at least part of those social systems, and namely science, performs Bayesian inference in order to learn which of the system's perceptions are illusory in nature, and which are veridical instead [31]. Because science is embedded in the social systems that we study, and because developing simulation capabilities is a specific function of the scientific component of a social system [32], we then assume that the perception of the social system towards the environment is veridical. We also assume the set of possible states of the world to be in total order [33]; and subsequently the perceptions by the agent, in this case the social system, to also be in total order. If we do that, we can identify an homeomorphic and order-preserving map $g : x \rightarrow p$ that provides to the agent an accurate method for performing comparisons between any two possible states of the world x_1 and x_2, by comparing the associated perceptual states p_1 and p_2. For the rest of this paper, we simplify the discussion by considering the states of the world, and not the associated perceptual states held by the agent. The same consideration applies for the perceptions by the agent of any changes that occurs in the environment, and also for the decisions made by the agents and the actions that the agents undertakes at the same time, which we consider to be equivalent, though they are often treated separately in the literature [34].

In response to the status of the world x_t, the agent takes an action a_t. The sequence of actions taken by an agent up to the time t comprise the sequence A_t, whereas the states of the world up to t comprise the sequence X_t. Consequently, of the actions by the agents and of the physical or chemical law that regulates it, the world then updates the state of the environment around the social system. We call W the world function, according to which the status of the world changes between two successive periods in the discrete time.

$$X_{t+1} = W(X_t, A_t) \tag{1}$$

We can now make some considerations regarding the nature of the world function, according to the relative importance that X_t and A_t have in determining the new state of the world in $t + 1$.

The first consideration that we make is that the world updates its own status, regardless of whether or not some agents are present who can perceive it. This is necessary, because we must allow the possibility for the environment to change its status regardless of the actions of the agent, notably if the agent suddenly disappears. A world without agents is thus characterized by the rule that the update of its status is independent of any agents' actions, for lack thereof. In this sense, the world function W is modeled as:

$$X_{t+1} = W(X_t) \tag{2}$$

This function describes a world that operates according to the rules of physics, mathematics, or computational rules of some kind, and that are independent from any agent's actions.

At the other opposite, we can also imagine a world in which the world status is entirely determined by the actions of the agent. We can call the world function associated with this world "the God function", because in it the agent determines the future status of the world in complete autonomy:

$$X_{t+1} = W(A_t) \tag{3}$$

This type of autonomy corresponds to the Greek notion of $\alpha\dot{v}\tau o\nu o\mu\acute{\iota}\alpha$, self-regulation, in the sense that the agent regulates the environment as a direct consequence of the actions that it undertakes, without there being any distinction whatsoever between the agent and the environment itself. In this world, physics, if it exists, is a reflection of the agent's internal cognitive processes rather than a collection of rules on the behavior of the environment. We can argue, in fact, that the environment does not exist in distinction from the agent. In this sense, the worlds described by both (2) and (3) are equivalent or indifferent. Both (2) and (3) do not represent well the relationship between the social system and the environment today, of course. However, it is still possible to consider (2) as a good model for a subinterval of the historical period that concerns the relationship between that social system and that environment. We will see shortly why this is the case, if the world undertakes a transition that allows agents to appear and disappear.

More in general, though, we can imagine a world in which the human agents in general, and the complex adaptive social systems that we discussed earlier in particular, have a certain level of agency. This is to mean, the world changes according to both its own rules of behavior, but also in response to the actions undertaken by those social systems. If this is the case, then the term A_t appears as an input to the world function:

$$X_{t+1} = W(X_t, A_t) \tag{4}$$

We call this function the "business-as-usual" function, because it describes a typical world in which an agent acts on the environment and causes effects upon it. The term A_t, in this context, can either indicate the individual action that immediately precedes the world update in $t + 1$, or the full sequence of actions $A_t = \{\emptyset, \emptyset, \ldots, a_j, a_{j+1}, \ldots, a_t\}$ that have taken place if the agent has appeared in the world at time j. If this is the general case, we can then study, as anticipated earlier, the transition between a world in which agents are not present and a world in which agents are active, and that operates according to the previous model.

A world function in which an agent appears at some time j can be described as the "creation of Adam" function, and can be formalized as follows:

$$\begin{cases} t < j & X_{t+1} = W(X_t) \\ t \geq j & X_{t+1} = W(X_t, A_t) \end{cases} \tag{5}$$

Further notice that, provided that the agent exists for at least a certain subinterval of the history of the world, there is however no requirement that the agent continues to exist indefinitely. Therefore, there can be some time k in which the agent stops existing in the world, or at least to cause effects on the world update as a consequence of its own actions. We can call the world function that is subject to that condition "the apocalypse function", and define it in a manner analogous to the previous world function:

$$\begin{cases} t < k & X_{t+1} = W(X_t, A_t) \\ t \geq k & X_{t+1} = W(X_t) \end{cases} \tag{6}$$

The considerations we made above for the content of A_t in relation to j are here valid for A_t in relation to k.

We can also notice how the expressions contained in (5) and (6) are related to one another. If the world is characterized by both the past appearance and the future disappearance of the agent, such that $j < k$, then the world function assumes this combined form:

$$\begin{cases} t < j & X_{t+1} = W(X_t) \\ j \leq t < k & X_{t+1} = W(X_t, A_t) \\ t \geq k & X_{t+1} = W(X_t) \end{cases} \tag{7}$$

We can call this world function the "complex adaptive social system" function, which models the adaptation of the complex social system from the moment of its emergence to its ultimate end.

4 A Changing Environment for an Adapting Agent

We can now consider the problem of fitness of the agent in relation to the environment, within the framework of the model defined above. To this regard we can imagine, as is suggested by the literature on computational sociology [35], that the agent operates in order to maximize some kind of objective function that we call "fitness". This function does not directly consider the internal states of the agent, but only the status of the world at a given time. The fitness F_t of the agent at time t is therefore a function of X_t but not of A_t, and can thus be expressed as:

$$F_t = f(X_t) \tag{8}$$

We do not make any specific assumptions on the shape of f, besides assuming its continuity and it being positive. Because the agent is, in this case, a social system, and because the social system is in turn composed by biological organisms, we can also argue that a minimal level of fitness must be maintained by the agent at all times, in order for

the agent to continue existing. The literature on human survival considers the extreme environments as those in which the human fitness is minimal [36], which is another way to say that a minimal fitness is needed for a human to exist in an environment. If this is valid for humans, we extend the same consideration to the group of humans that we call social system, and we conclude that a minimum fitness $F_{min} > 0$ must be preserved by the system at all times, lest the social system itself disappears. Notice that, because the concept of fitness is constructed around the agent and is not independent from it, we can infer some characteristics of the fitness function according to whether or not an agent is present in a given environment. Specifically, we can infer that any world in which the humans are present is a world where the fitness of the social system is higher than that threshold F_{min}. But also, that any world in which the humans are not present is a world in which the fitness of a social system that we hypothesize is under that same threshold. The same is valid for the temporal, not just the spatial, dimension. This is not to say that the conditions of a pre-human or pre-social world were unsuitable for the appearance of humans; simply, that the observation that humans are present in the world at a given moment implies that their fitness is sufficient for their survival.

If we consider the world functions that we described earlier, this implies that if the A_t term is appearing as a term in the world function at time t, then the fitness F_t that we compute at the same time t is:

$$\text{if } X_{t+1} = W(X_t, A_t), \text{ then } F_t = f(X_t) > F_{min} \tag{9}$$

Conversely, if the term A_t is not present in the world function W, this means that the fitness of the agent in that world is lower than the threshold:

$$\text{if } X_{t+1} = W(X_t) \text{ then } F_t = f(X_t) < F_{min} \tag{10}$$

If we apply (9) and (10) to the world function described in (7), we can see that the fitness of the agent transitions to $F_t > F_{min}$ for $j \leq t < k$, and is lower than F_{min} otherwise. The figure below shows the relationship between F_t and F_{min} with respect to j and k (Fig. 1)

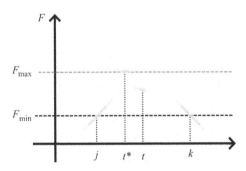

Fig. 1. The fitness function of a complex adaptive social system at time j and k.

If the function $F = f(X_t)$ is continuous, and if F is always greater than F_{min} for any $j \leq t < k$, then this implies that there is at least one value of t in the interval $[j, k)$ which

maximizes F. We can call that value t^* and we call F_{max} the maximum value or values assumed by F. There may or may not be a local or global maxima in f as a function, not in F as a distribution of values; and neither we, nor the social system whose behavior we model, know of any maxima in f. However, by observing the past values assumed by F, the social system can know whether it is approaching or deviating from a local maximum. This is because the agent can understand the general variation of its own fitness over time, on the basis of information that he holds on the status of X and on its recent variation. This consideration however requires a further argument.

Let us imagine that the social system possesses a certain number of methods for estimating which features of X are positively correlated with the value of F. Some of these methods can be simple, such as the comparison of the number of persons who are alive at any two given moments [37], with an increase in the demographic size of humans indicating an increase in aggregate fitness. Some can be more complex, and may relate for example to the capacity by a social system of extracting resources from the environment [38]. There are, in general, established methods for estimating the fitness of a single human individual [39], as well as the resilience of a society [40]; therefore, we can support the idea that the system is capable to assess the general direction of its evolutionary trajectory, and notably whether its fitness has increased between any two successive time periods $t-1$ and t. This information, which would be valuable in terms of adaptation, is part of the cognitive processes that determine the actions A_t taken by the agent, upon observing the status of the world in X_t and comparing it with X_{t-1}. In a sense, this type of comparison and its application to the decision-making process corresponds well to what we would call, in the literature on artificial intelligence, gradient ascent [41].

Now, we discussed above how climate change, though not climate variability, can be considered as a generalized decrease in the fitness of the social system. This means, within the framework of our model, that the social system observes today a general reduction in the value of its fitness F with comparison to its recent past. This observation tells the social system that whatever cognitive mechanisms the system is using need to be reassessed and modified, because the process of formulating decisions is currently leading to maladaptation. The system then observes that the maximum value assumed by F is in its past at some time t^*, and can therefore infer (not necessarily correctly) that a local or global maximum of f was reached for $t = t^*$, thus for $F = f(X_{t^*})$. The system may therefore reason that the status of the world in X_{t^*} is preferable to the current status of the world as observed in X_t, and may want to find the action a_t that would lead the world function W to produce a world X_{t+1} that would possess the same characteristics as the ones observed in t^*. In other words, it searches for:

$$a_t \in A_t : X_{t^*} = X_{t+1} = W(X_t, A_t) \tag{11}$$

There is the obvious consequence that, if a_t existed, it would imply the restoration of the world status to a prior state. The states X_{t^*} and X_{t+1} are, in fact, indistinguishable from one another. This would be problematic to justify in thermodinamic terms, because it would be impossible to discriminate between X_{t^*} (in the past) and X_{t+1} (in the future), even though the two worlds come from two world histories of different length. Specifically, the world history of X_{t^*} is a subset of the history of X_{t+1}.

We want however to clarify that a distinction is still possible. This is because the status of the agent is changed between t^* and $t + 1$, in a manner that is irreversible. The agent attempts to compute a_t in a manner that would produce an update $X_{t+1} = X_{t^*}$ by the world function; and not, in general, in order to maximize its own expected fitness, because the agent does not know the shape of f. In other words, the agent attempts to reverse the world function W, by finding:

$$a_t | A_{t^*} : X_{t^*} = X_{t+1} = W(X_t, A_t) = W(W(X_{t^*}, A_{t^*}), A_t) \tag{12}$$

Now, two cases are possible. In the first case, the agent reaches a new fitness value F_{t+1} that is higher than the maximum value $F_{max} = F_{t^*}$ recorded so far, and then continues to compute its actions as in (7). In the second case, if the new fitness value F_{t+1} is lower than F_{t^*}, the agent continues to search for its actions that would restore the world state to the one observed in t^*:

$$a_{t+1} | A_t : X_{t^*} = X_{t+2} = W(X_{t+1}, A_{t+1}) = W(W(W(X_{t^*}, A_{t^*}), A_t), A_{t+1}) \tag{13}$$

We can see that this process becomes increasingly self-referential as time passes, and only terminates upon reaching a higher value of fitness by the agent. This is valid whether reaching that value depends upon the agent's actions, the autonomous behavior of the external environment, or both.

5   Conclusions

We can now draw some generalizations from this particular formalization of the problem. If the maximum recorded fitness reached by a complex adaptive social system is in its past, the system then increasingly attempts to compute the consequences of its actions, and not the behavior of the environment, in order to maximize its own fitness. This is because, in absence of information on the general shape of its own fitness function, the system attempts to create a world state of higher fitness by attempting to restore the environment to its previous state. This attempt by the social system is an explanation, in terms of the adaptive fitness and the sustainability of an intelligent agent, for the development of social simulation capabilities that we observe today.

The type of framework that we proposed here opens the possibility, in future research, to study trade-offs between adaptive strategies by a population against climate change, where one of these strategies is the development of computational sociology. We do not currently have a framework for asking questions such as "how many less emigrants would there be, if the computational power of this particular society were to increase?"; even though we have reasons to believe that migration, as a social phenomenon, also comprises adaptation to climate change [42]. However, if we first frame any adaptive strategies as attempts to move upward along the fitness curve of a social system, as we did here, then this opens the possibility to compare alternative strategies for societal adaptation in terms of a unique adaptive process that encompasses them.

Acknowledgements. The authors received no financial support for the research, authorship, and/or publication of this article.

References

1. Patt, A., Siebenhüner, B.: Agent based modeling and adaption to climate change. Vierteljahrshefte zur Wirtschaftsforschung **74**, 310–320 (2005)
2. Oswald, S.M., et al.: Using urban climate modelling and improved land use classifications to support climate change adaptation in urban environments: a case study for the city of Klagenfurt, Austria. Urban Climate **31**, 100582 (2020)
3. Martens, C., et al.: Large uncertainties in future biome changes in Africa call for flexible climate adaptation strategies. Glob. Change Biol. **27**, 3440–4358 (2020)
4. Hassani-Mahmooei, B., Parris, B.W.: Climate change and internal migration patterns in Bangladesh: an agent-based model. Environ. Dev. Econ. **17**, 763–780 (2012)
5. Dignum, F., Dignum, V., Jonker, C.M.: Towards agents for policy making. In: David, N., Sichman, J.S. (eds.) Multi-agent-Based Simulation IX, pp. 141–153. Springer, Heidelberg (2009). https://doi.org/10.1007/978-3-642-01991-3_11
6. Thorngate, W., Tavakoli, M.: Simulation, rhetoric, and policy making. Simul. Gaming **40**, 513–527 (2009)
7. Troost, C., Berger, T.: Dealing with uncertainty in agent-based simulation: farm-level modeling of adaptation to climate change in Southwest Germany. Am. J. Agr. Econ. **97**, 833–854 (2015)
8. Conte, R., et al.: Manifesto of computational social science. Eur. Phys. J. Spec. Topics **214**, 325–346 (2012)
9. Wittek, P., Rubio-Campillo, X.: Scalable agent-based modelling with cloud HPC resources for social simulations. In: 4th IEEE International Conference on Cloud Computing Technology and Science Proceedings, pp. 355–362. IEEE (2012)
10. Botero, C.A., Weissing, F.J., Wright, J., Rubenstein, D.R.: Evolutionary tipping points in the capacity to adapt to environmental change. Proc. Natl. Acad. Sci. **112**, 184–189 (2015)
11. Donaldson-Matasci, M.C., Lachmann, M., Bergstrom, C.T.: Phenotypic diversity as an adaptation to environmental uncertainty. Evol. Ecol. Res. **10**, 493–515 (2008)
12. Gillings, M.R., Hilbert, M., Kemp, D.J.: Information in the biosphere: biological and digital worlds. Trends Ecol. Evol. **31**, 180–189 (2016)
13. Burkart, J.M., Schubiger, M.N., van Schaik, C.P.: The evolution of general intelligence. Behav. Brain Sci. **40** (2017). https://www.cambridge.org/core/journals/behavioral-and-brain-sciences/article/abs/evolution-of-general-intelligence/5AB1923F6D39AEED6AFB91E5 AACCEE8E
14. Miller, G.: Sexual selection for indicators of intelligence. In: Novartis Foundation Symposium, pp. 260–270. Wiley Online Library (2000)
15. Chollet, F.: On the measure of intelligence. arXiv preprint arXiv:1911.01547 (2019)
16. Keuschnigg, M., Lovsjö, N., Hedström, P.: Analytical sociology and computational social science. J. Comput. Soc. Sci. **1**, 3–14 (2018)
17. Chen, S.-H.: Agent-Based Computational Economics: How the Idea Originated and Where it is Going. Routledge, London (2017)
18. Gilbert, N., Ahrweiler, P., Barbrook-Johnson, P., Narasimhan, K., Wilkinson, H.: Computational modelling of public policy: reflections on practice. J. Artif. Soc. Soc. Simul. **21**, 1–14 (2018)
19. Preise, R., Biggs, R., De Vos, A., Folke, C.: Social-ecological systems as complex adaptive systems: organizing principles for advancing research methods and approaches (2018)
20. Simon, H.A.: The architecture of complexity. In: Klir, G.J. (ed.) Facets of Systems Science, pp. 457–476. Springer, Boston (1991). https://doi.org/10.1007/978-1-4899-0718-9_31
21. Holland, J.H.: Complex adaptive systems. Daedalus **121**, 17–30 (1992)

22. Miller, J.H., Page, S.E.: Complex Adaptive Systems: An Introduction to Computational Models of Social Life. Princeton University Press, Princeton (2009)
23. Weisberg, M.: Simulation and Similarity: Using Models to Understand the World. Oxford University Press, Oxford (2012)
24. Smit, B., Burton, I., Richard, J.T., Klein, J.W.: An anatomy of adaptation to climate change and variability. In: Kane, S.M., Yohe, G.W. (eds.) Societal Adaptation to Climate Variability and Change, pp. 223–251. Springer, Dordrecht (2000). https://doi.org/10.1007/978-94-017-3010-5_12
25. Badjeck, M.-C., Allison, E.H., Halls, A.S., Dulvy, N.K.: Impacts of climate variability and change on fishery-based livelihoods. Mar. Policy **34**, 375–383 (2010)
26. Kratz, T.K., Deegan, L.A., Harmon, M.E., Lauenroth, W.K.: Ecological variability in space and time: insights gained from the US LTER program. Bioscience **53**, 57–67 (2003)
27. Hendry, A.P., Schoen, D.J., Wolak, M.E., Reid, J.M.: The contemporary evolution of fitness. Annu. Rev. Ecol. Evol. Syst. **49**, 457–476 (2018)
28. Folland, C.K., Karl, T.R., Jim Salinger, M.: Observed climate variability and change. Weather **57**, 269–278 (2002)
29. Carley, K.M.: Simulating society: the tension between transparency and veridicality. In: Proceedings of Agents, p. 2 (2002)
30. Mark, J.T., Marion, B.B., Hoffman, D.D.: Natural selection and veridical perceptions. J. Theor. Biol. **266**, 504–515 (2010)
31. Nour, M.M., Nour, J.M.: Perception, illusions and bayesian inference. Psychopathology **48**, 217–221 (2015)
32. Larzelere, A.R.: Creating simulation capabilities. IEEE Comput. Sci. Eng. **5**, 27–35 (1998)
33. Harzheim, E.: Ordered Sets. Springer, Heidelberg (2006). https://doi.org/10.1007/b104891
34. Littman, M.L.: Markov games as a framework for multi-agent reinforcement learning. In: Machine Learning Proceedings 1994, pp. 157–163. Elsevier (1994)
35. Macy, M.W., Willer, R.: From factors to actors: computational sociology and agent-based modeling. Ann. Rev. Sociol. **28**, 143–166 (2002)
36. Piantadosi, C.A.: The Biology of Human Survival: Life and Death in Extreme Environments. Oxford University Press, Oxford (2003)
37. Rockwood, L.L.: Introduction to Population Ecology. John Wiley & Sons, Hoboken (2015)
38. Grafius, D.R., et al.: Estimating food production in an urban landscape. Sci. Rep. **10**, 1–9 (2020)
39. Maud, P.J., Foster, C.: Physiological assessment of human fitness. Human Kinetics (2006)
40. Maclean, K., Cuthill, M., Ross, H.: Six attributes of social resilience. J. Environ. Planning Manage. **57**, 144–156 (2014)
41. Zhang, Z., Wang, D., Zhao, D., Song, T.: FMR-GA – a cooperative multi-agent reinforcement learning algorithm based on gradient ascent. In: Liu, D., Xie, S., Li, Y., Zhao, D., El-Alfy, E.-S. (eds.) Neural Information Processing, pp. 840–848. Springer, Cham (2017). https://doi.org/10.1007/978-3-319-70087-8_86
42. McLeman, R., Smit, B.: Migration as an adaptation to climate change. Clim. Change **76**, 31–53 (2006)

Natural Interaction with Traffic Control Cameras Through Multimodal Interfaces

Marco Grazioso[1] , Alessandro Sebastian Podda[2] , Silvio Barra[1(✉)] ,
and Francesco Cutugno[1]

[1] Department of Electric and Information Technology Engineering (DIETI),
University of Naples, "Federico II", Naples, Italy
{marco.grazioso,silvio.barra,francesco.cutugno}@unina.it
[2] Department of Mathematics and Computer Sciences, University of Cagliari,
Cagliari, Italy
sebastianpodda@unica.it

Abstract. Human-Computer Interfaces have always played a fundamental role in usability and commands' interpretability of the modern software systems. With the explosion of the Artificial Intelligence concept, such interfaces have begun to fill the gap between the user and the system itself, further evolving in Adaptive User Interfaces (AUI). Meta Interfaces are a further step towards the user, and they aim at supporting the human activities in an ambient interactive space; in such a way, the user can control the surrounding space and interact with it. This work aims at proposing a meta user interface that exploits the *Put That There* paradigm to enable the user to fast interaction by employing natural language and gestures. The application scenario is a video surveillance control room, in which the speed of actions and reactions is fundamental for urban safety and driver and pedestrian security. The interaction is oriented towards three environments: the first is the control room itself, in which the operator can organize the views of the monitors related to the cameras on site by vocal commands and gestures, as well as conveying the audio on the headset or in the speakers of the room. The second one is related to the control of the video, in order to go back and forth to a particular scene showing specific events, or zoom in/out a particular camera; the third allows the operator to send rescue vehicle in a particular street, in case of need. The gestures data are acquired through a Microsoft Kinect 2 which captures pointing and gestures allowing the user to interact multimodally thus increasing the naturalness of the interaction; the related module maps the movement information to a particular instruction, also supported by vocal commands which enable its execution. Vocal commands are mapped by means of the LUIS (Language Understanding) framework by Microsoft, which helps to yield a fast deploy of the application; furthermore, LUIS guarantees the possibility to extend the dominion related command list so as to constantly improve and update the model. A testbed procedure investigates both the system usability and multimodal recognition performances. Multimodal sentence error rate (intended as the number of incorrectly recognized utterances even for a single item) is around 15%, given by the

© Springer Nature Switzerland AG 2021
H. Degen and S. Ntoa (Eds.): HCII 2021, LNAI 12797, pp. 501–515, 2021.
https://doi.org/10.1007/978-3-030-77772-2_33

combination of possible failures both in the ASR and gesture recognition model. However, intent classification performances present, on average across different users, accuracy ranging around 89–92% thus indicating that most of the errors in multimodal sentences lie on the slot filling task. Usability has been evaluated through task completion paradigm (including interaction duration and activity on affordances counts per task), learning curve measurements, a posteriori questionnaires.

Keywords: Control room · Multimodal interaction · Speech and gestures

1   Introduction

The user experience and the user interfaces often make a difference in the choice of a software rather than another. This because, during the years, the usability of a system has become the most powerful evaluation criteria, given the central role of the user within the life cycle of a software [32]. As a consequence, the user interfaces have accordingly modified, so to satisfy the user requirements, which suggest more natural interaction means [3]. Adaptive User Interfaces [6] are an evolution of the common UI, in which the interface adapts itself to meet the user interaction criteria; such kind of interfaces have furtherly narrowed the gap between the system and the user, since the interaction between the two components improves asymptotically after a partial experience of the system itself with that user [14]. A further step forward is identified in the Meta User Interfaces [7] which improve the UX by adding particular care to the environments which s/he acts within. The definition of such kind of interfaces is reported from the work in [8]:

> The interface is meta because it serves as an umbrella beyond the domain-dependent services that support human activities in this space. It is UI-oriented because its role is to allow users to control and evaluate the state of the ambient interactive space. By analogy, a metaUI is to ambient computing what desktops and shells are to conventional workstations.

These interfaces aim at supporting human activities in an ambient interactive space [7]; in such a way, the user can control the surrounding space and interact with it. Besides being very useful in smart environments [29], given their inner ability to allow interaction with the surrounding space, Meta User Interfaces develop their usefulness in contexts in which speed of action and reaction are fundamental, like in surgery scenarios [28,30] or in scenarios in which the user needs to have the entire surrounding under direct control.

In this paper, a Meta User interface for a video surveillance control room application scenario [1] is proposed; the interface is based on the Put That There paradigm [4], so to enable the user to a fast interaction by using natural language and gestures. Three interaction environments are considered:

- the *control room environment*: in this environment, the user is given both the ability to organize the views of the monitor s/he controlling, and the option to convey the audio of a specific monitor towards the headset or in the speakers spread into the room;
- the *video management environment*: the user can navigate a particular video, so to reach a specific minute or a particular scene showing a specific event. Also, s/he can zoom in/out a particular screen as well as applying a split-screen to compare a specific trait of the road from to different points of view (if proper ground cameras are placed). Finally, also the possibility to pan, tilt and zoom a particular camera is provided (if the camera is provided with such mechanics);
- the *road*; the user is offered the skill to act immediately whether an intervention is required on a road; in such sense, the operator is provided with interaction means for sending rescue vehicle in a particular street, in case of need.

The main contributions of the proposed paper are the following:

- an entity-extensible tool for gestures and vocal interaction with the surrounding environment;
- three environments act as the object of the interaction: the control room, the displayed video and the surveilled road;
- the system exploits the Kinect for modelling the operator joints and the **FANTASIA** framework for the rapid development of interactive applications.

The remainder of the work is organized as follows: Sect. 2 discusses the state of the art; Sect. 3 describes the entire system architecture and the related modules for gesture interactions and voice commands. Section 4 evaluates the system and reports details about the testbed. Finally, Sect. 5 concludes the paper and explores future directions.

2   Related Work

The user interfaces are nowadays strongly oriented to the improvement of the user experience, especially in those factors related to accessibility, adaptability and control. The accessibility finds its leading exponent in the multimodal interfaces [23], which provide several modalities of interaction with the system, thus resulting useful not only for normal users, which are able to choose the preferred interaction mode [9], but also for people with physical impairments whose interaction criteria are met by one or more of the provided modalities [34]. Few examples are described in [26] and in [24]: both are oriented to disabled users, in order to facilitate their interaction without using mice or keyboards; the first proposes an augmented communication system whose interface is controlled by different types of signals, like electromyography, electrooculography and accelerometer. In the second, instead, Voiceye is described, a multimodal interface system that combines voice input and eye-gaze interaction for writing code. Multimodal interfaces are also very useful in those environments in

which the user needs his/her hands for primary tasks, and therefore the interaction must take place in other ways. As an example, drivers have their hands on the steering wheel and therefore interactions with the surrounding cockpit must happen in other ways. An interesting multimodal interaction system oriented to drivers is presented in [13] which along with the common interaction modalities like touch and physical button, the authors proposes further modalities like *mid-air gestures, voice* and *gaze*.

AI-driven interfaces are more reliable for those systems which need to modify themselves during the interaction in order to further adapt to the user him/herself. An example of systems is described in [17] in which the application proposed delivers personalized and adaptive multimedia content tailored to the user. Other examples of adaptive systems are located in the field of robotics and Human-Robot Interaction, like in [33] and in [31]. In [27] an AI-driven approach for additive manufacturing is proposed. Many systems are also aimed at improving learning platforms, like in [12] in which the engagement and motivation of the students are inferred by analysing their implicit feedbacks.

A further step towards the complete environment automation and complete control for the user is given by the meta user interfaces whose main focus is to support the interaction with ambient intelligence systems [16]. Such kind of interaction earns much importance in the current era of IoT and smart environments, like highlighted in [18], in which the authors describe a 3D-based user interface for accessing smart environments. In particular, many works have dealt with the problem of interacting with Smart Homes [2,5], since, in such contexts, the user needs to control not a single item, but in some cases, the objects of the requirements are a bunch of sensors which have to cooperate in order to produce a result. From this point of view, there have been many works dealing with such issues [25] and the answers have been very controversial; in fact, in order to assist the user in the best way possible, some systems need to design both machine-to-human and machine-to-machine (M2M) communication systems [11].

3   System Architecture

The applications scenario is shown in Fig. 1; it depicts the control room in which the operator is immersed. The operator is located about 2 m from the monitor, so as to have a clear overall view of all the supervised streets. The interactive area consists of a 2,5 m high and 4,4 m long curved screen, which nine street camera views are displayed on in a $3 - by - 3$ configuration.

On the floor, on the basis of the screen, a Microsoft Kinect 2 [35] is placed, which captures user pointing and gestures so as to allow him a very natural interaction; simultaneously, a microphone capture the vocal command uttered by the user. The scenario is completed by a set of environmental speakers located in the upper corners of the room; also, the user can wear a headset provided with the microphone.

SYSTEM OVERVIEW

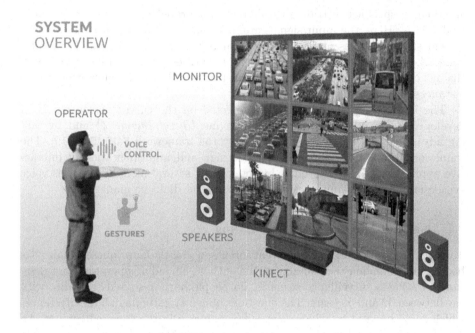

Fig. 1. The figure illustrates the application scenario. Note that the operator has a widescreen at his disposal, which he can interact with; the screen is logically divided in a 3-by-3 matrix-shaped form. At the basis of the monitor, the Kinect gathers the gestures of the operator. The speakers can be placed within the room, independently by the position of the monitor. Also, an environmental microphone (or a headset) gets the vocal commands of the operator.

The gestures data are acquired through the related module that maps the movement information to a particular instruction, also supported by vocal commands which enable its execution. The videos are synthetically generated and acquired from the *Grand Theft Auto V* video-game by Rockstar Games, by using a particular patch that enables specific tricks within the video-game, like the custom positioning of the cameras within the scene, creation of pedestrians for simulating crowds, traffic lights control and so on. Vocal commands are mapped by means of the **LUIS** (Language Understanding) framework by Microsoft, which helps to yield a fast deploy of the application; moreover, **LUIS** guarantees the possibility to extend the domain-related command list, to constantly improve and update the model. This framework receives a previously defined set of commands as input; there is no linguistic constraint to the possible utterances the user can produce, thus leading to a very natural way of producing commands. Speech and video features are acquired asynchronously and thus, a step of data fusion is necessary to perform the interpretation of the commands and make the system able to accomplish the users' requests. It has been adopted a multimodal fusion technique that makes it possible to easily incorporate expert knowledge

and domain-specific constraints within a robust, probabilistic framework: Open-Dial [10, 15]. Here, the input signals are modeled into random variables and the system state is represented as a Bayesian Network, where probabilistic rules define the way the system state is updated. Entities recognised into the speech channel compose the probabilistic rule together with labels signalling the gesture instances.

The interactive 3D environment projected on the curved screen has been developed by using the popular game engine *Unreal Engine 4*[1] and by taking advantage of the facilities provided by the framework FANTASIA [19]. The framework aims at supporting a rapid design and implementation of interactive applications for HCI studies by integrating several modules, such as voice synthesis, speech recognition, graph database and dialogue manager.

3.1   Natural Language Understanding

To gather data to model the different intents, a *Google Form* questionnaire has been released among the contacts of the authors, including operators that daily work in a video surveillance scenario. 25 people have answered the form, with age between 32 and 55 years. The questions aimed at gathering all the utterances which could be used to achieve a specific operation in the depicted context. An example of question has been related to all the possible ways to refer to a specific monitor. Recall that the screen is divided into nine monitors in a 3-by-3 shaped matrix. In this context, the single monitor can be referred in different ways; following some of the answers obtained:

- upper-right monitor - monitor in (1,1) - first monitor - north/west monitor;
- central monitor - monitor in (2,2) - monitor at the center - fifth monitor;
- lower-left monitor - monitor in (3,3) - last monitor - monitor number 9;
- ...

Analysing the utterance, a recurrent pattern emerges. In most of cases, people start the sentence by declaring the operation to do, followed by the target (or targets) of such operation. This observation has guided our choice for applying the intent-entity paradigm. This paradigm involves two steps: intent classification and entity recognition. The first task is in charge of the identification of the general purpose of the user query, e.g. the operation to be executed. The second one is responsible for retrieving from the query the objects which have a meaning in the specific domain. Typically, these tasks require the developing of neural networks and, as a consequence, the need to access a huge amount of labelled data; hence, starting from scratch is not a feasible solution. Therefore, the idea has involved the use of LUIS, so to take advantage of its capabilities. LUIS is a module included in the Microsoft cognitive services, which provides a simple way of loading sentences, defining intents and entities for data annotation, training models, even with a small amount of data, and exposing it as a service. Moreover, it gives the chance to define complex entities composed of a

---

[1] https://www.unrealengine.com/.

set of smaller sub-entities. This level of representation allows us to define more general concepts that could be composed differently, based on the contained sub-entities. For instance, the general concept *monitor* could be composed by the couple <MONITOR, REFERENCE NUMBER> or by the pair (X,Y) like in a matrix.

Starting by the matrix view shown in Fig. 1, the following intents have been identified:

- ZOOM_IN - ZOOM_OUT: for enlarging a specific monitor (zoom in) or for going back to the matrix view (zoom out);
- SPLIT_SCREEN: the matrix view is substituted by a view in which two videos are placed side by side;
- SWAP: swap the position on two videos into the matrix view;
- AUDIO_TO_DEVICE: the current audio is conveyed to a certain device (a headset or the room speakers);
- AUDIO_OFF: switch off the audio;
- REWIND - FORWARD: go back or forth to a particular minute in the video.

The involved entities are the monitor, which is composed by the sub-entities *ref*, *ref_x* and *ref_y*, and the device, in order to capture the reference to an audio device; in addition also deictic terms are modelled, so to allow the user to utter expressions like *this* and *that* (as an example, for referring to a particular monitor), in their singular and plural form.

3.2 Pointing Recognition

The pointing recognition module acts in an independent way, by asynchronously collecting the skeleton information provided by the Kinect sensor. Skeleton data consists of a set of 3D points representing the user's joints (see Fig. 2). The coordinates of such points refer to the Kinect coordinates system, where the axes origin corresponds to the sensor position. Since the 3D environment coordinates system does not match the Kinect one, the skeleton data has been transformed by rotating them according to the sensor inclination, so as to properly representing it in the 3D environment. Moreover, the user height, the distance from the Kinect and the lateral displacement are taken into account. Skeleton data, in combination with the Skeletal Mesh object provided by Unreal Engine, could be used to spawn an avatar of the user in the virtual space. The Skeletal Mesh consists of a hierarchical set of interconnected bones and it gives the chance to associate its joints with the Kinect one, obtaining a virtual representation of the user that follows his movements. Once obtained a good user representation, the next step is to estimate where the user is pointing at. This process could be divided into two sequential tasks:

- Pointing detection.
- Pointing recognition.

In pointing detection, it is important to distinguish between pointing and non-pointing gestures. Since, at the moment, the system does not recognize other

kind of gestures, it is possible to use the hand position and movement speed as discriminant. In particular, by computing the distance between the SpineMid joint and the hand one, such positions where the hand is very high or very low could be excluded, assuming that the user is not pointing to the screen. Moreover, an high speed movement of the hand suggests that the system is observing a transition movement and it must be excluded, too. Exclusion criteria are based on fixed thresholds empirically estimated. The detected gestures can now be processed to recognize the pointed object. To accomplish this task, a geometrical approach is used: it computes the line passing through the shoulder joint and the hand one, and stretches it forward until it collides with an environment object. In order to avoid errors caused by possible noise in the joints data and, eventually, transition movements that passed the first filtering step, our approach collects the pointed objects inside a window of 1 s and then, for each different object, computes the probability to be the current pointed object.

Fig. 2. Joints provided by the Kinect [20]

3.3   Multimodal Fusion Engine

The Natural Language Understanding (NLU) and the pointing Recognition activities, discussed in the previous subsections, have been fused into the Multimodal Fusion Engine here discussed, in order to provide the proposed concept of multimodality. Following the suggestions discussed in [21], the Multimodal Fusion Engine has been developed as an independent module. It receives asynchronous messages from the input modules related to the NLU and the gesture recognition, handled by the specific receivers:

- the NLU message consists of the sentence uttered by the user, together with the related intents and entities. Also, a confidence value is returned related to the recognition of the intents and the entities.
- the gesture recognition message consists of the pointed ojects, together with the related confidence values;

The OpenDial framework is in charge of receiving and managing the inputs from both the gesture recognition and the NLU modules. OpenDial has the peculiarity to be capable of managing the dialogue using a mixed approach based on both rule-based and probabilistic approaches. Indeed, whether it allows to integrate expert knowledge by defining rules, on the other hand it allows probabilistic reasoning by defining Bayesian networks. Given that, the received messages are mapped to the respective random variables encoded in a Bayesian network, so to derive a common interpretation. During the multimodal fusion, several aspects need to be considered and modelled to avoid wrong interpretation. According to the paper in [10], several OpenDial modules have been developed, that change the network configuration according to specific constraints. The *Multimodal Input Integrator* module aims at combining input variables coherently. In particular this module analyses verbal actions and pointed objects in order to understand the current request. Since the variables evolve in real-time, the *Multimodal Time Manager* is used to check the consistency and prune out-of-date variables. In particular, starting from time-stamps related to the input variables, once a new speech signal is captured, the module compares its time intervals with those computed for each pointing variable, pruning off pointing gestures whose occurrence was concluded more than 4 s before the start of the current speech signal. Pruning criteria were selected in accordance with the study discussed in [22]. In order to support multi target operations, (as an example"SWAP THIS MONITOR WITH THIS ONE) the system needs to keep in memory more than the last pointed object; to this regard, a linked list is implemented, so to keep trace the pointed objects from the most recent one to the last one respecting the previous criteria. Since the input variables come asynchronously, the *State Monitor* manages the entire operation by observing changes in dialogue state. Therefore, the unification methods are called by this component according to dialogue progresses. Once the system has derived the current request, the message containing the most appropriate action to be performed is sent to the game engine.

4   Evaluation

A test bed procedure is used to investigate both the system usability and multimodal recognition performances. Users have been involved in a task-oriented test consisting of six tasks. Since the main aim is to test the usability of the application in a real scenario, each of the tasks is composed of both a voice command and a gesture activity. In such a way all the involved modules are tested in their entirety.

After a brief introduction on both the scenario and the interaction modality, the task list was given to the users and they were then left free to interact as they preferred to accomplish their jobs. During the tests, users have been recorded in order to analyze their behaviour, so to obtain usability information. Each task was annotated with one of following labels: s for success, ps for partial success and f for failure. The explanation of each outcome is explained as follows:

- Success (s): the user has completed the task acting in an autonomous way and in reasonable times.

- Partial Success PS: the user has needed some suggestions or spent more time to complete the task;
- Failure F: the user was completely unable to proceed in the task completion.

Moreover, the fusion engine logged the information about the fusion process in order to compare its result with the recorded video, so obtaining an estimation of the precision and reliability of the fusion basing on the NLU and Gesture recognition inputs.

The tasks How cited above, the tasks have been built such that both the NLU module and the Gesture Recognition module are activated, so to properly execute the Multimodal Fusion Engine. Before starting a task, nine videos at the same time are shown, organized as a 3-by-3 matrix. The users are placed at 3m from the screen. The Kinect is placed at the bottom of the screen. The room is provided with four speakers at the top corners; however, the user is provided with a wireless headset with a microphone. In Table 1 the 6 tasks are reported; however, these have been defined to the user by not using keywords, like ZOOM, SWAP and so on, but paraphrases are used.

Table 1. The tasks involved in the test; the users are asked to accomplish the defined actions in sequence. When the action involves a "Random Monitor" this means that the user is free to choose the preferred monitors as objects of his/her action.

	Action 1	Action 2	Action 3	Action 4
T1	*Zoom <Monitor 1>*	*Zoom out*	*Zoom <Monitor 9>*	*Zoom out*
T2	*Zoom <Rand. Monitor>*	*Zoom out*	*Zoom <Rand. Monitor>*	*Zoom out*
T3	*Split <Monitors (1, 9)>*	*Zoom out*	*Split <Monitors (3, 7)>*	*Zoom out*
T4	*Split <Rand. Monitors>*	*Zoom out*	*Split <Rand. Monitors>*	*Zoom out*
T5	*Swap <Monitors (1, 9)>*	*Zoom out*	*Swap <Monitors (3, 9)>*	*Zoom out*
T6	*Swap <Rand. Monitors>*	-	*Swap <Rand. Monitors>*	-

Twelve participants have been hired for the testing phase; three considerations need to be done, in order to highlight the fairness of the test.:

- none of the participants works as operator in video surveillance field;
- all of the participants can be defined as average computer users;
- none of the participants have been invited to fill the form used for defining entities and intents.

4.1 Results

The data collected during the interaction between the system and the users have been used to generate the results in Table 2. This analysis represents a simple way of estimating the system usability by computing the *task completion rate*. This

measure has been computed by counting the total successes, the partial successes and the failures of the users in completing the assigned task and by making a weighted average of those values. In particular, the success has a weight equal to 1, the partial success has a weight equal to 0.5 and failures has a weight equal to 0. Proceeding as described, a total of 52 successes, 16 partial successes and 4 failures were obtained. Computing the task completion rate, a value of 0.83 emerged. Considering the few data used to train the NLU model this represents a good result; indeed, analysing the recorded test sessions, it was noticed that the most frequent cause of failure comes from the wrong intent or entity interpretation. This observation suggests that by increasing the amount and the variety of the examples used to train the model, it is possible to improve the results. Moreover, the success rate for both the NLU and Gesture recognition modules is computed. For the NLU model, the number of correct interpretation over the total speech interaction was counted, providing a success rate of 0.76. As said before, this value is strongly influenced by the amount and the variety of the examples. It is also important to say that answering an online survey is different from really interact with a system; in fact, the data could be not representative enough for the case study. In this regard, a pipeline of active learning would increase the NLU success rate and consequently the task completion rate. This activity would help to collect and interpret the misunderstood sentence, so improving the model. Regarding the Gesture recognition, the number of correct objects recognised by the system over the total interaction has reached an accuracy of 79%. As expected, most of the errors occur in the multiple object intent. Since this is a complex activity, several variables may influence the success of the action. In particular, wrong object selection doesn't come from an imprecise computation of the art direction, but comes from the users' movement speed

Table 2. Task completion table

| | | TASKS | | | | | |
		T1	T2	T3	T4	T5	T6
U S E R S	**U1**	S	S	S	S	S	S
	U2	PS	S	S	S	F	F
	U3	S	S	S	S	S	S
	U4	S	S	PS	S	PS	PS
	U5	S	S	PS	S	PS	S
	U6	PS	PS	S	S	S	S
	U7	S	PS	S	S	S	S
	U8	S	S	PS	S	F	F
	U9	S	S	S	S	PS	S
	U10	S	S	PS	S	PS	S
	U11	S	S	S	S	PS	S
	U12	S	S	PS	PS	S	S

from an object to another. If the movement takes long, the multimodal fusion starts before the user complete the activity. In most cases, this problem has regarded the users not so comfortable with technology. From this observation, it can be deduced that, to satisfy a larger part of the population, it is necessary to improve the recognition of multi-target intents by providing a time window large enough to consent to end the movement. It can be concluded that the multimodal fusion works properly under the assumption that both the NLU and Gesture recognition modules do their job correctly.

5 Conclusions and Future Work

In this paper, a meta user interface was proposed. The application scenario of the system is a video surveillance control room, in which an operator has an NLU module and a Gesture recognition module at his disposal, to issue commands to the environments by leveraging both on voice and pointing. The system has involved the use of a Kinect and the LUIS framework for gesture modelling and vocal command processing respectively; the OpenDial framework has been used for fusing information coming from the modules. The preliminary results are obtained by assigning six composed tasks to twelve participants; these show that the system is consistently reliable and usable, since the participants were not trained for the test, but they were only explained what the system was intended for.

Limitations. Given the good outcomes of the system results, many advances can be done: first, it would be possible to expand the use of the system also to other purposes, like event tagging, object annotations and so on. This would imply the definition of new entities (CAR, LAMPPOST, PEDESTRIAN, ...) and new intents (LABEL, PINCH, ...). This involves an enhancement also in the supporting devices, given the fact that some actions involving the single fingers, and not the hands, cannot be easily recognized by using the Kinect.

References

1. Atzori, A., Barra, S., Carta, S., Fenu, G., Podda, A.S.: Heimdall: an AI-based infrastructure for traffic monitoring and anomalies detection (2021, in press)
2. Balta-Ozkan, N., Davidson, R., Bicket, M., Whitmarsh, L.: Social barriers to the adoption of smart homes. Energy Policy **63**, 363–374 (2013). https://doi.org/10.1016/j.enpol.2013.08.043. https://www.sciencedirect.com/science/article/pii/S0301421513008471
3. Barra, S., Carcangiu, A., Carta, S., Podda, A.S., Riboni, D.: A voice user interface for football event tagging applications. In: Proceedings of the International Conference on Advanced Visual Interfaces, AVI 2020, Association for Computing Machinery, New York, NY, USA (2020). https://doi.org/10.1145/3399715.3399967
4. Bolt, R.A.: "put-that-there" voice and gesture at the graphics interface. In: Proceedings of the 7th Annual Conference on Computer Graphics and Interactive Techniques, pp. 262–270 (1980)

5. Bonino, D., Corno, F.: What would you ask to your home if it were intelligent? exploring user expectations about next-generation homes. J. Ambient Intel. Smart Environ. **3**, 111–126 (2011). https://doi.org/10.3233/AIS-2011-009910.3233

6. Browne, D., Totterdell, P., Norman, M. (eds.): Computers and People Series. Academic Press, London (1990). http://www.sciencedirect.com/science/article/pii/B9780121377557500017

7. Coutaz, J.: Meta-user interfaces for ambient spaces. In: Coninx, K., Luyten, K., Schneider, K.A. (eds.) TAMODIA 2006. LNCS, vol. 4385, pp. 1–15. Springer, Heidelberg (2007). https://doi.org/10.1007/978-3-540-70816-2_1

8. Coutaz, J.: Meta-user interfaces for ambient spaces: can model-driven-engineering help?. In: Burnett, M.H., Engels, G., Myers, B.A., Rothermel, G. (eds.) In: Proceedings of Dagstuhl Seminar End-User Software Engineering, Schloss Dagstuhl - Leibniz-Zentrum fuer Informatik, Germany, Dagstuhl, Germany, p. 07081 (2007). http://drops.dagstuhl.de/opus/volltexte/2007/1082

9. Grazioso, M., Cera, V., Di Maro, M., Origlia, A., Cutugno, F.: From linguistic linked open data to multimodal natural interaction: a case study. In: 2018 22nd International Conference Information Visualisation (IV), pp. 315–320 (2018). https://doi.org/10.1109/iV.2018.00060

10. Grazioso, M., Di Maro, M., Cutugno, F.: "what's that called?": a multimodal fusion approach for cultural heritage virtual experiences. In: CEUR-WS, vol. 2730 (2020). https://www.scopus.com/inward/record.uri?eid=2-s2.0-85096133907&partnerID=40&md5=46b2830cde8476d6e561254c280e6987

11. Kashyap, H., Singh, V., Chauhan, V., Siddhi, P.: A methodology to overcome challenges and risks associated with ambient intelligent systems. In: 2015 International Conference on Advances in Computer Engineering and Applications, pp. 245–248 (2015). https://doi.org/10.1109/ICACEA.2015.7164704

12. Kim, B., Suh, H., Heo, J., Choi, Y.: AI-driven interface design for intelligent tutoring system improves student engagement. arXiv preprint arXiv:2009.08976 (2020)

13. Kim, M., Seong, E., Jwa, Y., Lee, J., Kim, S.: A cascaded multimodal natural user interface to reduce driver distraction. IEEE Access **8**, 112969–112984 (2020). https://doi.org/10.1109/ACCESS.2020.3002775

14. Langley, P.: Machine learning for adaptive user interfaces. In: Brewka, G., Habel, C., Nebel, B. (eds.) KI 1997. LNCS, vol. 1303, pp. 53–62. Springer, Heidelberg (1997). https://doi.org/10.1007/3540634932_3

15. Lison, P., Kennington, C.: Opendial: A toolkit for developing spoken dialogue systems with probabilistic rules. In: Proceedings of ACL-2016 System Demonstrations, pp. 67–72 (2016)

16. Mostafazadeh Davani, A., Nazari Shirehjini, A.A., Daraei, S.: Towards interacting with smarter systems. J. Ambient Intell. Humanized Comput. **9**(1), 187–209 (2018). https://doi.org/10.1007/s12652-016-0433-9

17. Mozgai, S., Hartholt, A., Rizzo, A.S.: An adaptive agent-based interface for personalized health interventions. In: Proceedings of the 25th International Conference on Intelligent User Interfaces Companion, IUI 2020, pp. 118–119. Association for Computing Machinery, New York, NY, USA (2020). https://doi.org/10.1145/3379336.3381467

18. Nazari Shirehjini, A.A., Semsar, A.: Human interaction with IoT-based smart environments. Multimedia Tools Appl. **76**(11), 13343–13365 (2016). https://doi.org/10.1007/s11042-016-3697-3

19. Origlia, A., Cutugno, F., Rodà, A., Cosi, P., Zmarich, C.: FANTASIA: a framework for advanced natural tools and applications in social, interactive approaches. Multimedia Tools Appl. **78**(10), 13613–13648 (2019). https://doi.org/10.1007/s11042-019-7362-5

20. Ousmer, M., Vanderdonckt, J., Buraga, S.: An ontology for reasoning on body-based gestures. In: Proceedings of the ACM SIGCHI Symposium on Engineering Interactive Computing Systems. EICS 2019, Association for Computing Machinery, New York, NY, USA (2019). https://doi.org/10.1145/3319499.3328238

21. Oviatt, S.L., Cohen, P.: Perceptual user interfaces: multimodal interfaces that process what comes naturally. Commun. ACM **43**(3), 45–53 (2000)

22. Oviatt, S.L., DeAngeli, A., Kuhn, K.: Integration and synchronization of input modes during multimodal human-computer interaction. In: Proceedings of Conference on Human Factors in Computing Systems, CHI 1997, pp. 415–422 (March 22–27, Atlanta, GA). ACM Press, NY (1997)

23. Oviatt, S., et al.: Multimodal interfaces. The human-computer interaction handbook: Fundamentals, evolving technologies and emerging applications **14**, 286–304 (2003)

24. Paudyal, B., Creed, C., Frutos-Pascual, M., Williams, I.: Voiceye: A multimodal inclusive development environment. In: Proceedings of the 2020 ACM Designing Interactive Systems Conference, pp. 21–33 (2020)

25. Portet, F., Vacher, M., Golanski, C., Roux, C., Meillon, B.: Design and evaluation of a smart home voice interface for the elderly: acceptability and objection aspects. Pers. Ubiquit. Comput. **17**(1), 127–144 (2013). https://doi.org/10.1007/s00779-011-0470-5

26. Rocha, L.A.A., Naves, E.L.M., Morére, Y., de Sa, A.A.R.: Multimodal interface for alternative communication of people with motor disabilities. Research on Biomedical Engineering **36**(1), 21–29 (2019). https://doi.org/10.1007/s42600-019-00035-w

27. Röhm, B., Gögelein, L., Kugler, S., Anderl, R.: AI-driven worker assistance system for additive manufacturing. In: Ahram, T. (ed.) AHFE 2020. AISC, vol. 1213, pp. 22–27. Springer, Cham (2021). https://doi.org/10.1007/978-3-030-51328-3_4

28. Rosa, G.M., Elizondo, M.L.: Use of a gesture user interface as a touchless image navigation system in dental surgery: case series report. Imaging Sci. Dentist. **44**(2), 155 (2014)

29. Roscher, D., Blumendorf, M., Albayrak, S.: A meta user interface to control multimodal interaction in smart environments. In: Proceedings of the 14th International Conference on Intelligent User Interfaces IUI 2009, pp. 481–482. Association for Computing Machinery, New York, NY, USA (2009). https://doi.org/10.1145/1502650.1502725

30. Sánchez-Margallo, F.M., Sánchez-Margallo, J.A., Moyano-Cuevas, J.L., Pérez, E.M., Maestre, J.: Use of natural user interfaces for image navigation during laparoscopic surgery: initial experience. Minim. Invasive Ther. Allied Technol. **26**(5), 253–261 (2017)

31. dos Santos12, J.R.A., Meyer, T.S., Junior, P.T.A.: An adaptive interface framework for a home assistive robot

32. Wallach, D., Scholz, S.C.: User-centered design: why and how to put users first in software development. In: Maedche, A., Botzenhardt, A., Neer, L. (eds.) Software for people, pp. 11–38. Springer, Berlin, Heidelberg (2012). https://doi.org/10.1007/978-3-642-31371-4_2

33. Wijayasinghe, I.B., Saadatzi, M.N., Peetha, S., Popa, D.O., Cremer, S.: Adaptive interface for robot teleoperation using a genetic algorithm. In: 2018 IEEE 14th International Conference on Automation Science and Engineering (CASE), pp. 50–56. IEEE (2018)

34. Worsley, M., Barel, D., Davison, L., Large, T., Mwiti, T.: Multimodal interfaces for inclusive learning. In: Penstein Rosé, C., et al. (eds.) AIED 2018. LNCS (LNAI), vol. 10948, pp. 389–393. Springer, Cham (2018). https://doi.org/10.1007/978-3-319-93846-2_73

35. Zhang, Z.: Microsoft kinect sensor and its effect. IEEE Multimedia 19(2), 4–10 (2012)

Building Conversational Agents for Military Training: Towards a Virtual Wingman

Joost van Oijen[✉] and Olivier Claessen

Royal Netherlands Aerospace Centre, Amsterdam, the Netherlands
{Joost.van.Oijen,Olivier.Claessen}@nlr.nl

Abstract. In military training simulations there is an increasing need for synthetic agents as team members that are capable of replacing human role players. Human-agent interaction using spoken natural language is an important capability for such agents, especially when communication skills are part of training. Interaction technologies such as speech-to-text, text-to-speech, natural language understanding and dialogue management are maturing, though their use in human-in-the-loop simulations is scarce. In this paper we address design challenges for building conversational agents in the domain of military training. In this domain, agents often have to adhere to specific communication standards, protocols and use of jargon. We propose a data-driven design method to tailor conversational agents to a particular application domain, while minimizing human authoring. We demonstrate this method within an existing military training simulation capability and show that agents are able to effectively build shared situational awareness and coordinated their activities in a shared task environment.

Keywords: Conversational agents · Human-computer interaction · Training · Simulation

1 Introduction

Simulation-based military training benefits from agents that are capable of playing the role of adversaries or team-members. These so-called Computer Generated Forces (CGF) enable tactical training or mission rehearsal, allowing trainees to enhance skills and gain practice in applying doctrines, tactics or procedures, either individually or in teams [1]. The use of synthetic team members has the benefit of being able to replace human role-players, allowing for more flexible training opportunities, requiring fewer personnel, and thus saving costs [2, 3]. As team communication is an important aspect in many domains (e.g. in air combat [4]), synthetic team members should be able to communicate naturally with others (human or synthetic). When incapable of interaction, the use of such synthetic team members could be useless in certain domains, or potentially leading to negative training. In the military domain, speech-based interactive systems in general have a broad application [5].

An inherent capability for a speech-based conversational agent is its ability to express, recognize and understand interactions based on spoken natural language. Due

© Springer Nature Switzerland AG 2021
H. Degen and S. Ntoa (Eds.): HCII 2021, LNAI 12797, pp. 516–531, 2021.
https://doi.org/10.1007/978-3-030-77772-2_34

to advances in deep learning and natural language processing (NLP), technologies for these abilities, such as text-to-speech, speech-to-text and natural language understanding are becoming more and more mature. However, when they are to be integrated in conversational agents that represent synthetic team members in a simulation environment, this introduces several challenges:

- The technologies have to be coupled and integrated within a cognitive agent that operates in a particular domain in which it must be capable of multi-party, mixed-initiative communication with others in a shared task environment. In such an environment, the agent performs other tasks besides communication, as opposed to most conversational systems where interaction with the user is the primary task [6]. This introduces research & development challenges that are not easily catered to by off-the-shelf technologies.
- The technologies have to be tailored to the particular domain, i.e., must support communication in line with the specific activities of the agents. In military training this typically involves sharing situational awareness and coordinating goals and plans. This requires a highly domain-specific, task-oriented form of interaction that is often regulated by organization-specific operating procedures, specifying rules for interaction protocols, terminology and phraseology [4]. Such requirements has implications for the implementation of dialogue management and the NLP models. Further, example data that can benefit machine learning-based NLP is often scarce for military communication and difficult to obtain because of confidentiality and security reasons.

In this study we address the above challenges within the scope of building conversational agents for military training. In our approach, we employ available state-of-the-art interaction technologies and integrate these into an existing military training capability, with agents as role-players. The case that is studied in this research is fighter pilot radio communication, applicable for team-training for both fighter pilots and fighter controllers. Many of its characteristics are representative for other domains that involve military tactical communication.

In Sect. 2 a background is given on conversational agents. Afterwards, a general model for a conversational agent is presented where we illustrate the dependencies with domain knowledge (Sect. 3). Following we propose a data-driven design method to tailor a conversational agent towards a specific domain, focusing on automating domain configurations for individual interaction technologies, while minimizing human authoring (Sect. 4). The design method is demonstrated in the case study in Sect. 5, followed by a discussion (Sect. 6) and conclusion (Sect. 7).

2 Background

The use of conversational agents (CAs) is quickly growing within the field of human computer interaction (HCI). They are applied successfully in domains such as healthcare, business, education or the military, where they are used as tutors, personal assistance, chatbots or voice-controlled systems [7–12]. For training purposes, CAs can be used to train participants communication skills, to reduce costs, to increase training availability and to provide a standardized experience [2, 3, 13, 14].

The design of CAs varies per application, depending on the specific needs of the application. Some applications use text-based conversations where a keyboard is used to communicate with the agent and the agent's responses are printed on a prompt. Others allow spoken language interaction with embodied agents in simulations (e.g., [3, 8, 9]). Furthermore, the implementation of CAs can range from domain-expert handcrafted dialogue strategies (e.g. [15]), to automatically trained probabilistic models which can use natural language, to hybrid models using a combination of the two (e.g. [12, 13]). The exact implementation of the CA depends on several factors, including but not limited to naturalness, complex and robust dialog creation, scalability, reusability, applicability and data availability [7].

A lot of CA applications are geared towards one-on-one human-agent interactions [14, 16]. In such applications the agent is 'the system'. In this paper, we consider the agent to be embedded in a broader simulation of a shared human-agent team task environment, in line with for instance research efforts on building virtual humans for training applications [17, 18].

3   Model of a Conversational Agent

The scope of this study is visualized in Fig. 1. It highlights the two challenges identified in the introduction: the integration of interaction technologies to develop a conversational agent (top of the image), and the use of domain knowledge to tailor a conversational agent to operate in a desired application domain (bottom of the image). In the remainder of this section, we first briefly describe the involved technologies, followed by an analysis of domain knowledge requirements.

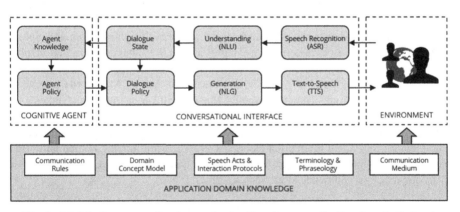

Fig. 1. Model of a conversational agent (top), tailored to an application domain (bottom)

The top of the figure shows a basic model for a conversational agent. A cognitive agent (left side) has the ability to communicate in a virtual environment (right side) through a conversational interface (center). The interface shown is based on a conventional model of a spoken dialogue system [19]. The development of a conversational agent requires the coupling and integration of different specialized technologies, each related to a research

field on its own. From left to right these include multi-agent systems, dialogue management systems, natural language understanding and generation, and speech recognition and text-to-speech. Whenever available, the use of mature, off-the-shelf technologies for one or more of these technologies allows one to employ state-of-the-art solutions, without the need to reinvent the wheel.

Besides a technical integration of technologies, a proper conceptual integration of domain knowledge is required to employ these technologies successfully in a target domain. In this study, the target domain is simulation-based military team training. Interactions in such a domain are typically well structured, regulated by standards, protocols or other underlying principles for effective communication. For instance, consider Grice's Maxims for effective communication which state that one should be informative, truthful, relevant, and clear [20]. Efficient and effective communication is of high importance, especially in time-critical missions in contested environments.

As a concrete target domain we consider fighter communication in air-to-air combat situations. Below we identify domain knowledge categories that can be used to analyze and capture a system's dependency on domain knowledge. These are the categories shown at the bottom of Fig. 1, described next from left to right.

Communication Rules: In task-oriented team communication, underlying motivations to interact with others can roughly be divided into (1) building shared situational awareness within the team, and (2) coordinating activities, in line with some shared team goal. Depending on their role, different team members have different responsibilities with respect to why, when and what to communicate with others. In fighter communication, such communication rules for individual team members are commonly defined by Standard Operating Procedures (SOP). For instance, a flight lead is expected to decide on and communicate engagement strategies and target assignments; or flight members are expected to communicate relevant updates on their progress during an engagement.

Domain Concept Model: For agents to understand the actual communicated content, a shared understanding of domain concepts is required. For instance, in a military domain, such concepts can include physical elements such as environmental features and properties of observed military units; abstract elements such as missions, tactics, formations or team roles; or events or activities such as firing a missile towards a target. In agent technology, domain concepts are commonly captured in ontologies. Ontologies can define the scope of situational awareness (what can be believed), decision-making (what can be decided), and for conversational agents, social interaction (what can be talked about). Shared ontology definitions between agents allow them to effectively operate in a shared domain [15].

Speech Acts and Interaction Protocols: At the dialogue level, speech acts describe a speaker's intention behind a spoken utterance and the desired effect it has on the listener. Speech acts can be related to actions that control the information flow between participants (e.g. request for information, inform, order, accept/reject, etc.) and acts to facilitate the interaction dynamics itself, such as initiating/ending a conversation, grounding or turn-taking acts [21]. Interaction protocols describe the turn-taking obligations between speakers to regulate a dialogue using speech acts. For example one is expected to answer a question.

The use of speech acts has been analyzed for fighter communication, see [4]. Examples include questions and informs about the enemy's activity, sharing tactics, giving orders, confirmations and informs about one's own intents and activities. Based on interviews with a subject-matter expert (SME), it was seen that interaction protocols can be effectively captured in rules. For example, initiation acts such as questions and orders require a response (inform, acknowledgement), and informative acts don't require a specific response. Protocols also state that callsigns (unique identifiers) should be used in utterances to explicitly denote senders and/or receivers.

Terminology and Phraseology: At the level of natural language, terminology is the set of terms or specialized words used in a particular field, profession or domain. Phraseology concerns rules on fixed expressions or multi-word lexical units. Military communication is often heavily based on terminology and phraseology. In fighter communication, these are defined by the SOPs. The majority of spoken utterances are based on structured, form-based phrases. Further, so-called Brevity Words are used which are standardized words that can represent rich semantics, such as specific strategies, tactics or activities. See [22] for an example list of operational brevity words used in air-to-air communications. Finally, there are more general rules based on principles for effective communication. For example, a heading is pronounced as a sequence of three numbers (e.g. 'heading zero five eight'); or unit types are mostly omitted in speech as these are based on standards known by all participants (e.g. nautical miles for distances).

Communication Medium: Finally we consider the environment and the medium used for communication. In the case study, we restrict this to voice-only, radio communication. In fighter communication, tactical team communication between a flight and fighter controller generally uses a UHF channel. Specific intra-flight communication (e.g. between a lead and wingman) can take place over VHF. The use of multiple channels limits the amount of chatter over a single channel and is more applicable in larger scenarios. We limit the scope to a single channel available to all team participants.

The above categories capture the domain requirements for the system as a whole. These translate to requirements for individual technologies. For instance, the domain's interaction protocols should be supported by the dialogue system; terminology and phraseology should be taken into account during language understanding and generation; and a speech recognition system needs to be tuned towards the domain's language to improve accuracy. In the next section we propose a design method that is geared towards data-driven solutions for configuring the technologies part of a conversational interface.

4   A Data-Driven Design Method

A data-driven design method is proposed for tailoring an agent's conversational interface to operate in a target application domain. In summary, the method employs domain knowledge to facilitate the implementation of Automated Speech Recognition (ASR), Natural Language Understanding (NLU) and Natural Language Generation (NLG). The method is described in the remainder of this section.

4.1 The Method

This basic principle behind the design method is that structured knowledge about the domain can be used to configure and/or optimize NLP processes automatically. In order words, if we have a semantic model of what an agent can talk about, this information can be used to better recognize and understand spoken utterances, assuming agents (and humans) employ the same semantic model. The method is illustrated in and is described below in four steps, as shown in the (Fig. 2)

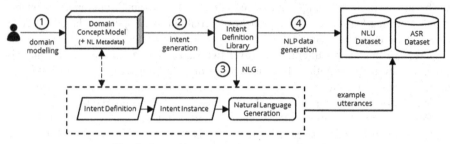

Fig. 2. Data-driven natural language processing

Step 1: Domain Modelling. In this step, a human modeler defines domain-specific semantic concepts that can be used to represent knowledge, reasoning rules or interaction content, denoted as the Domain Concept Model. The use of ontologies is a common way to define such concepts in agent systems [23]. Concepts can represent simple types (e.g. a name, direction or aircraft type), complex types (e.g. a radar picture), or events (e.g. a missile was fired). Additionally, the modeler associates the concepts with metadata that describes how they are expressed in natural language (NL). This is where a domain's use of terminology and phraseology is encoded. How NL metadata is used to create spoken utterances is described in step 3 (NLG).

During domain modelling, it is important to constrain the values of the concepts to possible or plausible values. As one of the features of the method is to generate example utterances from randomly generated knowledge (step 3), value constraints make sure this knowledge stays within logical bounds. For instance, a heading takes on values between 0 and 360; team-member identifications are based standard callsigns known in advance; or a fixed set of tactical maneuvers is applicable. Proper value constraints ensure that generated example utterances are better tuned towards the domain.

The above approach allows us to (1) capture the full knowledge space that an agent can conceptually have (and thus communicate), and (2) associate that knowledge space with corresponding representations in natural language.

Step 2: Intent Generation. In the intent generation step, a library of *communicative intent definitions* is generated. An intent in this scope represents a single, machine-understandable utterance that can be recognized or expressed in dialogue management.

An intent definition is a combination of a *speech act* and a *concept reference path*. A concept reference path can be seen as a query that points to one or more concepts from

the Domain Concept Model. Standard speech acts include a request for information (a query) and a transfer of information (the result of a query). At run-time, these speech acts represent internal actions to automatically query or update an agent's knowledge base, as will be described in the next section.

A library of the full scope of possible communicative intents can be generated in a fully offline fashion, based on generated combinations of speech acts and concept reference paths.

Step 3: Natural Language Generation. After generation of intent definitions, NLG can be utilized to create example utterances for each intent. Example utterances for intents enable machine learning-based approaches to configure and optimize the accuracy of ASR and NLU models (step 4).

Creating an example utterance involves two steps. First, an instance of an intent is generated from its definition. For intents that include knowledge (i.e. informs), this requires constructing a knowledge instance. Knowledge instances can be randomly generated from the Domain Concept Model, based on the concept reference path.

Second, for each created instance, natural language utterances are generated. For this study, a simple template-based approach towards NLG is employed [11]. A template represents an utterance in which variable-slots can be used to refer to the value of a knowledge instance. During NLG, slots (denoted with a '$') are resolved and replaced with the respective values part of the knowledge instance. For instance, knowledge about an *enemy formation* associated with the template *"$count group $shape"* is resolved to *"four group wall"* (four groups of enemies approaching in a side-by-side formation). Slot values can also refer to other complex knowledge concepts, which are then recursively resolved to generate the full utterance.

Using the above steps, one can generate as many example utterances as desired for a single intent definition. When performing this step for all entries in the intent library, a set of utterances of any configurable size can be built, covering the scope of possible natural language utterances in the application domain.

Step 4: NLP Data Generation. The final step involves the generation of NLU and ASR datasets. These datasets refer to data required by NLU and ASR models to configure them for a particular application domain. The exact form of this data depends on the underlying technologies used for NLU and ASR.

Data concepts that are shared by many NLU services are *intents* and *entities* [24]. Intents represent user intentions that can be recognized. Entities describe general concepts that have constrained values (e.g. locations, names, numbers, etc.). Entities can be embedded in intents to recognize and extract semantic concepts from utterances, sometimes known as Named Entity Recognition (NER). Finally, intents can be associated with a set of utterances to allow mapping of different expressions to a single meaning. Many services nowadays employ machine learning on example utterances to improve accuracy. In our method, intents from the intent library can be converted to NLU intents for a specific NLU service, including the intent's example utterances generated in the previous step. Entities can be extracted directly from the Domain Concept Model.

Data for ASR can be used in *language models* to capture probability distributions over sequences of words, or *dictionaries* to represent a vocabulary of words, their spelling

and pronunciation. In our method, data to facilitate such models can be obtained from the dataset of example utterances from the previous step. In this study, we have not explored ASR data further. In the NLU service that was used in the implementation (Sect. 5), ASR is a built-in capability and domain tuning is handled internally, indirectly based on data from the NLU Dataset.

4.2 Benefits of the Method

The above method has several benefits. First, human authoring is reduced to modelling of the Domain Concept Model. Based on this model, the data-driven approach automatically configures ASR and NLU models. Second, the user can directly associate modelled domain concepts with metadata about natural language usage, hereby facilitating NLG. Finally, the concept of a communicative intent enables automated back-end operations with an agent's knowledge base for storing and retrieving knowledge during communication. This is described in more detail in the next section where we apply the method to our case study, in the scope of an agent-based simulation system.

5 Case Study

In this section we describe our implementation of a case that applies conversational agents as synthetic team-members for military training. The domain chosen is team communication in air-to-air combat training. First we describe an example scenario that was implemented. Next we describe the system implementation that was used and show how the design method from the previous section was applied. We conclude with an evaluation based on a qualitative user experience.

5.1 An Example Scenario

A basic training scenario was considered that applies to the training of human fighter pilots and fighter controllers. The scenario chosen is a Defensive Counter Air (DCA) in a 2v2 configuration. The blue team consists of a 2-group (lead and wingman), supported by a fighter controller (FC). The hostile 2-group represents the red team. The use of agents for the blue team roles enables simulator-based training in a shared human-agent environment. I.e. each role can be fulfilled by human or agent.

In order to analyze and gain insight into the scope of blue team interactions, a story line was developed together with a SME. For illustration purposes, a simplified version of the resulting transcript is shown in Table 1. It shows spoken team interactions during a single successful engagement of blue versus red.

The second column shows simplified speech acts: QRY = query, INF = inform, ORD = order and ACK = confirmation. In the study, the speech act categorization from [4] was followed. The callsigns used in the utterances are *Magic* for the fighter controller, *Viper* for blue team consisting of *Viper1* for the flight lead, and *Viper2* for the wingman.

The transcript was analyzed based on the domain knowledge requirements identified in Sect. 3. All proactive interactions are the result of team-members' *communication*

Table 1. Transcript of blue team communication

Utterance	Act	Fighter controller	Lead	Wingman
(U-1)	QRY		*Magic, picture*	
(U-2)	INF	*Magic, two group wall, thirty wide*		
(U-3)	INF	*Magic, 50 miles*		
(U-4)	QRY		*Magic, declare all groups*	
(U-5)	INF	*Magic west group bullseye three-five-five at fifteen, Flanker, hostile. East group bullseye three-six-zero at fifteen, Flanker, hostile*		
(U-6)	INF		*Viper commit skate*	
(U-7)	ORD		*Viper2, target east group*	
(U-8)	ACK			*Viper2*
(U-9)	INF		*Viper1 fox 3 west group, crank right*	*Viper2 fox 3 east group*
(U-10)	INF		*Viper1 husky, out south*	*Viper2 husky, out south*
(U-11)	QRY		*Viper1 time out west group*	*Viper2 time out east group*
(U-12)	INF	*Magic west group vanished, East group vanished. Picture clean*		

rules (U-1, 3, 4, 5, 6, 7, 10 and 11), expressed depending on the evolution of the scenario. *Interaction protocols* state that queries and orders require a response (e.g. U-1-U-2 and U-7-U-8), and informs do not (e.g. U-3, U-9). When regarding callsign usage, for questions, the receiver's callsign is used (U-1, U-4), and for responses or informs, the sender's callsign is used (e.g. U-2). The flight lead uses its team's callsign when the interaction concerns the whole team (U-6). Callsigns are also used as a means of confirmation upon a given order (U-8). All utterances are subject to rules for *phrasings* and use of *terminology*, part of air-to-air communication standards. See operational Brevity words in [22].

5.2   System Implementation

The goal of the case study is demonstrate the integration of agents' communicative abilities into an existing military training simulation capability for fighter pilots. In

order to implement the conversational agents, a multi-agent system (MAS) was used. In the system, each of blue team's roles can be fulfilled either by a human-in-the-loop or an agent-based synthetic team-member. The system design is shown Fig. 3.

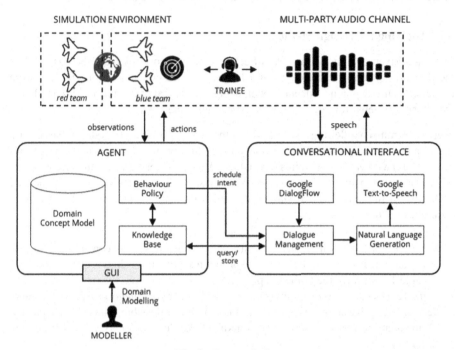

Fig. 3. System design

The MAS is based on an in-house developed toolkit for the construction of behavior models for CGFs in military simulations. The system can be connected to a variety of commercial- or government off-the-shelf simulation environments. Further, it allows end-users to model the behaviors of military units through a graphical user interface. Predefined behaviors models are available for blue and red team fighter tactics.

The conversational interface of an agent contains a Dialogue Manager that can interface with the agent's core components: the Knowledge Base (KB) and the Behaviour Policy. The access to the KB is regulated by speech acts and interaction protocols. When an *inform* act is received, the KB is automatically updated. When a *question* act is received, the KB is queried automatically to construct and formulate an answer, which in turn is an inform for the receiver. Received informs not directed to the agent but instead 'overheard' are also stored in the KB. The Behaviour Policy can proactively schedule communicative intents. These are translated by the Dialogue Manager to spoken utterances using NLG. Note that this is the same NLG process that was used offline to generate example utterances to optimize the NLP models.

Additionally, the Dialogue Manager interfaces with two commercial cloud-based services to employ state-of-the-art speech recognition and synthesis: Google DialogFlow

and Text-to-Speech respectively. DialogFlow[1] handles both ASR and NLU as a single process. All agents and humans communicate through speech over a single audio channel. There is no difference between human-agent or agent-agent communication (i.e. agents interact with each other through speech and no short-cuts are taken).

5.3 Applying the Design Method

The data-driven design method from Sect. 4 was used to implement the case. All of the domain concepts (and their natural language representations) from Table 1 were modelled in the Domain Concept Model, using a graphical user interface (GUI). Below we revisit the steps from the method by providing some examples.

Step 1: Domain Modelling: In the MAS, the Domain Concept Model is based on a frame-based ontology approach. Primary modelling primitives are simple data types (a.k.a. slots) with value constraints (a.k.a. facets) and concept classes with attribute-slots, which can refer to simple data types or other classes. Each primitive can be associated with custom meta-data which was used to provide natural language (NL) metadata. Examples include:

- A *heading* is based on an integer with value constraints between 0 and 360. NL metadata states that the value should be pronounced as a three digit sequence (see U-5 in Table 1), conform the domain's protocol.
- Named entities, such as *callsigns* (magic, viper1, viper2) or *enemy groups* (west group, east group) are based on strings, constrained by sets of possible values. Consequently, NL metadata defines their spoken versions and possible synonyms (e.g. western group, eastern group).
- Concept classes are for instance a *picture* with attributes *enemy group count, formation shape* and *distance* (see U-2); or *relative location* with a *heading* and *distance* from a reference point (a.k.a bullseye in jargon, see U-5). NL metadata encodes the specific phrasings that should be used for pronunciation (Step 3).

Step 2: Intent Generation: For each concept class, intent definitions are generated in two-fold: a query and an inform. For instance, intents for utterances U-1 and U-2 can be defined as *query(picture)* and *inform(picture)* respectively.

Step 3: Natural Language Generation: For each inform intent definition, example utterances are generated by first generating random values, and second resolving these to NL utterances using the metadata's NL template. E.g. the template *"$count group $shape, $space wide"* is resolved to *"two group wall, thirty wide"*. In the domain, utterances also include callsigns to denote a sender or receiver. These are also randomly generated and prefixed to the final utterances.

Step 4: NLP Data Generation: In the system, DialogFlow is used as the NLU service. DialogFlow provides both a web-based user interface and an application programming

---

[1] DialogFlow was only used for intent classification, as its features for dialogue management are now handled by the system's dialogue manager.

interface (API) to model entities and intents. Using the API, a converter was implemented to translate system intents to DialogFlow entities and intents, including so-called training phrases as example utterances. Upon updates, DialogFlow automatically retrains its model.

Any changes to be made for the way concepts are expressed as natural language can be performed in the Domain Modelling step using the provided GUI. Consecutive steps are automated. Therefore the conversational interface can be changed without the need for additional programming. Further, the multi-agent system offers end-users an editor to develop agent behaviour policies. As these policies are based on decision-making rules that also refer to modelled domain concepts, end users can create and extend agent models together with required interaction capabilities without any programming.

5.4 Evaluation

The implemented system was qualitatively evaluated using the 2v2 scenario described in the beginning of this section. Two experiments were conducted: one where all of the participants are non-human agents and one human-in-the-loop (HITL) experiment where a single human takes on the role of the flight lead, supported by two agents playing the role of fighter controller and wingman. For the former experiment, the focus was on system performance: are the agents capable of effectively sharing situational awareness and coordinating activities, purely based on speech interaction? For the latter experiment, the focus was on user experience and satisfaction, based on user interaction with the agents. It was conducted with the help of a former F-16 pilot who could provide valuable feedback in terms of shortcomings and value for training. The experiments were conducted in a controlled, low noise environment using a laptop's built-in speaker and microphone. It should be noted that the evaluation is preliminary with the goal to identify immediate shortcomings to address for further development.

In the all-agent experiment, no functional errors were found while running the scenario. Agents were able to successfully play out the scenario and each agent's situational awareness evolved according to the information being communicated (personally directed to or overheard). In the HITL experiment, we looked at user satisfaction based on similar metrics as used by [25]: TTS Performance (quality and understanding of agent speech), Task Ease (ease of interaction), User expertise (knowing what to say or do), Expected behavior (does the system work as to be expected) and Future use (viability of the system for operational use or use in other domains). No shortcomings were reported about TTS Performance, Task Ease and User expertise. This is most likely due to the advances in TTS technology in the past years, the task being easy to perform as the scenario is small and the SME having extensive experience in the domain. However, the system did not always match the expected behavior because of an occasional intent misclassification. When this happens, an agent requests for clarification, according to protocol ('e.g. Viper1, say again'). Also, when an agent does not recognize the sender or intended receiver in the utterance, there is the chance that all agents request for clarification. When regarding future use, the SME did see great potential for actual operational use in training if (1) unexpected intent misclassification is reduced to a minimum and (2) agents can cope with larger scenarios.

In conclusion, unwanted behaviors originate mainly from occasional intent mis-classification and the following need for repetition of unrecognized utterance by the human. We see room for improvements by means of further fine-tuning of aspects that have not been investigated in detail. For instance (1) the tuning of hyper parameters of DialogFlow, such as machine learning thresholds for intent matching; (2) the provision of more phonetic synonyms for words which are difficult to recognize (highly domain-specific words and jargon which is uncommon in English); (3) the use of scenario-phase specific contexts to configure NLU with expected utterances, rather than a global domain context; and (4) speaker identification from audio, rather than inference from recognized text. After improvements, a more extensive evaluation of the proposed system is required to judge its value for operational training. However, properly evaluating conversational agents is not an easy task as many measures for quality are subjective and good scores quantitative measures are not necessarily related to system performance or perceived quality by the user [26]. This is left for future work.

6 Discussion

In this paper, we have not presented particular new ideas for individual interaction technologies. Rather, we focused more on the analysis and design of the system as a whole, where off-the-shelf technologies have to be integrated and configured to create conversational agents that can operate in a desired application domain.

The presented data-driven design method has close ties to research on natural language interfaces to databases (NLIDB) [27]. As a concept, NLIDBs can understand and generate questions that can be answered by a knowledge base. For example, in [28], natural language questions are translated to formal queries that can be generated by a knowledge graph. Our approach employs similar ideas where the model definition of an agent's knowledge base is used to determine the variety of spoken utterances that the system is capable of understanding and producing. In addition, we showed how this top-down, data-driven approach can be taken further towards facilitating speech-based interaction through configuration of NLU and ASR models.

In our experience during development, many challenges remain to achieve a level of interaction that is indistinguishable from human interaction. In an integrated system, there can be shortcomings at any level, whether it is imperfect speech recognition, unnatural voices or intonations, unnatural timings, expressiveness, or limited cognitive capability (what can be understood). Many of these can be improved, however small improvements in fidelity can lead to significant additional development effort. More insight is required into what level of fidelity is sufficient to achieve user acceptance and training effectiveness for a particular domain. Also, a strategy that can be followed to cope with fidelity shortcomings is to align a user's expectation of system performance, capabilities and limitations, for instance through instructions prior to the use of the system. For instance, in [29] a set of design guidelines are proposed to enhance user experience for human-AI interaction. In future work we will address fidelity requirements and the effect of prior human instruction.

In our proposed design method, there is a strong focus on knowledge representations that closely match their counterparts in natural language. This has not only benefits

for interaction, but also for effective modelling of situational awareness and decision-making. As natural language has close ties to human thought, it is beneficial to base an agent's cognitive process on matching underlying knowledge representations: SME knowledge can more efficiently be translated to agents' belief and reasoning constructs, and agent models become more transparent and explainable. Interaction requirements (what can be said and understood) should be an integral part of the agent's design process, besides SA requirements (what can be known) and decision-making requirements (what can be decided). We experienced this first-hand in our attempt to add conversational interfaces to existing, non-communicative agent models. These models were developed prior to taking into account interaction requirements, and consequently led to mismatches in suitable knowledge representations that could have been prevented.

7 Conclusion

In this paper we addressed the challenges of building conversational agents to play the role of synthetic team-members for military training. These center around the integration of interaction technologies and how they can be tailored towards specific application domains in which communication is subject to domain-specific communication rules, standards, protocols and terminology and phraseology.

First, we analyzed domain-specific communication characteristics for a particular domain. These capture a domain's interaction requirements in terms of why, when, what and how agents should communicate with others. Second, a data-driven design method was proposed to automate the configuration of ASR, NLU and NLG models, based on human authoring of a domain concept model. Finally, we applied and demonstrated the method in a case study of air-to-air combat training simulation.

Although the evaluation was conducted in a very small scenario experiment, the system did successfully demonstrated team interactions in a shared human-agent environment. Agents were able to build a shared situational awareness and coordinate their activities. According to the SME subject in the human-in-the-loop experiment, the system has potential for operational use, provided that the agents' natural language understanding is robust, and agents have the cognitive capabilities to cope with variations possible in a scenario. For the former, we see room for additional optimizations to increase robustness. For the latter, this is more a matter of scaling the complexity of agents' domain models, rather than a communication limitation.

As for the applicability of the data-driven design method, the method has a strong reliance on formal knowledge representations. It assumes that such representations match the natural language concepts used during interaction. This assumption worked well in the case study, as fighter communication is designed for efficient and effective interactions, based on well documented standards. In this study we have not investigated the proposed method in domains that are less formal and support more freedom of expression of natural language (e.g. more socially-oriented interactions).

References

1. Dompke, U.: Computer Generated Forces - Background, Definition and Basic Technologies. TNO, The Hague (2003)

2. Hancock, P., Vincenzi, D., Wise, J., Mouloua, M.: Human Factors in Simulation and Training. CRC Press (2008)
3. Ball, J., et al.: The synthetic teammate project. Comput. Math. Organ. Theory **16**(3), 271–299 (2010)
4. Andersson, J., Svensson, J.: Speech acts, communication problems, and fighter pilot team performance. Ergonomics **49**(12–13), 1226–1237 (2006)
5. Noyes, J.M., Haas, E.: Military applications: human factors aspects of speech-based systems. In: Chen, F., Jokinen, K. (eds.) Speech Technology, pp. 251–270. Springer, New York (2010). https://doi.org/10.1007/978-0-387-73819-2_13
6. Jokinen, K.: Spoken language dialogue models. In: Chen, F., Jokinen, K. (eds.) Speech Technology: Theory and Applications, pp. 33–60. Springer, New York (2010) https://doi.org/10.1007/978-0-387-73819-2_3
7. Harms, J.G., Kucherbaev, P., Bozzon, A., Houben, G.J.: Approaches for dialog management in conversational agents. IEEE Internet Comput. **23**, 13–22 (2019)
8. Perez-Martin, D., Pascual-Nieto, I.: Conversational Agents and Natural Language Interaction: Techniques and Effective Practices, Information Science Reference - Imprint of. IGI Publishing (2011)
9. Kerry, A., Ellis, R., Bull, S.: Conversational agents in E-Learning. In: Allen, T., Ellis, R., Petridis, M. (eds.) Applications and Innovations in Intelligent Systems XVI, pp. 169–182. Springer, London (2009) https://doi.org/10.1007/978-1-84882-215-3_13
10. Klopfenstein, L., Delpriori, S., Malatini, S, Bogliolo, A.: The rise of bots: a survey of conversational interfaces, patterns, and paradigms. In: Proceedings of the 2017 Conference on Designing Interactive Systems (2017)
11. Gatt, A., Krahmer, E.: Survey of the state of the art in natural language generation: core tasks, applications and evaluations. J. Artif. Intell. Res. **61**, 65–170 (2018)
12. Graesser, A.C., Wiemer-Hastings, K., Wiemer-Hastings, P., Kreuz, R.: AutoTutor: a simulation of a human tutor. Cogn. Syst. Res. **1**(1), 35–51 (1999)
13. Talbot, T.R.A.: Virtual human standardized patients. In: Rizzo, A., Bouchard, S. (eds.) Virtual Reality for Psychological and Neurocognitive Interventions, pp. 387–405. Springer, New York (2019) https://doi.org/10.1007/978-1-4939-9482-3_17
14. Hart, J., Gratch, J., Marsella, S.: How virtual reality training can win friends and influence people. In: Human Factors in Defence, vol. 21, no. 1, pp. 235-249. Ashgate, Farnham (2013)
15. van Oijen, J., van Doesburg, W., Dignum, F.: Goal-based communication using BDI agents as virtual humans in training: an ontology driven dialogue system. In: Dignum, F. (ed.) AGS 2010. LNCS (LNAI), vol. 6525, pp. 38–52. Springer, Heidelberg (2011). https://doi.org/10.1007/978-3-642-18181-8_3
16. Graesser, A.C., Li, H., Forsyth, C.: Learning by communicating in natural language with conversational agents. Curr. Dir. Psychol. Sci. **23**(5), 374–380 (2014)
17. Traum, D., Rickel, J.: Embodied agents for multi-party dialogue in immersive virtual worlds. In: Proceedings of the First International Joint Conference on Autonomous Agents and Multiagent Systems: Part 2 (2002)
18. Kenny, P., et al.: Building interactive virtual humans for training environments. In: Proceedings of the Interservice/Industry Training, Simulation and Education Conference (I/ITSEC) (2007)
19. Chen, H., Liu, X., Yin, D., Tang, J.: A survey on dialogue systems: recent advances and new frontiers. ACM SIGKDD Explor. Newsl **19**(2), 25–35 (2017)
20. Grice, H.: Logic and conversation. In: Studies in Syntax and Semantics III: Speech Acts, pp. 41--58 (1975)
21. Traum, D., Swartout, W., Gratch, J., Marsella, S.: A virtual human dialogue model for non-team interaction. In: Dybkjær, L., Minker, W. (eds.) Recent Trends in Discourse and Dialogue. Text, Speech and Language Technology, vol. 39. Springer, Dordrecht (2008) https://doi.org/10.1007/978-1-4020-6821-8_3

22. AFTTP 3–1: Operational Brevity Words, Definitions, and Counter Air Communication Standards (2001)
23. Hadzic, M., Wongthongtham, P., Dillon, T., Chang, E.: Ontology-Based Multi-Agent System. Springer, New York (2009). https://doi.org/10.1007/978-3-642-01904-3
24. Braun, D., Hernandez, M., Matthes, F., Langen, M.: Evaluating natural language understanding services for conversational question answering systems. In: Proceedings of the 18th Annual {SIG}dial Meeting on Discourse and Dialogue, Saarbrücken (2017)
25. Walker, M., et al.: DARPA communicator dialog travel planning systems. In: 7th European Conference on Speech Communication and Technology, Aalborg, Denmark (2001)
26. Möller, S.: Assessment and evaluation of speech-based interactive systems: from manual annotation to automatic usability evaluation. In: Speech Technology: Theory and Applications, Boston, MA, pp. 301–322S (2010
27. Affolter, K., Stockinger, K., Bernstein, A.: A comparative survey of recent natural language interfaces for databases. VLDB J. **28**(5), 793–819 (2019). https://doi.org/10.1007/s00778-019-00567-8
28. Han, Y.J., Park, S.B., Park, S.Y.: A natural language interface concordant with a knowledge base. Comput. Intell. Neurosci. **2016**, 1–15 (2016). https://doi.org/10.1155/2016/9174683
29. Amershi, S., et al.: Guidelines for human-AI interaction. In: Proceedings of the 2019 HCI Conference on Human Factors in Computing Systems, Glasgow (2019)

Toward Automated Mixed Reality Interface Design: An Evolutionary Optimization Approach

Hongbo Zhang[1]([✉]), Denis Gracanin[2], Mohamed Eltoweissy[1], and Tianxin Zhou[1]

[1] Virginia Military Institute, Lexington, VA 24405, USA
{zhangh,eltoweissymy,zhout22}@vmi.edu
[2] Virginia Tech, Blacksburg, VA, USA
gracanin@vt.edu

Abstract. In this research, we propose an innovative capsule based optimization model toward the automated user interface (UI) design of mixed reality devices. Within the model, functional, spatial, and cognitive seams unique to mixed reality devices are considered as the constraints of the capsule, the unit cohort of UI elements. A genetic evolutionary optimization solution approach was used to solve the optimization problem. The results demonstrate that the number of optimal UI elements per capsule is linearly proportional to the initial estimate of the UI elements of the capsule. The total reward of the capsule design is also linearly proportional to the number of optimal UI elements per capsule. The identified linear relationship model suggests that it is feasible to predicate the optimal number of UI elements therefore achieving the automated design of the mixed reality interface. Experimental results have validated the proposed optimization framework suggesting that the optimization reward objective function is able to match app ratings from the app store. The seam constraints of the optimization function are also consistent with the time taken to complete the task which shows the validity of using seam to model the automated mixed reality interface design.

Keywords: Mixed reality · Automated user interface design · Evolutionary optimization

1 Introduction

The interaction between human and robot shows promises for practical applications [1–5]. While the benefits are clear, the challenges still exist. The interaction of mixed reality device interface is spatially distributed [17]. The spatial seam dictates that for completion of tasks may take longer to complete [18] lead to inconvenience of the interaction process [4, 19]. The cognitive seam is associated with the cognitive stress of awkward, limited visual cues, and non-intuitive gesture control [1]. Functional seam strongly related to the mixed reality device limitations e.g., voice recognition and environment mapping accuracy also exists [1].

© Springer Nature Switzerland AG 2021
H. Degen and S. Ntoa (Eds.): HCII 2021, LNAI 12797, pp. 532–540, 2021.
https://doi.org/10.1007/978-3-030-77772-2_35

For the functional seam, the field of view of the mixed reality devices is relatively small. The limited field of view determines that the displayed content on the mixed reality devices need to fit into the narrowed view space [15]. The limited view space does not impose significant challenges for simpler task of one-to-one interactions for example one operator interaction with one robot. Yet for one-to-many, e.g., one operator and multiple robots, the limited view space can cause impractical visualization and control. The second major functional seam is associated with the limited computational capacity of the mixed reality devices [16]. The computational capacity of mixed reality devices has been a limiting factor for computation intensive applications. It decides that the interaction between operator and robot needs to be lightweight task such as simple visualization of robot tasks.

In contrast to other types of devices, the spatial seam is special to the mixed reality device. The interaction of mixed reality device interface is spatially (e.g. 360° in free space) distributed [17]. The nature of such spatial distribution dictates that for completion of tasks, the interaction of the interface may take longer time, e.g. one group of UIs is distributed in one spatial location whereas the other groups of UI is located in the other locations [18]. Yet for completion of the task, all different groups of UIs need to interacted thus leading to the inconvenience of the interaction process [4, 19]. Majority of mixed reality devices utilize hand gesture for interaction, the spatial seam is thus also related to the hand gesture detection accuracy [20]. The detection of hand gesture in free space is often not accurate, thus leading to longer task completion time and potential frustrations of the interaction process.

The cognitive seam is associated with the cognitive stress embedded in the interaction process [1]. The cognitive seam is attributed to the cognitive stress that arises from following interactions such as the awkward and non-intuitive gesture control process - the moderate accuracy of mixed reality devices camera able to detect human gestures correctly, the limited feedbacks associated with the interactions. Most importantly, the limited visual cues of the interaction process are known associated with significant cognitive seams [21]. The cognitive seams can be attributed to the function seams, but not all functional seam contribute to the cognitive seams. Functional seam is more related to the mixed reality device, while the cognitive seam involves the user experiences of the mixed reality device and the interaction process [1].

Functional, spatial, and cognitive seams of mixed reality devices are associated with the context and resources of the system [1, 6–14]. The limited resource is attributed to the functional seam. For example, the limited field of view determines that there is only limited information able to be displayed by the mixed reality device. In the same vein, the limited camera detection range of the device also determines that there is very narrowed range where the device can detect gesture precisely so that it leads to spatial seam. Similarly, the low accuracy of gesture detection and the field of view can cause cognitive seams. Based on such observation, it is natural that we can formulate hypothesis that an appropriate management of the resources is likely able to reduce the seams thus increase the user interaction experiences.

Through the research, we propose an innovative capsule approach to address the spatial, cognitive, and functional seams to achieve an optimal design. Capsule is motivated by divide and conquer approach for solving the resource availability optimization

problem to reduce functional, spatial, and cognitive seams. More specifically, the inter-actions between human operator and robot are divided into solvable subspaces termed capsule to constrain the seams in an acceptable boundary to increase the overall efficacy and benefits of the design. The experiment has been conducted for the validation of the theoretical formulation of spatial, cognitive, and functional seams.

2   Methods

The concept of capsule is illustrated showing four capsules of typical task of robot visualization and control in Fig. 1. The capsule is further divided into more specific sub-capsules to manage robot visualization and control sub-tasks.

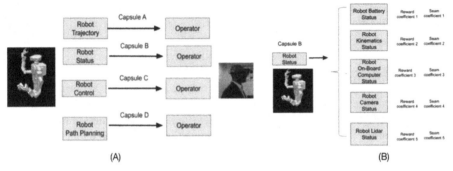

Fig. 1. (A) A representative example of single capsule design for human-robot interaction (B) hierarchical capsule design for human-robot interaction

The problem of the capsule design resembles a knapsack optimization problem with the following formulation.

$$\text{Max} \sum_{i=1}^{n} v_i x_i \tag{1}$$

$$\text{subject to } \sum_{i=1}^{n} \left(w_i^c x_i \right) \leq W^c$$
$$\sum_{i=1}^{n} \left(w_i^f x_i \right) \leq W^f \tag{2}$$
$$\sum_{i=1}^{n} \left(w_i^s x_i \right) \leq W^s$$

The objective function denoted by Eq. 1 maximizes the reward from the capsule design for n UI elements. Each UI is a binary variable ($x_i \in 0$ or 1). v_i is the UI reward coefficient for a specific UI i. In Eq. 2, w_i^c, w_i^s, w_i^f are the cognitive, spatial, and function seam coefficients bounded by W^c, W^s, W^f constants for all the UI elements. Reward coefficient v_i is sampled from a uniform random distribution between 1–10. Seam coefficient w_i^c, w_i^s, w_i^f are sampled from normal distribution with mean 0 and standard deviation 1. A genetic evolution approach is adopted for solving the constrained optimization problem. In the genetic model, a population of 30 are generated as the initial condition. Crossover and mutation are used to enhance the healthiness of the gene offspring. The evolution of the genes is performed until the converging conditions when the number of UI elements stop decreasing.

3   Simulation Results

3.1   Single Capsule Simulation

A strong linear relationship between optimal and initial UIs per capsule is shown in Fig. 2. The initial and optimal number of UIs per capsule are in the range of [2–10] and [2–7].

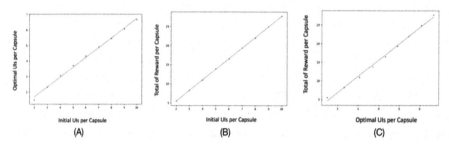

Fig. 2. (A) The initial number of UIs versus the optimal number of UIs (B) The initial number of UIs versus the total reward per capsule (C) The optimal number of UIs versus the total reward per capsule.

3.2   Hierarchy Capsule Simulation

The hierarchy capsules consists of multiple layers of capsules. As evident from Fig. 3 and 4, the optimal number of UIs remains similar between 5-6 optimal UIs regardless of the changes of the number of UIs per capsules. Similarly, the associated total of reward also remains within a small range between 19 to 22.

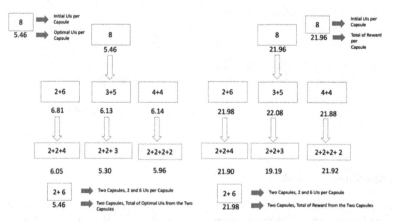

Fig. 3. (Left) The hierarchy of optimal number of UIs and initial number of UIs per capsule (Right) The hierarchy of total of reward and initial number of UIs per capsule (8 initial UIs used as the representative example)

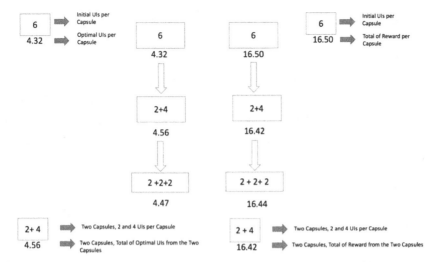

Fig. 4. (Left) The hierarchy of optimal number of UIs and initial number of UIs per capsule (Right) The hierarchy of total of reward and initial number of UIs per capsule (Both left and right use 6 initial UIs as the representative example)

The optimal number of UIs and the total of reward are summarized in in Fig. 5 Illustrating that one capsule with 10 UIs is further decomposed to two capsules consisting of 2, 8; 3, 7; 4, 6; 5, 5 UIs. For this condition, the total number of optimal UIs and total of reward remain similar while the number UIs changes per capsule.

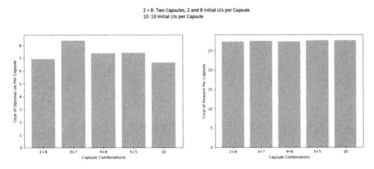

Fig. 5. The optimal number of UIs (left) and total of reward (right)

3.3 Experimental Results

In the experiment, five participants were recruited to asses different Smart Phone based augmented reality apps. The participants are healthy young adults with moderate use of Smart Phone for their college studies. Due to COVID-19, it becomes infeasible to access and share our mixed reality device with participants. As such, personal Smart Phone

based augmented reality app was used for mocking up the mixed reality device. The mixed reality apps are divided into the following categories respectively entertainment, educational, and room furniture Participants were instructed to become familiar with the experimental setup. Participants were also instructed to become familiar with functional, spatial, and cognitive seams through illustration of demonstrative augmented reality app. Following that, participants were required to perform the tasks of environment scanning using the Smart Phone augmented reality apps. The three apps chosen are AR Vid, Sol AR, and Augment. The results of the time taken and associated seams for completion of the environment scanning task are shown below in Table 1.

Table 1. Time (second) taken to complete the task and associated seams for Smart Phone based augmented reality app

	Name of app	Participant 1	Participant 2	Participant 3	Participant 4	Participant 5
Time	AR Vid	29	46	30	85	84
	SOL AR	37	71	9	7	14
	Augment	32	16	43	44	22
Seams	AR Vid	1. Functional Seam 2. Spatial Seam 3. Cognitive Seam	Spatial Seam	1. Functional Seam 2. Spatial Seam	1. Functional Seam 2. Spatial Seam	1. Functional Seam 2. Spatial Seam 3. Cognitive Seam
	SOL AR	Functional Seam	Functional Seam	Functional Seam	Functional Seam	Functional Seam
	Augment	Spatial Seam	Spatial Seam	Spatial Seam	Cognitive Seam	1. Function Seam 2. Spatial Seam 3. Cognitive Seam

Following the preliminary assessment of the augmented reality Smart Phone app, more extensive apps were adopted for counting the number of UI components for different scenes. For each scene, the reward is also computed through the user rating of each UI. The sum of the reward is equal to the multiplication of user rating with the number of UI presented in the equation, which is the objective function. Same as the simulation, the rating is between 1 to 10. In total, three scenes were used for the evaluation. The three scenes include the app landing interface, the environment scanning interface, as well as the object selection interface. App store ratings for the apps and total of reward are shown in Table 2 to show the correlation between the value of reward and app store rating for the apps.

Table 2. App store rating and total of reward for evaluation of smart phone based augmented reality apps

	Name of app	App store eating	Total of reward
Entertainment	Roar	3.2	12.6
	Dino's AR	4.4	24.3
	AR Vid	4.5	26.6
	Eugene's pet's store	4.7	22.6
Educational	Planetarium	4.0	12.6
	Sol AR1	4.4	24.3
	Sol AR2	4.6	26.6
	Solar walk lite	4.7	22.6
Furniture setup	Inferior design	4.3	26.0
	Augment	4.5	23.0
	Triple	4.6	29.6

4   Conclusion

Our research proposed an innovative method for addressing the functional, spatial, and cognitive seams embedded in the human-robot interaction process using mixed reality devices. The method is able to divide complex interactions and associated functional, spatial, and cognitive seams into a solvable subspace called capsule. An evolutionary optimization approach is used to solve the optimal design of the capsule given the constraints of the functional and cognitive seams. Results suggest that the number of optimal UI elements per capsule is linearly proportional to the initial estimate of the UI elements of the capsule. The total reward of the capsule design is also linearly proportional to the number of optimal UI elements per capsule. The number of optional UIs does not change throughout the hierarchy chain of capsule decomposition process indicating that the merging or decomposition behavior does not alter the optimal number of UIs significantly.

The experiment has also validated that the proposed method optimization objective function is able to match with real world settings. The reward obtained from evaluations of three scenes of the Smart Phone based augmented reality apps is able to match with the app store rating of these apps. While for Eugene's pet's store app, it does not seem that the app rating is able to match with the total reward, however, participants do suggest that the rating from the app store is relatively limited therefore the app store rating of the app is likely too high. The time taken to complete the environment scanning task does also match with the overall feedbacks of the seams. The SOLAR app has relatively smaller number of seams as such the time taken to complete is indeed shorter for participant 3, 4, and 5. Such finding indicates that the seam, which serves as the constrains of the optimization function, is able to model the general usability of the app.

There are limitations associated with the work. Due to COVID-19, we have not been able to utilize mixed reality device in the experimental setup. While the Smart Phone based augmented reality app is able to approximate the mixed reality app, the gap between them needs to be considered. Another limitation of the research is that the optimal number of UIs needs to further examined. Future work involves using different generations of specific mixed reality apps and compare the number of UIs per scene should be conducted. The work is in particular useful to address large scale and complex human computer interaction problems such as massive human-robot interaction and collaboration applications.

References

1. Billinghurst, M., Hirokazu, K.: Collaborative mixed reality. In: Proceedings of the First International Symposium on Mixed Reality, pp. 261–284 (1999)
2. Hoenig, W., Milanes, C., Scaria, L., Phan, T., Bolas, M., Ayanian, N.: Mixed reality for robotics. In: 2015 IEEE/RSJ International Conference Intelligent Robots and Systems (IROS), pp. 5382–5387. IEEE (2015)
3. Al-Barhamtoshy, H.M., Al-Ghamdi, A.: Toward cloud-based mixed reality e-learning system. In: Informatics, Health and Technology (ICIHT), International Conference, pp. 1–6. IEEE (2017)
4. Atsali, G., et al.: A mixed reality 3D system for the integration of X3DoM graphics with real-time IoT data. Multimedia Tools Appl. **77**(4), 4731–4752 (2017). https://doi.org/10.1007/s11 042-017-4988-z
5. Fairchild, A.J., Campion, S.P., García, A.S., Wolff, R., Fernando, T., Roberts, D.J.: A mixed reality telepresence system for collaborative space operation. IEEE Trans. Circ. Syst. Video Technol. **27**(4), 814–827 (2017)
6. Kiyokawa, K., Takemura, H., Yokoya, N.: A collaboration support technique by integrating a shared virtual reality and a shared augmented reality. In: 1999 IEEE International Conference Systems, Man, and Cybernetics, 1999. IEEE SMC'99 Conference Proceedings. vol. 6, pp. 48–53. IEEE (1999)
7. Monahan, T., McArdle, G., Bertolotto, M.: Virtual reality for collaborative e-learning. Comput. Educ. **50**(4), 1339–1353 (2008)
8. Pouliquen-Lardy, L., Milleville-Pennel, I., Guillaume, F., Mars, F.: Remote collaboration in virtual reality: asymmetrical effects of task distribution on spatial processing and mental workload. Virtual Reality **20**(4), 213–220 (2016). https://doi.org/10.1007/s10055-016-0294-8
9. Gonzalez-Franco, M., et al.: Immersive mixed reality for manufacturing training. Frontiers **4**(3), 1 (2017)
10. Jacoby, D., Coady, Y.: Perspective shifts in mixed reality: persuasion through collaborative gaming. Int. Work. Pers. Persuas. Technol. (2017)
11. Seth, N.S.: Real time cross platform collaboration between virtual reality and mixed reality. Arizona State University (2017)
12. Brigham, T.J.: Reality check: basics of augmented, virtual, and mixed reality. Med. Ref. Serv. Q. **36**(2), 171–178 (2017)
13. Gonzalez-Franco, M., et al.: Immersive mixed reality training for complex manufacturing (2017)
14. Howard A.M., Roberts, L., Garcia, S., Quarells, R.: Using mixed reality to map human exercise demonstrations to a robot exercise coach. In: 2012 IEEE International Symposium Mixed and Augmented Reality (ISMAR), pp. 291–292. IEEE (2012)

15. Boudoin, P., Domingues, C., Otmane, S., Ouramdane, N., Mallem, M.: Towards multi-modal human-robot interaction in large scale virtual environment. In: Proceedings of the 3rd ACM/IEEE International Conference on Human Robot Interaction, pp. 359–366. ACM (2008)

16. Costanza, E., Kunz, A., Fjeld, M.: Mixed reality: a survey. In: Lalanne, D., Kohlas, J. (eds.) Human machine interaction. LNCS, vol. 5440, pp. 47–68. Springer, Heidelberg (2009). https://doi.org/10.1007/978-3-642-00437-7_3

17. Green, S.A., Billinghurst, M., Chen, X., Chase, J.G.: Human robot collaboration: an augmented reality approach—a literature review and analysis. In: ASME 2007 International Design Engineering Technical Conferences and Computers and Information in Engineering Conference, pp. 117–126. American Society of Mechanical Engineers (2007)

18. Sato, S., Sakane, S.: A human-robot interface using an interactive hand pointer that projects a mark in the real work space. In: Robotics and Automation, 2000. Proceedings. ICRA 2000. IEEE International Conference, vol. 1, pp. 589–595. IEEE (2000)

19. Donalek, C., et al.: Immersive and collaborative data visualization using virtual reality platforms. In: 2014 IEEE International Conference on Big Data (Big Data), pp. 609–614. IEEE (2014)

20. Lehner, V.D., DeFanti, T.A.: Distributed virtual reality: supporting remote collaboration in vehicle design. IEEE Comput. Graph. Appl. **17**(2), 13–17 (1997)

21. Chen, I.Y.-H., MacDonald, B.A., Wünsche, B.: Designing a mixed reality framework for enriching interactions in robot simulation. In: GRAPP, pp. 331–338 (2010)

Estimation of Learners' Engagement Using Face and Body Features by Transfer Learning

Xianwen Zheng[1(✉)], Shinobu Hasegawa[1], Minh-Tuan Tran[1], Koichi Ota[1], and Teruhiko Unoki[2]

[1] Japan Advanced Institute of Science and Technology, 1-1 Asahidai, Nomi, Ishikawa 923-1292, Japan
{s1810428,hasegawa}@jaist.ac.jp
[2] IMAGICA GROUP/Photron, 1–105 Jinboucho, Kanda, Tokyo, Chiyoda 101-0051, Japan

Abstract. In recent years, online learning plays an essential part in education due to distance learning technology development and control of COVID-19. In this context, engagement, a mental state to enhance the learning process, has been brought into the limelight. However, the existing engagement datasets are of a small scale and not suitable for education time-series research. We proposed an estimation method on time-series face and body features captured by built-in PC cameras to improve the engagement estimation on small and irregularly wild datasets. We designed upper body features using the facial and body key points extracted from OpenPose. To reduce the influence of the extracted noises from OpenPose, the moving average, the average value of a fixed period in the videos, is used to process the training data. Then, we compose a time-series dataset of online tasks with 19 participants. In the composed dataset, there remained self-reports of participants' mental state and external observation to confirm the different engagement levels in the answering process. The combined self-reports and external observation results were used as the engagement label. Finally, the transfer learning was used to solve the insufficient data issue. We pre-trained a long short-term memory (LSTM) sequence deep learning model on a big dataset and transferred the trained model to share learned feature extraction and retrain our dataset. Our proposed method achieved 63.7% in experiments and could apply to estimate and detection engagement in future works.

Keywords: Engagement · Online learning · Transfer learning · Time-series data · Long short-term memory

1 Introduction

Online learning sets learners free from space and time constraints. Coursera is one of the online learning platforms that provide flexible and individual online

H. Degen and S. Ntoa (Eds.): HCII 2021, LNAI 12797, pp. 541–552, 2021.
https://doi.org/10.1007/978-3-030-77772-2_36

learning opportunities and high-quality courses to the masses [1]. Besides, the outbreak of the pandemic Covid-19 forces schools to shift to an online mode of teaching [2]. Those circumstances, therefore, have fostered the development and popularization of distance education. However, there seem to be some problems in distance education.

Learners, when joining online learning, seem not be able to concentrate on learning courses, and therefore, cannot maintain high learning efficiency [1,4]. Furthermore, since learners and instructors do not meet or contact directly in online learning contexts, the lack of face-to-face communication and interaction [2] made instructors hard to grasp the student's engagement and confirm the lectures' contents have been absorbed by students or not. Thus, the research on engagement has been brought to the limelight.

Engagement is a mental state that can helps students feel positive and realize high-quality learning. Simultaneously, in online learning, self-regulated learning (SRL) support is another critical topic related to learner engagement [3]. Still, the studies of SRL in online learning are relatively inadequate comparing to the studies of traditional classroom settings. Simultaneously, online learning platforms need to be equipped with necessary SRL support to improve and maintain learners' engagement [1] and achieve personal learning goals. Therefore, if we can estimate learners' engagement in the online learning process, the teaching progress and lecture content can be adjusted by the instructors, and students' self-regulated become available.

This article proposed improving the engagement estimate performance of small and irregular wild datasets through time-series facial and body information recorded by a built-in PC camera to restore the traditional scene in real-time online learning. A vital issue in engagement estimation research is the insufficient scale dataset [4,5]. However, collecting new time-series dataset with engagement labels requires lots of expenditure and time, which is not cost-effective. For this reason, we adopt transfer learning, a machine learning method that has been recognized can effectively improve the performance of small data, to transfer a long short-term memory (LSTM) sequence deep learning model to retrain the model on a small dataset.

The remainder of this article is organized as the followings. In Sect. 2, we introduce the related work for engagement estimation and detection in online learning. Then, in Sect. 3, we compare two datasets for the transfer source and target models. Section 4 will illustrate the experiment settings and discuss the results. Finally, we present implications and suggest further work in Sect. 5.

2   Related Work

Dewan et al. [7] proposed a deep learning-based approach to online learners' engagement detection, in which on the Dataset for Affective States in E-Environments (DAiSEE) they applied the Local Directional Pattern (LDP) to person-indepemdemt edage features to extract different facial expressions and used Kernel Principal Component Analysis (KPCA) method to capture nonlinear correlations among the extracter features [6]. DAiSEE dataset includes 9068

snippets of videos from 112 participants. Each video is approximately 10 s with 1920*1080 pixels and 30fps.

Chang [5] proposed a series of methods for detecting engagement at different levels in the EmotiW Challenge. The methods include a cluster-based framework for fast engagement level predictions, a neural network using the attention pooling mechanism, heuristic rules using body posture information, and a model ensemble for more accurate and robust predictions. OpenFace was used to track the head pose, gaze directions, and action units (AUs), and OpenPose to directly track head, body, and hands as the model training features. The record videos in the EmotiW dataset were 197 videos collected on non-constrained environments and approximately 5 min in one video. The participants in the dataset were watching Korean Language videos.

Hasegawa et al. [4] proposed a couple of machine learning and deep learning models to estimate learners' mental states such as difficulty, interest, fatigue, and concentration using the emotion, eye gaze, face orientation, face position features that change with the time-series between learners and their learning tasks. They extracted the features using an existing face image recognition library Face++ by each frame of the videos. The dataset was recorded learners' videos with facial expressions using a PC built-in camera and retrospective self-report from 19 participants who are doing the online task.

Nomura et al. [9] used a pressure mat and a web camera to record students' postural data of upper body pressure distribution and upper body pose during e-learning lectures to grasp the students' engagement. They extracted 38 features from upper body pressure distribution and 33 features from upper body pose for every minute and trained a Support Vector Machine (SVM) classifier model to estimate engagement levels. The engagement labels are from the reported questionnaire displayed on the screen every minute during e-learning and answered by 7 participants aged 20–22.

Monkaresi et al. [10] proposed supervised learning (Updateable Naïve Bayes, Bayes Net, Logistic Regression, Classification via Clustering, Rotation Forest, Dagging) for detection of concurrent and retrospective self-reported engagement through the three sets of features, heart rate, Animation Units, and local binary patterns in three orthogonal planes (LBP-TOP) that were extracted using computer vision techniques and an ECG signal. The participants were 23 students (14 male, 9 female) from a public university aged from 20 to 60 who completed a structured writing activity. An auditory probe was produced every two minutes during the essay writing activity to collect their engagement level verbally, and a week after their writing session, they would fill out a questionnaire after viewing each segment to annotate the engagement labels.

In previous research, the training features in computer vision-based methods for estimating or detecting engagement usually have three aspects, facial and body expression, body postures and motions, and eye movement. Other information is also obtained through external devices like heart rate collected by the ECG signal and pressure mat. From affective computing research, affect expression occurs through combinations of verbal and nonverbal communication channels, and bodily expressions have been recognized as more critical for

nonverbal communication than was previously thought since the body may better communicate some affective expressions than the face [11]. Besides, some biological features require external devices, which is difficult to collect in real-time online learning. Facial and body information is no intervention that will not disturb the learners' learning process and cost-effective. Therefore, our research used time-series facial and boy information, including head pose, mouth shape, eye movement, and body distance from the screen as the training features.

There remain some datasets for engagement estimating and detecting studies, DAISEE (Dataset for Affective States in E-learning Environments) [6], and EmotiW2019 (Emotion Recognition in the Wild Challenge) [12]. However, some limitations exist in these datasets. All these two datasets are single race-based participants. The female and male rate is imbalanced. DAISEE and EmotiW2019 datasets were labeled by crowdsourcing. Another issue in current datasets is the bias of data. For instance, the labels in the DAiSEE: very low: 61, low: 459, high: 4477, very high: 4071 (10 s each video), and for EmotiW2019: very low: 9, low: 45, high: 100, very high:43 (5 min each video). We can observe the low-level engagement data is very limited in these datasets. For this research, we collect videos from 19 participants who take the answer process and get self-reports of their mental state after recording videos. Moreover, combine the self-report and external observation to annotate the recorded videos.

3   Dataset

To improve the estimation accuracy on small dataset, we trained a model with a big dataset that reused the trained model to train a model with an insufficient dataset. This research uses the DAISEE dataset as the source domain and the newly created dataset as the transfer target.

3.1   Source Dataset

DAISEE dataset [6], an affective states dataset in the e-environment for engagement recognition, will be used in our experiment as the source domine. This dataset contains 9086 snippets of 10s videos captured by a full HD web camera from 112 (80 male, 32 female) participants when they watch learning videos. All of the participants are 18-to-30-year-old Asian learners. The collected videos are in dorm rooms, crowded lab spaces, and libraries with different illumination settings to restore the real-world environments. The video labels are crowd-sourced based on a gold standard created using expert psychologists. Thus, each video in the dataset is labeled with one affective state of boredom, confusion, engagement, and frustration, and each label has been arranged four levels from low to high intensity.

Besides, from the observed in the DAiSEE dataset, very low and low label videos are very similar. In other previous research, they changed the labels into three levels and proved feasibility. Thus, in our experiment, we rearranged the dataset provided four engagement levels into three: (0–1) low engaged, (2) normally engaged, and (3) high engaged.

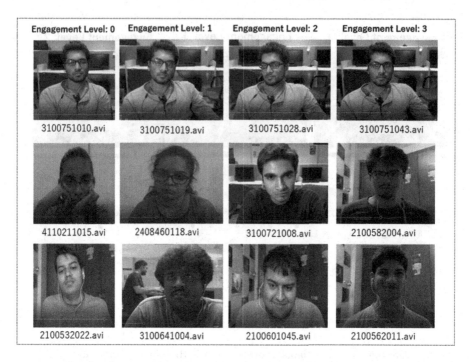

Fig. 1. DAiSEE dataset. From left to right the engagement levels are from Level 0 to level 3

3.2 Target Dataset

Based on the limitation on existing engagement research datasets, we made a time series dataset of online tasks with 19 participants as the transfer learning target dataset. The videos are recorded by a built-in PC camera facing the learners in the test answering process. The online learning content is the Cognitive Assessment Battery (CAB) test, a test for checking user cognitive speed/attention, episodic memory, visuospatial functions, language, and executive functions [13]. The participants take the CAB test maximum of 30 questions and within 12 min. Based on the different individual speeds, the length of recorded videos is also different. After completing the CAB test, the participants write a self-report about their mental state in order to confirm the engagement levels. To guarantee engagement labels' accuracy, we combined the self-report from 19 participants and external observation results from several study members as the engagement labels. This dataset contains the engagement change regularity.

Our collected dataset is of a small scale, which is not appropriate for further time-series studies. At the same time, collect more data is not cost-effective. Therefore, we apply transfer learning to improve the created dataset's performance and find a method for time-series research on small scale datasets (Tables 1 and 2).

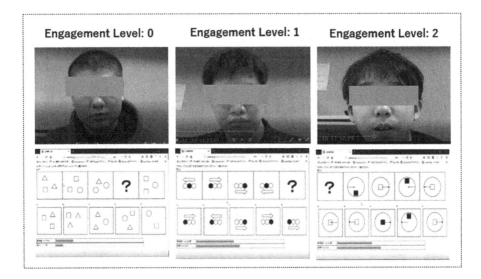

Fig. 2. New collected dataset

Table 1. Engagement labels.

Dataset	Low label	Normally label	High label	Total label
DAiSEE	**520**	**4477**	**4071**	**9068**
New Dataset	**158**	**195**	**128**	**481**

3.3 Features

Verbal and nonverbal communication channels are the external manifestation of emotion [11]. Some studies have shown that engagement is directly related to emotion [14], so we assume that engagement can be analyzed by observing nonverbal communication channels such as facial and bodily expressions. Furthermore, in different situations, facial and bodily expressions play a dominant role, respectively.

OpenPose and OpenFace are the existing popular libraries for computer vision research. The advantage of these libraries is that they can extract facial, body, and hand keypoints on images and videos from a built-in PC camera in real-time [15]. Considering the importance of facial and bodily expressions, we design up body features such as head pose, eye movement, mouth shape, and body movement using the facial and body key points extracted from OpenPose.

Table 2. Comparison table of DAiSEE (source) and new collected (target) dataset.

Heading level	DAiSEE	New collected dataset
Device	A full HDD	Built-in PC camera
Videos	9086	19
Length	10sec	Within 12 min
Participants	112	19
Race	Asian	Chinese, Japanese
Space	Rooms, lab spaces, libraries	Dorm rooms, lab spaces
Engagement level	4 levels	5 levels
Labels	Crowd-sourced	Self-report, external observation

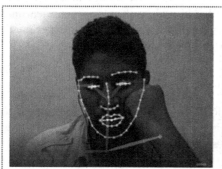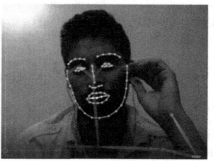

Fig. 3. Extracted features by openpose

3.4 Data Pre-processing

Calculate Moving Average. In our previous experiment, the existing library OpenPose generated noise and lost video frames when extracting external information harmful to the final estimation result. On the other hand, for the video format, bandwidth, and low-spec PC, frames per second (fps) may not keep the same, and each sample video's length may also be unstable. It is hard to set a standard for unprocessed wild data. In order to satisfy as many situations as possible, the moving average method, which calculates the average value of a fixed period in the videos, is used to process the data and reduce the extracted noise effect.

Over Sampling. As we mentioned, the number of engaged/unengaged data in the DAiSEE dataset is imbalanced, which is harmful for estimation. To deal with such problems, we applied oversampling to the low engaged data. In the DAiSEE dataset, a one-second video has thirty frames, which is more than our needs. Engagement is a continuously changing mental state, and it cannot present as

a moment of external expression. Therefore, we set 5 frames as the fixed period and calculated the fixed period average value in all the settings. Besides, we got six times sampling data by shifting one by one average value in the low engagement level data.

4   Experiment

4.1   Development Environment

In this paper, to improve the flexibility and compatibility in the further real-time online learning process, we used Google Colaboratory service, which allows us to write and execute Python from a web browser and no need to build an environment with free access GPU. Under this unrestricted condition, the trained model would be running on any devices that only need a built-in camera and connecting internet, like mobile and iPad. Meanwhile, Keras, a deep learning API written in Python would build sequence models, was used in our experiment. Keras API runs on top of the machine learning platform TensorFlow.

4.2   LSTM Classification and Transfer Learning

In some research, they applied Convolutional Neural Network (CNN) models to classify the learners' engagement. CNN is an image classification deep learning model that is popular for engagement estimation and detection. The training input face images are detected from the dataset videos [7,16]. However, engagement is a continuously changing mental state, but face images cannot show the change and the state of learners in online learning on engagement.

Therefore, our research proposed a long short-term memory (LSTM) sequence model using Keras API to train models on DAiSEE and a new collected dataset about learners' engagement estimation in online learning tasks. LSTM network models are appropriate for classification, processing, and forecasting based on time series data, and one LSTM cell includes input, forget, and output gates that can store information over extended time intervals. Consequently, long short-term memory (LSTM) network models can process images and audio or video time-series data.

Because our collected dataset is on an insufficient scale, we apply transfer learning to improve the performance of the new dataset. Transfer learning has been recognized can effectively improve the performance of deep learning models for Time Series Classification (TSC) [8]. For the sequence of deep learning, more time-series data is better. Our dataset is on a small scale. It is insufficient for sequence deep learning models. We use transfer learning, which pre-trained a model on big datasets DAISEE then transfer a model on our datasets. To improve transfer learning's interpretability and flexibility, we build an inner model with one or two nature network layers in the pre-training model for transfer.

4.3   Experiment Setup and Results

In the experiment setting, we first trained a model on the DAiSEE dataset and built an inner model with one nature network layer in the pre-training model to improve interpretability and flexibility. Then we take the trained inner model from the pre-trained model and freeze it to avoid destroying the information learned before. Next, we add a new trainable LSTM nature network at the bottom of frozen layers to share pre-trained feature extraction and train the new model. Finally, fine-tuning the new built LSTM model.

Table 3. Model structure

Model	Source model (DAiSEE Dataset)	Target model (New collected dataset)
LSTM layer 1	**32** transfer layer	**32** trained layer
Dropout	**0.1**	**0.1**
LSTM layer 2	**32**	**8** new layer
Dense layer	**3**	**3** new layer
Activation	**softmax**	**softmax**
Batch Size	**32**	**32**
Epoch	**10**	**early stopping**

Table 3 shows the experiment model structure. In data processing, we normalized both pre-processed data of the source and target dataset range from -1 to 1 and divided it into 8:2 for training and validation set. The source model contained two hidden LSTM layers with 32 cell nodes and a dense output layer with 3 cell nodes with a SoftMax function. We built an inner model with the first hidden LSTM layer in the pre-training model for improving transfer earning's interpretability and flexibility. After trained the source model, we reused the frozen inner model to build a new LSTM nature network and add a new LSTM layer as the target model for training on the new dataset. The previous experiments found that the fully trained source model will be harmful to the transfer target model. Thus, to avoid overfitting, the source model was trained for 10 epochs. On the other hand, the transfer target model sets early stopping since it trains on the new dataset but easy to overfit.

Table 4. Results

Model	Low engaged	Normally engaged	High engaged	Accuracy
Source Model Result (1)	**92.6**	**60.9**	**47.2**	**65.1**
Target Model Result (2)	**64.5**	**73.7**	**41.6**	**62.6**
Transfer Learning (3)	**68.4**	**68.9**	**51.4**	**63.7**

Table 4 shows the results of our experiments:

(1) Source Model Result: The estimation result of a fully trained long short-term memory (LSTM) sequence model trained on the DAiSEE dataset containing one LSTM layer with 64 cell nodes and one dense output layer with three nodes.
(2) Target Model Result: The structure in this model trained on our collected new dataset is the same as the source model structure, but the LSTM layer with 32 cell nodes.
(3) Transfer Learning Result: This is the result of our proposed transfer learning method (Table 3).

(1)

Source		Engagement Labels			
		0	1	2	Σ
Estimate	0	498	42	13	553
	1	66	478	258	802
	2	40	283	334	657
	Σ	604	803	605	2012

(2)

Target		Engagement Labels			
		0	1	2	Σ
Estimate	0	51	27	1	79
	1	18	90	14	122
	2	5	37	30	72
	Σ	74	154	45	273

(3)

Transfer		Engagement Labels			
		0	1	2	Σ
Estimate	0	54	21	4	79
	1	23	83	16	122
	2	9	26	37	72
	Σ	86	130	57	273

Fig. 4. Confusion matrix

The confusion matrix in our experiments is shown in this figure. The blue grid is the correct answers, and the incorrect answers no color.

4.4 Discussions

The result of the transfer learning model and confusion matrix proved our proposed method improved the collected new dataset performance and overall engagement levels. The ±1 range estimation error is 35.1% in the target model, and 31.5% in the transferred model, approximately 4% has been reduced. However, the ±2 range estimation error is improved by around 2%. The learned information from the pre-trained model improved the transferred model performance trained on the new collect dataset. However, the shared information also brought some partially redundant information, which caused the error rate of ±2 to increase. Thus, how to choose the source dataset is another critical issue that we need to consider. To avoid negative transfer, the source dataset's features need to be similar to the target dataset.

The transferred model accuracy has been improved, but not obvious. The source dataset is also on a small scale. A more engagement recognition dataset needs to be collected for increasing the proposed method performance in further work. Furthermore, we observed eye information is more important than other features. Japanese and Chinese students are reserved and not many body movements in the learning process. In this situation, eye information like eye gaze, eye movement, and wink are essential features to distinguish different engagement levels.

5 Conclusions and Further Work

5.1 Conclusions

In conclusion, we introduced the transfer learning method to estimate learners' engagement in online learning. We collected a dataset with 19 participants and proposed some methods to improve the estimation performance. 1) we used time-series facial and bodily expressions to estimate engagement. 2) designed the features use the extracted up-body key points by OpenPose. 3) to relieve the widespread noise issue generated by OpenPose, and solve the wild data irregularity, we implied the moving average method. 4) To balance the number of engaged/unengaged data, we oversampled the unengaged data. 5) To let the deep learning sequence model work well on our small-scale dataset, we implement transfer learning and pre-trained an inner model on the DAiSEE dataset. As a result, we achieved the engagement estimation correct rate of 0.637. The achieved correct rate is 0.01 higher than the model fully trained on our new dataset.

According to our research experiments, we proposed that transfer learning is effective for small scale dataset. However, the experiment's effect is not apparent but has excellent potential. Thus, we will merge the existing public time-series datasets to collect a massive dataset for transfer in further work.

5.2 Application in Further Research

As we mentioned in the previous part, learners' engagement needs to be feedback in online education to retain their attention. To solve this problem, we plan to develop an engagement support system that can feedback the student's engagement at an opportune time and saved the valuable history for further research. The system will use the engagement analyzed result from the recorded time-series expressions and body features from learners to improve online learning quality. The support system will feedback the engagement analysis results to both teachers and learners immediately. At the same time, save the results for checking leaners' engagement results history anytime. Also, analyze the collected engagement records, proposing plans to increase learners' engagement. Depended on the feedback, professors can adjust teaching progress and lecture content difficulty. Also, students can improve their engagement and self-regulation.

References

1. Wong, J., Baars, M., Davis, D., Van Der Zee, T., Houben, G.J., Paas, F.: Supporting self-regulated learning in online learning environments and MOOCs: a systematic review. In: Int. J. Hum.-Comput. Interact. **35**(4-5), 356–373 (2019)
2. Dhawan, S.: Online learning: a panacea in the time of COVID-19 Crisis. J. Educ. Technol. Syst. **49**(1), 5–22 (2020). https://doi.org/10.1177/0047239520934018
3. Tran, M.T., Hasegawa, S.: Self-regulated learning recognition and improvement framework. In: The Asian Conference on Education 2020: Official Conference Proceedings, pp. 449–465. (2021)

4. Hasegawa, S., Hirako, A., Zheng, X.W., Karimah, S.N., Ota, K., Unoki, T.: Learner's mental state estimation with PC built-in camera. In: International Conference on Human-Computer Interaction, HCII 2020: Learning and Collaboration Technologies. Human and Technology Ecosystems, vol. 12206, pp. 165–175 (2020)
5. Chang, C., Zhang, C., Chen, L., Liu, Y.: An ensemble model using face and body tracking for engagement detection. In: The 20th ACM International Conference on Multimodal Interaction, pp. 616–622 (2018)
6. Gupta, A.., D'Cunha, A., Awasthi, K., Balasubramanian, V.: DAiSEE: towards User Engagement Recognition in the Wild. J. Latex Class Files 14(8), (2015) https://doi.org/10.10007/1234567890
7. Dewan, M. A. A., Lin, F., Wen, D., Uddin, Z.: A deep learning approach to detecting engagement of online learners. In: 2018 IEEE Smart World, Ubiquitous Intelligence Computing, Advanced Trusted Computing, Scalable Computing Communications, Cloud Big Data Computing, Internet of People and Smart City Innovation, pp. 1895–1902 (2018)
8. Fawaz, H.I., Forestier, G., Weber, J., Idoumghar, L., Muller, P.: Transfer learning for time series classification. arXiv:1811.01533v1 [cs.LG] 5 Nov 2018
9. Nomura, K., Augereau, O.: Poster: estimation of student's engagement based on the posture. In: Proceedings of the 2019 ACM International Symposium on Wearable Computers, pp. 164–167 (2019)
10. Monkaresi, H., Bosch, N., Calvo, R.A.: Automated detection of engagement using video-based estimation of facial expressions and heart rate. IEEE Trans. Affect. Comput. 8(1), 15–28 (2017)
11. Kleinsmith, A., Berthouze, N.B.: Affective body expression perception and recognition: a survey. IEEE Trans. Affect. Comput. 14–18 (2019)
12. Dhall, A.: EmotiW 2019: automatic emotion, engagement and cohesion prediction tasks. In: 2019 International Conference on Multimodal Interaction (ICMI 2019), pp. 546–550 (2016)
13. Nordlund, A., Pahlsson, L., Holmberg, C., Lind, K., Wallin, A.: The cognitive assessment battery (CAB): a rapid test of cognitive domains. Int. Psychogeriatr. 23(7), 1144–1151 (2011)
14. Fredricks, J.A., Blumenfeld, P.C., Paris, A.H.: School engagement: potential of the concept, state of the evidence. Rev. Educ. Res. 74(1), 59–109 (2004)
15. Cao, Z., Hidalgo, G., Simon, T., Wei, S., Sheikh, Y.: OpenPose: Realtime Multiperson 2D pose estimation using part affinity fields. IEEE Trans. Pattern Anal. Mach. Intell. 43(1), 172–186 (2019)
16. Dewan, M.A.A., Lin, F.: Engagement detection in online learning: a review. Smart Learn. Environ. 6(1), 1–20 (2019)

Correction to: Tool or Partner: The Designer's Perception of an AI-Style Generating Service

Kyungsun Kim, Jeongyun Heo, and Sanghoon Jeong

Correction to:
Chapter "Tool or Partner: The Designer's Perception
of an AI-Style Generating Service"
in: H. Degen and S. Ntoa (Eds.): *Artificial Intelligence in HCI,*
LNAI 12797, https://doi.org/10.1007/978-3-030-77772-2_16

In the originally published version of chapter 16, the corresponding author was not correctly marked. This has now been corrected.

The updated version of this chapter can be found at
https://doi.org/10.1007/978-3-030-77772-2_16

Author Index

Printed in the United States
by Baker & Taylor Publisher Services

HCI International 2021 Thematic Areas and Affiliated Conferences

Thematic Areas

- HCI: Human-Computer Interaction
- HIMI: Human Interface and the Management of Information

Affiliated Conferences

- EPCE: 18th International Conference on Engineering Psychology and Cognitive Ergonomics
- UAHCI: 15th International Conference on Universal Access in Human-Computer Interaction
- VAMR: 13th International Conference on Virtual, Augmented and Mixed Reality
- CCD: 13th International Conference on Cross-Cultural Design
- SCSM: 13th International Conference on Social Computing and Social Media
- AC: 15th International Conference on Augmented Cognition
- DHM: 12th International Conference on Digital Human Modeling and Applications in Health, Safety, Ergonomics and Risk Management
- DUXU: 10th International Conference on Design, User Experience, and Usability
- DAPI: 9th International Conference on Distributed, Ambient and Pervasive Interactions
- HCIBGO: 8th International Conference on HCI in Business, Government and Organizations
- LCT: 8th International Conference on Learning and Collaboration Technologies
- ITAP: 7th International Conference on Human Aspects of IT for the Aged Population
- HCI-CPT: 3rd International Conference on HCI for Cybersecurity, Privacy and Trust
- HCI-Games: 3rd International Conference on HCI in Games
- MobiTAS: 3rd International Conference on HCI in Mobility, Transport and Automotive Systems
- AIS: 3rd International Conference on Adaptive Instructional Systems
- C&C: 9th International Conference on Culture and Computing
- MOBILE: 2nd International Conference on Design, Operation and Evaluation of Mobile Communications
- AI-HCI: 2nd International Conference on Artificial Intelligence in HCI